THE OXFORD HISTORY
OF ENGLAND

Edited by G. N. CLARK

THE OXFORD HISTORY OF ENGLAND

Edited by G. N. CLARK
Provost of Oriel College, Oxford

The titles of some volumes are provisional

* *These volumes have been published*

THE
EARLIER TUDORS
1485–1558

BY

J. D. MACKIE
C.B.E., M.C., Hon. LL.D. (St. Andrews)

PROFESSOR OF SCOTTISH HISTORY
AND LITERATURE IN THE
UNIVERSITY OF GLASGOW

OXFORD
AT THE CLARENDON PRESS

Oxford University Press, Amen House, London E.C.4

GLASGOW NEW YORK TORONTO MELBOURNE WELLINGTON
BOMBAY CALCUTTA MADRAS KARACHI
CAPE TOWN IBADAN NAIROBI ACCRA SINGAPORE

FIRST PUBLISHED 1952
REPRINTED WITH CORRECTIONS 1957
PRINTED IN GREAT BRITAIN

PREFACE

My thanks are due to Miss Anne Robertson, Curator of the Hunter Coin Cabinet in the University of Glasgow, for the Appendix on Tudor Coinage; to Sir Ernest Bullock for help with the section on Music; and to J. C. Thomson, Esq., late of Charterhouse School, for advice and assistance in the section on Architecture.

I acknowledge, too, with great gratitude the generosity of the Council of the Royal Historical Society in permitting me to make use of the valuable Lists of Officers contained in the *Handbook of British Chronology*, and of the Saint Catherine Press for giving me leave to use the Lists of Officers, especially of Lords Admiral and Earls Marshal, published in the *Complete Peerage* (volume ii).

My indebtedness to the many scholars who have worked upon this period, which will be obvious to every reader, have been expressed in the footnotes and the Bibliography.

To the Provost of Oriel College I am grateful for a most helpful and considerate editorship, and to the staff of the Clarendon Press for their labours in seeing the work through the press.

Finally, I must express my thanks to a succession of Research Assistants, Miss Lyndall Miles (Mrs. John Luce), Miss Jill Walker (Mrs. Edwin Drummond), Miss Margaret Burnet (Mrs. William Brown), Miss Moira Bruce (Mrs. James McIntosh), and Miss Anna Dunsmore, for care and patience which have survived many changes of arrangement.

Arrangement, in a period when so many conflicting factors entered, and when each reign must needs be treated as a unity, has been a great difficulty. In the main I have adhered to chronology, but it has been necessary sometimes to present the various aspects of a single chapter in separate subsections, dealing respectively with foreign affairs, domestic political affairs, constitutional development, religion, economic and social questions.

It has not been possible to signalize the structure of each chapter by the use of subheadings, but it is hoped that the table of contents and the page-titles will help to guide readers where the text departs from exact chronology, and will prove convenient if the book is used for reference.

J. D. M.

GLASGOW, 1951

CONTENTS

CONTENTS ix

CHAPTER VI. FOREIGN AFFAIRS

CHAPTER VII. THE ACHIEVEMENT OF HENRY VII

CHAPTER VIII. SPLENDOUR OF YOUTH

Chapter IX. THE CARDINAL

Chapter X. ROYAL SUPREMACY

CHAPTER XIII. ECONOMIC DEVELOPMENT

Chapter XIV. THE YOUNG JOSIAH

Chapter XV. THE REIGN OF MARY

LIST OF MAPS

I
THE NEW MONARCHY

THE Renaissance has been described as the transition from the age of Faith to the age of Reason, as the reaction of the individual against the universal, as the victory of a spirit of criticism over a spirit of acceptance. All these definitions are good and all make clear the essential fact that the Renaissance was not an event but a process. It owed something to the capture of Constantinople by the Turks, no doubt, and certainly it owed much to a rediscovery of Greek thought. But Manuel Chrysoloras had taught Greek in Italy before the fourteenth century was ended, and even remote Oxford had Decembrio's translation of Plato's *Republic* by 1443 at latest.[1] Birth is not really a sudden affair: it is the result of long silent processes, and the Renaissance was carried in the womb of the middle ages for centuries. It did not spring to life like Athene in full panoply: like most other births it moved from infancy to splendour and to decay. It did not come to all the countries of Europe at the same time, and it did not develop in the same way and at the same rate in different atmospheres. The Italian Renaissance in art was decadent before the English Renaissance in letters reached its full glory. Yet wherever it was felt, and whatever form it took it represented the same thing. It was a rebellion of the facts against the theories.

To regard the thought and the institutions of the middle ages as static and uniform is absurd: there was progress and there was variety. But none the less, the basic theories of church, of state, of economics, of philosophy, of life generally were set in the frame of that universalism which survived amongst the ruins of the Roman empire. The world was the special creation of God and the centre of the universe. It was an ordered unity, reflecting the divine harmony of the New Jerusalem where Christ presided over the holy angels. Every individual, man or institution or idea, had being as part of the great whole from which it was derived. This great whole had several aspects, and

---

[1] See *English Historical Review*, xix. 511, and Dr. R. Weiss, *Humanism in England during the Fifteenth Century*, p. 59, for books of the *Republic* and other Greek books obtained by Duke Humphrey in 1441.

so its derivatives could be, indeed must be, arranged into several categories. But throughout the whole structure harmony prevailed. Human society was in all points regulated by a divine, universal law. There was a single church ruled by the pope, in which all ecclesiastical authority originated, and though the implications of the theory were never recognized in England, a single state governed by the emperor, from which all temporal authority was derived. In philosophy there was one single truth from which proceeded all particular truths. In morality there was one single code of righteousness; legislation was the enunciation of the eternal right rather than the formation of anything new. In the realm of economics every article had its *justum pretium*, and the customary rents and the customary wages represented the divine institution concerning these matters.

With these complete and satisfying theories, the actual facts had at no time tallied, and as the centuries passed the discrepancies became more and more apparent. They were not unnoticed by the thinkers of the times,[1] but, as a rule, they were either ignored or explained away by a subtle philosophy. The whole genius of the age was for harmonization and reconciliation.[2] Even at a time when sailors were constructing, for very necessity, maps which showed things as they were, there seemed to be nothing incongruous in the production by clerics of a *mappa mundi* which showed the earth as a disk with Jerusalem at the centre. Even after the complete failure of the Crusades[3] it was still possible to regard the Kingdom of Light as coextensive with the habitable world. The scholastic philosophy was, in essence, an attempt to reconcile eternal verity with particular verities which might seem to contradict it or to contradict one

---

[1] See Professor Jacob in *History*, N.S., xvi. 214. It is pointed out that 'Humanism is much too complex a phenomenon to admit of simple qualification like "the emergence of the individual"'; diplomacy, strong centralized administration, Reason of State, foreign exchange, and many other features generally associated with the Renaissance had their roots in the middle ages; that, to a medievalist 'if there is any break in the continuity of European tradition, it is the break between the later Roman-early medieval epoch and the Middle Ages proper . . . rather than between the Middle Ages and the Renaissance'. See also A. S. Turberville, 'Changing Views of the Renaissance', in *History*, N.S., xvi. 289. None the less it is evident that the Neoplatonists and Erasmus were conscious that their age was making a new approach to truth.

[2] See Clemens Bäumker, 'Die europäische Philosophie des Mittelalters' (*Die Kultur der Gegenwart* (vol. ii), in *Allgemeine Geschichte der Philosophie*, 1909).

[3] In that they stemmed the Moslem attack the Crusades succeeded.

another. There was a hierarchy in truth as in church and state. To pope, archbishop, and bishop, to emperor, king, and baron, there corresponded the order of the Bible, the fathers, and the schoolmen. Derived from the great fountain-head, and derived by a right logic from an authoritative interpretation, particular truths could not be incompatible one with another. The names of the great medieval books speak for themselves. The *Sententiarum Libri* of Peter the Lombard, the *Sic et Non* of Abelard, the *Concordia Discordantium Canonum* of Gratian—all these were essays in reconciliation, and it was the supreme achievement of Aquinas that he reconciled the Aristotelian Θεός with the Christian God. Against even the hard realities of economics the theories asserted themselves. The good old doctrine was that since the barren could not breed, the taking of interest for the loan of money was a monstrous thing. Yet when capitalism was rendered necessary as business ceased to be local and retail, its existence was reconciled with the orthodox view by various logical devices. In the realm of political thought the theory of the feudal pyramid was itself merely a 'device' for reconciling with the ideal world-state those national and local powers whose actual being could not be disputed.

So ignoring, pretending, and philosophizing, the middle ages went on their way until the discrepancies between theory and fact grew too wide to be ignored by minds well practised in the search for truth. To the greatest thinkers of the middle ages it was clear enough that all questions could not be settled out of hand by an appeal to authority. As early as the twelfth century Adelard of Bath had said that to accept authority in the face of common sense was the action of a senseless brute, and when Bernard of Chartres compared the men of his own day to dwarfs mounted upon the shoulders of giants, he had claimed, at least by implication, that his contemporaries saw more than the great ones of old. For men who argued thus, philosophy was plainly more than an exposition of long-known truth, and as the middle ages were receptive—they learnt, for example, from the Arabs—the sum of knowledge steadily increased.

While men's critical faculty thus developed, the established ideas and the established institutions became steadily less able to endure criticism. The journeys of the explorers and the experiments of the physicists played havoc with the *mappa mundi*, the Babylonish Captivity and the Schism shook the

authority of Rome, and the struggling Italian state which was
the papacy obviously bore little resemblance to a world-church.
The empire was become an 'imponderous rag of conspicuous
colour',[1] tossed upon the whirlwind of German politics, and
power had passed to the national monarchies, the great princedoms, and the city states. Wholesale enterprises were beginning
to flourish; already wealth was leaving the country-side and
multiplying itself in the towns. The impact of a critical spirit
upon theories far removed from actuality is what is called the
Renaissance. The inquiring spirit turned itself upon the universe,
and Copernicus (Nikolaus Koppernigk of Thorn: 1473–1543)
discovered that the stable earth rotated daily on its own axis.
It turned itself upon the earth and Columbus discovered
America; Vasco da Gama sailed round the Cape to India
(1497–9) and back; Magellan's ship sailed round the world
(1519–22). It turned itself upon human society and it perceived
that the world was not really divided by the horizontal lines of
caste or tenure, or at least not by these alone; it was divided by
the vertical lines of geography. The world-state was a fiction,
but the nation states were real and their relations were governed
not so much by divinely appointed law as by human opportunism. It turned itself upon man's body and it saw that the
body was not a mere clog upon the spirit but a thing of beauty
and of worship in itself. It turned itself upon man's soul and it
perceived that the soul was possessed of an infinite yearning for
God and that this yearning could not be satisfied by the mere
acceptance of the authority or even by the experience of others;
the soul must find a personal satisfaction and a real security of
its very own. To criticize is easier than to create, and when the
triumphant Renaissance began to rationalize its experience—
to consolidate its gains—it too began to theorize; and some of its
theories were not unlike those which it had discarded with scorn.
The doctrine of hereditary divine right, which was later used to
support the national monarchy, was not far removed from the
medieval conception of kingship after all; and in more than one
country new presbyter turned out to be but old priest writ
large—or, in some cases perhaps, small. But with these later
theories we are at the moment not concerned. The essential
feature of the early Renaissance was its reliance upon facts. Its
genius was to reveal and to accept the thing which was actually

[1] Carlyle, *Frederick the Great*, ii, ch. xiv.

there. Many of the facts were self-evident. The world was round, and Magellan's ship had sailed round it. The human body was beautiful and the painters showed it so. The national state was actually in being, governed upon a system to which there has been given the name of the 'new monarchy'. The France of Louis XI, the Spain of Ferdinand and Isabella, the Scotland of James IV, the England of Henry VII—all these are examples of this new monarchy. In all of them the new régime did not spring suddenly into being by the creative act of some tyrant, but owed its existence to a new handling of old institutions. There was no conscious adoption of a new conception of the meaning of the word 'state', but there was an unconscious pursuit of realism, and an instinctive ability to pick out the new and the growing from the unnecessary and the effete.

Produced by the working of the same spirit upon material, which was throughout western Europe very much the same, the states which came into being as the middle ages crept to their end have very much in common. The essential features are these. The state is regarded as the expression of some local group—perhaps an incipient nation—to which a common experience over a long period has given definite self-consciousness. The grand object of this state is self-sufficiency, and in pursuit of this self-sufficiency it should consider no interest but its own. The centre of its existence is the prince, necessarily a person of great ability, who finds the business of being a prince an end in itself. He is not immoral; but he is amoral, and practises a *Realpolitik*. He cultivates, if not the poor, at least the middle classes, partly because they, no less than he, believe in a 'mercantilist' economy; he keeps the nobility and even the church in order, but he commends himself to the people as a whole by a justice which gives the impression of being impartial or at least less partial than the old feudal justice. He does not encourage 'representative institutions' but favours a small council, composed of the 'new men', many of whom are hardworking officials. In a broad sense it may be asserted that his authority rests upon good government, but his *ultima ratio* is force. He has a standing army of professional soldiers, he has perhaps a few ships, and above all he has powder and guns. He is wise concerning men; he may have personal charm and the art of acquiring and keeping popularity. He may be the idol of his subjects. But in the end he stands, not for liberty, but for authority.

Judged by the theory of the middle ages, the rule of this sort of prince might well seem to be a shocking innovation: but it was not, in fact, far removed from medieval practice. It was the service of the Renaissance to tear away the decent sheepskin which had covered the medieval wolf, and incidentally to justify his existence on the ground that one great wolf was better than a pack of lesser carnivores.

In England, as elsewhere, the new monarchy did not emerge fully developed either in the year 1485 or in any other year. This fact was not always recognized. Henry Hallam, for example, in his *Constitutional History from Henry VIIth's Accession to the Death of George II*, gave the impression that it was the Tudors who removed 'the essential checks upon the royal authority' which had operated during the middle ages, and for many people the appearance of the Tudor dynasty has been associated with the establishment of a new system of government. But John Richard Green, who first gave to the system the name 'The New Monarchy', had other ideas about its origin. In *The History of the English People* (1877–80), book V is called 'The Monarchy 1461–1540'; in a smaller edition, however, published in 1876, the title 'The New Monarchy' is given to the whole period 1422–1540, and specifically to section iii (1471–1509). Green plainly took the view that the year 1485 was not of supreme importance in the history of the English constitution. He accepted, however, the idea that there was a constitution in England, and that this epoch is one of 'constitutional retrogression', witnessing 'something strange and isolated in our history', 'a sudden and complete revolution' in which parliamentary life was 'almost suspended'.[1]

Recent authority has pointed out that the idea of a definite breach in the constitution is somewhat of a mistake, that the fifteenth century hardly conceived the idea of a 'constitution'; that 'constitutional history of England' was virtually a new phrase when Hallam made it the title of his great work; that the so-called 'constitutional experiment' represented less a constitutional or liberal tradition than a breakdown in the governmental system whereby two of the essential elements, the king and the parliament, were set in opposition. On this show-

---

[1] A. F. Pollard, *Parliament in the Wars of the Roses*, Glasgow University Publications, xlii (1936), p. 15.

ing, 'the novelty was a return to an older and more orderly system of government than the anarchy of the Wars of the Roses or even the Lancastrian experiment. The crown recovered the initiative in public legislation exercised by Edward I . . .',[1] and one of its features was a restoration of confidence between Crown and parliament.[2] Edward IV, perhaps because like the commons he was interested in trade, recreated that confidence, and when, after the troubles of Edward V and Richard III, the nation resumed its ordered life, the Tudors developed the system in which Crown and commons were in alliance.

This view might be criticized on the ground that the partnership was not always cordial, that Henry VII dispensed with parliaments towards the end of his reign; that Wolsey did not like them, and that it was only in the stress of the Reformation that Henry VIII made full use of parliament to back him in his policy. But in its main lines it is undoubtedly correct. There was no great breach in constitutional tradition in the year 1485.

In support of this view it may be pointed out that although Bacon and later chroniclers lay stress on the fall of a tyrant, contemporary historians drew no such moral from the battle of Bosworth. The *Chronicle of London*[3] refers to it somewhat casually between the mention of a sheriff's death and a notice of the appearance of the sweating sickness in the city. The reticence may be due to the prudence engendered by uncertain times, but the evidence of the *Chronicle* as a whole suggests indifference rather than prudence. Before, throughout, and after the year 1485 the same tale goes placidly on—the succession of mayors and sheriffs, the notices of 'benevolences' and executions. For the author, plainly, the battle of Bosworth was not a historical landmark, the beginning of a new age.

His attitude is easily comprehensible. England had witnessed in three decades several violent transferences of authority, and these transferences had been accomplished without any change in the 'constitution'. There was no reason to suppose that this fresh alteration of the dynasty would cause any general upheaval in the life of England, and no reason to suppose that the dynasty would endure for long. If such were the prognostications of the chronicler they were to some extent realized in fact. The struggle of the Roses continued. In 1487 there was fought

---

[1] Ibid., p. 15.   [2] Ibid., p. 28.
[3] *Chronicles of London*, ed. by C. L. Kingsford (1905), p. 193.

at Stoke a pitched battle in which two of the king's 'battles' did not engage until they saw what would be the issue; from 1491 to 1497 the Yorkist cause found a leader in the person of Perkin Warbeck, and the ignominious end of that pretender must not conceal the fact that the chances of the 'duke of York' were weighed seriously by European politicians; throughout all his reign Henry was apprehensive—not without cause—of some treachery or rebellion undertaken in the name of the extruded dynasty.

When it is added that the faction-victory which set Henry upon the throne was very like other faction-victories gained during the past thirty years, and that the 'constitutional' and 'economic' expedients adopted by the new king resembled those of his predecessors in principle and even in detail, it is easy to see how a case can be made for the view that the battle of Bosworth was not of vital importance in the story of England's development, and the year 1485 of no particular significance.

But there is another side to the matter. Bosworth was not the last battle of the Roses, but it was the decisive battle. Its decision was never reversed. The new dynasty kept the throne which it won for over a hundred years—as long as its own life lasted—and every attempt to overthrow it served only to hasten the ruin of its opponents. Yorkists intrigued, intriguers covered their designs beneath the Yorkist mantle, but every unsuccessful venture strengthened the Tudors. For the Tudors held the crown, and their rivals, branded by their failure as traitors and rebels, suffered the doom of forfeiture and death. The wars of the Roses were over, not in theory only, but in fact.

Henry VII stood for no particular principle. He did not even stand entirely for one particular faction, for he was pledged to marry a daughter of York before ever he set sail for England. He was the man who by his own good fortune, skill, and courage ended in his own favour a long and profitless struggle, and who was able by his power to keep the position which he had won. As became a 'new monarch' he rested upon the realities.

The triumph of the new dynasty may be ascribed to two factors—the attitude of England towards the civil wars, and the personality of Henry, who in himself possessed both a strong position in the dynastic quarrel and gifts of character which peculiarly fitted him for the task in hand. England was heartily sick of the war. For thirty years it had gone on, its fires replenished by the fuel of hate which it created as it burned, but there was no great principle involved. It was merely the struggle of two noble factions for the crown, and by the year 1485 this had become obvious to all observers. Even to foreigners it had become a by-word for its folly and its ferocity. The Scottish historian, Robert Lindsay of Pitscottie, has an odd tale to tell.[1] According to him there was in Richard's camp on the eve of Bosworth a Scottish ambassador, the bishop of Dunkeld,[2] who as he took his formal leave, witnessed a curious scene. Calling for the crown, the king had it set upon his head, declaring that he would wear it in the battle which was soon to join and that he would win or die 'crownit king of England'. So it befell that the crown remained unguarded in the royal tent when the king was called forth suddenly to still a tumult among his people, and it was stolen by a Highland servant of the bishop, McGregor by name. Arrested and taxed with his crime the Highlander made a remarkable defence. He had realized early in life, he said, that like his fathers before him he would be hanged for theft, and he had resolved that he would not die for sheep or cattle but for some thing of great price—and it would be a great honour to his kin and friends that he should be hanged for the rich crown of England, for which so many honourable men have lately died, some hanged, some beheaded, some murdered, some killed in battle, and for which the king had offered himself, within that hour, to die for, ere his enemy Harry got it off his head. Pitscottie was an incurable story-teller, and his tale must rank as a γενναῖον ψεῦδος. Its moral, at least, is true. To Scottish observers of the late fifteenth century, the wars of the Roses were a dispute for 'the rich crown of England'.

Another foreign observer took the same view. Commines in his *Memoirs* mentions more than once the bloody civil wars of England, which he regarded as a divine punishment for the ravaging of France by English invaders.[3] In these wars, he says, 'threescore or fourscore persons of the blood-royal of the king-dom were cruelly slain' and others reduced to want and poverty so that no common beggar could have been poorer. He had seen

---

[1] *Croniclis of Scotland* (Scottish Text Society), i. 196.
[2] George Brown, bishop of Dunkeld, represented James III at Henry's corona-tion on 30 October, but his safe-conduct seems to date only from 22 September.
[3] Scoble's translation of Commines, 1904, i. 182.

the duke of Exeter barefoot begging his bread from door to
door. Yet, he adds, 'England enjoyed this peculiar mercy above
all other kingdoms, that neither the country, nor the people,
nor the houses, were wasted, destroyed, or demolished; but the
calamities and misfortunes of the war fell only upon the soldiers,
and especially on the nobility'.[1]

The English evidence confirms that of the foreigners. In the
speech attributed to him in More's *History of King Richard the
Third*[3] Buckingham argues that war is never so mischievous as
when it is intestine. In England, he says,

'what about the getting of the garland, keeping it, losing and winning
it again, it hath cost more English blood than hath twice the winning
of France. In which inward war among ourselves hath been so great
effusion of the ancient noble blood of this realm that scarcely the
half remaineth, to the great enfeebling of this noble land, beside
many a good town ransacked and spoiled by them that have been
going to the field or coming from thence. And peace long after not
much surer than war. So that no time was there in which rich men
for their money, and great men for their lands, or some others for
some fear or displeasure, were out of peril. For whom trusted he
that mistrusted his own brother? Whom spared he that killed his
own brother?'

It may be contended that Buckingham was making a case,
and that the towns were in fact little affected. Certainly in his
narrative, More makes his common folk remark 'that these
matters be King's games, as it were stage-plays, and for the
more part played upon scaffolds. In which poor men be but
the lookers-on. And they that wise be will meddle no further.'[2]
More's interpretation of the attitude of London is supported
by the *Chronicle*. There the events of the civil war are catalogued
along with notices of 'goodly and costious bankettes', of the prices
of commodities, of the vagaries of the weather. No opinion is
expressed as to the justice or injustice of the numerous executions which are recorded—'There were some hanged and some
were heded', and the mercy of Jesus is besought for the souls
of all. Civic disorder is mentioned with implicit, provincial
disorder with explicit, condemnation, but the quarrels of the
nobles are plainly a matter for the nobles themselves. It is 'the

---

[1] Scoble's translation of Commines, 1904, i. p. 394.
[2] *The English Works of Sir Thomas More*, i. 441.
[3] Ibid., p. 448.

gentilmen of England' who were 'so dismayed that they knew not which party to take but at all adventure'. The part of the citizens was, usually, to stand aloof. At one time their mayor is seen, with 5,000 well-arrayed men at his back, riding 'abowte the cite daily' to keep a close eye upon the rival parties; at another time they are found changing their allegiance with the greatest complacency. The Fellowships are shown 'in scarlet and the comonys in Grene' bringing Edward IV in triumph through London, and greeting Edward V as he entered the city. But the same paragraph which tells how they donned their 'violet clothynge' to meet King Richard tells how they put on their violet eighteen days later to lead Henry through the city in triumph.[1]

Other cities were no less prudent than London. York, at one time, is shown as very anxious to keep trouble at a distance; Norwich sent men to both sides, and the evidence of Leland makes it plain that the wars had left few scars upon the English towns. One of his few references is significant. In a note upon Wakefield and its battle he remarks, 'the commune saying is there that the erle (of Rutland) wold have taken ther a poore woman's house for socour, and she for fere shet the dore and strait the erle was killed'. That poor woman's attitude was the attitude of the essential England.

But to suppose that 'the people' were utterly uninterested in the wars of the Roses would be wrong. The support of London meant much to the Yorkists, and the commons of certain areas at times took the part of their lords. It is true enough that in these cases the commons may have been moved rather by local grievances than by an acute interest in national affairs; but it is equally true that a great part of England must have been vexed by the constant necessity for vigilance and by the interruption of trade. It can have been no pleasure to the citizens of London to pay for the strong watch which must ride about the town in troublous times. And even though the economic life of England as a whole continued on an even course there must have been endless dislocations in town and country. With the dynastic issue itself England felt little concern, but she was very anxious to see the end of uncertainty, and prepared to support any authority which seemed likely to bring the purposeless quarrel to an end. For her, the proper authority was the

[1] *Chronicles of London*, ed. by C. L. Kingsford (1905), pp. 176, 190, 192–3.

Crown. This conviction is illustrated by the story noted by Camden:

'When Richard the Third was slain at Bosworth and with him John Howard Duke of Norfolk, King Henry the seventh demanded of Thomas Howard, Earl of Surrey, the Duke's son and heir, then taken Prisoner, how he durst bear Arms in the behalf of that tyrant Richard. He answered: "He was my crowned King, and if the Parliamentary authority of England set the Crown upon a stock, I will fight for that stock: And as I fought then for him, I will fight for you, when you are established by the said authority." And so he did for his son King Henry the eighth at Flodden field.'[1]

Camden's story may be apocryphal; but it certainly expresses the conviction that England must have a king, and that the king should command the allegiance of every Englishman. The reference to parliament is something of an anachronism, though Richard III is said to have claimed that the crowns of England and France had been 'by the high authority of parliament entailed'[2] to his father. Fortescue had contended that the monarchy of England was not absolute, but it did not occur to an Italian observer[3] either to doubt the reality of the royal power, or to suppose that it was founded on consent. Certainly the judges had recognized that parliament made the law, in accordance with which the king should reign, but the king himself could do no wrong which the judges could remedy, and parliament was still his high court. The great officers of state, the judges, the sheriffs, and the mayors all exercised their authority in the name of the king, and although, during the troubles of the Roses, irregularities had crept into the system, the system itself was still intact. Government was a function of the royal power, and the wearer of the crown not only enjoyed a high prestige, but also controlled a quite efficient machine. He controlled the royal property too, and the royal rights of taxation. Whoever assumed the crown assumed also a real

[1] *Remains concerning Britain* (ed. 1870), p. 294.
[2] *The English Works of Sir Thomas More*, i. 443.
[3] Rawdon Brown, in his introduction to *Four Years at the Court of Henry VIII*,—a selection of dispatches written by the Venetian Ambassador Sebastiano Giustiniani (London, 1854)—ascribes the collection of the material to Andrea Trevisano who represented the Signory in England from 1497 to 1502. The compilation must have been made by one of his gentlemen, who makes it clear that his magnificent and most illustrious lord had himself been in England, and the *Relation* would be the basis of the formal *Relazione* which Trevisano must present to the Signory on his return. (Translated for Camden Society by Charlotte Augusta Sneyd, 1847, with useful notes, and generally cited as *The Italian Relation*.)

authority, and under the shadow of that authority all the waverers might well find courage to make up their minds.

The apathy of England towards the dynastic quarrel of two noble factions; the general relief, amongst the nobility at least, at the disappearance of the hard-handed Richard; the general desire for stable government—these things provided some sort of guarantee that any rule which promised peaceful order would be generally accepted. The obvious failures of a succession of baronial cliques and the weakness of the church inclined men to place their hopes of peaceful order in the monarchy. The Crown was in a better position than either of the rivals who might possibly have disputed its power.

To say that the baronage had been destroyed by the civil wars would be to exaggerate. But certainly, as Buckingham had said in his oration, it had been much weakened. Some of the greatest names had disappeared, and although old blood trans-ferred by heirs female ran in the veins of men who bore new titles, yet the transference had meant a certain loss of prestige, and, in most cases, either a division amongst co-heirs or a disputed succession. 'For where is Bohun? Where's Mowbray? Where's Mortimer?' The questions asked in 1625 by Chief Justice Crewe, contemplating the possible end of the De Veres of Oxford, might well have been put in 1485. Bohun had been absorbed in Lancaster and Buckingham; Lancaster was extinct in the male line, and Buckingham had been attainted. Mowbray had gone, after a division between co-heirs, to Norfolk; and Norfolk died, the king's enemy, at Bosworth. The main line of Mortimer had ended in 1424, and though scions of the female descent still existed in the defeated Yorkist party they were far removed from the parent stem. Edward's sons were dead; Richard's son was dead; Clarence's son Warwick still lived, but Clarence had been attainted. Some of the old families which survived were now represented by minors, and amongst the noble houses old or new there had been so much attainder and restoration according to the fluctuating fortunes of the civil war, that few can have felt secure in their estates. Only eighteen nobles subscribed the oath taken in Henry's first parliament, and of the titles represented, several had disappeared before the reign was ended.[1] Those who had opposed a reigning king were

[1] See A. F. Pollard, *The Reign of Henry VII*, iii, app. ii, for the diminution of the old families.

in constant danger of proscription and were, by now, hardly strong enough to threaten his position. Those who had befriended him were in no case to embarrass him, even if they had been inclined to do so, by demanding gratifications so large as to sap the royal power upon which they themselves depended. John De Vere, thirteenth earl of Oxford, for example, represented the really old nobility, but his return to fortune was so closely bound up with the success of the Tudors that he never showed the slightest desire to waver in his allegiance. It would be idle to pretend that the over-mighty subject no longer existed. During the civil war the baronage had profited from the weakness of the central power not only to gain control over high politics but also to establish itself in might throughout the country-side, and while Edward IV and Richard III had regained for the Crown the direction of national affairs, they had had neither the opportunity nor the inclination to interfere with the local authority of the magnates. With the far-reaching privileges of their supporters they had interfered little, and while they overthrew the baron who offered an unsuccessful opposition, their general policy was to grant his lands and authority to some other nobleman whom it was desired to win or to reward, or to entrust them to one of the royal servants. No doubt the royal courts, never ineffective, nibbled away at the feudal privileges, but in his own area each great noble was still a petty king— 'Get you lordship,' wrote an anonymous correspondent to John Paston, '*quia ibi pendent tota lex et prophetae.*'

Where he had power, certainly, the lord could sway the law in his own interest. Maintenance, the supporting of a magnate's dependents in courts of law, was common; courts were venal, royal officers were terrorized or bribed. But it is necessary to distinguish between the influence improperly enjoyed by the over-mighty subject, which would vanish with the might which gave it birth, and the legitimate authority of a baronial court. In the north, where the old 'liberties' endured until the reign of Elizabeth, the feudal courts preserved a criminal jurisdiction commensurate with that of the royal justices, together with a civil jurisdiction which covered almost every aspect of landholding, and regulated the whole fabric of local economy. But elsewhere they had over a long period slowly lost their authority before the unrelenting pressure of the royal justice. Local justice was sadly interrupted during the wars of the Roses, and trial by jury was often a contest in perjury. But the king's courts continued to sit at Westminster, cases and arguments to be reported in the 'Year Books'. Over the greater part of England the law, in all important matters, was the king's law, and its officers were the king's officers. For long particularism had had its way, and doubtless public opinion had accepted the fact that upon lordship hung the law. But, for all that, much of the local authority enjoyed by the great noble was illegal and rested solely upon his possession of actual power. That power had been weakened by the frequent changes consequent upon the alternate proscriptions of the civil war, and by the constant presence of the more efficient justice of the Crown. Once let the king gain real power and he would find that feudal justice, despite its ostensible strength, was less firmly entrenched than it seemed to be; let him offer royal justice as an alternative to baronial justice and it would become obvious which was the more popular. Already the Yorkist kings had opened the attack upon maintenance—the outer perimeter of the aristocratic stronghold had already been breached.

While the local authority of the magnates thus crumbled away they were less able in another way to make trouble for the Crown than their fathers had been. Too much emphasis has been laid upon the fact that the king possessed the only train of artillery in the kingdom, for in England artillery was only just becoming a decisive factor in warfare. Nevertheless the point is of importance. It was not long before the baronial castle ceased to be a sure refuge to its owner, whose military power had also been diminished by other developments in the conduct of war. By this time the feudal magnate had come to rely largely upon hired men. It is said that 'every baron and gentleman of estate kept great horses for men-at-arms. Some had armories sufficient to furnish out some hundreds of men', and although the baronial resources have no doubt been exaggerated, it seems to be true enough that the nobles did in fact raise 'their power' by equipping friends and mercenaries from their private stores when they wished to take the field. From More's account of the arrest of Rivers it appears that the harness was kept in barrels and moved from house to house with its owner; and it is obvious that a force raised according to this system would depend very largely on the ability of the magnate to pay wages. Now the barons as a class were poorer than they had been. The French wars were long since ended and they had ceased to be profitable long before their end. Apart from the accidents of attainder and legal process and the mere wastage of civil war, estates were less remunerative than they had been, and the nobles as a class did not take to merchandise. Their incomes, based largely upon customary rents from land, were fixed, while prices were already on the up-grade. They felt the pinch of this the more keenly because they were great spenders. It was an age of display and lavish housekeeping, and it was a man's duty to keep his 'countenance'—to live, that is, in the style proper to his rank.

An Italian observer who described England about the year 1500 was struck with the evidence of splendour which he saw about him. The English wore fine clothes and spent their money lavishly if selfishly: 'They would sooner give five or six ducats to provide an entertainment for a person than a groat to assist him in any distress.'[1]

The Venetian's picture of the English nobility is confirmed by a study of household books of the great families. Not only were sumptuous feasts given on special occasions, but the ordinary day-to-day expense was on a scale not incomparable with the royal court. When Edward Stafford, duke of Buckingham, the son of Richard's victim, kept his Epiphany feast on 6 January 1508, what with yeomen or valets, *garçons* or grooms, he had 519 persons for dinner, and 400 for supper. The invitations seem to have been issued to the important guests with a casual generosity—the duke's sister bringing fifteen guests with her, and smaller people nine, six, three, or two as the case might be. There were minstrels, trumpeters, waits, and players, besides special cooks; and the expenses in beef from 'the lord's store', fresh purchases—'achates'—of poultry and game, wine from the cellar, and 521 quarts of ale from the buttery, were tremendous.[2]

Most revealing of all is the story of the fifth earl of Northumberland (1478–1527) as it is revealed in his household books. He was the son of the earl who was killed in 1489 in attempting to collect a royal tax. Skelton blames the lord's servants for cowardice, but whether they behaved well or ill it is plain that they were unable to cope with the 500 'uncourteous carls' who confronted them. To generalize would be dangerous, for an

---

[1] *The Italian Relation*, p. 22, n. 31 and app., for the splendours of a gentleman's wardrobe.   [2] Ibid., n. 60.

accident may happen at any time; but it seems fair to argue
that the 'power' of the house of Percy was not what it had been.
And it may be significant that the fifth earl played no great
part in politics or war. Yet his magnificence was extreme. On
occasions of display, as for example when the Princess Margaret
went north to her marriage, the earl of Northumberland appears
as a dashing figure and a generous host, splendid in his dress,
his house, and his retinue. If he spent freely abroad it is clear
from the famous Northumberland Household Book[1] that he had
little opportunity to save at home. Ten years later he accompanied his royal master to France, and it was left to the competent Surrey to command the army of Flodden, which included
in its ranks representatives of the name and vassalage of Percy.
It is possible that despite his gallantry the splendid earl was
not too much trusted by the king; it is certain that his service
overseas cost him dearly, for the army, as described by Hall,
must have been one of the best dressed that ever left our shores.

Where did the money come from?

His domestic expenditure was carefully supervised by his
council, and economies were attempted, but there is ample
evidence of lavish spending. The maintenance of a household
of 166 persons of various degrees involved a yearly outlay of
£933. 6s. 8d.—nearly one-fifteenth of the cost of the royal
household—and it may well be doubted whether the revenues
of lands in Northumberland, Yorkshire, and Cumberland left
any real margin. At all events, when the earl died he left debts
amounting to £17,000 and cash assets of £13. 6s. 8d. He may
have been more than usually extravagant. He cultivated the
Muses, at least vicariously. His accounts show some provision
for 'My Lords Lybrary' and 'My Ladies Lybrary'; he patronized
Skelton; he paid for a beautifully illuminated transcript of
some poems; he decorated his walls and ceilings with poetical
inscriptions; he maintained a secretary, William Perris, who
produced a metrical 'Chronicle of the Family of Percy'.

His was not the only family to maintain its songsmiths.
There were minstrels in the service of the house of Stanley, who
gave in the allegory of 'The Rose of England' and in the epic
ballad 'The Song of the Lady Bessie' versions of Henry Tudor's

---

[1] *The Regulations and Establishment of the Household of Henry Algernon Percy, the fifth
earl of Northumberland, at his castles of Wressle and Leckonfield, in Yorkshire* (1905,
usually cited as *The Northumberland Household Book*).

successful venture designed to enhance the glory of their
patrons. No doubt other great lords had other hobbies, and
plainly all had, as a matter of course, very heavy expenses.
When it became necessary for my lord to 'make his power'
money might well be short. The case of Rivers, already cited,
reveals the fatal delay that might elapse before the household
was put on a war footing, just as the success of Henry's invasion
reflects the provision of money by the competent Reginald
Bray. If funds could be obtained from some outside source the
great lord could still raise a considerable power. But all
depended on money, and the nobles, as a class, were already
finding it difficult to make ends meet. If they wished to indulge
their warlike proclivities, their easiest course was to take money
from the king, and if the king, using the old system of indentures,
were able to supply the necessary cash, the barons would cease
to be his rivals and become but the agents of his power.

In yet another way was the power of the magnates crumbling.
Their morality had been shaken. The long and brutal struggle
had sapped the ancient honour and the ancient sense of duty.
In the words of More, 'the state of things and the dispositions
of men were then such, that a man could not well tell whom he
might trust or whom he might fear'.[1] The careers of some of the
magnates of this time are eloquent of changing loyalty and
broken faith, justified sometimes by subtle casuistry. Buckingham himself, who spoke so feelingly of the mutual mistrust,
could find specious arguments for preferring Richard's title to
that of Edward's children, and for abandoning Richard in
favour of Henry Tudor. In setting forth the reasons for his
change of heart, however, he explained quite naïvely to Morton
that the king had cheated him of his reward, refusing him a
part of the earl of Hereford's lands and the 'Highe Constable
Shyppe' of England which were patently his due. From the
failure of his first suit he had drawn an evil augury and at one
time thought of claiming the crown himself.[2] The Italian
Relator represents that behind the outward affability and good
address of the English noblemen there lay an essential selfishness; for them even love and marriage were matters of computation. Possibly the Italian was deceived, for the Englishman
does not wear his heart on his sleeve. But if the portrait painted

---

[1] *The English Works of Sir Thomas More*, i. 423.
[2] Ibid., p. 452, and Hall's *Chronicle* (1809), pp. 382, 387.

by Titian which is usually called 'An Englishman' is truly
named, then another Italian made an estimate not dissimilar.
We are shown a well-made, well-dressed, fair-haired young man
whose face, though handsome, is marked by a calculating eye.
He appears competent, avid of life perhaps, but with no
illusions.[1]

While the nobles were in no case to challenge the power of
a king who knew how to make use of his resources, the church
in England was at this time ill fitted to act as a check upon the
royal authority and still less to be a substitute for it. Its position
in the field of politics was not unlike that of the nobles; it had
real strength, yet that strength could be employed in alliance
with the royal power more readily than in opposition. Admittedly its power was great, far greater than its subsequent history
suggests. The suddenness of the Reformation seems to point to
inward collapse, and with this phenomenon always at the back
of their minds, historians have been apt to employ their industry
in detecting elements of weakness.

To most Englishmen, it is clear, religion was a necessary part
of life; it is not without significance that Richard III tried to
win popular support by putting up Doctor Shaw and Friar
'Penker' to preach popular discourses, and that Henry VII,
especially after victory, showed himself punctilious in public
acts of devotion. The church was accepted as an essential
element in an ordered society. It commanded, if not general
enthusiasm, at least a general allegiance, and it must not be
forgotten that there was enthusiasm too. It was in the days of
Henry VII that notable additions were made to the fabric and
the ornament of Canterbury Cathedral[2] by Prior Goldsmith
and Cardinal Morton, and pious foundations were being made
right up to the Reformation. The church seemed to stand firm
in its ancient strength with its 2 archbishoprics, its 19 bishoprics,
its 600 religious houses,[3] its stately parish churches,[4] and the

---

[1] See H. L. Gray, 'Incomes from the Land of England in 1436', in *English
Historical Review*, xlix (1934), 607, for a summary of the numbers and wealth of the
nobility and gentry. The peers had £500 to £1,000 a year; the gentry, perhaps
9,000 families, from £10 to several hundreds a year.

[2] See *The Italian Relation*, p. 84, for a brief note on the enrichment of Canterbury
at this time.

[3] Gairdner enumerated as suppressed, or surrendered, or confiscated, 606
houses, including nunneries and commanderies or Preceptories of the Knights of
St. John. *The English Church in the Sixteenth Century*, pp. 419–28.

[4] The number of parishes in England was usually over-estimated. The Italian

'college' of clergy in every fair-sized town. It enjoyed wealth which astonished foreign observers, and privileges which shocked them, or possibly aroused their envy; its leaders, as the Venetian observed, played a great part in the government of the country. 'The clergy are they who have the supreme sway over the country, both in peace and war' he asserts; and after noticing the great privileges of the churchmen as regards sanctuary and benefit of clergy, he remarks 'nor is the common saying of this country without cause that the priests are one of the three happy generations of the world'. Certainly the clergy played a great part in affairs. The chancellor was invariably a cleric,[1] and many of the other great offices and what we should call now the civil service were mainly in clerical hands. Only with the advance of the Renaissance did there appear laymen with the technical skill needed for administration, and the English Crown, like other European governments, found it convenient to pay its officials by the grant of church livings. The bishops were not really canonically elected, though the forms were doubtless observed; promotions were in the hands of the Crown, which made with Rome one of those undefined working arrangements which are of the essence of English politics. There was no 'concordat', or 'pragmatic sanction', but the king saw to it that the statutes of Provisors and Praemunire were not really observed; the pope saw that the king's nominees were not opposed, and each party was at pains not to upset an arrangement which suited both.[2] It is fair to say that there was no great scandal; the nominees were in the main good men, and some of them, like Castelli, commended themselves by sound Latinity and good letters. None the less the usual ground for advancement was political utility. The pope supported the men whom the king delighted to honour.

It will be obvious then that much of the political authority which the church seemed to enjoy really depended upon its understanding with the Crown, and that it had not in itself the

---

Relator gives 52,000. The guileless lay financiers of 1371 reckoned on 40,000, and discovered from the chancellor that there were fewer than 9,000.

[1] See E. F. Jacob, 'Changing Views of the Renaissance' in *History*, N.S., xvi. 228. There were a few instances of lay chancellors in the fifteenth century, but until Sir Thomas More became chancellor in 1529 the chancellor was always a cleric at this time.

[2] Gairdner, *The English Church in the Sixteenth Century*, pp. 2–4; F. M. Powicke, *The Reformation in England*, pp. 8, 9.

force to assert its independence. Despite the survival of devotion its moral position was less secure than it had been. In England there was, as yet, little of the philosophic scepticism which had manifested itself in Italy,[1] and had begun to show itself north of the Alps; though it is possible that wealthy men who had visited Italy may have come home with some doubts. But, as the trials for heresy attest, the old Lollard opinions still survived in the lower ranks of society;[2] and it is probable that the old notion of confiscation survived too. It is certain that clerical wealth was attracting the eyes of a laity thinking more and more in terms of trade and industry. A spirit of indifference was manifest. What is worse, this spirit was operative in the church itself and, behind the imposing façade of its institutions, weakness was evident to those in a position to see. In the splendid houses of religion there were only 7,000 monks and 2,000 nuns all told,[3] and it was becoming difficult to provide for the proper singing of the offices. Some of the parishes were attached to the great foundations—to the bishoprics and the monasteries—and the devout in the country-side had to do as best they might in the hands of ill-paid vicars.

An old-established institution whose machinery still functioned regularly, the church may not have been conscious of its weaknesses, though it must have been aware of the criticisms levelled against it. It had not in itself the inherent strength to overcome these weaknesses, and while it might be able at times to exercise some moral pressure on the government, it was ill fitted to offer an effective opposition. Rather was it disposed, if only the king would be reasonable, to be a good ally of the Crown.

All stood upon the Crown. The barons were impoverished and divided. The church was tending to become a royal instrument. The people were indifferent to the dynastic dispute, and, if the Venetian is to be believed, indifferent to the old

---

[1] The ironical reference to the Jubilee of 1500 mentioned in Hall (p. 492) is little more than a verbal translation from Polydore Vergil (ed. 1950, p. 120). Either the papal collector himself or his collaborator Federigo Veterani, the learned librarian of the dukes of Urbino, must bear the responsibility. See F. A. Gasquet, 'Some Materials for a New Edition of Polydore Vergil's History' in *Transactions of the Royal Historical Society*, N.S., vol. xvi.

[2] The Venetian remarks that though the English do not 'omit any form incumbent upon good Christians, there are, however, many who have various opinions concerning religion'. *Italian Relation*, p. 23.

[3] G. Baskerville, *English Monks and the Suppression of the Monasteries*, p. 285, n. 2.

sanctions. According to him they were 'though somewhat
licentious, never in love', and though they kept a jealous guard
upon their wives anything might be compensated in the end
by the power of money; they were hard on their children, and
family life was at a low ebb. 'They are so diligent in mercantile
pursuits that they do not fear to make contracts on usury.'
Despite severe laws, 'there is no country in the world where
there are so many thieves and robbers,' and the common 'people
are held in little more esteem than if they were slaves'.

This gloomy picture of English morality must be discounted
by the consideration that the Venetian was not only surprised
but jealous of the mercantile prosperity which he saw about him.
The advance of the Turks was robbing Venice of her control of
the gorgeous East, and his city was not what she had been.
None the less, his picture of an amoral self-interested people is
supported by other evidence, and it is hard to resist the conclusion that the England of 1485 was realist in its outlook on
life. 'Get Lordship'—in politics it was power which mattered.
The barons and the church were losing their power, but the
Crown possessed both a prestige and a governmental machinery
which made it the natural expression of national power. If there
could arise a claimant to the throne whose title was good enough
to satisfy the conservative instincts of England, and if that
claimant had power enough to gain the crown, then he could
keep it—as long as he had the power to do so. If he gave the
good order which would allow a practical people to realize their
practical ends he would meet with little opposition on the part
of outraged principle. It was because Henry Tudor had the
power, if only just the power, to gain the throne, and because
he and his successors had the power to keep it, that the year
1485 is justly reckoned a turning-point in English history.

England was sick of the war; yet there seemed to be no
reason why the war should stop. The dynastic issue was not
settled, the overmighty subjects, though shorn of some of their
might, were still with power, anxious to recover old lands which
had been lost, greedy to hold new lands which had been won.
The fires of hatred were not slaked and might have smouldered
on to burst into flame at the breath of random wind. It was
a sadly battered crown which Henry donned at Bosworth; but
he kept it, and made it a splendid diadem. Saving himself he
saved his country too—it is the epitome of Tudor greatness—

and his accession may well be accounted the beginning of a new age.

Yet in one sense Green was right. It is wrong to suppose that in the year 1485 any great change took place in the means whereby England was governed. The 'new monarchy' was the political expression of the Renaissance, and the Renaissance represented the triumph of the facts. No new theory of kingly power emerged, and any new machinery which came into being was clearly a development of the old and known. The whole atmosphere was impregnated with medieval ideas, and the institutions followed medieval practice. That there was a transition to the modern world is not to be disputed, but the transition was gradual, especially at first, and only with the great movement known as the Reformation did it become spectacular.

It is the task of this book to trace the history of a change which was not, in all its aspects, apparent to the generations which experienced it, and which cannot be presented as a concrete, clear-cut thing without a distortion of the truth. Sometimes the changes which resulted were not those which had been designed by the innovators, sometimes the innovators themselves were not aware that they were making any change of consequence. But it was the gift of the statesmen of this time that they had an almost unconscious, empiric, feeling for the realities, and it is as apostles of the realities that the makers of the Tudor monarchy stand before us. To the problems of their day they offered not an ideal but a practical solution; founding upon the incontrovertible fact and the acted deed they built a government which was to stand the tests of time and circumstance.

The Roads of England in the Tudor Period.

II

THE FACE OF ENGLAND

WHAT manner of country was it, the England of the early sixteenth century? The question is not so easy to answer as might be supposed. It is true that interest in geography and topography were growing, and that 'Descriptions of England' were printed by Caxton in 1480 and by Wynkyn de Worde in 1497, but both these 'descriptions' are little more than borrowings from the fourteenth-century *Polychronicon* of Ranulph Higden, and Higden, who referred to his sources with care, relied on authorities like Giraldus Cambrensis, Bede, and Pliny. To dismiss his work as unimportant would be foolish as well as ungrateful, for his general observations on geography and climate are sound, and he emphasizes certain features of importance. England is a rich country and her wealth is not derived from one source only. Her soil is fertile and yields good crops of grain and fruit. Animals are plentiful—cattle, horses, and sheep, and beasts of the chase besides. Fowls of all kinds are to be found and the many rivers and the surrounding seas teem with fish of great variety. There is abundance of metals and of the useful minerals—copper, iron, tin, lead, and silver, salt, chalk, marble, and white clay. Altogether this is a rich and noble country, sufficient for its own needs and indispensable to the rest of the world.[1]

Descriptions of this kind, though they are of interest as illustrating the pride of the Englishman in his native land, say nothing of many things about which the modern inquirer is most curious. To the men of every generation the commonplaces of life do not seem worth recording, and to the men of the late fifteenth century it did not occur to describe in detail the towns or the country-sides in which their lives were set, or the manner in which everyday life was lived. Everyone knew the parts of the country and the social conditions which concerned himself; other places and other conditions were of little importance. Before the day of Leland there is not to be found in English sources anything like a comprehensive survey of Tudor England.

[1] *Polychronicon Ranulphi Higdeni*, ii. 12–20 (Rolls Series, 1869).

Something, however, is to be gained from the accounts written by observant foreigners. To them the strange rainy lusty island on the edge of civilization was full of interest. Erasmus, to be sure, tells us little. Convinced though he was that the seeker after the good life should be secure in physical amenities, he seems to have concerned himself little as to the means whereby these amenities should be produced, and his letters throw little light upon economic or social conditions. In the first flush of rosy hope he finds England so delightful that he writes to an Italian humanist in Paris, Fausto Andrelini, that 'did you but know the blessings of Britain you would clap wings to your feet and run hither'. The kisses, so freely given, of the English ladies were delightful. A measure of disillusionment came a few months later when, on leaving Dover, he was relieved of all his money—about £20—in accordance with an old law, recently re-enacted by Henry VII, which forbade the export of bullion from England. Two subsequent visits, begun also in hope, ended in disappointment, and as the passing years proved the 'mountains of gold' to be illusory, the tone of the references to England in Erasmus's correspondence becomes more critical.[1] England is a place where the carriers rob you of your wine, so that you are compelled to drink the health-destroying beer; where the dirtiness of the people is hateful, where the manner of life is such that an Italian, unable to dwell with some merchant from his own country, might well wonder whether it was possible to go on living in England at all. There is mention of the 'inhospitable Briton', of the plague, of the many robberies which vexed the country. The sailors, or possibly the porters, on the Dover–Calais transit despoil passengers in various ways; their atrocious conduct reflects the greatest discredit upon the whole nation and sends travellers away with a very poor opinion of the country.

A picture far more satisfactory was left by Polydore Vergil of Urbino, who came to England as a sub-collector of papal revenue in 1502, became a naturalized Englishman in 1510, and, except for occasional visits to Italy, remained in his adopted country until the year 1551, when he departed to spend the evening of his life in Italy. Before he set foot on

---

[1] Erasmus visited England several times—from June 1499 to January 1500, from April 1505 to May or June 1506, from June–July 1509 to July 1514, in March–June 1515, in August 1516, and from March to April, 1517.

English soil he had already a good repute in the world of letters, and a few years after his arrival he was asked by Henry VII to write a history of England. He set to work with industry and acumen, and it is plain that, when he left on a visit to Italy in 1514, he had already brought his narrative down to the battle of Flodden.[1] But it was not until 1534 that his *Anglica Historia* appeared in print, and then it went only as far as the death of Henry VII. Its appearance was greeted with a storm of obloquy, for in the course of his researches the author had convinced himself that Geoffrey of Monmouth had mingled fact and fiction in a manner quite unjustifiable, and that the story of Brute, the alleged founder of the British kingdom, was a mere fable.

> Virgilii duo sunt; alter Maro, tu Polydore
> Alter. Tu mendax, ille Poeta fuit,

wrote one enemy. Other accusers alleged that, having completed his own history, he caused a wagon-load of ancient manuscripts to be burnt, lest his own errors should be detected; that he shipped old manuscripts to Rome; that he 'borrowed' books from the public library at Oxford and never returned them. Polydore was, in fact, a historian of the new 'Renaissance' school, who criticized as well as chronicled, and a century passed before this heretical assailant of the antiquity of Britain found a defender. But his book was of obvious value; it was copied by some, and read by many—by Edward Hall and Richard Grafton, for example—and to this day it speaks for itself. It is the work of a man who saw the necessity of seeking reliable authority and of balancing evidence. Granted that the Italian may have over-emphasized the features in which his adopted country differed from his native land, his description of England may be reckoned as worthy of close attention.

'The whole countrie of Britaine (which at this daie, as it were in dowble name, is called Englande and Scotlande) . . . is divided

---

[1] For a definitive account of Vergil and his work see the Introduction to the edition of the *Anglica Historia* (for the years 1485–1537), with a translation, edited by Denys Hay for the Camden Society (1950). Vergil obtained ecclesiastical promotion, became archdeacon of Wells in 1508 and a prebend in St. Paul's in 1513. In 1515 he was imprisoned at the instance of Wolsey, but was released soon after Wolsey was made a cardinal. See also Denys Hay in *English Historical Review* (April 1939), p. 240; E. A. Whitney and P. P. Cram on 'The Will of Polydore Vergil' in *Transactions of the Royal Historical Society* (1928), p. 117; and C. L. Kingsford, *English Historical Literature*, pp. 191 and 254.

into iiij partes . . . whereof the one is inhabited of Englishmen,
the other of Scottes, the third of Walleshemen, the fowerthe of
Cornishe people. 'Which all differ emonge themselves, either in
tongue, either in manners, or ells in lawes and ordinaunces.
'Englond, so called of Englishmen the inhabitauntes, beinge farre
the greateste parte, is divided into xxxix Shiers, which commonlie
men call cownties.'

The counties he divides into four groups—the ten between
the Thames and the sea, the sixteen between the Thames and
the Trent, the six 'towards Walles and the west partes', and the
seven lying in the middle or navel of the country, the norther-
most of which lie over against the Scots. After describing the
organization of the ecclesiastical government the author goes
on to make some observations on the land as a whole. Although
he criticizes Caesar in the matter of the circumference of the
island he himself falls into some of the errors of the old writers
in the matter of geography; he puts Ireland between Britain
and Spain, and makes the eastern coast of Britain incline
towards Germany. Yet even in the field of geography he makes
a new contribution by contrasting the lack of ports on the east
coast with the abundance of havens on the west, and when in
describing the interior economy of the island he relies upon his
own observation, he presents his reader with the fruits of an
active and intelligent inquiry.[1] Plainly, he regards it as a
wealthy land 'most frutefull on this side of the river of Humber,
for on the other side it somewhat to muche abowndethe in
mountaynes; for notwithstanding to the beholder afarre off it
[apparently the country as a whole] appearethe very champion
and plain, neverthelesse it hathe manye hills, and such as for
the moste parte are voyde of trees, with most delectable valleys'.
In these valleys are the homes of men—the houses of the nobles
and the little towns, for the people, 'according to their ancient
usage [Is this a recollection of the *Germania*?] do not so
greatlie affecte citties as the commodious nearness of dales and
brooks'. 'The grownde is marvelous frutefull, and abundantlie
replenished with cattayle, wherebie it commethe to pass that of
Englishemen moe are grasiers and masters of cattayle then
howsbandemen or laborers in tilling of the fielde, so that allmoste
the third parte of the grownde is lefte unmanured, either for their
hertes, or falowe deere, or their conies or their gotes—(for of

---

[1] He seems to have used Higden, however, and *The Italian Relation*.

them allso are in the northe partes no small numbers);—for almoste everie where a man may se clausures and parckes paled and enclosed fraughte with such venerie, which as they minister greate cause of huntinge, so the nobilitie is much delited and exercised therein.'[1] None the less he notices the crops of wheat, barley, and pulse, and while he remarks upon the absence of the olive, he comments upon the numerous vines which, to be sure, seldom produce wine, though they serve for ornament. He observes that the English geld their animals with the result that they are tractable. Of the riding horses many amble instead of trotting, and the oxen may be yoked many to a plough. He notes too the superb sheep which—this seems strange to an Italian—are never watered, but slake their thirst from the dew of the air, conveyed no doubt through the rich pasture provided not only by meadows but by the common fields which are left open after the crops have been lifted. Of fish and fowl there is an abundance, and the Englishmen are great eaters. Their chief food is flesh; their oxen and wethers are 'beasts as it weare of nature ordagned for feastinge', and their 'beafe is peereles, especiallie being a fewe dayse poudered with salte'.

Albeit the other cities are of little account, yet London is worthy of commendation, for in the wonderful gulf, 60 miles long, the tide ebbed and flowed twice in 24 hours, so that merchandise had ready access to the city, and commerce flourished. Across the river, admired by all beholders, stood London Bridge on its 20 piles of squared stones joined with arches, 60 feet high, 30 feet broad, and, by reason of the houses which it carried, more like a street than a bridge.

The general impression is one of prosperity. The climate is healthy and the Englishmen who, though valiant and void of fear in battle, do not over-exert themselves in labour,[2] live to a great age, to 110, or 120 perhaps. The men are tall, fair, grey-eyed, and good companions; the women are of excellent

---

[1] The quotations are from an English version of the *Historia* which was made about the middle of the sixteenth century, 1546; the three books which cover the period from 1422 to 1485 were edited by Sir Henry Ellis for the Camden Society in 1844, and the first part, from the beginning to 1066, in 1846. These parks, however, are mentioned in the Latin version of which the first draft was made about 1512–13. It seems that the inclosures were for sport, not for agriculture. (*English Historical Review*, 1939, p. 240.)

[2] 'Manie men live in divers places an hondred and tenne yeares, yea some sixe skore, albeit emonge artificers and husband men it is receaved as a prescripte that thei should sweate by no meanes.' Camden Society, 1840, p. 19.

beauty and 'in whitenes not muche inferior to snowe'. Silver is
to be seen on many tables, and everywhere there seems to be
ample food. As for the country people, largely because they
eschew cities and in their rural dwellings have 'entercourse and
daylye conference' with 'the nobilitie confuselie dwellinge
emonge them' they are 'made verie civill'. Too much has been
made by some writers of the good estate of the English peasantry
in the fifteenth century, but plainly, in the eyes of an acute
observer from south of the Alps, the English countryfolk were
very different from the peasants of Italy.

Much the same impression of general well-being is given by
the famous *Italian Relation* composed about the year 1500. The
country is definitely to be accounted rich; it has an equable
climate and possesses various sources of wealth. The soil is
fertile and carries most of the trees known to Italy, though the
olive is not to be found, and the vine is of little account. Wine,
however, is freely imported from Candia, Germany, France,
and Spain, and the natives brew from wheat, barley, and oats
two beverages, one of which is called beer, and the other ale.
'These liquors are much liked by them, nor are they disliked
by foreigners after they have drank them four or six times; they
are most agreeable to the palate, when a person is by some
chance rather heated.'

'Agriculture is not practised in this island beyond what is required
for the consumption of the people; because, were they to plough and
sow all the land that was capable of cultivation, they might sell a
quantity of grain to the surrounding countries. This negligence is,
however, atoned for by an immense profusion of every comestible
animal, such as stags, goats, fallow-deer, hares, rabbits, pigs, and
an infinity of oxen, which have much larger horns than ours, which
proves the mildness of the climate, as horns cannot bear excessive
cold; whence, according to Strabo, in some northern countries, the
cattle are without horns. But above all they have an enormous
number of sheep, which yield them quantities of wool of the best
quality. They have no wolves, because they would immediately
be hunted down by the people; it is said, however, that they still
exist in Scotland.'

Birds of all sorts abound. Special notice is taken of the flocks of
swans which grace the Thames and of the rooks, ravens, and
jackdaws which, abhorred by the Italians as of evil omen, are
recognized in England as useful scavengers. Fish, too, are

plentiful, both in the numerous rivers and in the surrounding seas. The sea-fish are preferred by the inhabitants, but they prize most of all the salmon, which is indeed a most delicate fish.

Amongst the resources of this fortunate island mineral wealth is to be accounted. There is a certain amount of iron and silver, and an infinity of lead and tin, from which are produced pewter vessels as brilliant as if they were of fine silver. Yet silver itself is surprisingly common, for 'the riches of England are greater than those of any other country in Europe, as I have been told by the oldest and most experienced merchants, and also as I myself can vouch, from what I have seen'. The country needs to import practically nothing except wine, whereas it sells tin and wool abroad, and as a law of long standing prohibits the export of bullion in any form, it follows that there is a vast accumulation of the precious metals. 'There is no small innkeeper, however poor and humble he may be, who does not serve his table with silver dishes and drinking-cups; and no one, who has not in his house silver plate to the value of at least £100 sterling, which is equivalent to 500 golden crowns with us, is considered by the English to be a person of any consequence.' The wealth of the religious houses in gold, silver, and jewels was prodigious, and London offered a display of wrought silver which was astounding. 'In one single street, named the Strand,[1] leading to St. Paul's, there are fifty-two goldsmiths' shops, so rich and so full of silver vessels, great and small, that in all the shops in Milan, Rome, Venice and Florence put together I do not think there would be found so many of that magnificence.' The street must have been Cheapside where, on the south side, opposite to the cross and extending to the east as far as Bread Street, Thomas Wood, Goldsmith, Sheriff of London, had erected, in 1491, 'Goldsmiths' Row', described by Stow (1598) as 'the most beautiful frame of fair houses and shops that be within the walls of London, or elsewhere in England'. It contained 'ten fair dwelling-houses and fourteen shops, all in one frame, uniformly builded four stories high, beautified toward the

---

[1] See Miss E. Jeffries Davies in *History*, N.S., xvii, p. 48. This is a mistranslation of the 'Strada' of the original. The street in question cannot have been the Strand which does not, in fact, lead to St. Paul's, and which was outside the city proper—'part was in the duchy of Lancaster, part in Westminster'. The traditional goldsmiths' quarter was immediately east of St. Paul's churchyard, and the 'Strada' in question must be Cheapside. Besant's *Tudor London*, app. vi.

street with the Goldsmith arms and the likeness of Woodmen (in memory of the founder's name) riding on monstrous beasts, all of which is cast in lead, richly painted over and gilt'.

London, indeed, is a luxury-city, and its citizens, who serve a long apprenticeship in youth, rank as highly as gentlemen in Venice. Its total population will compare with that of Rome or Florence. But the other cities are of little consequence. Bristol, a great seaport to the west, and York have some celebrity, and Oxford and Cambridge have universities, but of the other towns even those which contain great churches and religious houses, are of little importance, and even the country-side seems to Italian eyes to be sparsely populated: 'The population of this island does not appear to me to bear any proportion to her fertility and riches.'

In reading the encomiums pronounced by Italians it is well to remember that English visitors thought Venice far more magnificent than anything they had seen in their own land,[1] and that the laudatory tone of the descriptions may reflect the surprise of men who had expected to find little save barbarism at the end of the earth. None the less the essential accuracy of these pictures of England given by intelligent foreigners is established by a comparison between them and the first detailed and personal survey made by an Englishman. Between the years 1534 and 1543 John Leland,[2] the keeper of the king's libraries, undertook several journeys of investigation up and down England, and collected notes which he recorded in carefully kept Itineraries. So perceptive was his eye and so sound was his method that he may justly be regarded as the father of English topography.[3] His professed motive, indeed, at all events after the 'Suppression of the Monasteries' in 1536, was largely bibliographical. His object was to rescue and to preserve in the king's libraries the treasures of the monastic houses which had been mishandled or scattered at the time of the suppression of the monasteries, and to use the evidence they contained to

---

[1] *The Pylgrymage of Sir Richard Guylforde* (Camden Society, 1850, p. 8).

[2] Much the best edition of *Leland's Itinerary in England and Wales* is that of Lucy Toulmin Smith (5 vols., 1906–10). It will be cited as *Leland*. The volume which is third in sequence was printed first, and does not bear the title volume iii. It will therefore be cited as *Wales*.

[3] William Botoner or William of Worcester, whose *Itinerary* (ed. Nasmyth, 1778), written in the late fifteenth century, includes interesting antiquarian notes, may be called a predecessor of Leland in a small way; but he did not, like Leland, begin a tradition.

refute the impudent scepticism of Polydore Vergil in a majestic
De Antiquitate Britanniae of some fifty books. He was mistaken,
no doubt, in his implicit faith in Geoffrey of Monmouth, but
he was absolutely right in his belief that unless they were
resolutely guarded the books and the manuscripts of the
monastic libraries would speedily disappear. The accusations of
vandalism made against Polydore himself were probably ill
founded, but there was some truth in Leland's complaint that
'the Germans, perceiving our desidiousness and negligence, do
daily send young scholars hither that spoileth them, and cutteth
them out of libraries, returning home and putting them abroad
as monuments of their own country'. The object of the Germans,
perhaps, was less to secure credit for their own country than to
procure evidence as to the state of the medieval church.
Certainly this was the end which Flacius Illyricus of the
'Magdeburg Centuries' had in view[1] when he organized, at a
somewhat later date, a painstaking search of the monastic
libraries, and though the charges of purloining made against
his accredited agents have been disputed, the very naïve account
given by Marcus Wagner of his own depredations at St.
Andrews and elsewhere sheds a lurid light upon the methods
of the German *antiquitatum inquisitor*.[2] In one way and another
the literary treasures of the past were disappearing, and England
may well be grateful to this resolute investigator, who, as he
wrote to the king, 'to the intente that the monumentes of
auncient writers as welle of other nations, as of this yowr owne
province, mighte be brought owte of deadely darkenes to lyvely
lighte . . . conservid many good autors . . . of the whiche parte
remayne yn the moste magnificent libraries of yowr royal
Palacis', though part remained in his own custody.

No less grateful should we be that Leland's enthusiasm soon
outran the limits of his self-imposed task. Inspired, like others
of his time, with a love for the 'worlde and impery of England'
he made descriptive notes of every road upon which he travelled
and of every place which he visited. These notes are invaluable.
It is true that Leland was concerned, in theory at least, with
gentle families and noble antiquities rather than with topography

---

[1] According to his *Newe Yeare's Gyfte* to Henry VIII the vindication of truth
against old error was one of Leland's motives in making his investigations.

[2] See the *Copiale Prioratus Sanctiandree*, edited by J. H. Baxter (1930), Introduction, p. xx.

and systems of land-economy, and that at the very time of his journeyings the face of England was undergoing the changes consequent upon the 'agrarian revolution'. It is true, too, that his work has survived in a fragmentary form; for a mental breakdown prevented him from completing that formal 'description of the Realm of England' which he promised to Henry VIII, and his notes, scattered after his death, in 1552, were only brought together two centuries later by Thomas Hearne. But when every discount has been made, it remains indisputable that Leland produced what is really a detailed survey of a great part of England, and that this survey, though it takes account of changes, yet portrays a civilization which is in its essentials old and stable.

To one who follows Leland on his well-marked path certain features make themselves obvious at once. There is a manifest distinction between the fertile lands of the south and the midlands, and the great 'mores' and 'craggi and stoni montaines' of Wales and the north. The distinction is not absolute. Even in the 'mountainiouse ground' good pasture might be noted. Yorkshire, of course, had good pasture and corn. 'Good rye, barly and otes, but litle whete' were to be found in the Welsh valleys. Conversely, in the fertile area 'hethes' were not uncommon, and in the lowlands of Lincolnshire and Nottingham, as well as in the Fens proper, there was 'low wasch and sumwhat fenny ground'. Forests were to be found everywhere, and although some of them were 'forests' in the technical sense of being hunting-grounds, much woodland still remained. But the trees were already vanishing. Even the great forest of Windsor supplied timber for wharves on the Thames. The forest of Dean, where the ground was 'frutefull of yron mynes', contained much meadow and corn-land besides its woods, and the same thing was true of the imposing forest of Rockingham, which was twenty miles long by four or five across and housed in its lodges the keepers of the fallow deer. The forests of Shropshire harboured roes, but though Cannock Chase or 'Cank Wood' was still a forest, much of the wood about Lichfield had completely disappeared, and the ground, under the influence of time and culture, 'waxithe metely good'.

Between Knaresborough and Bolton in Craven, however, was a forest twenty miles long, which was in some places eight miles across, and though 'the principal wood of the forest is decayed' it is not stated that meadow or arable land has replaced the timber. Charnwood in Leicestershire, which belonged mainly to the marquess of Dorset, but partly to the king and the earl of Huntingdon, was a real forest; twenty miles in compass, it contained within its precinct 'plenty of woode', but 'no good toune nor scant a village'.

Leicestershire, in fact, was divided. 'Marke that such parte of Leircestershir as is lying by south and est is champaine, and hath little wood. And such parte of Leircestershir as lyith by west and north hath much woodde.'

But in the main the distinction holds, and the Trent, perhaps, may be taken as a boundary:[1] 'After that I cam a litle beyond Trent I saw al champaine grounde *undecunque* within sight, and very litle wood, but *infinita frugum copia*.' This was Leland's impression as he returned south after a tour of Lancashire and Yorkshire.

Throughout the country, and especially in the south-east, communications, if not good, were numerous. Leland did not confine himself to the great highways which radiated from London—to Exeter (via Salisbury), to Bristol, to Gloucester, to Chester, to Lancaster, to York and the north, to Walsingham in Norfolk, to Ipswich, and to Dover. He used many cross-roads besides, and it is plain that, although there were bad patches, especially where clay soil must be traversed, most of these roads were meant to carry wheeled traffic. The number of bridges is surprising, and most of them were of stone, borne on one or more 'fair arches' and approached in some cases by raised causeways through the 'wasche' and low ground about the stream. In the thirty miles between Pately Bridge and Skip-bridge there were no fewer than nine bridges, and between Skipbridge and York a small length of the main road, carried upon a causeway, had 'nineteen small bridges on it for avoiding and overpassing carres [pools] cumming out of the mores there-by'. From York, Boroughbridge, it must be remarked, might be approached by water as well as by land, for the rivers were important highways. Not only Reading but Staines and Maidenhead were important river-ports. There was a bridge at Gloucester but sea-going vessels sailed up the Severn as far

---

[1] Henry VII was at York on more than one occasion and in 1487 actually spent a few days at Newcastle. Henry VIII came to York upon one occasion only (1541); none of the other Tudor sovereigns came so far north.

as Tewkesbury, and smaller craft might go as far as Shrewsbury, passing on their way the ports of Bewdley and Bridgnorth. The Trent, also, was a great artery for commerce, and the smaller rivers were useful too.

The greater streams were kept free for traffic. There was no bridge across the Ouse between York and the sea, and none across the Trent lower than those at Newark; London Bridge was a much-admired rarity. Yet, where bridges were not, ferries, and in some places fords, were used, and altogether the internal communications of England must be reckoned as more than moderately good. The surfaces of the roads were poor, and in winter the difficulties of travel must have been great, but on the whole the country was not ill provided.

The country-side to which the roads and rivers gave access, though sparsely populated in Italian eyes, must have given the general impression of quiet prosperity. There was no great distance between the various little towns which lay along the roads, and each of these towns would be surrounded by its own arable land and pasture—for all towns at this time were in part agricultural. Dotted about between the towns were the villages, each in the midst of its fields, most of them still open fields. Certainly Leland notices the existence of the numerous 'parks' of the gentry which had attracted the notice of the Italian, and it is true that his itinerary did not include some of the areas—Kent for example—where inclosure had long been common. It is possible, too, that he may have thought it prudent not to say too much about rural depopulation. But, although there is mention of pasture, there is frequent reference to good crops of wheat, oats, barley, and sometimes rye, and the general impression given by his survey is that, despite the presence of moors and heaths, much of the cultivable land was devoted to a busy champaign farming.

Apart from the active agriculture there were other signs of industry. The mineral wealth of England, which had attracted the attention of foreign observers, is noticed also in Leland's survey. Quarries of stone, of 'coarse marble', and of slate are frequently mentioned. Alabaster was quarried at Axholm in Lincolnshire and also at Chellaston in Derbyshire, and Tutbury and Burton in Staffordshire. 'Cole' or 'secole' was to be found in Yorkshire, Durham, Lancashire, Staffordshire, and Shropshire; iron in the Mendips, the forest of Dean, Staffordshire,

Warwickshire, Lancashire, and Weardale. Lead was mined in Yorkshire and in Weardale; and it had been mined extensively in Cardiganshire too, though there the industry had been depressed by lack of wood.

The general appearance of the country told a tale of long-continued peace. The bitter wars of the Roses had left few scars. Leland tells of the great battles, of St. Albans, of Northampton, of Wakefield, and of Towton, for example, and he duly records the names of the places where the dead were buried. At Towton they had lain 'in five pittes', still to be seen in his day, before the bones were transferred to the churchyard of Saxton. But of great devastations there is no trace. The careful observer who notes the effect upon Leicester of the 'Barons wars' of Henry III's day, and who hazards the view that Lincoln had been destroyed rather by the Danes than by 'King Stephan', has little to say of the ravages of recent war. Most of the castles mentioned in the itineraries are already ruinous. The Tower of London, of course, was at once citadel and arsenal, but London was not included in Leland's survey, nor was Berwick, a fortress comparable to Calais, and equipped by the time of Henry VIII with the most up-to-date defences. Carlisle, however, appears as a strong town girt with a very good wall and with its castle in order, and elsewhere along the Border were strengths which were designed for use as well as for show. In that vexed country-side, which Leland did not penetrate, the four-square stone-built villages were designed for defence. Farther south, Durham castle, much restored by Bishops Fox and Tunstall, was strong and stately, and in Wales some castles were still intact. Pembroke castle, for example, where Henry VII had been born, and where he, as a child, had long been protected, was 'veri larg and strong, the chymmeney new made with the armes and badges of King Henry VII', and the town itself was walled. In the more settled parts of the realm a few royal castles still retained some remnant of their departed glory. At York, though the castle was partially decayed, some of it remained. The city walls, too, were intact, garnished with towers in one of which there was a great 'chein of yren to caste over the Ouse'. Pontefract, 'of sum caullid Snorre Castelle', with its many towers, was in fair condition; 'bullewarkes of yerth' had been made before the castle gate of Northampton, where the 'kepe' was still of power; Fotheringhay was 'fair and meately strong

with doble diches, and hath a kepe very auncient and strong';
Kimbolton, too, in Huntingdonshire, though modernized, is
'meately strong' also.

But these places are exceptions. Again and again Leland tells
of castles which are ruinous. Some are mere shells, some are used
for ignoble purposes, some are become mere heaps of building
material; of some the very site is hard to trace. The great castle
of Richmond in Yorkshire is 'in mere ruine' and the town walls
are 'ruinus'. Chartley in Staffordshire is 'yn ruine' and 'a good
flite shot' from the 'goodly manor place' which is now in use.
The interior buildings of Rockingham in Northamptonshire are
roofless and crumbling, and within the ruins of the neighbouring
castle of 'Berengarius Moyne' is now 'the meane house for a
fermar'. The castle of Leicester is 'of smaul estimation' and
there is 'no appearaunce other of high waulles or dikes'. No
stone-work at all remained upon the hill upon which the keep
of Groby had once stood. Lord Zouch at Haringworth had
'a right goodly manor, buildid castelle like', preserving, as it
seems, with an eye to defence some of the features of an older
strength. It must be remembered that Leland was interested in
ruins and that his picture was painted towards the end of Henry
VIII's reign, when feudal privilege had withered for some
decades before the 'damn'd disinheriting countenance' of the
Tudor monarchs. But Polydore Vergil, writing much earlier in
the century, had already remarked on the paucity of fortresses,
and it is clear that the gentlemen of England at this time
fashioned their houses for comfort rather than for security. The
day of private local wars was ended. It is plain too that the
king, though he kept some real fortresses for the protection of
his coast and borders, had few castles in a state of defence in the
heart of his land. The Tudor monarchs were of the same counsel
as Machiavelli in the matter of fortresses. They were not without
their uses, but in themselves they were no sure props of kingly
power.

Writ large on Leland's pages is the immensity of the contribution made to medieval building by the church. Almost
every structure which he commends as being well built or well
supplied with water is of ecclesiastical origin. The recent or
contemporaneous 'dissolution of the monasteries' was producing
changes about which it was prudent not to say too much, and
concerning the fate of some of the religious houses the observer

remained discreetly silent. In Northampton 'the Gray freres house was the beste buildid and largest house of all the places of the freres, and stoode beyond the chief market place'. Here, as in other instances, the past tense is used without comment, but sometimes it is made evident that houses recently suppressed are already crumbling to decay. On the other hand, the walls of St. Mary's abbey at Leicester, 'three quarters of a mile aboute', are still intact, and the 'chapelle' of St. Thomas at Oundle is apparently still in use, though it is now called 'of our Ladie'. At Fotheringhay what had been a nunnery survived as a collegiate church, though here the change had been made long before the quarrel of Henry VIII with Rome. That quarrel, though it played havoc with the religious houses in many places, yet made little difference to the parish churches. Every little town had its parish church and the big towns had several. Northampton, for instance, had seven parish churches within the town proper and two in the suburbs. All Hallows, 'the principale, stonding yn the harte of the toune', is large and 'welle-buildid'. The churches generally earn the commendation of Leland for the excellence of their building, and it may be remarked that in some cases the inns—the English inns had a good reputation—were later developed from the town houses of ecclesiastical foundations. Besides its church and its inn, the town of any size would boast, in most cases, an almshouse, a hospital, and a grammar school usually founded by private benefaction. For during the fifteenth century the towns of England had thriven fairly well, and the prosperity had expressed itself sometimes in the construction of larger build-ings less beautiful than the old, sometimes in the addition of new wards. For themselves wealthy traders built fine new houses, picturesque in line and graced with good timber, and to the general good of the community, either as individuals or in company with their fellows, they made handsome dona-tions.

The choice of building material was determined largely by the nature of the local supplies, but there is some evidence that wood was actually preferred to stone. Where stone was abun-dant, it was freely used. 'Sleford is buildid for the most part al of stone, as most part of al the townes of Kesteven be: for the soil is plentiful of stone.' But Doncaster, not so far away, 'is buildid of wodde, and the houses be slated; yet is there great plenty of stone there about'. At Burford, in the Cotswolds, was a notable quarry of fine stone, but near Henley a castle which had been converted into a 'maner place' was 'buildyd about with tymbar and spacyd withe brike'. The 'hole toune' of Leicester at this time was built of timber; so was Loughborough. All the old building at Northampton was of stone, the new was of timber. Much brick was used at Hull, where clay was easily come by, and the greater convenience and smaller cost of brick probably led to a steady increase in its use, though timber was often employed to give additional strength. The gables, even of stone houses, were reinforced or ornamented with wooden beams.

Some of the 'pratie uplandisch tounes' mentioned by Leland were evidently very small; indeed the word meant a village as well as a town in the modern sense. The possession of four paved streets is regarded as worthy of special mention and some towns had but a single street—'Hepworth is the best uplandisch toun for building in one streate in the isle' (of Axholm in Lincolnshire).

It is possible that to the antiquarian eye of the itinerist new constructions might appear as 'modern churches', 'commodious dwellings', and 'housing-schemes' appear to an antiquarian to-day, and that he dismissed too lightly innovations which, now hallowed by time, stand forth to us as monuments of modest affluence and of civic piety. But he was by no means blind to the significance of the towns which, large or small, were fairly numerous along the roads he followed. The ports like 'Lyrpole' had their 'good marchandis'; the inland towns had their local industries. 'In Rotherham be veri good smithes for all cutting tooles'; 'ther be many smithes and cuttelars in Halamshire'; 'Al the hole profite of the toun' of Wakefield 'stondith by course drapery'. 'There was good cloth making at Beverle but that is nowe much decayid.' 'The toun (of Bath) hath of long tyme syns bene continually most mayntainid by making of clothe.' Yet it is essential to remember that the inhabitants of a town are stated to be 'summe marchauntes, sum artificers, sum fermers', and that in describing a town Leland often makes mention of the ground about it. Still largely agricultural, the towns of England were, in comparison with those of other countries, very small. Exact figures cannot be given, but York, the second city of the realm, had perhaps 25,000 inhabitants, and Norwich and

Bristol, which were placed next, may have boasted only 15,000 apiece. There was, however, one city which was large even by continental standards; London with its 75,000 inhabitants[1] was bigger than any German city, though it was but one-third of the size of Paris, the giant of the west. It is not hard to gain a picture of sixteenth-century London, although it has no place in the itineraries of Leland—at least in those itineraries as we have them today—because the city entered so much into national history, because it kept records of its own busy life, and because it was described by admiring foreigners. London throve because it summed up in itself all the various assets which gave to various towns a local importance. It had a good port situated on an estuary over against the prosperous Netherlands; it had its native industries; it had a rich agricultural *hinterland*; it stood upon the great waterway of the Thames, and was, as it had been since Roman days, the nodal point of a great road system. Dowered with these blessings London dominated the country. The Venetian 'Relator' is almost rhapsodical about the city's wealth, and considered it equal to Florence or to Rome in population. The judicious John Major held that it was bigger than Rouen, the second city of both Gauls, and enlarged upon its prosperity which he attributed to three factors. Within her walls sit the supreme courts of justice; she enjoys the almost constant presence of the king, who at his own expense provides for a great household and supplies to them all their food; and, most important of all, she harbours a great concurrence of merchants.

Another Scot has left a more famous tribute. In December 1501 there came to London certain ambassadors to discuss the marriage of James IV with Margaret Tudor. Along with them was the poet, probably William Dunbar, who sang: 'London, thou art of Townes A *per se*.' Lauding the wealth and beauty of this 'lusty Troynovaunt . . . Strong Troy in vigour and in strenuytie', he praised the river and its great ships, its barges, and its swans; the bridge with its pillars white; the Tower with its great artillery.

[1] These figures appear to be highly speculative. See Thorold Rogers, *Six Centuries of Work and Wages*, 2 vols., pp. 62, 117, 121, 463, and Aeneas Mackay's additional notes to his edition of *Majoris Britanniae Historia*, p. 395 (Scottish History Society, vol. x, 1892). See also E. F. Gay in the *Quarterly Journal of Economics* (1902, xvii. 575–9). The best modern authorities decline to commit themselves to figures for the population of England at this time.

Strong be thy wallis that about thee standis;
  Wise be the people that within thee dwellis;
Fresh be thy ryver with his lusty strandis;
  Blithe be thy chirches, wele sownyng be thy bellis;
  Riche be thy merchauntis in substaunce that excellis;
Fair be their wives, right lovesom, white and small;
  Clere be thy virgyns, lusty under kellis:
London, thou art the flour of Cities all.

By this time London had spread beyond her wall, but the
wall was still essential. Beginning on the east at the Tower it
ran north to Aldgate, then north-west by Bishopsgate, Moor-
gate, and Cripplegate, south-west to Aldersgate, west again to
Newgate, and finally south to Ludgate to find its western end
where the unsalubrious Fleet issued into the stately Thames.
The strong Bridgegate at the southern end of London Bridge,
backed by a drawbridge and another tower, completed the
defences. The bridge itself was believed, not by the citizens only,
to be one of the wonders of the world. Begun in 1176 and
finished in the reign of John, it remained, thanks to timely
repairs, for centuries a monument to its builders; only in 1832
was it pulled down to give place to its successor of today. The
old bridge was longer than the new one, as much of the old
foreshore has been reclaimed. It had 19 arches with spans from
10 to 30 feet. Each pier was supported by a ring of encircling
piles, within which stones were heaped, and the resultant
'stradelings' or 'starlings', which were islands at low tide, so
constricted the channels that the rush of water beneath the
arches was tremendous. 'Shooting the bridge' was a dangerous
business; prudent folk landed at the 'Old Swan' above the
bridge and re-embarked at Billingsgate (Bolin's Gate) if they
meant to continue their journey by water. This they might well
do, for the river was one of the great arteries of traffic in a city
whose narrow streets, far from straight, were rendered more
narrow still by 'pentices', projected from the houses at first-
floor level and supported upon pillars. Most of the houses were
of wood, with roofs of tiles; though thatching, albeit forbidden,
was not unknown.

Yet the general impression cannot have been one of squalor.
Apart from the grim Tower there were some important secular
buildings, the Guildhall, the Custom House, near Billingsgate,
and the three markets—Blackwell Hall for cloth in Basinghall

Street, the Leadenhall, and the Stocks Market on the site of the present Mansion House.[1] Where Cannon Street station now stands the Hanse Merchants had their depot—the Steelyard, strongly built of stone, and the halls of the Great Companies were handsome buildings. Those of the Merchant Taylors, the Grocers, the Goldsmiths, the Skinners, the Haberdashers, the Vintners, and the Clothworkers stood in the days of Henry VII upon the very sites which they now occupy, or occupied until the aerial bombardments of the recent war. Magnates, both lay and clerical, maintained town-houses or 'inns', built of fair stone, and some of the great merchants, too, had magnificent dwellings, of which Crosby Hall in Bishopsgate, now removed to Chelsea, where it is used by the International Federation of University Women, may serve as an example.

Along the river-front were wharves and steps and the hum of commerce, but there were some big buildings too, conspicuous among them Baynard's Castle, which Henry VII rebuilt, not after the former manner with embellishments and towers, but as a gracious palace surrounding two courts, and, save that it lacked other gardens, not unlike the other great houses which lined the Strand farther west.

But in London, as elsewhere, most of the great buildings were ecclesiastical. St. Paul's, set upon its hill with a spire 500 feet high, dominated the city; it boasted two cloisters, and round its precinct clustered the houses of the bishop of London and of the clergy. To the south lay the famous churchyard, at the east end of which Colet erected his school. Inside the walls were no fewer than ninety-seven parish churches, each with its yard about it. The traveller approaching the city must have beheld a very forest of spires and on entering must have been greeted with a cheerful clamour of bells. Besides the parish churches there were 'a score of great religious houses', each in its own precinct with graveyard, gardens, and cloisters surrounding a beautiful church.

London, however, had expanded far beyond its wall. Across the stream and about the end of London Bridge lay the important suburb of Southwark containing the churches of St. Olave and St. Mary Overy, the hospital of St. Thomas, and the palace

---

[1] See C. A. J. Armstrong, *The Usurpation of Richard III*, pp. 159–60, for some useful notes.

of the bishop of Winchester, whose liberty 'the Clink' housed some doubtful places of amusement. Elsewhere the city had moved into the surrounding fields in a kind of 'ribbon development' along the great roads. Hoxton, Islington, Clerkenwell, Shoreditch, and Whitechapel were garden suburbs. These 'towns' already had begun to inclose the common fields, so that neither could the young men of the city shoot, nor the ancient persons walk for pleasure without the risk of their bows being broken and of themselves being indicted. In 1514 the process was arrested, for a time at least, by the resolution of a turner who, crying 'shovels and spades', inspired the citizens to destroy the obnoxious hedges and ditches.[1] But 'the fields' were doomed to disappear. One street, which issued from the east, took its busy way to Wapping, but the more aristocratic roads which debouched to the north-west and west ran past the houses and the gardens of the great. Most important was the road which led from Ludgate to Westminster. Between Fleet Street and the river lay Bridewell, a royal residence until, after a period of disuse, it became in 1553 a house of correction; Whitefriars, before long too, became a dangerous 'liberty' where licence throve, and the Temple where the lawyers had been established —it is uncertain how—ever since the middle of the fourteenth century. Beyond Temple Bar, then made of wood, the Strand continued past an array of episcopal palaces, past the Savoy, where Henry VII had erected, or revived, a hospital for the poor, to Charing Cross. There the main road, leaving the open fields upon its north-west and west, ran southwards alongside the river to advance to Whitehall, the house of the archbishops of York until 1529, the great hall of Westminster, the palace which lay to the south of it, and the abbey to which Henry VII added the magnificent chapel which bears his name.

Westminster was the true seat of government. It was there that the parliament met, and there that the king most often resided, though he might upon occasion occupy the Tower or Baynard's Castle, or at a later date Bridewell. But Westminster, after all, is very close to London, and from his palace there, which backed upon the river, the king had easy access by water to the city and

---

[1] Hall's *Chronicle* (ed. Whibley, 1904), i. 119. Cf. Stow, *Survey of London* (ed. Kingsford, 1903), ii. 77. Much of the material used in the description of London is derived from Stow.

to the great stronghold which dominated it. From Westminster he could easily gauge the temper of the city, could borrow of its wealth, could if need be threaten or coerce. In its half-rural seclusion at Westminster the government could and did maintain the closest relations with the city of London. It needs must, for London was the heart of England.

III

THE NEW KING AND HIS RIVALS

'The crown which it pleased God to give us with the victory of our enemy in our first field.' *Will of Henry VII.*

THE battle of Bosworth was fought on 22 August 1485. 'Ricardus inter confertissimos hostes praelians interficitur.' Richard III died like a king with his sword in his hand. He was carried stark naked from the field and hugger-mugger buried in the Grey Friars at Leicester.[1] The power and the terror of him had vanished with his breath. His conqueror Henry Tudor was made king by virtue of the crown which was found under the hawthorn and placed upon his head, and by the acclamations of the soldiery. 'The coronation ceremony is older than any act of parliament', and in that ceremony the verbal assent of the people is an essential part. To Bacon, enumerating the 'three several titles' of the new king, the 'title of the sword or conquest' took its place after that of the Lady Elizabeth, secured by the 'precedent pact' of marriage, and after the 'ancient and long-disputed title' of the house of Lancaster. Certainly the 'kind of military election or recognition' which took place on the battlefield was later confirmed by more formal acts; but it was none the less decisive. The essential thing was that Richard was not only beaten but killed. Had he lived neither the crown nor the triumphant shouting would have secured Henry in his new estate, and the formal acts of confirmation might never have taken place. Though the titles enumerated by Bacon might well be placed in a different order, Henry had, apart from his personal gifts, the three assets necessary for kingship in fifteenth-century England—a dynastic claim, some support from the rival house, and force enough to make his pretension good.

His title by inheritance was weak. His mother Margaret Beaufort was great-great-granddaughter to Edward III through John of Gaunt; but if a claim on the distaff side were once

---

[1] In the words of a *continuator* of the Croyland chronicle 'inter pugnandum et non in fuga, dictus Rex Richardus, multis letalibus vulneribus ictus, quasi Princeps animosus et audentissimus in campo occubuit'.

admitted the Yorkists possessed a better right in their descent from Gaunt's elder brother Clarence, and the cause of Lancaster was utterly lost at the outset. The Beaufort claim, moreover, was marred by a stain of illegitimacy which had not been quite clearly removed.[1] On the paternal side too Henry's descent was obscure as well as illustrious. The Tudors were an old-established family in Anglesey which claimed Cadwaladr as its progenitor, and a scion of this house, Meredith by name, had been butler—some would give him the dignity of *scutifer*—to a bishop of Bangor and escheator of Anglesey[2]. Owen Tudor, son of Meredith, became clerk of the wardrobe in the household of Catherine, widow of Henry V, and so commended himself by his good looks and good address that he succeeded in marrying his royal mistress. Of obvious impropriety, and perhaps of doubtful legality,[3] the match was not unnaturally kept secret, and after Catherine's death in 1437 Owen was summoned before the Council. After some doubtful passages he was committed to Newgate, from whence he twice escaped; he was pardoned in 1439, fought for the Lancastrians in the civil war, and was executed by the victorious Yorkists after the battle of Mortimer's Cross. The legitimacy of his issue, however, was never in dispute, and Henry VI was kind to his half-brothers. The younger son, Jasper, raised to the dignity of earl of Pembroke (1453), showed himself an unwavering supporter of the

---

[1] See S. Bentley, *Excerpta Historica* (1883), p. 152. The Beauforts, children of John of Gaunt by Catherine Swynford before the marriage of their parents, had been legitimated by a patent of Richard II which was confirmed by parliament. When in 1407 Henry IV confirmed this patent he added the words *excepta dignitate regali*, and these words were added, apparently at the same time, to the patent of 1397. The earlier patent, however, had been ratified by parliament in its original form, and a royal patent in itself could not avail against act of parliament. In fact the Beauforts had played the part of princes of the blood in the politics of the fifteenth century.

[2] For the Tudor pedigree the best account is still that of 'J. W.' in *Archaeologia Cambrensis*, 1869. Meredith, however, does not appear among the sons of a well-established Tudur-ap-Gronw (d. 1367) in the earliest pedigrees, though the story giving him this paternity goes back to 1710 anyhow (*Arch. Camb.*, 1849, 268, 9). The obvious necessity of magnifying Henry VII's ancestry after his accession makes the account of Meredith's descent a little uncertain though his relation to the Anglesey Tudors cannot be doubted in view of contemporary bardic references. Henry himself evidently valued his Welsh lineage highly; Bernard André made much of it. W. Garmon Jones, *Welsh Nationalism and Henry Tudor* (Cymmrodrion Transactions, 1917–18).

[3] H. T. Evans, *Wales and the Wars of the Roses* (C.U.P., 1915), p. 67. It seems likely that Owen had not received letters of denization when he married Catherine.

cause of Lancaster throughout the civil war; the elder, Edmund, who was created earl of Richmond in 1453, married Margaret Beaufort and became the father of Henry, later Henry VII, born in Pembroke castle on 28 January 1457. The vicissitudes of Henry's youth were remarkable. A posthumous child whose mother was fourteen years of age when he was born, he was brought up in Wales by his uncle Jasper, until, when Harlech castle surrendered in 1468, he fell into the hands of Lord Herbert, soon to become the new earl of Pembroke. A marriage was projected between the little boy and Maud, daughter of his captor, but in 1470 he was recovered by Jasper who managed to get him across to Brittany when the alliance between the Lancastrians and the King-maker came to its disastrous end. In Brittany he remained—somewhat under restraint—the pivot or the victim of Lancastrian or Yorkist conspiracy, until he was hunted out by the vigorous pursuit of Richard, who corrupted the venal chancellor of the ageing Duke Francis II, Pierre Landois. He took refuge in France, and it was from Harfleur that he set out with a motley following on his venture for the English throne on 1 August 1485. Not without justification might Richard have described him as 'an unknown welshman (whose father I neuer knew, nor hym personally sawe)'.[1] The speech put into his mouth by Hall is obviously a rhetorical exercise, but it reflects the contemptuous arrogance of Richard's proclamation against Henry of June 1485.[2] Henry's title by descent was important because it was the only 'Lancastrian' title. There were better titles to be found in the house of York, and he was wise to support his slender claim by alliance with the stem of his rivals.

Happily for him this was the easier because the scions of the White Rose were at variance among themselves. Edward IV had alienated the old aristocracy by his marriage to Elizabeth Woodville. He had broken the king-maker, he had executed Clarence. Richard III had declared the issue of Edward IV illegitimate, had murdered the princes, and had seized the

---

[1] Hall's *Chronicle* (1809), p. 415.

[2] See Pollard, *The Reign of Henry VII from Contemporary Sources* (henceforth cited as Pollard, *Henry VII*), i. 3, quoting from *Paston Letters*, iii. 883. Richard treats the alleged claim of this Henry Tydder with no respect. The vicissitudes of Henry's career attracted great interest. They were noted by Commines. After he was securely upon the throne they were quoted by his supporters as an example of divine providence.

Crown himself. Each successive act of violence had left its legacy of hate, and the once powerful Yorkist party was shivered into fragments, mutually distrustful but all ill disposed to the king. Richard had alleged that Edward IV's children were illegitimate; but in fact they were not; and after the murder of the princes there survived several daughters of whom Elizabeth was the senior. He had declared that Clarence's attainder had corrupted his blood; but Clarence had left a son Edward, earl of Warwick, and a daughter Margaret, who became countess of Salisbury. His own sisters had left issue too, and when, in 1484, his only son died, it was plain that this issue might pretend an interest in the Crown.[1]

Richard himself was conscious of the danger of his position and had sought to ally the scattered forces of York. Early in 1484 he induced Elizabeth Woodville to come out of sanctuary promising good marriages to her daughters, and on his son's death he proclaimed as his heir John de la Pole, the son of his sister Elizabeth. After the death of his queen in 1485 he thought of marrying his niece Elizabeth, sister of the murdered princes, and it was even suspected that he had poisoned his ailing wife to that end; although, warned even by his most unscrupulous councillors, he decided that this unnatural marriage was impossible and publicly contradicted the rumour as a gross slander, it seems that the project was certainly in his mind. His interest in Elizabeth was due in part to the knowledge that there was afoot a design to marry her to Henry Tudor. The exiled party of Lancaster had already been reinforced by Yorkist malcontents. Buckingham's abortive rising in 1483 had been made partly at the instigation of the silver-tongued Morton, bishop of Ely, and had been meant to coincide with a landing by the earl of Richmond which was prevented by ill weather. Henry's mother, now Lady Stanley, had been involved in the plot, along with the Courtenays and other gentlemen whose allegiance to Lancaster was traditional; but an essential element of the conspiracy was that Henry should marry Elizabeth, and many of the Yorkists had been involved too. The facile but ambitious Elizabeth Woodville had given her consent; her brother Edward Woodville, a notable soldier,

---

[1] See genealogical table—Pretenders of the House of York. If attainder were held to vitiate a claim, then the claim of Henry was barred by his own attaint.

and her son, the marquis of Dorset, had come in. Buckingham's family had always supported the White Rose, and he
himself had been a close confidant of Richard; amongst his
supporters were a representative of the house of Bourchier
and Sir Thomas St. Leger, who had married Richard's own
sister.

When the rising collapsed most of the conspirators who
survived took refuge in Brittany, and Henry had been quick to
consolidate his position by a formal act of union with the traditional enemies of his house. On Christmas Day, 1483, the whole
party went together to Rennes Cathedral where they pledged
themselves to be true to one another and swore allegiance to
Henry as if he were already king; Henry for his part gave his
corporal oath to marry the Princess Elizabeth as soon as he
obtained the Crown.

Richard's reaction was prompt. He demanded from the duke
of Brittany the surrender of Henry and the exile saved himself
only by a hasty flight into Anjou disguised as a page among his
own servants. He won back the uncertain queen-dowager and
at her suggestion Dorset tried to break away from his new allies,
though he was arrested and brought back. He raised money by
forced loans, set the coasts in a state of defence, and posted
himself at Nottingham ready to meet the invasion on whatever
coast it should land. He was still confident; he took no severe
measures against Henry's mother, and still trusted her husband
Stanley. But in fact many of the old adherents of the White Rose
had given their sympathies to Henry.

Sympathy was not enough. The problem for Henry was to
find the necessary force. His resources were not large, but
various considerations urged him to prompt action. His narrow
escape in 1484 and the defection of Dorset reminded him that
his enemy's arm was long and would reach to Brittany. The
story of Richard's intention to marry Elizabeth so alarmed him
that he contemplated marriage with a sister of Sir Walter
Herbert and actually began negotiations through Northumberland, who had married another of Sir Walter's sisters. Delay
was dangerous, and he received at this time a new supporter
whose advice was probably in favour of a bold venture. This
was John de Vere, the thirteenth earl of Oxford, whose spirit
was unbroken by a long imprisonment; he brought to Henry's
service his keeper James Blount, who had released him from the

castle of Hammes and fled along with him. Henry determined to use at once the small means available to him. He equipped half a dozen ships at Rouen. The French regency lent him 60,000 francs and 1,800 mercenaries under Philibert de Chandé, later created earl of Bath. It has been alleged that Henry had some understanding with Scotland and that, included in his force, were some of the good Scots troops which the French king kept about him; but it is clear that the French Crown would not employ the competent professionals of the Scots Company of Men-at-Arms or the *corps d'élite* of the Archer Guard upon so doubtful an enterprise. According to a French account his soldiers were the worst rabble one could find, and in spite of the very definite assertions of John Major and Pitscottie it is improbable that any large body of Scots fought at Bosworth. A few competent captains there certainly were; and there may have been some vague promises from Scotland, but that is all.[1]

His hopes were based on the Stanleys whom he could approach through the agency of his mother, and upon Wales. It has been represented that the spirit of Welsh nationalism was strong, and that, disappointed in the leadership first of Lord Herbert and later of Buckingham, it turned instinctively to the young Tudor. This argument may be exaggerated, for Welsh sentiment was still in some sense tribal, and unity was hard to come by. What is certainly true, however, is that circumstances had tended to drive all Welshmen into the same camp. Those who had given to the Yorkists the homage due to the brood of Mortimer had been shocked by the murder of the princes, true children of the royal line. Jasper, earl of Pembroke, had a great reputation, and it is true enough that beneath the local jealousies there survived the sense that all Welshmen were akin after all and might well give their allegiance to a son of the old British line. Throughout the land men were studying the obscure vaticinations of the

---

[1] Miss Conway, in *Henry VII's relations with Scotland and Ireland 1485–1498*, was inclined to attribute the good relations between Henry VII and James III to the help given by the Scots at Bosworth, and quotes (p. 6) the evidence of Major and Pitscottie. As Pitscottie numbered amongst his informants Sir William Bruce of Earlshall whose father, Sir Alexander, certainly fought at Bosworth, there is some foundation for the story. It seems certain, too, that Bernard Stuart of Aubigny, who had recently been in Scotland to renew the auld alliance, accompanied the expedition and may have been in command of a contingent. *The Stuarts of Aubigny*, p. 27. Lady Elizabeth Cust, quoting good evidence. But Henry was having trouble from the Scots as early as October. Pollard, *Henry VII*, i. 19–21.

bards who promised great things to the scion of the famous kings of old:

> Jasper will breed for us a Dragon—
> Of the fortunate blood of Brutus is he—
> A Bull of Anglesey to achieve;
> He is the hope of our race.[1]

This great bull should conquer the snarling boar.

Henry heard from a lawyer, Morgan of Kidwelly, that Rhys ap Thomas and Sir John Savage were ready to take his part, and it was in Wales that he made his venture. He left Rouen on 1 August and landed at Milford Haven on 7 August. Immediately on landing he knelt down, crossed himself, and sang 'Judica me, Deus, et decerne causam meam'. Bacon is sceptical about the sincerity of Henry's acts of devotion, but without cause; Henry had ventured all upon an ordeal of battle, and he was well aware what he was doing. It has been argued that a man so prudent as Henry would not have launched his attack until he was confident of success, but, as has been shown, Henry's venture was almost thrust upon him, and the issue of the trial by combat was uncertain.

He landed with what Richard described (according to Hall) as 'a nomber of beggerly Britons and faynte harted Frenchmen',[2] and Wales rallied to him less quickly than he had hoped. Afterwards it was said that his friends deliberately refrained from declaring themselves in order to deceive Richard, but their delay may have been due to other reasons. Rhys ap Thomas, the hero of tradition, did not openly join him until he had reached Shrewsbury, and even after his force had been swelled by English adherents, including some deserters from Richard's army, it amounted to barely 5,000 men on the day of battle as against his opponent's 10,000. It is easy to believe Polydore Vergil's story that Henry, separated by accident from his army near Tamworth, spent the whole of a dreadful night 'non magis ob praesens quam ob futurum periculum perterritus'.[3] Richard displayed a confidence which may have been only in part assumed. More's picture of the restless, suspicious tyrant, which survives in Shakespeare's play, was painted after the king was dead, though it must be observed that the Croyland *continuator*,

---

[1] See W. Garmon Jones, *Welsh Nationalism and Henry Tudor*, pp. 32 and 33.
[2] Hall's *Chronicle*, p. 415.
[3] Polydore Vergil, *Anglica Historia* (1534), xxv. 555.

who finished his work on 30 April 1486, mentions the terrible dreams which tormented the last night of the doomed monarch. Richard's crimes were notorious, but the England of his day was not easily shocked; his forced loans had alienated the merchants, and Collingbourne's famous distich[1] shows that he and his confidants were bitterly disliked in some quarters:

> The Rat, the Catte and Lovell our dogge
> Rule all England under the hogge.

Yet if we judged by the record of his parliaments he would rank as an enlightened ruler; he was popular as a good soldier, and one who had retaken Berwick; even after the murder of the princes the bishop of St. David's could write of him 'God hathe sent hym to us for the wele of us all'.[2] Even after his fall the town council of York spoke of the 'grete hevynesse of this Citie' on hearing of the death of King Richard 'late mercifully reigning upon us'.[3] No doubt he knew that there was disaffection in his ranks, but he knew how to deal with that. Lord Strange, Stanley's son, was in his hands as a hostage, and his reputation for ruthlessness made men chary of deserting his cause. When the news of Richmond's landing came to him his forces gathered quickly, but he may have been taken by surprise by the rapid advance of Henry who reached Shrewsbury unopposed, for although the keeper of the Tower of London came to him hot-foot on his summons there is no word of any artillery in the royal host. When the battle was joined Northumberland stood aloof and the Stanleys came in on the Tudor side, though only after the fighting had begun. In the first encounter Richard broke into Henry's battle, killed his standard-bearer, Sir William Brandon, with his own hand, and tried to bring his rival to a personal encounter. It may well be surmised that, had it been Henry himself who fell, the issue of the battle would have been other than it was. But the Stanleys advanced; Richard's followers, mutually suspicious of each other's loyalty, did not back their king; the attack was repulsed; it was Richard who fell.

The ordeal by battle was ended. The first act of the victor was to thank Almighty God upon his knees for the victory and to pray for grace to rule in justice and in peace his 'subjectes

[1] Hall's *Chronicle*, p. 398.
[2] See A. F. Pollard, 'Sir Thomas More's Richard III', in *History*, xvii. 319–20.
[3] Pollard, *Henry VII*, i. 17.

and people by God now to his gouernaunce committed and assigned'.[1] Already Henry regarded himself as king and before long Lord Stanley set upon his head the battered crown of England which Richard had worn when he went into action. The representative of Lancaster, pledged to the Rose of York, had found the force needful to secure the crown, and with the crown went the governance of England.

Henry behaved like a king from the first. Indeed it is possible that he had already used the royal title in summoning his Welsh adherents; and no sooner was the victory won than he displayed his prerogative by knighting eleven of his supporters on the field itself. Immediately he sent Sir Robert Willoughby to Sheriff Hutton in Yorkshire to secure the persons of the Princess Elizabeth and of the luckless Edward, earl of Warwick. He moved slowly towards London, well received as he went along perhaps because he had commended himself by a circular letter wherein he had commanded his people on pain of death 'that no manner of man rob or spoil no manner of commons coming from the field; but suffer them to pass home to their countries and dwelling-places, with their horse and harness'.[2] Probably he rested a few days at St. Albans before he made his formal entry, while a great welcome was prepared for him. At a meeting of the common council held on the 31st it was resolved that the sixty-five 'mysteries', represented by delegations which reflected their importance, amounting in all to 435 persons, should welcome him in robes of 'bright murrey'. The mayor and the aldermen were to ride in scarlet, accompanied by sword-bearer and sergeants, and the aldermen's servants 'in tawney medley'; there were to be as many armed men as could be mustered. When Henry arrived on 3 September he was met at Shoreditch and proceeded in triumph to St. Paul's—*laetanter* not *latenter*[3]—where he offered his three standards, the banner of St. George, the red fiery dragon of Cadwaladr on a white and green ground, and a dun cow painted upon 'yelowe tarterne'. Then, after prayers and a *Te Deum*, he went to stay for a few days in the palace of the bishop of London.

---

[1] Hall's *Chronicle*, p. 420.

[2] See Pollard, *Henry VII*, i. 12, quoting *Letters of the Kings of England*, i. 169.

[3] Speed, relying on some misreading of Bernard André, paraphrased *latenter* as 'covertly, meaning, belike in a horse litter or close chariot'; and Bacon went on to interpret his action as that of a man who, having been a proscribed person, 'chose rather to keep state and strike a reverence into the people than to fawn upon them'.

The council had voted him a sum of 1,000 marks as a present, and that their goodwill was shared by the citizens appears from the records of processions and pageants. These were cruelly interrupted by the outbreak of the 'sweating sickness' brought by the foreign mercenaries, which slew the lord mayor, his successor, and six of the aldermen in the course of a few days. But the plague ceased as suddenly as it began, and on 30 October Henry was crowned with a magnificence 'which had probably never at this time been equalled at the coronation of any English monarch'. The purchases made by Robert Willoughby, knight of the king's body, later as Lord Willoughby de Broke, steward of the household, are impressive. Cloth of gold in purple or in crimson for the king cost £8 per yard. White cloth of gold for the henchmen was obtained at 33s. 4d. to 40s. per yard, blue cloth of gold at 46s. 8d., and green cloth of gold at 60s. Purple velvet for the king's robe cost 40s. per yard; Lord Oxford's crimson velvet 30s., and crimson satin 16s. The king's confessor was clad in russet cloth at 13s. 4d. a yard. Scarlet cloth was bought in large quantities from several 'citezyns and taillors' at a price which varied from 14s. to 7s. a yard. Fine blue cloth, red cloth, and black cloth were obtained at lower prices. Ermine and miniver, sarcenet, buckram, and worsted were liberally bought. Payments were made for velvet for red roses and dragons, and for embroidering Cadwaladr's arms. Fringes and tassels of gold and silk, ostrich feathers for the henchmen, swords and saddles and horse-gear, all were provided with a lavishness which seems at variance with the alleged parsimony of the first Tudor king. It must be noted, however, that the accounts were carefully kept; and while the variation in the cost of material reflects the varying importance of the wearers, the different prices paid for scarlet cloth suggest that the king's officer bought where he could at the best market available. It seems possible, indeed, that the demand for this particular commodity was so great that the price began to rise against the king.[1]

Bacon's suggestion of the almost furtive king is wide of the mark, but it is true enough that behind the outward display the king was quietly concentrating his power. The small group

---

[1] For the royal purchases and the promotions of the king's supporters see W. Campbell, *Materials for a History of the Reign of Henry VII* (2 vols., Rolls Series, 1873, 1877), ii. 3–29, *et passim* (henceforward cited as Campbell, *Materials*).

of intimate advisers who had accompanied him from France provided him with the core of a council which was reinforced after Bosworth by Stanley and soon by Fox from Paris, and by Morton who returned from Flanders before the year was out. Jasper, duke of Bedford, the earl of Oxford, Lord Strange, Sir William Stanley, Giles Daubeney, Lord Dynham, Sir Richard Edgecombe, Sir Edward Poynings, and above all Reginald Bray were among the members of the council. Administrative and financial posts were given to trusty adherents; John Alcock, bishop of Worcester, became chancellor in October 1485; Thomas Lovell was made chancellor of the exchequer, Edgecombe comptroller of the king's household, Fitzherbert king's remembrancer, and Robert Bowley master of the coinage of groats in Dublin and Waterford. Later, as the king's authority became better established, a closer liaison was made between the unofficial advisers of the Crown and the formal administration; in July 1486 Dynham became treasurer; in the following February Fox was made keeper of the privy seal, and in March Morton succeeded Alcock as chancellor. From the first, however, Henry had a competent council[1] which could supervise the work of government. In a meeting held soon after his arrival in London the marriage with Elizabeth had been discussed, but, like a later king, he had no mind to be his wife's 'gentleman usher', and he was at pains to assume the crown and have his title recognized in parliament before the marriage was completed.

Meanwhile he saw to it that actual power passed from Richard's supporters into the hands of men on whom he could rely. To mark his coronation he bestowed a few honours. Pembroke became duke of Bedford, Stanley earl of Derby, and a new peerage was created for Sir Edward Courtenay, who became earl of Devon. Twelve new knights banneret were also made. But this meagre list of honours is only the outward symbol of a steady process whereby his friends were rewarded and his enemies abased. A glance at the Patent Rolls and the Register of the Privy Seal shows that before and after his coronation Henry made a whole series of grants which established his authority throughout the length and breadth of the

---

[1] Hall's *Chronicle*, p. 424. The list there given is obviously incomplete. Probably Hall omitted some of the names which might be 'taken for granted' but the 'council' was a somewhat indefinite body. Pollard, *Henry VII*, iii, app. i.

land. The patronage at his disposal, including that of the rich duchy of Lancaster, was surprising, and by its use Woodvilles and Stanleys, Nanfan, Edgecombe, Morley, and other trusty friends who had served him 'by yonde the see as over this side' took the place of the followers of Richard, always described as 'in dede and not of right' king of England.[1] Some of the posts which changed hands were of the first consequence. The castles went into safe keeping. Sir William Willoughby obtained Norwich, Sir Giles Daubeney Bristol, Lord Welles, the king's uncle, Rockingham; Cardiff, with authority over the counties of Glamorgan and Morgannock, went to Roger Cotton, Launceston to Sir Richard Edgecombe; and the castles of Salisbury, Gloucester, Penrith and Worcester, Carisbrooke, and others were given to new custodians. Sir Thomas Bourchier was restored to his constabulary of Windsor castle, and the faithful De Vere, earl of Oxford, became high admiral of England and constable of the Tower. Besides these great offices there were many others which brought power, influence, and remuneration. Oxford, for example, was made 'keeper of the lions, lionesses and leopards' in the Tower of London with wages of 12 pence a day for himself and 6 pence a day for the support of each animal; and he was to be examined from time to time regarding the number of animals in his care. For less exalted servants of the Crown the castles offered the humbler posts of porters and watchmen. The crown lands and lands in the king's hands presented other opportunities for the confirmation of the royal power; there is a long list of appointments of stewards and bailiffs. The forests, too, provided a host of offices great and small which could be used to reward the king's friends. The king's stepfather, Lord Stanley, was created 'master forester and steward of all the game northwards beyond Trent', and at the other end of the scale John Grey, gentleman, was made a rider, for life, of the forest of Dean. The rolls and registers abound with the names of the king's supporters who were made constables, rangers, parkers, walkers, and warreners.

One way and another the royal authority was quietly extended throughout the country; there was apparently no serious opposition; some of Richard's stoutest supporters—Norfolk,

---

[1] The phrase has been held to exemplify Henry's 'settled' aversion to the house of York as noted by Bacon, but it was in fact used by the Commons in Edward IV's reign. *Rotuli Parliamentorum*, v. 509.

Ferrers, Ratcliffe, Brackenbury, Catesby—had been killed in the field, though Northumberland was a prisoner, and Lincoln and Lovell had disappeared. The Yorkist party had lost its leaders, and without them it could do nothing; the apprehensions of a great body of their supporters had been allayed by a general pardon to 'persons of the north parties' issued on 24 September.[1] It is probable that loyalty to the White Rose was less enthusiastic than has sometimes been represented, for the military force in the king's hands was surprisingly small. He controlled, of course, the great arsenal of the Tower, but he dismissed his foreign mercenaries as soon as he might; and, except for some gunners, the only 'regular force' at his disposal was the small guard which he created at the time of his coronation, formed, no doubt, upon the French model. This consisted of 'a certain nomber as well of good Archers, as of diverse other persons, being hardy, strong and of agilitie, to geve dailye attendance on his person, whom he named Yeomen of his Garde'.[2] This force, however, though it impressed an England unused to a permanent establishment of paid soldiery, was very small; it mustered only fifty at first, though the Italian Relator, writing about 1500, made it three or four times as strong. It served its purpose in guarding the king, and in later days it was occasionally useful in *émeutes* in London or about the court; but it cannot seriously have augmented the military strength of the Crown. The secret of Henry's easy triumph was that his authority was generally regarded as the only guarantee of good order. Even the Croyland chronicler, who commended the valour of Richard, regarded the victory of Bosworth as divinely given and said that the victor 'began to be lauded by all men as an angel sent from heaven'.

In the eyes of the sixteenth-century Englishman, however, good order must have a formal basis. It might be empirically true that Henry had gained the crown by battle, and would keep it by the power and the patronage which he enjoyed. As says a great authority: 'It was a case of Aaron's rod swallowing all the rest: the greatest feudal magnate gained the Crown.'[3] It is

---

[1] Williams, *England under the Early Tudors*, p. 3.

[2] Samuel Pegge, *Curialia* (1791). There were yeomen of the guard about Henry from the moment of his advent to England and evidently before that. See Campbell, *Materials*, i. 8, 46, 71, for grants to yeomen in September and October.

[3] Pollard, *Henry VII*, i. xxx, n. 1. See *Rotuli Parliamentorum*, vi. 241, where the importance of a parliamentary declaration, even of a right founded on the laws

true enough that the limitations upon the *praerogativa regis* known
to the middle ages applied to the prerogative only in its feudal
sense, and that the meaning of a 'royal authority' was well
understood by king and by people alike. Yet was there a general
belief that authority must express itself in the traditional way,
and it is significant that, only twelve days after his arrival in
London, Henry had issued writs of summons for a parliament.
It was with the aid of parliament, he realized, that he could best
do certain necessary things—some of them unpopular. He could
set his finances upon a proper footing; he could reward his
friends and punish his enemies in the most correct legal fashion;
above all he could invest his new-found crown with the
semblance of a properly constituted authority. Not without
intent, it may be supposed, he had arranged that his coronation
should precede his meeting with his parliament, and when
parliament assembled on 7 November its business was conducted in the time-honoured way. The king was seated upon
the royal throne in the 'chamber commonly called of the Cross
within the palace of Westminster'; the lord chancellor, Alcock,
delivered the oration with which it was proper that 'an archbishop, or a bishop, or a great clerk discreet and eloquent'
should open the proceedings, and the oration, graced with
quotations from Livy and from the Fathers, followed the familiar
lines, lauding righteousness as the foundation of a happy state,
but distinguishing between the duties of the prince and those of
his subjects. The commons were dismissed with instructions to
choose a Speaker, and then the king, himself nominating
receivers and triers of petitions for Gascony and Scotland as well
as for England and Wales, made patent his claim to all the
rights his predecessors had enjoyed. The Speaker who was
presented next day turned out to be Sir Thomas Lovell, one
of the Bosworth men, a trusty councillor, who had already been
made chancellor of the exchequer,[1] and who had evidently
been designated for his new office before parliament had even
met.[2] To his formal prayer for the accustomed privileges of the
Speaker, it was the chancellor who replied, but later the king
himself addressed the commons, telling them he had come to

---

of God, of nature, and of England, is firmly asserted in the act about the title of
Richard III.
[1] 12 October.
[2] Campbell, *Materials*, i. 84; Pollard, *Henry VII*, ii. 11.

the throne by just title of inheritance and by the right judgement of God given in battle, and promising to maintain all his subjects in their rights and possessions except such persons as had offended against his royal majesty who would be punished in parliament.

The grateful commons replied by a grant of the subsidies of tunnage and poundage to be used especially for the keeping of the sea, and of the usual wool subsidy for the defence of the realm. The caveat that the grant was not to be made a precedent has been noticed by some historians, but the same caveat had been made in the grant to Richard III; it was not a novelty.[1]

Everything was represented as normal; yet behind the seeming normality lay three awkward questions to which an answer must needs be given. Why should England now recognize as her king a person who was until recently a proscribed exile?; how could he and other attainted persons now sit in parliament and attaint their enemies?; what was to be the effect of a change of monarchs upon grants made by earlier monarchs? For the solution of the first of these questions there might seem to be precedents—the crown had passed from head to head in bewildering fashion during the preceding three decades—but the precedents did not exactly meet the case. The long act of 1461[2] showed the accession of Edward IV as the proper consequence of a decision already taken in parliament; the long act of 1483 rehearsed the 'Rolle of Perchement' presented to Richard III 'in the name of the three Estates of this Reame'—as was alleged—before he assumed the crown; but the only parliamentary proceeding relative to the claim of Henry VII was an act of attainder made against him. The Gordian knot was loosed with decisive simplicity. A very short bill was exhibited that

to the pleasure of Almighty God, the wealth, prosperity and surety of the realm, to the comfort of all the king's subjects and the avoidance of all ambiguities, be it ordained established and enacted by authority of the present parliament, that the inheritance of the crowns of the realms of England and France . . . be, rest, and remain and abide in the most royal person of our new Sovereign Lord King Henry the VIIth and in the heirs of his body lawfully comen perpetually . . . and in noon other.[3]

<hr>

[1] *Rotuli Parliamentorum*, vi. 238 *b*.
[2] Ibid. v. 463 and vi. 240.   [3] Ibid. vi. 270 *b*.

The king with the assent of the lords spiritual and temporal and at the request of the commons gave his assent—*le voet en toutz pointz*—and the thing was done.

What was done is not immediately clear. An act was passed establishing by the authority of parliament the position of a king who was already upon the throne, and who had played the part of a king in making the act in question. It appears to be a complete *petitio principii*. Yet the procedure, if lacking in logic, abounded in political wisdom; it was an exemplification of the principle laid down in the act of Richard III, that for the people of England a 'manifestation and declaration of any truth or right made by the three estates of the realm assembled in parliament' was of sovereign efficacy.

It may be that the king's advisers thought it imprudent to declare the exact nature of the title by inheritance, and that men generally, as is hinted by the Croyland *Continuator* felt that that in any case defects in the title would be remedied by the marriage to which the king was pledged; but it is not unreasonable to suppose that it was Henry himself, a practically minded prince, who declined to rake up old controversies, and decided to take his stand simply upon the accomplished fact. If that was the attitude of his mind it accorded well with the wisdom of the judges as revealed in their opinions upon the other two great questions. As to the attaint the judges pronounced that Henry's attainder needed no reversal, having been automatically cancelled by the very fact that he had assumed the crown. Further, resting upon the convenient fact that the condemnation of Lovell had proceeded upon a personal arrest by the king—declared illegal in the time of Edward IV—they held that Lovell had never been really attainted and could properly act as Speaker. Other lords and commons now in parliament must have their attainders reversed in parliament provided that they themselves did not sit while this was done. On the question of existing rights of subjects they declared that the old grants were still valid despite the new settlement of the crown.

The stage being thus set, the drama proceeded with a smooth celerity. The attainders of Henry VI and his family, of Bedford, and of many supporters of the Tudor cause were reversed, and the queen-dowager Elizabeth—future mother-in-law of the king—was reinstated in her dignity. The sensible provision was made that no person restored should have any action arising from his restoration until parliament was ended; the king did not wish a rather attenuated assembly to be depleted by the departure of nobles and gentlemen who might hurry down to the country to make good their claims.

The king's friends being thus remembered, his enemies were not forgotten; the dead Richard and more than a score of his chief supporters were attainted. There was an ugly side to the business. In order to facilitate condemnation it was held that Henry was already king, and that those who had fought against him at Bosworth were rebels. There was great consternation. As Thomas Betanson wrote to Sir Robert Plumpton 'ther was many gentlemen agaynst it, but it wold not be, for yt was the Kings pleasure'.[1] The Plumpton interest was in the north, and Betanson may be representing the perturbation felt by those who had supposed themselves covered by the general pardon of 24 September, or who had expected that that pardon would be still more widely extended; but it is more probable that the discontent was general, that all over England men were sick of the constant uncertainties, had hoped that, under a king who was pledged to unite the rival roses, proscriptions would cease, and were now disappointed. Many of the anxieties were no doubt personal, but some at least realized that a great principle was involved, as appears from the remark of the Croyland chronicler: 'Oh God! What security shall our kings have henceforth that in the day of battle they may not be deserted by their subjects.' If men knew that by obeying the dread summons of a ruling king they would be liable to death and forfeiture if that king were defeated, what would become of loyalty and allegiance?

The king pushed the measure through, but either because he was moved by the resistance, or because he felt that a display of severity would suffice, he did not push matters to extremes. He had already issued a general pardon, it is to be supposed the 'coronation pardon' of Bacon, which had brought out of sanctuary many men who were prepared to submit and swear fealty;[2] now some individual pardons were granted, and in two cases at least a 'proviso' was used to mitigate the effect of a

---

[1] *Plumpton Correspondence* (Camden Society, 1839), p. 49.
[2] Hall's *Chronicle* (ed. 1809), p. 424. De Giglis writes as if a general pardon in parliament had been granted: Pollard, *Henry VII*, i. 27.

restoration in the interest of a noble lady.[1] Northumberland was soon set at liberty. Surrey himself, a prisoner in the Tower since Bosworth, was pardoned in March 1486, and before long was given honourable office in the north.[2]

That the king's mind was set upon a policy of what Cromwell might have called 'healing and settling' appears from the unusual expedient adopted on 19 November, when an oath to preserve the peace was tendered first to the commons and to the members of the royal household specially summoned to the parliament-chamber, and then, after a commendatory speech by the chancellor, to the lords spiritual and temporal. All were asked, or made, to swear from retaining or aiding any felon; from retaining or giving their livery contrary to law; from making any unlawful assembly; from interfering with the king's justice, and from giving protection to fugitive criminals.[3] The measure bears the appearance of a pious aspiration only, and events were to show that something more than promises, even on oath, were needed to produce the desired effect; but meanwhile the king was laying the foundation of a strong government by reorganizing the royal finances.

Bacon's statement that he thought it not fit to demand any money or treasure from his subjects at this time, means only that the king asked for no subsidy beyond those already granted almost as of course; the fact is that he found means to enrich the Crown by an enormous act of resumption.[4] This, though it was clogged by many 'provisos', put him in possession of all the crown lands held by Henry VI on 2 October 1455, and as many of the old charges upon these lands were not renewed he found himself in possession of a steady income which he was at pains steadily to increase. He was, however, prepared to regulate as well as to acquire, and with the advice of his council drew up a plan to check the abuses of 'purveyance' by assigning £14,000 from his revenues to the expenses of the royal household. The expedient was not new; it had been tried in 1439,[5] but a comparison between Henry VII's arrangements and those of Henry VI sheds a useful light upon the methods of the 'new monarchy'. In 1439 the household expenses—of amount unspecified

[1] Campbell, *Materials*, i. 166, 172, 323, 540, *et al.*; i. 127; *Rotuli Parliamentorum*, vi. 311.

[2] Campbell, *Materials*, i. 392; ii. 480.   [3] Pollard, *Henry VII*, i. 26.

[4] *Rotuli Parliamentorum*, vi. 336–84.   [5] Ibid. v. 7a, 32b.

—were to be met from the revenues of the duchies of Lancaster
and Cornwall after certain charges, also unspecified, had been
met. It is no surprise to find that the plan did not work and that,
before the parliament which made it was dissolved, it was
decided to apply the fourth part of a newly granted fifteenth
and tenth to meeting the necessary outlays. Far other were the
arrangements made by the first Tudor.[1] The total required was
estimated at £14,000, and provision was made from more than
four-score sources of revenue—lands, leases, and customs—scattered all over England, every one of which was charged with an
exact quota. Some of the contributors paid very small amounts,
one the awkward sum of £16. 16s. 10½d.; the duchy of Cornwall
was charged with £2,700, and the duchy of Lancaster with
£2,303. 13s. 5½d., obviously because it was convenient to make
up an even sum from a convenient and well-provided source.
For the maintenance of the great wardrobe a similar provision
was made. The annual expense was estimated at £2,105. 19s.
11d. and appropriations for this exact amount were made in
precise sums from definitely stated sources. Not only did Henry
fix the amount he was prepared to spend upon his household,
but he obviously endeavoured to limit his expenses to the
assigned sum. There is in the Public Record Office a manuscript,[2] incomplete and of uncertain date, which includes an
'estimate of the daily diet and expense of the king's household'.
This provides for everything which can be foreseen, for the
'diets' of the king and queen, of their 'bordes', of their 'chambers', and of their servants generally; for the expenses of the
stable; for the cost of fuel and light and carriage; for the royal
alms and offerings in the chapel. Provision, in round sums,
is even made for contingencies such as the entertainment of
ambassadors and the 'enlarging of the principal feasts'. The
gross total of the estimate is shown as £14,365. 10s. 7d.;
comment is unnecessary.

Oddly interspersed with the great matters were a few items
of other business. The king consulted his own interest by making
hunting in the royal forests a felony, and the importunity or
influential clamour of his subjects by a few concessions. Thanet
was to have a bridge to replace the ferry ruined by the silting-up

---

[1] *Rotuli Parliamentorum*, vi. 299–304.
[2] *Exchequer Accounts*, 416. 10 (Public Record Office; see Lists and Indexes,
xxxv. 254).

of the channel; Winchester was to be paved; alien merchants who had received letters of denization were to pay dues as aliens, and their encroachments on the privileges of the native-born were checked by other statutes; a kind of 'navigation act' prohibited the import of Gascon wines in foreign ships and, in the interests of good leather, curriers were forbidden to be tanners, and tanners to be curriers. One great matter, however, was kept to the very end. On 10 December the commons with their Speaker appeared in parliament with a petition that, since the crowns of England and France were now settled upon himself and his heirs, he would take to wife Elizabeth, daughter of King Edward IV. The lords spiritual and temporal, rising in their places, supported the request in reverential tones, and Henry declared himself content to act according to their desire. Then the chancellor, with a final admonition to all to seek the peace of the church and of the realm, and to promote good order in their own countries when they arrived there, announced that it would be impossible to finish business adequately before Christmas and in the name of the king prorogued parliament until 23 January. It did not in fact reassemble.

The king had got all that he needed for the time; he had confirmed by parliamentary action the 'three several titles' of which Bacon afterwards wrote. Having made it appear that his claim to the throne was valid in itself, and that a marriage with Elizabeth was not a political necessity, he saw no need longer to delay the fulfilment of his promise. He may, indeed, have felt that delay would be dangerous, and on 18 January 1486 he wedded his kinswoman Elizabeth without waiting for the papal dispensation, for which he had applied only after his title to the throne had been declared in parliament. He had, however, secured a dispensation from the papal legate in England, the bishop of Imola, and he must have been confident that he stood well with Rome. Before parliament was ended John de Giglis, the papal collector in England, had written to Innocent VIII commending the moderation of Henry and representing that in it, and in the projected marriage, lay the hopes of future peace. Plainly his cause had been pleaded with effect. On 6 March the pope issued a Bull commending the legate's action. On 27 March he issued another[1] in which he not only gave his own dispensation and confirmed the dispensation

---

[1] Pollard, *Henry VII*, i. 35.

already granted in his name, but recognized categorically the title of Henry to the English throne and denounced any who should oppose him as rebels against whom the sentence of excommunication would inevitably be pronounced. Henry had called to his aid the distant but still potent thunders of the church. It may fairly be supposed that Rome would not have spoken in such uncompromising terms, after so long a civil war, unless she had felt sure that the new king, fortified by marriage with the rival house, would establish a durable authority; and the very fact that the Bull was so resolutely worded is in itself a proof that to acute observers Bosworth was something more than a mere incident in the intestine strife.

Henry's position seemed secure. Yet despite the fair prospects there were many elements of danger. England was in a most unsettled state. No doubt it is true that the rigours of attainder and resumption would be modified for individuals by the ties of kinship, by the accidents of tenure and by local sentiment; but the wholesale transferences of power, land, and office from one party to another must have left sore hearts and restless spirits throughout the length and breadth of the land. What was to become of all those castellans, porters, and foresters whose employment was gone, of those landowners whose acres had passed into other hands? Among the churchmen, too, there were those whose sincere sympathies were with York, and those who felt that their chances of promotion were gone. Long ago it was observed that it is possible to detect some clerical influence behind most of the conspiracies against the new king.[1] Altogether there must have been many whose hopes of future prosperity were centred upon a Yorkist restoration, and as the stem of the White Rose still bore many lively branches it was certain that the vengeful and the desperate would know where to look for leaders. The event was to show that Henry by his prompt arrests and by his eternal vigilance was usually able to account for the true leaders of the rival line; but in those days of secrecy, when even the murder of the princes in the Tower was only a hypothesis, it was easy for baffled ambition to produce some supposititious claimant, and the king found himself threatened not only by authentic scions of the house of York, but also by 'apparitions'. What made the situation dangerous was that the new dynasty had enemies abroad. Ireland

---

[1] Sir Frederick Madden in *Archaeologia*, xxvii. 153 ff.

remembered Richard Plantagenet with affection; although the
French government was not usually hostile there were many in
the newly invigorated France who recalled with animosity the
story of the English invasions; and in the Low Countries the
dowager duchess of Burgundy, Margaret, sister to Edward
IV, was always ready to play the kindly aunt to Henry's
enemies. For the first twelve years of his reign the new king was
constantly engaged in putting down rebellions, checking conspiracies, and passing laws which made it dangerous as well as
difficult to foment local disorder. It was his settled policy to
represent the risings against his authority as things of small
account, and it was partly because of the supreme confidence
that he always showed that he won through; but it was not till
after 1497 that he established a clear ascendancy and stood
forth in the eyes of Europe as the founder of a dynasty which
had come to stay.

The completeness of the victory which he gained in the end
has blinded later historians as to the reality of his difficulties,
but these were very obvious to his contemporaries. Not without cause did Bernard André[1] compare his labours with those
of Hercules, and the chronicler Edward Hall contrast his
'troublous' reign with the 'triumphant' reign of his son.

There was trouble almost from the start. Determined to show
himself as a king to his people, Henry set out on progress in the
spring of 1486, and took his way towards the north where the
influence of his enemies was strongest. As he kept Easter at
Lincoln he had word that Lord Lovell and the two Staffords,
Humphrey and Thomas, had left their sanctuary at Colchester
and disappeared. When he reached Nottingham he heard that
there was a rising in the north of Yorkshire, about the great
castle of Middleham where Richard held sway, and that the
Staffords were preparing an attack on Worcester. Undismayed
he moved on towards York, gathering with surprising speed a
force in which the newly pardoned earl of Northumberland was
conspicuous, though Hall says that many of the men were ill
armed, with breastplates made of tanned leather. The resolute
Bedford, however, was in command, and he advanced boldly,
showing no hesitation but at the same time promising pardon
to all who would submit. Resolution had its reward. York

---

[1] 'Les Douze Triomphes de Henry VII', in *Memorials of King Henry VII* (Rolls
Series, 1858), p. 133. All citations of André's work are from this volume.

greeted the new king with enthusiasm, at pageants[1] wherein 'King Ebrancus', Solomon, David, 'Our Ladie', Prudence, Justice, and others paid their practical tributes. An attempted *coup de main* by Lovell came to nought, and its leader went into hiding, first in Lancashire, where he was protected by Sir Thomas Broughton, and later in the neighbourhood of Ely.[2] The rising in Worcestershire collapsed, and the Staffords took sanctuary again at Culham, near Oxford, where they were soon arrested, on the ground that even the protection of a holy place did not avail against high treason. Humphrey was hanged at Tyburn, but the younger Thomas was spared, and against the rank and file of the rebels Henry showed a wise clemency.[3] Through the agency of the sheriffs of Yorkshire and Cumberland proclamations were issued offering pardon, with a very few exceptions, to all who would submit within forty days, and by August the work of receiving submissions was in full swing. Meanwhile the king had continued his progress by way of Worcester, where on Whitsunday the useful papal Bulls were publicly declared by Alcock at the end of his sermon, to the west country, where he visited Hereford and Gloucester. Arrived at Bristol he won the goodwill of its people by his obvious interest in their trade, and before long the lord mayor of London and the citizens in their gaily decked barges, accompanied from Putney to Westminster this king who had ventured into the country of his foes and had returned triumphant. Some months later, on 20 September, at Winchester, the queen bore a son, to whom, in honour of the British race from which the Tudors had sprung, the name of Arthur was given. Bernard André hymned his rejoicings over the birth of this prince who united the red and the white rose, and the event was hailed by all as an omen of good success.

Henry's first year had ended well; but before long there were rumours of a new rising against him. As early as November 1486 men were saying that more would be heard of the earl of Warwick before long, and early next year they heard that a claimant professing to be Warwick had appeared in Ireland. This was Lambert Simnel, son of an Oxford tradesman who

---

[1] Leland, *De Rebus Britannicis Collectanea* (ed. Hearne, 1774), iv. 187 (from MS. Cottonian Julius B. XII).

[2] Pollard, *Henry VII*, i. 41 (Countess of Oxford to John Paston, 19 May 1486).

[3] Ibid., i. 43, 45.

was so obscure that he was variously described, in his own day, as a joiner, an organ-maker, a baker, or a shoemaker.[1] At the age of ten years the boy attracted the notice of a 'subtile' priest named William Symonds, who, impressed by his natural graces and ready wit, coached him to play the part of a Plantagenet prince. At first he was to impersonate Richard, the younger of the murdered sons of Edward IV; but later, perhaps upon a rumour that Warwick was dead in his prison or had escaped from it, he was represented as that luckless boy, and it was as Edward, of Warwick, son of the duke of Clarence, that he made his début in Ireland. Even today it is impossible to trace the exact origins of the plot. What is certain is that behind the apparition were the solid figures of certain great Yorkists who, well knowing him to be an impostor, meant to get rid of him when he had served his turn, and that the conspiracy was widespread. It was, in fact, more dangerous than has been generally believed.

It is safe to suppose that the queen-dowager, a flighty woman, was involved, for after a council held at Sheen in the beginning of February 1487 Henry deprived her of her lands, which he afterwards settled upon her daughter his wife, and sent her with a moderate pension into the convent at Bermondsey. It was alleged against her that by her surrender to Richard of a daughter already pledged to Richmond she had imperilled the lives of her friends and the success of their cause; but this was an old story, and that the council was really concerned with the new conspiracy appears from the other measures which it took. It decided to exhibit the real Warwick in London, and next Sunday the prisoner was taken to St. Paul's at the time of High Mass and allowed to speak openly with those who knew him. It also decided—and this bespeaks the gravity of the situation— to offer pardon, throughout all England, to all offenders, even to those who, like Sir Thomas Broughton, had been guilty of high treason. If it believed that in so doing it had scotched the conspiracy the council was sadly mistaken, for sitting in its very midst was one who was, in all probability, the author of the plot, although it is just possible that he was driven into rebellion

---

[1] The act of parliament against Lincoln and his supporters makes him 'joynour'. (*Rotuli Parliamentorum*, vi. 397a); André either a baker or a shoemaker (*Memorials of King Henry VIII*, p. 49). Bacon later called him a baker, but Symonds, who should have known, described him, before convocation, as an organ-maker (Pollard, *Henry VII*, iii. 247; cf. Morton's Register, f. 34).

by fear of unjust accusation. This was John de la Pole, earl of Lincoln, son of Richard's sister Elizabeth and of John de la Pole, earl of Suffolk. So far father and son had appeared to be on Henry's side. Both had attended his parliament, and taken the oath to promote good order; in July 1486 Lincoln was one of the justices of oyer and terminer appointed to deal with treasons, murders, conspiracies, and unlawful assemblies in the city and suburbs of London.[1] How long he himself had been conspiring is a matter of surmise. It is curious that the fugitive Lovells should have sought refuge near Oxford, where the new conspiracy was being hatched, and significant that Simnel was sent to Ireland, of which Lincoln had been titular lord lieutenant in the days of Richard III. On the other hand, it seems unlikely that well-informed conspirators, who must have known that Warwick was alive, would have put forward an obvious impostor when they had an obvious claimant of their own—Lincoln had been named as his successor by Richard—and it is possible that the real leaders of the Yorkists came into the Simnel plot only after it was already a going concern. Lincoln may all along have cherished secret hopes of the crown. At all events, soon after the council at Sheen he took flight to the Low Countries, where he met Lovell, and, according to one account, assured him that, to his own knowledge, Warwick had indeed escaped. There he received help from his aunt Margaret who, if she was not, as Henry seemed to assert,[2] the real author of the whole business, was always ready to play Juno to the Tudor Aeneas, and prepared to make a descent upon England. The east coast seemed to be the obvious target for his attack, and it was there that Henry prepared to meet him;[3] but meanwhile the cause of the impostor had thriven marvellously in Ireland, and it was from Ireland that the invasion ultimately came.

For a political adventure, especially for an adventure in the cause of York, Ireland offered a most attractive field. Theoretically it contained the 'soft, gentle, civil and courteous' people who lives inside the English Pale and the 'wild, rustical, foolish and fierce' people who lived outside it. But the Pale itself enclosed only a small area between Dundalk and Dublin,

[1] Campbell, *Materials*, i. 482.

[2] Sir Henry Ellis, *Original Letters*, 1st ser., i. 20. In a letter to Sir Gilbert Talbot, July 1487. Cf. Hall, *Chronicle*, p. 430.

[3] Pollard, *Henry VII*, i. 47, quoting from the *Paston Letters*; Hall, *Chronicle*, p. 433.

rarely as broad as thirty miles, and as a line of demarcation
between two different kinds of population it was not so effective
as might be imagined—it was in the towns which valued their
trade, especially in Waterford far outside the Pale, that the
English interest was strongest. Despite the statute of Kilkenny
(1366) which tried to keep the Anglo-Irish distinct from the
'wild' Irish, there had been much intermarriage, and the
'feudal system', especially in its later form, was more closely
akin to the 'tribal system' than has sometimes been supposed.
Easily might an Anglo-Norman baron become a tribal chief,
and a tribal chief vest himself in the panoply of a feudal magnate. The old families retained their old power, and their
internecine feuds resembled very much the aristocratic feuds
which throughout western Europe marked the end of the middle
ages, though it seems that the 'tribesman' entered into the
quarrel of his lord more heartily than the 'tenant' who sometimes preferred to leave the actual fighting to the baron and his
mercenaries. The Irish chiefs, however, no less than the English
barons, had been quick to seize the struggle between Lancaster
and York as an excuse to realize ambition or to gratify a lust for
battle, and the main effect of the wars of the Roses in Ireland
had been to ornament old controversies with new names. During
the struggle the party of the White Rose had gained the
ascendancy. It enjoyed a popularity which descended from
Richard of York, a well-loved lieutenant, to his sons[1] Edward
and Richard, and this popularity was backed by the strength
of the great house of Fitzgerald, boasting two earldoms, Kildare
and Desmond. Under the dynasty of York the lieutenancy was
held first by Clarence and later by Lincoln, but the executive
power had lain in the hands of two successive earls of Kildare
(the seventh and eighth earls), who held the office of lord-
deputy. The Butlers, the great rivals of the Geraldines, had
given their support to Lancaster; but the sixth earl had died
at Jerusalem upon pilgrimage, his brother the seventh earl
had been little in Ireland and the leadership of the family in
Ireland had been in dispute between various branches. The

---

[1] Richard's parliament made it high treason to bring any writs, privy seals, or
commandments over from England to attach any person remaining in Ireland.
His object was to cover himself and his friends whilst he prepared his revolt against
Henry VI. By a generous interpretation the act was held to grant immunity to all
English rebels in Ireland. Pollard, *Henry VII*, iii. 296, for the annulment of this
act and *infra*.

seventh earl had seemed to compromise with Richard at one time, but he had rallied at once to the Tudor banner, no doubt in hope of recovering his ancient power. Henry could do no more for him than admit him to the council and make him chamberlain to the queen. He advanced Bedford to the honorific post of lieutenant, and may have tried to call Kildare to account, though it is possible that the summons to the royal presence, generally believed to have been sent by him, may really have been sent by Richard III.[1] Whether or not he had tried to take control of the situation, Henry had been compelled to leave things as they were in Ireland. At first he even seems to have hoped to make use of Desmond, to whom he issued a commission to arrest traitors, only six weeks before he commanded Edgecombe to pardon them,[2] and when Simnel arrived there the Fitzgeralds were still supreme. Kildare was lord-deputy; his brother Thomas was chancellor; and the friends of York were able to use official authority to oppose the king in whose name the authority was, or should have been, exercised.

Whether they believed in the authenticity of this 'Warwick' is uncertain; Hall writes as if they did and there is other evidence to the same effect; they may even have been forewarned of his coming from the Low Countries.[3] At all events when Lincoln and Lovell landed at Dublin on 5 May, along with 2,000 German mercenaries paid for by Margaret and commanded by the redoubtable Martin Swart or Schwarz, they speedily declared themselves. King Edward VI was crowned in Christ Church in Dublin, with a diadem borrowed from a statue of the Virgin, and shown to the people on the shoulders of Darcy of Platten, the tallest man of his time.[4] A parliament was summoned in his name and coin was struck. The Irish clergy, though the Florentine archbishop of Armagh and the bishop of Clogher resisted, adopted his cause with enthusiasm, and even voted a subsidy to be employed in reversing any ecclesiastical censures which might be obtained against them. In the southeast where the Butler influence was strong there was some opposition in Kilkenny, Clonmel, and a few other towns, and

---

[1] Gairdner, *Letters and Papers Illustrative of the Reigns of Richard III and Henry VII*, i. 91. Gairdner dates the letter 1486?; but the instructions to John Estrete recognize the earl's good service to Edward IV, and there is no reference at all to Richard III.

[2] Campbell, ii. 291, April 1488. Ibid. ii. 315, May 1488.

[3] Bagwell, *Ireland under the Tudors*, i. 103 (cf. G. E. C. in *Peerage*, s.v. Kildare).

[4] Ibid. i. 104.

the stout city of Waterford, whose mayor was a Butler, offered a bold defiance to Kildare. But despite their valiant words the townsmen could not quit their stronghold to take the field, and with Ireland securely held, the venturers hastened across to England. They took with them many Irishmen, 'beggerly' according to Hall, and armed only with skenes and mantles, who, whether they believed in Simnel or not, obeyed the word of Kildare and hoped for spoil in wealthy England. Hoping that Broughton's influence would raise Lancashire, they landed at the Pile of Fouldry, near Furness, on 4 June; they gathered few recruits, probably, as Bacon says, because the people had no mind for a king 'brought in to them upon the shoulders of Irish and Dutch'. The stout-hearted Lincoln, however, hoped for better things, and, forbidding his men to plunder lest the hearts of the people should be alienated, marched resolutely south-east to seek his enemy.

Henry was ready to meet him. If he had been powerless to achieve much in Ireland, he had been active enough in England, though his preparations had been hampered by uncertainty as to where the blow would fall, and perhaps by uncertainty as to the inward mind of some of his professed supporters. On mere suspicion he had arrested Dorset, arguing that if he were a true friend he would not take offence, and sent him to the Tower, from which he was afterwards released 'without a stain upon his character', as they would say nowadays. He had visited Norwich, and prayed at Walsingham, but while he had kept his eye upon the east he had at the same time 'dispatched certeyne horsemen throughout all the west partes', to give prompt notice of any landing and to pick up any spies who might come from Ireland.

Meanwhile he had gathered his forces. If the evidence of the *Paston Letters* is not misleading there was some hesitation amongst the English gentry, men arguing as to the exact nature of the summons sent to them, and concerting joint action among themselves instead of answering at once the call of the king's lieutenant, Oxford. None the less he had collected a considerable force which had as its leaders reliable Bedford and Oxford, that *miles valentissimus*; when he was sure that the attack would not be delivered against East Anglia he moved into the Midlands—as Richard had done in his hour of peril—and made his headquarters at Kenilworth with his army about Coventry. He had issued proclamations forbidding his people to plunder or otherwise misbehave; he had warned the neighbouring counties to be ready at an hour's notice, and when the news came that the enemy was landed he was able to concentrate a powerful army at Nottingham without undue delay. Lincoln had hoped to take him unawares by the rapidity of his advance, but Henry 'was in his bosome and knewe every houre what the Erle did', and when the invaders approached the great artery of the Fosse Way, south of Newark, they found themselves confronted by the royal host arrayed in the three customary 'battles'. Lincoln did not refuse the challenge—according to the act of attainder later passed against him he had a force of 8,000 men— and on the morning of 16 June he attacked with great courage. The German mercenaries were found to be as good as Englishmen, and their leader Schwarz better than most Englishmen in valour, strength, and skill at arms; but the half-naked Irish, though they 'foughte hardely and stuck to it valyauntly', were no match for armed men, and were slain like 'dull and brute beastes'. For three hours the struggle lasted, but numbers told in the end, though, if Hall be rightly interpreted, only Henry's main battle was seriously engaged. Lincoln, Schwarz, Broughton, and Sir Thomas Fitzgerald were killed fighting with a valour which commended itself even to their enemies; Lovell disappeared, drowned in the Trent as he fled, it was believed; the priest Symonds and his pupil were taken, the one, after a public confession, to suffer lifelong imprisonment, the other to become a turnspit in the king's kitchen, to be promoted falconer, and ultimately to enter the service of Sir Thomas Lovell.[1] If Henry knew the value of clemency he knew the value of derision too. 'My masters of Ireland, you will crown apes at last' he told Kildare and other Irish lords when he gave them audience in 1489; one day when the visitors were dining they were informed that their 'new king Lambarte Simnel brought them wine to drink and drank to them all'. Only the Lord of Howth, a merry gentleman who appreciated the joke the more because he had never acknowledged the impostor, accepted the challenge; and he drank, 'for the wine's sake,' saying 'as for thee, as thou art, so I leave thee a poor innocent'.[2] Of Celtic blood himself Henry knew how to deal with Celts. But the English seem to have

---

[1] He seems to have survived till 1525.
[2] *Carew MSS.* vi. 190 (Book of Howth).

indulged in some mockery too, if we may judge from a contemporary ballad.

> Martin Swart and his men, sodledum, sodledum,
> Martin Swart and his men, sodledum, bell!

Derision is all very well after the event, and in any case when the English use mockery against their foes—as against 'Boney' and 'Adolf'—it is usually a sign of some inward apprehension. Before the field was stricken there had been ugly rumours, spread by the king's enemies as was believed, that Henry's army had been beaten, and that he himself was in flight.[1] Rumours were of great importance in a society where few could read and communications were bad, and earlier in the year Henry had issued a proclamation appointing the pillory as the punishment of offenders.[2] It is said that during the crisis an evil rumour prevented some from joining Henry's host, and the failure of two of his battles to engage before they saw the issue of the day bears an ugly resemblance to the conduct of a part of Richard's army at Bosworth. Hall's language on this point is obscure, but there are other evidences of hesitation in the royal host.[3] The king and his devoted followers may not always have been as confident as they pretended. The battle of Stoke, the last battle of the wars of the Roses, was a real crisis in the affairs of Henry, and he had good cause for the solemn thanksgivings for his victory, which he offered at Lincoln without delay.

The victory, however, was complete. Henry did not misuse it; but he used it to the full. He had given orders that Lincoln's life should be spared, in hope of learning from the leader the names of his confederates. That plan having failed he instituted a strict inquiry to discover not only those who had supported the invasion, but also those who had spread false tidings of its success, and though as a result there were relatively few executions there were very many fines. Henry believed in making war pay for itself. In accordance with his practice of showing himself in the disaffected areas as in others, he progressed northwards by way of Pontefract, York, and Durham to Newcastle, where he stayed, not as Hall says for 'the remnaunt of the somer', but only for four or five days (14–18 August). From

---

[1] Henry's letter to Pope Innocent VIII, 5 July 1487. Gairdner, *Letters and Papers Illustrative of the Reigns of Richard III and Henry VII*, i. 94.

[2] Pollard, *Henry VII*, ii. 110; cf. ibid. ii. 66.

[3] Fisher, *Political History of England*, v. 18.

there he sent Fox, now bishop of Exeter, and Sir Richard Edgecombe to try to conclude a negotiation with Scotland which had been interrupted by the Simnel affair.

In spite of some vague threats of attack from Scotland, which emanated probably from the restless nobles rather than from the peace-loving king,[1] Henry's relations with James had been good. Scottish ambassadors had been present at his coronation, and at the triumphant reception which greeted him[2] on his return from his first progress. In June–July of 1486 there had been arranged a truce of three years between England and Scotland, and one of the articles had provided that commissioners should meet in March 1487 to discuss not only a prolongation of the truce, but a triple marriage alliance whereby James and his two sons should wed the queen-dowager and two of her daughters. This meeting had not taken place, and it was in order to keep his engagement at the earliest opportunity that Henry now sent his emissaries north. They met with a good reception; the marriages they proposed were 'thought expedient' by the Scottish representatives, and James may, as Polydore Vergil says he did, have made a secret promise to prolong the truce as long as he could. Not until November, however, was anything formally concluded, and then the truce was prolonged for only a few additional months—it was to expire in September instead of in June 1489. Arrangements were made for further negotiations, and the possibility of a meeting between the two monarchs was discussed; but when the Scottish parliament met in January 1488 it made the surrender or the destruction of the fortress of Berwick[3] the condition of a continuance of the marriage negotiation. This England could not consider and Henry's reply was to strengthen the garrison in Berwick.[4] Before long the Scots lords, partly upon the pretext of their master's 'inbringing of Englishmen', rose in rebellion, and the death of James III at Sauchieburn on 11 June brought the whole transaction to a close, although, in the following October, the truce was renewed for three years by the new king. Yet it

---

[1] James IV referred to his father as *Princeps togatus* (*Epistolae Regum Scotorum*, i. 89.

[2] Leland, *De Rebus Britannicis Collectanea*, iv. 203.

[3] *Acts of the Parliament of Scotland*, ii. 181–2. See *Rotuli Scotiae*, ii. 471–82, for the whole negotiation.

[4] *Rotuli Scotiae*, ii. 483: cf. Conway, *Henry VII's Relations with Scotland and Ireland*, p. 11.

may have served a useful purpose in showing the restless spirits on the Border that the Tudor arm was long, and to historians it is important as an early example of Henry's policy of friendship with Scotland.

While Henry was occupied in treading out disaffection in the north, his determination to secure good order revealed itself in a very different place; on 6 August Pope Innocent VIII issued what Bacon calls 'a very just and honourable Bull' to limit the abuse of sanctuary. Well aware that the hope of secure retreat in case of failure was an incentive to rebellion, and feeling perhaps a little uneasy about his treatment of the Staffords, Henry had resolved to seek the help of Rome in order to regulate the matter, and the Holy Father now declared that a 'sanctuary man who emerged from his refuge to commit crime should have no further protection; that, even when the persons of offenders were sacrosanct their goods were not, and that, in case of high treason the offender was to be under the supervision of the royal officers, even when in the holy place'. For the moment the king was content to rest upon papal authority in this delicate matter, but when after his triumphant return to London he met his second parliament on 9 November he made plain, not only his intention to establish good order, but the means whereby his purpose should be carried out. John Morton, who had succeeded Bourchier as archbishop of Canterbury in 1486, and Alcock as chancellor early in 1487, preached the opening sermon upon the text 'Depart from evil and do good; speak peace and pursue it'. After the manner of his time he divided his oration into heads, each with its quota of sub-heads, and garnished it with quotations; yet he managed to drive home the lessons that peace was not a state of lethargy but a positive thing, and that the pursuit of peace was a duty to be actively undertaken by all.[1] Some of the business which followed may be described as normal; twenty-eight of Simnel's followers were attainted, the name of Lovell being accidentally omitted from the list; various measures were passed to regulate trade, industry, and currency in accordance with the theories of the age; but acts were passed also for the maintenance of good order and these emphasized the need for good local officers and the need of a strong central authority. The king's own servants were evidently to be brought under discipline. Arrangements were

---

[1] *Rotuli Parliamentorum*, vi. 385–408; *Statutes of the Realm*, ii. 509–23.

made for the execution of justice in the royal household, and local officers and holders of the king's land were forbidden to be unlawfully retained, to retain others, or permit others to be retained. The acts against murder and the carrying-off of women laid stress on the duties of coroners, sheriffs, and justices of the peace, and a very practical act made it unprofitable for an offender to delay justice on an ill-founded writ of error. The famous act, afterwards entitled *Pro camera stellata,* did not create anything new, but strengthened the discretionary jurisdiction of the council by empowering an unusually strong committee to deal with acts against the public peace, and emphasizing the power of making inquisition.[1] There were some jarring notes; the defences of Calais and of Berwick were reorganized, and the king was granted two fifteenths and tenths 'for the necessary and hasty defence of his realm'. Already the bright prospect of peace was clouded by the shadow of approaching war.

For the moment, however, Henry stood before his people as a king established by the favour of heaven towards a just title, and by his strong right arm, who was winning good opinion by a just and clement rule, according to the traditional way. He could afford to crown Elizabeth now, and knowing well the value of display, he made the ceremony the occasion for a display of the first magnificence.[2] London did her part; the streets were cleaned, the citizens arrayed; the queen and her mother-in-law were conducted from Greenwich to the Tower by water in a gay pageant, and next day a stately procession escorted her to the palace of Westminster. The coronation, which took place in the abbey on the morrow, Sunday, 25 November, was followed by a great feast in Westminster Hall conspicuous alike for the number and splendour of the guests and the abundance and excellence of the fare. Knights of the Bath were created on this happy occasion, and in 1488 the king kept with great ceremony the feast of St. George, as well as the feasts of Easter and Whit Sunday. Plainly he wished to appear in the eyes of England, and of Europe too, as a monarch who ruled at his ease in power and splendour; but while he maintained his outward state he by no means neglected the business of government, and in 1488 he added to his labours an effort to settle the affairs of Ireland.

---

[1] See Ch. VII *infra* for a full discussion of these matters.
[2] Leland, *De Rebus Britannicis Collectanea,* iv. 216–33.

In the act of attainder there had been no mention of Simnel's Irish supporters. Possibly the king thought that the nobles had really been deceived; possibly he saw no use in pronouncing sentences which he could not execute; possibly he believed that the failure of the revolt, if judiciously handled, gave him a chance to assert his power effectively. Certainly, as a realist who put practice before exact logic—always a useful thing to do in Ireland—he meant to achieve his end in the manner most convenient for himself. As usual, he adopted from the start a tone of complete mastery, and in thanking the citizens of Waterford for their loyalty, instructed them to bring Kildare and the city of Dublin to complete submission. This was obviously beyond their power, but the confidence of Henry's enemies had been shaken by the news from Stoke—how many of the 'wild' Irish ever got home?—and the citizens of Dublin, whose trade was now exposed to eager English ships, made haste to surrender, laying the blame upon Kildare and upon the archbishop of Dublin and his clergy. After some delay Henry decided not only to pardon all the offenders, but even to retain Kildare as lord-deputy provided that he and the other lords took the oath of allegiance in proper form. To administer the oath he sent Sir Richard Edgecombe, who set sail from Mounts Bay on 23 June, equipped with four ships, 500 soldiers, and a Bull which Innocent VIII had issued against rebels. This simple apparatus sufficed to give at least the appearance of success. Distant Kinsale, where Edgecombe first landed because the vain pursuit of a pirate had taken him off his course, gave him a good reception, and, after administering the oath which the local chiefs readily took, he returned to Waterford. There he did his best to calm the apprehension naturally felt by the townsmen when they learned that Kildare was to be continued in office, and having done what he might for a city whose loyalty—not on this occasion only—received little reward, he went on by sea to Dublin. There obstreperous Kildare kept him waiting for eight days, but he repaid the discourtesy by declining to bow when he entered the presence of the lord-deputy, and his firm demeanour proved effective. After some talk the great Fitzgerald took the oath, as did the other lords, though none of them would enter into recognizances for the forfeiture of their estates if they broke it—'they would rather become Irish every one of them'. Edgecombe replied by having the oath expressed in terms as stringent as he could, and by seeing that the Host on which the vow was made was consecrated by his own chaplain. Then after pardoning the conspirators in the name of the king he took his way home, less than five weeks after he had landed at Kinsale.

On the face of it his mission had accomplished little. Kildare was still in power; his representative in England was the bishop of Meath who had preached Lambert's coronation sermon; before long the archbishop of Armagh, who no doubt wanted the office of chancellor for himself, informed Henry that the lord-deputy was engaged in fresh conspiracy. None the less the stout-hearted Edgecombe had achieved some real success. He had asserted the power of the English king, and, when in 1489 Henry summoned the principal Irish nobles to court, they all came except Desmond and Fitzmaurice of Kerry. Moreover, when in 1491 Perkin Warbeck appeared in Ireland, he got relatively little help from Kildare, and nothing like the general support which Simnel had received.

'Ireland was as it were a theatre or stage on which masked princes entered, though soon after, their visors being taken off, they were expelled the stage.'[1] The drama of Perkin Warbeck, however, was a long-drawn-out affair, played against a background of European wars, and before it can be understood the foreign policy of Henry must be examined. Here it suffices to say that the triumphant monarch of 1487, who seemed to have laid so successfully the ghost of the Yorkist claim, was to be confronted before long by yet another 'apparition'. His troubles were not over; and already in 1488 he had begun to drift into war.

---

[1] Bagwell, *Ireland under the Tudors*, i. 109 (quoting Ware, *Rerum Hibernicarum Annales*).

IV

FOREIGN POLICY

THE DIPLOMATIC SCENE

ENRY VII has always been regarded as a pacific monarch, and that he really valued the good peace which his chancellor commended there is no reason to doubt. Whatever were his ethics in the matter, his abundant common sense must have shown him that a king seated uneasily upon a newly acquired and uncertain throne would be wise to keep out of war, and his first movements in the field of diplomacy prove that he tried to maintain good relations with all neighbours.

As early as 12 October 1485 he proclaimed a year's truce with the old enemy of France, and on 17 January 1486 this truce was extended to last for three years;[1] to France's old ally, Scotland, he made pacific overtures; he concluded a commercial treaty with Brittany in July 1486;[2] early in his reign he entertained ambassadors from Maximilian, king of the Romans, though he did not renew Edward IV's treaty till January 1487, and then for one year only;[3] and in March 1488 he began negotiations for the marriage of his son Arthur with Katharine, daughter of Ferdinand and Isabella.[4] All promised peace; and yet within a few years he was at war.

It has been supposed that his hand was forced by a prevailing war-spirit in England—he still claimed to be king of France and it was only some thirty years since Normandy and Guienne had been lost; this supposition has been rebutted on the ground that there was not, in fact, a great clamour for war on the part of his people, and that, however much England may have wished to vex France, she was most reluctant to pay for a full-dress war. The counter-suggestion that Henry was dragged into hostilities at the heels of his new ally, Spain, takes too little account of his native caution and of his diplomatic ability; granted that he did wish to please Ferdinand he could have found some way of doing so without waging a grand campaign on his behalf. The truth is that his action resulted, not from one

[1] Till 17 January 1489. Rymer, *Foedera*, xii. 277, 281.
[2] *Foedera*, xii. 303.   [3] Ibid. xii. 318–21.
[4] Pollard, *Henry VII*, iii. 2; *Spanish Calendar*, i. 3,

clear and compelling cause, but from a series of circumstances, some of which were beyond his control; that he went to war because, in the end, it seemed the best thing to do; and that, having gone to war, he considered first his own interest, which, after the manner of the time, he identified with the interest of his country. The ends which he had in view are obvious; his diplomacy was part of his general policy. He wished, above all things, to secure his throne, and in order to do this he must not only suppress all possible rivals, but also endear himself to his own people. His aims, therefore, were to deprive his rebels of foreign support; to show himself a true king of England by maintaining the high pretensions of his predecessors; to win the regard of his subjects by promoting their economic welfare. When Bacon says that he 'bowed the ancient policy of this estate from consideration of plenty to consideration of power' he tells at best a half-truth. Henry was by no means the first king of England to adopt a mercantilist policy, and the object of that policy was plenty as well as power. No man understood better than he that plenty is the mother of power. In its ends, therefore, his foreign policy resembled the traditional policy of earlier English kings; it was in the realization of the ends that new features presented themselves. There was a restless spirit in the air.

The diplomacy of the period reflected the individualism and the realism which were of the essence of the Renaissance. Treaties were still fortified with threats of apostolic censure; but the papal thunder was not infrequently disregarded as a *brutum fulmen*, solemn promises were readily broken, and each prince sought his own ends by all the means in his power. Just titles were valuable—if they could be made good in fact—and, since the state was identified with the person of the prince, royal marriages, and even personal idiosyncrasies were matters of high diplomacy. The result was a political kaleidoscope where the patterns would 'change and mingle and divide' with baffling uncertainty, a melodrama very different from the old-fashioned mystery where established characters played their recognized parts upon a familiar stage. The diplomatic stage of north-west Europe no longer presented only the simple drama wherein England, perhaps in alliance with the Netherlands or with Burgundy, was opposed to France and her remote ally, Scotland, and wherein outside interference, except for that of the pope, was almost unknown. Recent developments in the political world had not only altered the old balance, but had brought the protagonists into relation with distant powers whose collective interests embraced the affairs of all Christian Europe; and the diplomacy of an English king was now affected by events which occurred in the Mediterranean, in the Baltic, and at Vienna.

To express the situation in terms of a simple formula is impossible, but it may be regarded thus. In certain areas—in France and Spain, for example—the state had made itself the expression of a rising spirit of nationality and had, in consequence, become not only strong in itself but also aggressive towards its neighbours. In other countries—notably Italy and Germany, haunted by the ghosts of old imperialisms and by the lively menace of the Turks—the national idea, though it expressed itself in city princedoms and in a common sentiment, had not become identified with a single unitary state; these countries continued to suffer from internecine war and tended to be the targets for the aggression of their neighbours. The formula cannot be applied without modification. Venice was not Italy, but she was a power of the first magnitude; Naples, Milan, and Tuscany were not to be despised, and, in the dawn of the sixteenth century, Cesare Borgia came near to making the Papal States the core of a united Italy. Charles the Bold, whose forebears, by politic marriages, had added the rich provinces of the Netherlands to the French duchy of Burgundy, nearly succeeded in realizing the dream of so many statesmen in the creation of a 'middle kingdom'. Maximilian, busy here and there and finishing so few of the things which he began, yet by his endless activity restored the prestige of the empire and tried to create an administrative machinery. There was some local patriotism in the German princedoms, and in Brandenburg the germ of the later Prussian state; and Germany was, as always, a mother of fighting men—the landsknechts, the 'fellows from the plains', were as ready as the Swiss mountaineers to serve a master for pay. Yet, when every modification has been made, the significant fact remains that the countries most able to direct the course of European politics were those in which the national sentiment took a political expression.

Among these states was Spain. Spain was not formally a 'unitary' state, for the marriage of Ferdinand of Aragon and

Isabella of Castille in 1469 had effected only a personal union, and old jealousies died hard. Yet the 'catholic' monarchs had, with the support of the church, developed a strong centralized power; they had called to their aid the crusading zeal of a haughty people, somewhat cut off from Europe psychologically as well as geographically, and much disposed to set their own things first. The potentialities which were to make Spain the great imperial power of the sixteenth century were not yet plain to view, but already she stood forth as a successful state, expelling Islam from the peninsula, and possessed of claims against her neighbours which she might soon make good. Already mistress of Sicily, Aragon had some title to Naples, and she was particularly anxious to recover the provinces of Cerdagne and Roussillon, pledged to Louis XI in an hour of difficulty (1462), which would open the door to France.

France, fortunate alike in her geographical position, her rich soil, and her progressive *bourgeoisie*, was the supreme example of the new-fashioned monarchy. Charles VII and Louis XI had created a professional army in which the *corps d'élite* of *gardes* was backed by the heavy cavalry, the *ordonnances*,[1] the hired Swiss infantry, and the mobile artillery produced by the brothers Bureau. Armed with this striking force the monarchy had expelled the English, reduced to submission all the semi-independent princedoms except Brittany, thwarted the ambitions of Charles the Bold, and brought the whole of the kingdom beneath the rule of a centralized, authoritarian government. The ebullient life of this rejuvenated France presaged a policy of aggression, and her kings possessed, by inheritance, two useful claims in Italy—the Visconti claim to Milan and the Angevin claim to Naples. Her immediate purposes, however, were to secure Brittany and to exploit a very promising opportunity which had presented itself on her north-eastern frontier. Charles the Bold had been killed by the Swiss at Nancy (1477); his daughter Mary died as the result of a riding accident in 1482. Mary had been married to Maximilian, and had left a son, Philip, who would succeed in due course; but he was only an infant, and during his minority his subjects refused to accept Maximilian as regent. The dying Louis saw his chance. By the

---

[1] So called after the 'Ordonnance' of 1439, which had substituted the *taille* for military service; the revenue obtained was used to establish about fifteen companies of men-at-arms—*gens d'armes, gens d'ordonnance*, or simply *ordonnances*.

treaty of Arras (1482) he betrothed the dauphin to Margaret, daughter of Maximilian and Mary, and secured as her portion the counties of Artois and Burgundy (*Franche Comté*); he made some parade of removing French Flanders from the jurisdiction of the *parlement de Paris*, but he quietly assumed that, in default of heirs-male, the great duchy of Burgundy had already reverted to the French Crown. The treaty was not fully carried out but his own death, in August 1483, did not arrest, though it delayed, the triumphant progress of his country. During the minority of his son Charles VIII, the government was entrusted to his capable daughter Anne, who was married to Pierre de Beaujeu, brother and heir of the ailing duke of Bourbon; and though the king's second cousin, first prince of the blood, Louis of Orleans, who had hoped himself to be regent, concerted resistance with other discontented nobles, the regency not only maintained itself in power, but was able to improve the position of the French monarchy both in the north-east and the north-west.

In the north-east its task was easy. The provinces of the old Burgundian dominion had never been closely knit; there was a standing antipathy between the towns and the country nobles; two rival parties, the *Hoeks* (Hooks) and the *Kabeljauwschen* (Cod-fish) maintained a feud as devoid of principle as that of the old Guelfs and Ghibellines, and the whole county was given over to atrocities such as those recorded—with some error of date—in *Quentin Durward*, and in *Les Chroniques* of Molinet,[1] where the wholesale hanging of a defeated garrison is recorded almost as a matter of course. Maximilian, whose father was an emperor without an empire—Vienna itself was lost to the Hungarian Matthias Corvinus between 1485 and 1490—had to rely on his own resources, and though he was not without friends he had many enemies too. In the welter of strife it was inevitable that some of those enemies should declare for France, and though they were rather partisans than a party, it was easy for France to promote her own interests either by secret intrigue or by open war. Under the capable Philippe de Crèvecœur, Seigneur d'Esquerdes, who had once been in the service of Charles the Bold but was now governor of Picardy, her troops gained some success, capturing St. Omer and Thérouanne.

As regards Brittany the position of France was hardly less advantageous, even although here she encountered a genuine

---

[1] *Les Chroniques de Jean Molinet* (Best edition 1935–9).

'nationalist' opposition. Duke Francis II was old; he had no son, and the opponents of the insidious French influence strove to gain allies by promising the hands of his two daughters, Anne and Isabel, to foreign princes, who, ignorant of the speech and the sentiment of Brittany, were ill-fitted to champion her independence. In 1484 the corrupt treasurer of the duchy, Pierre Landois, brought the disgruntled Orleans to Nantes and promised him a marriage with Anne. This scheme broke down, and Landois was hanged by a party of nobles who professed to support France (1485); but the estates of Brittany arranged in March 1486 that the elder girl should marry Maximilian and the younger his son Philip. The situation was complicated still further when Francis promised his elder daughter to Alain d'Albret, head of a powerful Gascon house. The net result was that an unfortunate child of ten was pledged, or half-pledged, at the same time to a French prince, who was already married—lovelessly—to Jeanne of France, to the king of the Romans, who was a widower of thirty-one, and to a southern noble of forty-five, ugly in face and ugly in reputation. None of the suitors had any real sympathy with the aspirations of the Bretons, and it is not surprising that in the *guerre folle* which followed the grand coalition, which had hoped to dismember France, was hard put to it to defend the duchy. The French had the best of some confused fighting in 1487, and though they failed in an attempt on Nantes they captured several places, including Vannes.

This was the news brought by French ambassadors who, officially sent to congratulate Henry VII on the victory of Stoke, met him at Leicester as he came south in September. They added to their good wishes a vindication of their doings in Brittany—their king, they said, like Henry himself, was putting down rebellious princes, and subduing the remote parts of his realm.

Henry was in a sad quandary. Brittany had given him shelter when he was an exile, France had helped him to gain his crown, and he owed gratitude to both. He was in no case to undertake a great war; yet no king of England could view with equanimity the steady aggrandisement of the secular enemy France, especially when her advance threatened to cut off Calais on one flank, and on the other to give her control of Brittany which guarded the entrance to the Channel and was—as still she is—a great nursery of ships and sailors.

Wisely, he tried for a compromise and sent his almoner, Christopher Urswick, to visit both France and Brittany in the hope of arranging some settlement. His envoy, however, though a very able negotiator, met with scant success at either court, and in 1488 occurred a series of events which not only brought his temporizing policy to nought but even robbed him of his status as a mediator. In February, Maximilian, who had rashly ventured into hostile Bruges, was imprisoned by the citizens, who held him in captivity until May; he could do nothing in defence of Brittany. Yet Brittany was in sore need of defence for the French forces, under the competent La Trémoille, were making steady progress. The upshot was an unofficial English attempt to check them which ended in complete disaster. Edward Woodville, Lord Scales, uncle of the queen, who was governor of the Isle of Wight, begged Henry's leave to try a fall with the old enemy and on being refused collected a few hundred[1] stout fellows and made the venture himself. In May he landed at St. Malo, having plundered a French ship *en route*, and the Breton leaders, glad of even a small reinforcement, decided to give battle. On 28 July the armies met at St. Aubin du Cormier. Anxious to magnify the extent of the English assistance, and perhaps to exploit the old prestige of the English bowmen in France,[2] the Breton leaders dressed 1,300 of their own men in jerkins bearing the red cross of St. George and placed them with Woodville's men in the van. For some time the day was stiffly contested, and where the English were it at first went well enough; but the motley host of Germans, Spaniards, Navarrese, and Bretons, though perhaps 10,000 strong, was in the end no match for the well-appointed French army, in which the artillery and the heavy cavalry were conspicuous. The accounts of the action are not entirely clear—Molinet's confused version seems to suggest that Orleans was unlucky or unsuccessful—but there is no doubt about the result. The allied army was shattered; Orleans was taken, and, as the French gave no quarter to the red cross, Scales and his followers died almost to a man. The French followed up their victory,

---

[1] Polydore Vergil says there were only 400 in all; other accounts say 700.
[2] The Italian Relator says that the English were great men of war, and that the French feared them.

and on 20 August forced upon Francis II the treaty of Sablé,
whereby the defeated duke acknowledged himself to be vassal
of France; promised not to marry his daughter Anne without
the consent of Charles; undertook to clear his territory of
foreign soldiery and handed over four towns as pledges of his
integrity. When three weeks later, on 9 September, the duke
died, leaving Marshal de Rieux as guardian of his children, the
real objective of French policy at once declared itself. Charles's
government claimed wardship in his name and disputed the
right of Anne to take the title of duchess until the position of
France was made clear: Brittany it seemed was to go the same
way as Burgundy. The Bretons would not tamely submit.
Rennes itself had never surrendered; the fighting was renewed;
Anne and her guardian applied to Henry for help.

Henry had apologized for Scales's invasion, made at a time
when his own ambassador was speaking peace in Paris, and on
14 July at Windsor he had extended his existing treaty with
France to make it last for an extra year—till January 1490.[1]
But now English blood had been shed; now it was plain that
France meant to engulf Brittany altogether. Henry realized that
something must be done; but he was unwilling to act alone, and
at once began to concert measures with his new ally, Ferdinand,
who, though he had no interest in Brittany, was most willing to
embarrass France and to seek in troubled waters the coveted
provinces of Cerdagne and Roussillon. He hoped, in fact, to
use the newly begun marriage negotiation as an instrument to
push Henry into war without going to war himself.

The English king's first proposal was that Anne should marry
his own subject, the duke of Buckingham, but he abandoned the
project when Ferdinand, who of course did not like it, urged
that its adoption would alienate both Rieux and Albret. It was
necessary to try some other plan, and the steady advance of the
French in Brittany urged him to prompt action. In November
he held a great council to consider how Brittany could be
saved, and on 11 December authorized a whole stream of
embassies.[2] Dr. Thomas Savage and Sir Richard Nanfan were
accredited to Ferdinand and Isabella, and also charged to
deliver the insignia of the Garter to the king of Portugal; Dr.

---

[1] *Foedera*, xii. 344, 345; Campbell, *Materials*, ii. 334. The truce with the Scots
was ratified twelve days later.

[2] Campbell, *Materials*, ii. 376-8; *Foedera*, xii. 347-55.

Henry Ainsworth and Sir Richard Edgecombe went to the Duchess Anne; and other envoys were sent to Maximilian and to his son Philip. Christopher Urswick, it is true, was dispatched to France with the object of negotiating for a firm peace; but it is significant that in the same month commissions were issued to take musters of archers in preparation for the king's expedition to Brittany, and that in the following months ships and crews were got ready.[1] On 1 March Sir Robert Willoughby and Sir John Cheyne were named as commanders of the expedition, and it is obvious that Henry was preparing for war, though he still spoke peace and may still have hoped for it.

The proceedings of the parliament, which met on 13 January 1489,[2] seem to reflect the general uncertainty. Morton's opening speech said nothing of any war, but was a scholastic—and not uninteresting—disquisition upon the various kinds of justice. Of the measures passed many dealt with economic matters; some of these sought to regulate in the familiar way, the conduct of butchers, cap-makers, wool-merchants, and dealers in bullion, but others, such as the act forbidding the import of Gascon wines in ships other than English, and the act aimed at checking the practice of 'inclosure', have more than a passing interest. Several acts, again, exemplify the king's determination to secure good order. Yeomen of the guard and justices of the peace were firmly reminded of their duty; full 'benefit of clergy' was restricted to those in holy orders—other offenders were to be branded on the thumb with M for murder and T for theft, and were to have no protection if they were arraigned a second time. This was an important act, and no less important was the so-called 'Statute of Fines', though it was neither so novel nor so subtle as has sometimes been supposed. Richard III had passed a similar act, and its object was not to deliver a sly attack on baronial power by limiting entails, but to put an end to the many disputes about land consequent upon the civil war; henceforth open and undisputed possession during four legal terms would constitute a good title.[3]

Yet amid all this legislation, which plainly continued the

---

[1] Campbell, *Materials*, ii. 384, 403, 409, 419; *Foedera*, xii. 355; *De Sagittariis pro Relevamine Partium Britanniae Providendis*.

[2] *Statutes of the Realm*, ii. 524–48, and *Rotuli Parliamentorum*, vi. 409–39. Several of the acts are reprinted in Pollard, *Henry VII*.

[3] Minors and persons barred by infancy, imprisonment, or 'coverture' were given five years after the removal of their disability in which to make a claim.

programme presented by the king to the previous parliament, there were several items which spoke of war. An act was passed to protect the legal interests of those who went to Brittany upon the king's service, and parliament was asked to vote £100,000 to provide an army of 10,000 archers for one year, against the ancient enemies of the realm.[1] For the purpose in view the sum asked was not excessive—even at 6*d.* a day, the lowest rate for an unmounted archer, 10,000 men would cost in pay alone more than £91,000 in a year; but it is quite possible that parliament was staggered by the demand of a sum equal to about three-fifteenths and tenths, and only after a long delay did it vote supply. Even then it adopted a new method which was likely to prove ineffective and certain to be unpopular; the clergy were to find one-quarter of the total amount, and the laity were to provide the rest partly by an income-tax of 10 per cent. on all incomes greater than 10 marks a year, partly by a levy of 1*s.* 8*d.* on every 10 marks ($\frac{1}{80}$, that is) of personal property. Some attempt was made to prevent dissatisfaction by exempting the northern counties, already burdened with the defence of the Border, and by providing that the special assessments, to be made by special commissioners, were not to be matter of record. Implicitly, as well as explicitly, the commons made it clear that the grant was not to create a precedent.

The whole of the money was never got in, and there were difficulties from the very start. By the month of April the dissatisfaction in the restless north had issued into open revolt. The insurgents, doubtless with the act about benefit of clergy in mind, protested that they were resisting 'suche unlawfull poyntes as Seynt Thomas of Cauntyrbery dyed for',[2] and when the earl of Northumberland,[3] his existing powers fortified by a special commission, came up to restore order he was met near Thirsk by an angry mob. Their leader was John a Chambre who, though he was described as a 'simple fellow' or 'a very boutefeu who bare much sway among the vulgar and popular',

---

[1] *Rotuli Parliamentorum*, vi. 421.   [2] Pollard, *Henry VII*, i. 70.
[3] The fourth earl, who had been taken at Bosworth, had been restored to the office of lieutenant-general of the Middle and East Marches which he had occupied under Richard III, and by a special indenture had undertaken for a fixed sum (£3,000 or m. 3,000) to defend these marches along with the town of Berwick. Conway, p. 34. He became sheriff of Northumberland in February 1488 (Campbell, *Materials*, ii. 240), and along with others had a special commission to deal with insurrections, &c., in the city of York on 10 April 1489 (Ibid. ii. 443).

had hitherto shown himself a useful servant of the king. Bacon alleges that native harshness led Northumberland into peremptory action which was attributed to his own asperity though he was carrying out his master's orders; whether this is so or not is uncertain. At all events a scuffle ensued, he was deserted by his servants in a manner which aroused the disgust of Skelton, and miserably slain.[1] Having sued in vain for pardon the rioters took as their leader Sir John Egremont, who may have had Yorkist sympathies, and so serious did the situation appear that the king prepared to come north in person. Before he arrived, however, his work had been done for him by the competent Surrey whom he had sent on in advance. The rebels had been dispersed; Egremont had fled to Flanders; John a Chambre had been hanged at York. It remained for Henry only to play his accustomed role of the clement prince who spared the deluded.

While the king was receiving this rude lesson in the difficulty of making England pay for a war, his diplomatic efforts had gone on apace. To us of today the course he followed may seem both illogical and dishonest, for he tried to behave as if he were still at peace with France, and at the same time to make war against her in Brittany. In those days, however, men distinguished between waging war as a principal and taking part as an ally of one of the combatants; and Henry, if taxed with unwisdom or double-dealing, would probably have been much surprised. The object of both his policies, he would have explained, was the same—he wanted to maintain the *status quo* in Brittany. He hoped to do this by inducing France not to press her attack, but if he failed to do so he was prepared to resist her both by his own arms and by the creation of a league to maintain Breton independence. There was nothing underhand in his procedure. The exact details of his treaties were not published abroad, but there was no great concealment of the fact that he was dealing with Brittany, with Maximilian, and with Spain. In the first half of 1489 treaties were made with all three powers. An arrangement with Brittany presented no difficulty, because the luckless Anne and her advisers were cajoled to seek English help upon any terms, and the treaty of

---

[1] 28 April. Pollard, *Henry VII*, i. p. 72, for Skelton's poem on the 'Doulourus dethe and muche lamentable chaunce of the most honourable Erle of Northumberlande'. Ibid., p. 79 n., for John a Chambre.

Redon, concluded on 10 February 1489, illustrates less Henry's generosity than his bargaining ability. The duchess was pledged to give support, if England ever sought to recover her lost dominions in France; to arrange no marriage without Henry's permission; to make no alliance without the consent of England, except with Maximilian and Ferdinand, and with them only if England were included; to pay the expenses of any English force sent to her assistance, and to hand over to that force as soon as it arrived two strong places, fully equipped with guns and munitions of war. In return Henry undertook to send 6,000 men to serve the duchess until All Saints' Day of the same year. Plainly, by representing his soldiers as servants of the duchess he meant to keep out of the war 'as a principal'; his aid would be strictly limited within the terms of her promise. He drove a hard bargain.[1]

To come to terms with Maximilian was less easy. Apart from the fact that the cautious Tudor distrusted the flighty Habsburg, the relations between the two monarchs had for some time been far from good. The Low Countries were the breeding-ground of Yorkist conspiracies; Flemish pirates preyed upon English shipping, and Henry, who did not believe the excuses offered, had shown his displeasure by restricting English trade with the dominions of Maximilian. In July 1488 he had been quite outspoken to the Spanish ambassador Puebla upon the matter.[2] Ferdinand, however, was most anxious that the king of the Romans should be used against France, partly because he did not wish England to gain too great an ascendancy in Breton affairs, and urged the English king to forget his grievances. Henry accepted his advice, and on 14 February[3] concluded with Maximilian a treaty whose general effect was to re-establish the good relations set up by the alliance made between Edward IV and Burgundy in 1478.

Henry's deference to Spain in the matter of Maximilian is illustrative of his general attitude. He could impose his will upon Brittany, he could deal equally with the king of the Romans, but as regards Spain he was the suitor, and though he bargained resolutely he was content to accept the best terms he could get. Evidently he valued highly the alliance with this rising monarchy whose shipping would be useful to his trade, whose power

---

[1] *Foedera*, xii. 362–72.   [2] *Spanish Calendar*, i. 10, 11.
[3] *Foedera*, xii. 359; Gairdner, *Letters and Papers*, i. 52–54.

would limit overmighty France, and whose ambitions would be directed to fields in which England had little concern. Very early in his reign he had encouraged Spanish merchants and Spanish captains,[1] even permitting the export of wheat and tin, and when the Breton question took first place upon the European stage he hoped to engage the interest of Spain not less because Ferdinand was Anne's kinsman than because he too must regard with apprehension a great increase in French power. Ferdinand, however, was, or affected to be, little dependent upon the goodwill of England; when early in 1488 Henry sent an embassy to propose a mutually defensive treaty and a marriage alliance[2] he at once sent Puebla and Sepulveda to negotiate, but it is plain both from the formal instructions of his ambassadors, and from their private comments,[3] that he thought himself in a position to dictate terms. His ambassadors enlarged upon the might of Spain, and upon the uncertainties of the island monarchy, where kings were made and unmade so easily; it was generous in Spain, they said, to consider even the possibility of giving a daughter to England. Evidently Ferdinand had hoped to take complete control of Henry's dealings with Brittany and with Maximilian, to detach him from Portugal, and to involve him in an open war with France, to be waged without help from Spain; moreover, though he laid less emphasis on this, he had hoped also to reduce the amount of the necessary dowry. How far he was disappointed in his hopes appeared from the provisional treaty concluded in July 1488 and from Puebla's account of his activities. Henry had shown himself extremely anxious for the alliance. He doffed his bonnet whenever he mentioned the names of Ferdinand and Isabella; he opened his eyes for joy and said *Te Deum laudamus* when he found that the ambassadors were armed with powers to conclude a marriage alliance. But the English commissioners were adamant about the dowry, insisting on 200,000 crowns at 4s. 2d. each, and they demanded that the Infanta should be sent to England to be educated long before she became of marriageable age; they declined to abandon the old English understanding with Portugal; they refused to commit their master to a war

---

[1] *Spanish Calendar*, i. 3.
[2] The instructions of the English ambassadors said nothing of a marriage alliance (*Spanish Calendar*, i. 3), but later the Spaniards pointed out that it was England who had made the first proposals (ibid. i. 6) and their assertion was not contradicted.   [3] *Spanish Calendar*, i. 5–12.

with France, though they were prepared to swear on the mass-book that, after the marriage treaty was signed, Henry would make war on Charles VIII, at the bidding of the Spanish monarch. Ferdinand was inclined to grumble because better terms had not been obtained, but Puebla had apparently found that Henry's power was more secure than had been believed, and he was quite definite in his opinion that an alliance should be concluded. The result was that when Savage and Nanfan, accompanied by the Richmond herald, arrived in Spain after a very stormy voyage,[1] they were able to obtain a treaty which, though it left Ferdinand with an advantage, gave him much less than he had at first expected.

The treaty, which was signed at Medina del Campo near Valladolid on 27 March 1489,[2] covered both the questions previously discussed, the marriage being regarded as a means of strengthening the political alliance, which took pride of place in the actual document. As regards the marriage, England fared well enough. She got the portion which she had demanded, though there was some doubt as to whether the jewels and ornaments to be brought by the Infanta should or should not count as one-fourth of the 200,000 crowns—the oaths of Fox and Gunthorpe were to be taken upon a promise they were said by Puebla to have made, but which they denied. The jointure offered, one-third of the revenue of the principality of Wales, the duchy of Cornwall, and the county of Chester, amounting to at least 23,000 crowns, was accepted, though it was to be increased when Catherine actually became queen. One half of the dowry was to be given as soon as Catherine came to England; Spain still insisted that she must come only on the eve of her wedding, but a loophole was left by provision for a marriage *per verba de futuro* as soon as the children attained the necessary age. The date of the bride's arrival was left to be fixed later, and by a subsequent treaty (8 March 1493) it was agreed, or at least proposed, that she should come when she reached the age of twelve.[3] The political alliance, however, was less satisfactory to Henry. Some of its terms, indeed, expressed a complete parity. The customs were to be reduced to what they had been thirty years before; neither monarch was to aid the other's rebels or to grant letters of

---

[1] Gairdner, *Memorials*, pp. 157, 328.   [2] *Spanish Calendar*, i. 21.
[3] *Spanish Calendar*, i. 21, 48.

marque against the other's subjects. If either sovereign were attacked, the other was to assist him by sending aid within three months if he were asked to do so, such aid to be at the expense—fairly assessed—of the requesting party.

All this was fair enough, but the inclusion of certain clauses, aimed specifically at France, gave a definite advantage to Spain, both because Henry had not wished to signalize France as the great enemy, and because Henry was asked to contribute to the common action upon very unequal terms. Neither side was to make peace, alliance, or treaty with France without consulting the other; and each was to make war upon her when the other did so. If such war did break out neither was to make a separate peace unless his territorial claims were made good, unless, that is, France voluntarily surrendered Normandy and Aquitaine to England, or Cerdagne and Roussillon to Spain. Apart from the fact that Henry did not want to declare open war on France, these terms were quite inequitable. France could easily surrender two small frontier provinces which in fact she held only in pledge,[1] but she could not possibly give up two large and wealthy territories, one of them of indefinite extent, which were essential to her very being. England, in effect, was promising to continue a war with France during Ferdinand's good pleasure. There has been much discussion[2] as to whether a clause, permitting a renewal of Henry's truce with Charles, did or did not give the English king a loophole for escape. Technically, if Henry could, within a given time, include Ferdinand along with himself, he could prolong indefinitely the truce with France which he had already extended to last until 17 January 1490;[3] but it is obvious that Spain might refuse so to be included, and it seems likely that Ferdinand, confident that actual hostilities between England and France would break out in any case, was merely providing an excuse for his own abstention for a year or two on the ground that, while the truce lasted, England was not engaged in war as a principal, but only as the ally of Brittany.

In any case there is small profit in discussion. Stipulations as to existing truces were made in many treaties, and were, as a rule, little regarded. In those days it was even easier than it is

---

[1] She surrendered the provinces without much demur in January 1493. *Spanish Calendar*, i. 43.

[2] For his discussion with Gairdner see Busch, *England under the Tudors*, pp. 53, 331, 435.

[3] *Foedera*, xii. 344.

now to adduce some 'incident' as a proof that the truce had already been broken whenever one wished to break it. How little attention was given to formal documents will appear from the history of the main treaty itself. It was duly ratified by Ferdinand and Isabella on 28 March,[1] the day after its conclusion, and a copy, signed by both sovereigns, was sent to England. Henry, however, did not at once ratify it as it stood. His formal reason was that certain matters about the dowry and the date of Catherine's arrival were still unsettled;[2] but the main object of his delay was to secure an alteration of the terms upon which the peace with France, once broken, could be restored. Only on 17 September 1490 did he publicly proclaim that the political treaty had been concluded, and then in a somewhat casual form, coupling it with an announcement of a new-made alliance with Maximilian.[3] Six days later he gave his formal ratification[4] to the treaty as it stood, but coupled it with amendments which he presented in a form designed to give the impression that they were necessary adjuncts to the treaty itself. These he embodied in two documents, both executed on 20 September. In the first of these,[5] which was entitled *Declarationes* upon the treaty, he proposed that, as there should be equality between the two contracting parties, neither should make truce or treaty with France without consulting the other; in the second,[6] which was styled *Ratificatio* of the treaty and of certain amendments, the proposition was that the war should be continued until the territorial claims of both parties were satisfied—until, that is, Normandy and Aquitaine, Cerdagne and Roussillon were all surrendered. There is no evidence that Spain ever accepted these amendments. The Spanish monarchs evidently lost patience, for the copy of the treaty which they had so expeditiously sent to England was cancelled by the simple process of cutting away the signatures. Both sides continued to quote the treaty when it was convenient to do so, but Spain asserted, and Henry apparently agreed,[7] that it had never been fully

---

[1] *Spanish Calendar*, i. 24.   [2] Ibid. i. 27, 33, 37.   [3] *Foedera*, xii. 410.
[4] Ibid. xii. 417.   [5] Ibid. xii. 411.   [6] Ibid. xii. 413.

[7] In November 1494, in their belated report to Henry of their peace with France (January 1493), Ferdinand and Isabella, in pointing out that their action was in accordance with the treaty of Medina del Campo, since they had gained Cerdagne and Roussillon, noted that the treaty had no force since Henry had never ratified it. Next year Puebla, in forwarding Henry's stiff reply, added a note that Henry agreed that the former treaties no longer existed, and in 1496, in the preliminaries

ratified, and in fact an entirely new treaty was negotiated in
1496.[1]

From the documentary evidence, then, it appears that the
famous treaty of Medina del Campo, though quoted by both
contracting parties when convenient, had at no time a complete
legal validity. The fact remains that it defined the relations of
England and Spain during two decades, and that its essential
provisions were eventually carried out. Justly has it been regarded as a landmark in the history of English foreign policy;
it is an example of the *Realpolitik* of the time, an expression of
the common interest of England and Spain. Henry and Ferdinand understood the situation very well; both knew well the
value of a correct form, but, to both, reality was more important
than form. When the legal status of the whole treaty is thus
uncertain, it is idle to dispute the exact meaning of the clause
concerning Henry's truce with France. The facts rather than
the words were significant.

In 1490, to which year we now return, Henry, truce or no
truce, was actually at war with France and he had to do all, or
nearly all, himself. His 6,000 men had landed in Brittany in
April 1489, occupied Guingamp, which the French had hastily
abandoned, and made for Concarneau; the French garrison in
Brest had been attacked by the local countryfolk and before
long formally besieged by Rieux; as appears from his letter to
Oxford,[2] the English king at first thought that all was going
extremely well. Disillusionment came speedily. There was a
bitter quarrel between Anne and Rieux, who still urged her to
marry Albret, and, when the little duchess was persuaded—
wrongly as it seems—that the English also favoured the suit of
the ogre in hope of advantage in Gascony, she shut herself
up in Rennes, and declined to co-operate with her deliverers.
Eventually after taking Concarneau in September the English
retired in disgust to Guingamp; in October Rieux lost his guns
before Brest; in November Anne accepted the treaty of Frankfurt,[3] but her efforts to make peace were foiled by the continued
opposition of Rieux and Albret, and by the fact that the English
persisted in retaining Guingamp and Concarneau. The French

---

for the new treaty, Ferdinand and Isabella note Henry's dissatisfaction that, after
so long negotiation, no alliance had been concluded. *Spanish Calendar*, i. 51, 52,
54, 84.   [1] *Foedera*, xii. 636.
   [2] Pollard, *Henry VII*, i. 68 (22 April 1489).   [3] See on p. 99.

continued their attack, and exerted diplomatic pressure on
Henry who was now the sole obstacle to their complete
success.

For Maximilian, characteristically, had already abandoned
his English ally, even though that ally had given him real help
in the Netherlands at a time of sore need. Occupied himself by
his war against the Hungarians, he had left the defence of the
Netherlands to Albert of Saxony, who, though a good general,
soon found himself in straits. He had few troops and civil
dissension was rife. On the ground that he had favoured the
Cod-fish the Hooks flew to arms under Philip of Cleves, lord of
Ravenstein, and held out a welcoming hand to Esquerdes in
Picardy who sent some help and promised much more. Early
in 1489 an army of 4,000 rebels, largely from Bruges and Ghent,
sat down before Dixmude, whose fall seemed imminent. Maximilian's partisans could count only upon their own limited
resources and upon a few mercenary garrisons amongst whom
600 Germans at Nieuport were conspicuous. As a last resort they
appealed for help to the captain of Calais, Lord Daubeney, who
at once asked Henry for instructions. Henry's reply shows the
Tudor government and the Tudor spirit at its very best.
Realizing that, if Dixmude fell, the French would soon occupy
the surrounding coast and cut Calais off from the Netherlands,
the English king determined on a bold stroke. Lord Morley was
sent across with 1,000 archers, professedly to strengthen the
Pale; Daubeney collected what he could from his own garrison
and that of Sir James Tyrrell at Guisnes, and, only a few days
after the receipt of the appeal, the doughty captain of Calais
marched out in the darkness with 2,000 archers, 1,000 pikemen,
and sixteen guns. His left flank was guarded by six or eight
ships of war and his advance was swift and sure; he made
contact with the garrison at Nieuport, and by his valiant words
induced it, not to defend, but to join with him in attack.

The rebels had fortified their camp very strongly and the
main approach was along an exposed causeway half a league
long, but the English had been told of a weakness in the rear-
ward defences by a Ghentish spy whom Tyrrell saved from the
gallows, and it was probably with the aid of his information that
Daubeney found a way in whilst the main assault was being
made at the north gate. This was a desperate business. Morley,
who had refused to dismount, was killed with a gunshot;

Tyrrell was wounded with a bolt. According to Molinet[1] the headlong English rush was beaten off with heavy loss and the day was won by the disciplined valour of the Germans, who lost all their officers before they gained the wall. Hall,[2] of course, gives the credit to the 'crewe of valiaunt archers and souldiers' who had marched from Calais and says that the English bowmen silenced the French guns. Neither is an impartial witness, but it is significant that the only real victory of Maximilian's party was won when the English fought for him. The victory was marred by a massacre, for when it was known that Lord Morley was dead, the English, mindful perhaps of St. Aubin, killed all their prisoners; but the booty was immense and Daubeney was able to get it safely into Calais before Esquerdes came up with a large army.[3] That 'covetous' (and perhaps unimaginative) 'Lord Cordes' who 'would commonly saye that he would gladly lye seven years in Hell so that Calais were in the possession of the Frenchmen' at once attacked Nieuport with great fury. The town was so hard pressed that even its women were mustered for the defence, but it was saved by the arrival of a small reinforcement from Calais. The garrisons of Dixmude, Ostend, and Furnes were strengthened, and for the moment the situation in the Netherlands saved.

The action at Dixmude was fought on 13 June;[4] less than six weeks later Maximilian deserted his English ally. Perhaps he should not be too much blamed. Despite his high pretensions he had little real power anywhere; the Netherlands were in desperate need of peace; the Germans were concerned about their eastern frontier, had no interest in Brittany, and saw no need for a war with France. The diet of the empire met at Frankfurt on 6 July and there on 22 July 1489 was signed a formal treaty between Maximilian and Charles.[5] The contracting parties were to preserve perpetual peace on the basis of the treaty of Arras of 1482; Charles was to behave liberally about the duchy and the counties which France had then detached from the Burgundian inheritance; pending a final settlement Madame Anne—the Duchess Anne, not Anne of Beaujeau—

---

[1] Molinet, *Chroniques*, ii. 135.

[2] Hall, *Chronicle*, p. 445; cf. Leland, *Collectanea*, iv. 247.

[3] Hall says 20,000 men; Molinet 18,000. Even the lower figure may be too high, but it is plain that the army was big.

[4] Pollard, *Henry VII*, i. 80 n., quoting the *Chronicle of Calais*.

[5] Dumont, *Corps Diplomatique*, iii. 2, 327; *Spanish Calendar*, i. 28.

NE. France and the Low Countries.

was to hand over four towns, to be held in neutrality, by a representative of France (Bourbon), and a representative of Maximilian (Orange); meanwhile she was to dismiss the English troops from her service. Whether Maximilian, always optimistic, really hoped to recover any of the losses of Arras may be doubted; but he knew very well that the cessation of French interference in the Netherlands would allow him to recover his authority there, and for him this gain far outweighed the loss of Brittany.

Brittany, it seemed, must inevitably be lost. Maximilian had deserted her; Anne had accepted the treaty of Frankfurt; the peasants were sick of war; Ferdinand had sent no real help. Meanwhile France redoubled her efforts to induce England to abandon the war. The fact that the truce would expire in January 1490 gave her the opportunity to resume negotiations, and towards the end of 1489 she dispatched two very distinguished embassies to induce Henry to accept the settlement of Frankfurt. Her efforts were seconded by Pope Innocent VIII, who sent Henry the sword and the cap of maintenance,[1] and in the spring instructed his nuncio in France, Lionel Chieregato, to speak peace at the English court. He wished to obtain French help against the Aragonese king of Naples,[2] and knew that it was the Breton war which kept France from a projected invasion of Italy; but he said—and it was true—that there was great need of Christian unity in the face of the Turkish advance and that there was a golden opportunity for a grand crusade since he had in his hands Djem, the brother and rival of Sultan Bayazid. The great coalition against France was broken up. Everyone was speaking peace, and it seemed as if Henry must abandon the war without even thanks for all his pains. Failure stared him in the face.

But there was that in the Tudor which was slow to accept defeat, and calculation must have shown him that if it was difficult for him to continue the war it was dangerous for him to make peace. If he did so he would alienate his remaining ally Ferdinand; he would lose the prestige so important to a new dynasty and possibly also the 'tribute' claimed by English kings

---

[1] *Venetian Calendar*, i. 548. Henry received this mark of papal approbation on two subsequent occasions—from Alexander VI in 1496 and from Julius II in 1505. Kingsford, p. 211; Gairdner, *Letters*, i. 243.

[2] *Venetian Calendar*, i. 560; *Milanese Calendar*, p. 400.

since the treaty of Picquigny; he would leave unhampered in Spain, France, and the Netherlands those rebels who so easily became aspirants for his crown. So, in spite of all discouragements, he held firmly to his old policy. He continued to treat with France to whom he accredited commissions to discuss terms of peace in February[1] and again in June 1490.[2] He was represented at what resembled a European peace conference convoked by Chieregato at Boulogne and Calais, and it was not until August that negotiations were broken off.[3] They were resumed in October, and again in February 1491; Henry was still trying for a peaceful settlement.[4] Yet he was marshalling his forces for a continuation of the war, and it soon appeared that his assets were greater than his enemies had supposed. With instinctive skill he took his own people with him. He recalled his prorogued parliament twice, in October 1489 and again in January 1490, persuaded it that the successive French offers were inadequate, and, at the price of remitting the uncollected portion of the last grant, induced it to vote him on 27 February a fifteenth and tenth with which to carry on the war.[5]

Abroad, as well as at home, he was active. In August 1489 he had renewed the old treaty with Portugal, which dated back to 1387, and in December he secured the ratification of King John.[6] In January 1490 he amplified into a political and commercial alliance a treaty which he had made with Denmark in the preceding August.[7] In April he made a treaty with Florence.[8] His motives were largely economic. In the interests of English trade he was challenging the commercial supremacy of the Hansa and of Venice, but his policy had a diplomatic value in showing the world that the king of England did not lack friends. Meanwhile he had taken steps to ensure that the struggle should go on. On 15 February 1490 Duchess Anne had appointed a fresh commission to deal with him, and he used the opportunity to obtain the town and castle of Morlaix as well as Concarneau in pledge for his expenses;[9] when the treaty was finally concluded on 26 July he exacted a promise of 6,000 gold crowns per annum

---

[1] *Foedera*, xii. 449; see Bacon, *History of King Henry VII* for Gaguin's eloquence.
[2] Ibid. xii. 453.
[3] Busch, *England under the Tudors*, i. 58 n.; *Venetian Calendar*, i. 571, 574, 593; *Foedera*, xii. 454.
[4] *Foedera*, xii. 431, 435.
[5] *Rotuli Parliamentorum*, vi. 438.
[6] *Foedera*, xii. 378, 380.
[7] Ibid. xii. 373, 381.
[8] Ibid. xii. 389.
[9] Ibid. xii. 387, 394.

in lieu of the revenues of Morlaix.[1] He did his best to keep warm the friendship of Spain, and took no umbrage at a Spanish proposal that Anne should marry the son of Ferdinand and Isabella. The proposal was not seriously meant, being advanced mainly to exclude some undesirable bridegroom, perhaps the duke of Guelders,[2] but Ferdinand showed his interest by sending into the duchy 1,000 men. These did not co-operate with Henry's troops, alleging that[3] they did not wish to share in the unpopularity of the English, who were suspected of being too friendly to Albret, but their presence was a proof that Spain was still interested in Brittany. Ferdinand had, in fact, discovered that Maximilian might enter the matrimonial lists again,[4] and he meant to keep control in his own hands. The inconstant king of the Romans was veering to the English alliance once more; freed from French interference by the treaty of Frankfurt, he had at length (30 October 1489) persuaded the Netherlanders to acknowledge him as guardian of his son, and he was casting his restless eye on Brittany once more. On 22 May 1490 he instructed his ambassadors to approach Henry with proposals for an attack upon France, and on 11 September a formal treaty was made.[5]

Whilst his negotiations with Maximilian were in progress the English king had signed a treaty with Ludovico Sforza, Il Moro, duke of Milan, who knew himself threatened by France.[6] The great coalition against France was in being again, and at Christmastide 1490 its revival was signalized in dramatic fashion. At Neustadt Maximilian was invested with the insignia of the Garter with the greatest pomp imaginable and in Brittany his handsome proxy, Wolfgang von Pelhain, was married to the Duchess Anne, who early in 1491[7] publicly took the title of queen of the Romans. Henry, who a year before had seemed to stand alone, was now a member of a strong coalition which encircled France.

Its strength was all in the seeming. Even today it is not easy

---

[1] Ibid. xii. 456.   [2] *Spanish Calendar*, i. 27, 31.
[3] Ibid. i. 29, 32.   [4] Ibid. i. 34.
[5] *Foedera*, xii. 397–400: characteristically, Maximilian had renewed his treaty with France at Ulm in July 1490, in the midst of these negotiations.

[6] 4 October. Ibid. xii. 430.

[7] In powers to English ambassadors of date 26 February Anne is described as *Ducissa Britanniae*; in powers of 29 March she is styled *Romanorum Regina: Foedera*, xii. 436, 438.

to co-ordinate the action of allied powers, and in the days of
bad communications and slow transport concerted operations
were impossible. The diplomatic history of the sixteenth century
is an essay upon the impotence of coalitions, and this coalition,
more than most, concealed in its midst the germs of dissolution.
Maximilian was involved in a war against Ladislas Jagiello,
king of Bohemia, for the crown of Hungary; the imperial diet
at Nuremberg did, eventually, vote 12,000 landsknechts for a
campaign in Brittany, but they had no means of sending them
to the duchy even if they had been ready. Ferdinand, as usual,
was playing a double game. In the summer of 1489, with
complete disregard for any obligations under the treaty of
Medina del Campo, he had entered into secret negotiations
with France, discussing the possibility of a match between
Charles VIII and his daughter Juana, and, although this
project did not go very far, he next year concluded a six-months[1]
armistice without any reference to England. There was even
talk of a personal meeting between the king of France and the
Spanish monarchs. In the autumn he withdrew his troops from
the duchy, save for a small garrison at Redon,[2] and though he
promised to send them back in the spring he did not do so. He
wanted them for the assault upon Granada.

In Brittany itself things were in confusion. During the year
1490 Anne and Rieux had been reconciled, and the French had
made little progress; but the marriage of the duchess with
Maximilian at the end of the year threw the disappointed Albret
into the arms of France. Although Charles had made a truce with
Brittany, and was actually in negotiation with England and the
duchess upon the eternal question of Henry's expenses,[3] he was
quick to seize the opportunity. On 4 April 1491 Albret surrendered Nantes to him and before long his troops were over-
running the duchy. In May the king and queen of the Romans
appealed to England for aid,[4] but though Henry dispatched a few
troops he could do no more; Maximilian could send no help;
Anne's mercenaries were mutinous for want of pay. The French
took Redon from the Spaniards, Guingamp and Concarneau
from the English, and sat down before Rennes. Henry, who continued to hold Morlaix,[5] levied fresh money, 'from the more able

---

[1] *Venetian Calendar*, i. 195; Busch, *England under the Tudors*, p. 60.
[2] *Spanish Calendar*, i. 34, 35.   [3] *Foedera*, xii. 436.   [4] Ibid. xii. 443.
[5] The date of the evacuation of Morlaix is not ascertained.

sort', after consultation with his council, by way of benevolence,[1] and offered to assist Anne to join her husband. He continued his preparations for war; but Anne had had war enough. When her jewellery no longer sufficed to pay her soldiers, when defence was no longer possible, she still abode in her capital. The king of France marched in, made an armistice with his gallant opponent, offered his hand in marriage, and courteously retired to Touraine. He had already procured a dispensation to cover any bar of consanguinity, and though this said nothing of Anne's wedding by proxy, or of Charles's pre-contract to Maximilian's daughter by the treaty of Arras, her confessor made short work of the lady's scruples. He knew that the pope, who was anxious to see the French in Italy, would grant a full dispensation, as indeed he did on 15 December. It served only to ratify an accomplished fact. Anne had come to see her new suitor and on 6 December had married him at Langeais. The duchess of Brittany had become queen of France, the last semi-independent province had been absorbed into the French kingdom. It was the realist answer of the new unitary state to the airy cobwebs of the far-flung coalition. It was a victory for the New Monarchy in France.

From another 'new monarchy' came the counterstroke. Of all the members of the coalition Maximilian had the greatest grievance, for he had been robbed at once of his bride and of his prospective son-in-law; but, entangled in the east and confronted by fresh civil war in the Netherlands, he could do nothing. Spain would do nothing; her thoughts were centred upon Granada, which fell at last in January 1492; for her the Breton adventure had been only a means of recovering her lost provinces from France, and she had no further use for a broken lever. England, however, was resolute to act, and to act alone. To many observers, and not least to Anne, the Tudor must have appeared as an importunate tradesman for ever presenting his account, and it is probable that, if his demands had been met he would have withdrawn from Brittany, though there is no proof that he would have done so without including Spain in any arrangement that he made. He still wanted his money, for he

---

[1] As a means of raising money the 'benevolence' had been levied by Edward IV, but it had been condemned in an act of Richard III. See Tanner, *Tudor Constitutional Documents*, p. 619.

felt that of all the allies he alone had done any real fighting, and he continued his negotiations with France. But for him it was no longer simply a question of money. English blood had been shed, and there was nought to show for it; there were murmurings of Yorkist discontent and in the autumn of 1491 another 'masked prince' appeared in Ireland. His prestige was at stake. He was determined to fight, or at least to show himself ready to fight. He did not abandon hope of gaining his ends by a bargain with France,[1] and he continued to negotiate for allies abroad,[2] but he realized that in the end he must rely upon his own resources, and the winter of 1491-2 was filled with bustle of military preparation. The parliament which assembled in October was as Bacon says 'merely a parliament of war'. The text of Morton's opening address was significant—'Expectavimus pacem et non est bonum, et tempus curationis et ecce turbacio'. After animadverting upon the Jugurthine deceits of the French king the speaker argued that peace was better than war provided that it was not *fantastica, sophistica vel diabolica*, and concluded that one might wage, not only with the apostle a spiritual war, but also a corporal war, provided that the cause was just, that the war was publicly proclaimed, and the rules of humanity observed.[3]

According to Bacon Henry signalized the existence of a new situation by himself addressing the assembly. When he proposed to wage war in Brittany 'by my lieutenant', he had allowed his chancellor to tell them of his purpose, but now that he meant to go to France in person he would himself declare the matter to them. 'That war was to defend another man's right, this is to recover our own.' Charles's ambition had alienated the whole world, and England could count on many allies but 'God forbid but England should be able to get reason of France without a second'. He recalled old victories; claimed that England, now happily reunited under himself, could vindicate her ancient rights; asserted that, against a country so rich as France, war would pay for itself, and adjured them to lose no time in playing their proper part. The speech, put by Bacon into

---

[1] Henry continued to deal with Charles in 1492—5 February, 12 June, 26 July. *Foedera*, xii. 470, 481, 497.

[2] In January 1492 he planned to capture Brest with the aid of disaffected nobles; he dealt with the pope, December 1491, and Milan, January 1492. *Venetian Calendar*, i. 207, 210. Maximilian promised to serve him with 10,000 men, but in spite of high-sounding proposals had nothing ready. Hall, p. 455.

[3] *Rotuli Parliamentorum*, vi. 440.

sonorous prose, is a fairly exact paraphrase of the arguments used by the commons to justify the granting of supply to the extent of two fifteenths and tenths, with a promise of a third if Henry remained abroad for more than eight months. Provision was made for keeping intact the crown lands in the event of the king's death.[1] An act was passed to punish deserters and captains who cheated men of their pay.[2] Legal privileges were granted to 'persons being in the king's warrs'. On the ground that open war was better than a feigned peace all Scots who were not denizens were ordered to leave the country without delay.

Outside parliament things moved apace. In December commissions were issued for the hiring abroad of 500 ships and 300 crossbowmen. Men were raised by indenture, and William Paston was not the only man who bestirred himself to procure good arms and a good horse. By February it was reported that the king daily dispatched ordnance to the sea-side; that very many tents were being made, and that all the gentlemen were busy providing themselves with military equipment; it was thought that the expedition would start at Easter.[3] But Henry took his time, and though the fleet sailed in June it soon returned. As to the reasons for his delay it is possible only to speculate. He may have hoped for some help from his allies, though his solemn celebration of the fall of Granada on 6 April had produced no answering gesture from Ferdinand, and his envoys to Maximilian, Urswick, and Risley soon returned with the news that the king of the Romans was 'sore sicke of a flux of the pursse' and unable to move. He may have believed that the French army, which relied on hired foot-soldiers, would be unprovided with infantry if he delivered his attack late in the year. He may have hoped he would gain his ends without attacking at all; he had appointed a fresh commission to deal with Charles on 12 June[4] and negotiations were still in progress. Meanwhile a considerable force was collected at Portsmouth, and three great breweries were erected to supply its need— without beer the English soldier of the period was not able to do his best.[5] In May[6] indentures were made with magnates for

---

[1] *Rotuli Parliamentorum*, vi. 443. There is no mention of any speech from the throne.   [2] *Statutes of the Realm*, ii. 549.
[3] Pollard, *Henry VII*, i. 89; *Plumpton Correspondence*, p. 102.
[4] *Foedera*, xii. 481; Hall, p. 456.
[5] Several instances can be adduced to show that this view was held officially.
[6] *Foedera*, xii. 477.

provision of lances, demilances, archers on horse and on foot, and billmen, who were to serve for a whole year. In August patents were issued to the sheriff of Kent, the lieutenant of Dover castle, and the Cinque Ports, bidding them have their people ready at an hour's notice,[1] and in the same month a minor operation was undertaken against Sluys.[2] This port had been seized by the count of Ravenstein, the leader of the Hooks, who had turned it into a den of pirates, garrisoned it with Danish mercenaries, and successfully defied Maximilian's general, Albert of Saxony. The advent of Henry's ships and men under the efficient Poynings turned the scale, and when after a desperate defence Sluys capitulated on 13 October, the castle, though not the town, was handed over to the English admiral.

By this time Henry was in France. He had concentrated about London a puissant army, to which many great nobles had brought contingents raised by indenture. Bedford and Oxford were in command, and it was with a force of 25,000 foot and 1,600 horse that, leaving Arthur as regent in his stead, he sailed from Dover to Calais early in October. There he tarried awhile, and it was not until 18 October that he sat down before Boulogne, which was well defended. The siege was not pressed hard, and save that Sir John Savage was killed nothing of note occurred. Nine days later the business was ended; Fox and Daubeney had been dealing with Esquerdes and on 27 October Henry presented to his counsellors proposals for peace which had been sent to him at Étaples. Along with these he produced reasons advanced by his captains against the continuance of the war in an inclement season; the king, they said, was offered better terms than those given to Edward IV at Picquigny, and it would not hurt Henry's honour if he, like his predecessor, made peace without winning a battle or capturing a town. The reasoning, or the king's will, prevailed, and the treaty was signed at Étaples on 3 November. Its main provisions[3] followed

---

[1] Gairdner, *Letters and Papers*, ii. 371-3; *Foedera*, xii. 482.

[2] Gairdner, *Letters and Papers*, ii. 373, for commission to Sir Edward Poynings, 24 August.

[3] Pollard, *Henry VII*, iii. 6, cf. i. 91; *Foedera*, xii. 497 ff. The 'franc' was to be equal to 20 'sous tournois', and the crown to 35. In England the franc was reckoned at 2s. (Fabyan and Arnold). Henry would then get an annual income of about £5,000. In crowns at 3s. 6d. this would be 28,571 crowns; if an equation provided by the City Chronicle (Pollard, *Henry VII*, i. 93) be accepted the crown would equal 3s. 5d. only, and £5,000 would equal over 29,000 crowns. In fact the sum paid for a half-year in 1511 was only 13,793 crowns (*Foedera*, xiii. 310); this would

common form in providing for peace to endure until one year
after the death of the king who should live the longer, and for
the avoidance of *casus belli*. Henry's ally Maximilian was given
the option of being included, and if he refused could count on
English help only if he were attacked by France; custom duties
were to be equitable; each king was to abstain from supporting
the other's enemies, from the condoning of piracies, and from
the issue of letters of marque. But in a separate instrument
Charles also undertook to repay Henry's outlays in Brittany
(620,000 crowns) and the arrears of eight payments due under
the treaty of Picquigny (125,000 crowns)—745,000 gold crowns
in all, at the rate of 50,000 francs a year. By another document
Charles undertook not to assist Henry's rebels.[1]

The peace was announced at Boulogne on 4 November; five
days later the lord mayor proclaimed it in London, and Morton
announced it in St. Paul's where *Te Deum* was sung. The troops
came home by way of Calais, and on 22 November the king
entered his capital in triumph. *Ridiculus mus!* After all the talk
and preparation his campaign had lasted about three weeks,
and his troops had been in contact with the enemy for about
nine days.

Henry was blamed as one who 'did but traffic with that war',
one who 'cared not to plume his nobility and people to feather
himself'. To many people it must have seemed that he had
never meant to fight at all, that he had been sure of a speedy
peace before ever he set sail. The gentlemen who had wasted
their time and energy for months, and who had spent good
money on equipping themselves for war, felt that they had been
defrauded. Their losses were perhaps less than has been supposed, for they would be exactly paid the sums promised under
indenture for their daily wages and those of their men;[2] but
they had reckoned on booty as well as adventure and glory, and
their hopes had been deceived. As for the aldermen and other
rich men who had contributed to the benevolence, they felt that
they had had very little for their money. Henry himself was
probably conscious that his action would be ill received at home.

make Henry's gain less than £5,000 per annum—about £4,900 at 3s. 6d. a crown.
But with fluctuation in money value and 'deductions' it is hard to be accurate.
Hall took a 'crown', which he calls a 'ducat', to equal 4s. and Fabyan adopted the
same valuation (1494).

[1] *Foedera*, xii. 508.
[2] Gairdner, *Letters and Papers*, ii. 86, where Henry points this out.

The treaty, according to its terms, was to be confirmed by the English parliament within one year, but by a special instrument[1] Charles agreed to extend the time-limit to three years, on the inadequate ground that to summon the estates would plunge his cousin into expense. Parliamentary confirmation was not given till 1495.[2]

The king of England appeared to have cheated his own people, and not them alone; he appeared to have cheated his allies too, for in the making of his peace he had consulted neither Ferdinand nor Maximilian. The king had 'gone forth to Normandy', or at least to Picardy, and there, as it seemed, he had bartered English honour for French gold. There is another aspect of the matter. Ferdinand himself alleged, when it suited him, that the treaty of Medina del Campo had never been ratified. He had made no serious effort to attack France. He had, as Henry may have known, begun private negotiations with the enemy, and in January 1493 he recovered his lost provinces by the treaty of Narbonne. As for Maximilian, he had played fast and loose with his word and deserted his English ally completely at Frankfurt and at Ulm. Henry, on the other hand, had not only provided for the inclusion of Maximilian and his son in the treaty of Etaples, but had reserved the right to support them by war, even against France, if France attacked their territories. On 23 May 1493 the king of the Romans did, in fact, secure at Senlis an advantageous treaty by which he regained his hold on Artois and the Franche Comté. Henry's allies had little cause to grumble.

The king's conduct towards France, though it displeased his people, was really in accordance with the policy which he had followed all along; throughout the whole of the Breton imbroglio he had always, even when he opposed France in arms, kept the door open for negotiations. Only when his schemes were defeated in Brittany did he declare war upon Charles, and there is no proof that his invasion was all a sham. Had it not been made, his negotiations with France might well have gone the way of many earlier negotiations. As it was, his arrival at a time when France was not well prepared—her field-army did not put in an appearance—served to bring his opponents to the point and to give him what he had so long demanded.

He had, it is true, made no attempt to recover Normandy and

---

[1] *Foedera*, xii. 508. [2] *Statutes of the Realm*, ii. 635.

Aquitaine; but he, and all serious politicians, must have known all along that reconquest was impossible, and the experiences of his troops before well-defended Boulogne must have convinced the hotheads that France was no longer a land where great spoil might be won at little cost. It is true, too, that he had failed to save Brittany; but the cause of Breton independence, in the hands of its aristocratic champions, had lost much of its native force; his allies had failed him and single-handed he could not have done more than he did.

To set off these failures, he had made substantial gains. He had secured a solid peace, and peace, he knew, was better than war. He had closed France to his Yorkist enemies and he had obtained a useful annual income which was paid to himself and his successor until the year 1511. He had shown Europe that England was not to be despised; that, with her internecine wars quelled by his strong hand, she could once more send a strong army overseas. When it is added that, even in the midst of hostilities, he had found time to promote, by commercial treaties, the trade of his subjects in the Baltic and in the Mediterranean, it becomes clear that his foreign policy must be regarded as extremely successful. It was not splendid in its action, but its results were real.

V

PERKIN WARBECK

THE treaty of Étaples was a turning-point in the political history of western Europe. Along with the companion treaties of Narbonne and Senlis it marked the reorientation of French policy. France was clearing her feet for the stride into Italy which she took in 1494, and for the next half-century[1] the interest of Europe centred upon 'the Italian wars'. Italy invited invasion; her wealth, her beauty, and her helplessness made her the natural victim of aggression. In her population were numbered some of the cleverest merchants and financiers in the world, and in her great cities—notably Florence—large-scale industry was already organized on a 'capitalistic' basis. She excelled in the arts, and during the fifteenth century her excellence was revivified by the stimulating breath of the Renaissance. But she lacked political cohesion; at this time, as at others, Italy was only a geographical term. Of the city-states some had become princedoms—in Mantua the house of Gonzaga held sway, in Ferrara the house of Este; others such as Genoa (which fell under the control first of France and later of Milan), Lucca, and Siena had become considerable republics; but many cities remained independent under their own princes, though they stood in constant danger of being annexed by some acquisitive neighbour. The politics of Italy, however, were dominated by five great states—the papacy, the republic of Venice, the kingdom of Naples, the republic of Florence, the duchy of Milan.

None of these states was very stable. The Papacy had lost much of its European authority and, though its political weight was still considerable, its energy was largely directed towards the creation of a temporal principality, which Pope Alexander VI (1492–1503) and his son Caesar hoped to make hereditary in the house of Borgia. To the south the kingdom of Naples had been held since 1435 by the Aragonese who had gained it after a long struggle with the house of Anjou. Since 1458, how-

---

[1] There was little actual fighting in Italy after the 'Ladies Peace' of Cambrai (1529), but Italian questions played a great part in the politics of west Europe until the peace of Cateau-Cambrésis (1559).

ever, an illegitimate prince, Ferrante,[1] had been reigning, vexed by endless conspiracy, unpopular for his gross cruelty, and apprehensive not only of Aragon but also of France, which had absorbed the rights of Anjou on the death of Charles of Mayenne in 1481. In Tuscany Florence, though still in theory a republic, had virtually become a princedom in the hands of the Medici, but when Lorenzo the Magnificent died in 1492 it became apparent that his son Piero was unequal to his destiny, and the volatile citizens, scourged by the eloquence of Savonarola, cast their eyes back to their old 'liberty'. The duchy of Milan had been made great by the Visconti whose princesses had married into the houses of Habsburg and Orleans, but on the death without true issue of Filippo Maria in 1447 the inheritance had been seized by Francesco Sforza who had married an illegitimate daughter Bianca. The new house had formed a marriage alliance with the Neapolitans, but this alliance was broken when in 1494 Ludovico il Moro imprisoned his nephew, the true duke, and took the title to himself. Venice, great in her good government, her unrivalled diplomacy, and her rich sea trade, still maintained her old reputation, but her power was being sapped by the steady advance of the Turks. It has been pointed out that it was not until the Ottomans seized Egypt in the days of Selim (1512–20) that the great trade-route was cut, but it is obvious that the whole position of the great republic had been endangered long before. If, as in 1479, she made peace with the Infidel her neighbours raised their hands in pious horror; if she fought she was left to fight alone; and when, endeavouring to recoup herself for her losses in the east, she turned her face to the west she made herself unpopular as a power which profited by the misfortunes of others.

Unstable within, Italy was threatened from without. The Turks, who occupied Otranto for a while in 1480 and nearly captured Rome in 1481, were an ever-present menace, and the grand crusade, so much upon the lips, if not in the hearts, of European princes, served only as an excuse for foreign intervention in the peninsula. Both France and the empire had claims upon Milan. Charles VIII, it is true, would not bestir himself on behalf of the restless Orleans, and Maximilian, who

---

[1] Ferrante (Ferdinand) died in 1494 and was followed by his son Alfonso II (1494–5), by his grandson Ferrantino (Ferdinand II, 1495–6), and then by his second son Federigo (1496–1501).

in 1493 had been given Ludovico's niece Bianca with an immense dowry, was not disposed to interfere; but both powers were interested in the duchy and when, in 1498, Orleans mounted the French throne as Louis XII the fate of Milan became at once a matter of European concern. As for Naples, it was in immediate danger. The young king of France was anxious to assert the old Angevin title and his intervention in southern Italy was bound to provoke the opposition of Ferdinand of Aragon, never cordial to his Neapolitan kinsmen and, by his conquest of Granada, free to pursue his ambition outside Spain. It was from France that the blow came.

The interest of France in Italy had long been recognized. During the second half of the fifteenth century she had been invited to interfere at one time by Venice and later by the pope, and now the adventurous Charles VIII found his aid solicited from several quarters. Ludovico, conscious of his misconduct towards the Neapolitans, suggested that Charles should make good the old Angevin claim; the San Severino princes in Naples sought his help to avenge the atrocities of Ferrante; Cardinal Della Rovere (later Pope Julius II) joined the king of France at Lyons to urge the deposition of the wicked Borgia. Charles VIII was not the man to refuse such enticements. Already he dreamt of a grand crusade, summoned Postel, the orientalist, to his court and bought up the rights of the Palaeologi.[1] Réné of Anjou had been titular king of Jerusalem and in 1493 the king of France began to call himself king of Jerusalem as well as king of Naples; 1494 was filled with preparation and in the late summer of that year the French army crossed the Alps. The Italian princes, probably Ludovico himself, were taken by surprise, and the French advance was a triumphant procession. On 17 November Florence was occupied and Piero driven to flight; on 31 December the French army was in Rome, and on 22 February 1495 it was in Naples. Charles had conquered Italy with a piece of chalk—the quarter-master's chalk. There was universal alarm, not only in Italy but throughout west Europe. At first the princes had watched the proceedings of Charles with complacence. Already they had gained from the hasty treaties which Charles had made whilst he prepared his attack; no doubt they hoped that the French king

---

[1] The eastern emperor in whose day Constantinople fell to the Turks was Constantine XI, of the house of Palaeologus.

would waste his strength on the fevered plains of Italy; in any
case they were confident that they might profit from his absence
from his own country. Now they realized with a shock that
Charles might return victorious and they hastily concerted
measures to meet the danger. On 30 March 1495 a league for
the rescue of Italy was signed at Venice by the pope, Ferdinand, Maximilian, Venice, and Milan, and Charles, who had
neglected to take counter-measures, was driven out as fast as
he had come in. Only by a lucky victory at Fornovo on 6 July
did he make good his escape, and the garrisons which he had
left in Naples were next year reduced by the Spaniards under
Gonsalvo de Cordoba. Having managed, however, to retain
a few fortresses in north Italy he hoped to renew the attack,
and the vigilance of his opponents was unceasing. Ferdinand
was particularly active. He set himself to unite the foes of
France in a great combination dependent upon himself. In
August 1496 he dispatched his daughter Juana to marry Philip
of Burgundy; on 1 October he concluded a new treaty for the
marriage of Catherine with Arthur; he curbed the extravagance
of Maximilian who, quite unjustly, had a grievance against Henry
for the Breton affair, and did his utmost to establish peace between
Scotland and England. The confirmation of the English marriage treaty on 18 July 1497 and the Anglo-Scottish treaty of Ayton
on 30 September of the same year mark the triumph of his plan.

They mark also a real success for Henry, a success apparently
obtained without effort. While the princes of Europe had been
busy here and there Henry had stood aloof. He felt that England
had no concern in Italy. It had been hoped in some quarters
that England would seize the opportunity to attack France;
indeed Maximilian, optimistic as ever, had proposed that Henry
should not be allowed to enter the League of Venice[1] unless he
pledged himself at once to cross the Channel in force. By threats
and by promises alike the Tudor was unmoved. Late in 1495
the treaty of Étaples was formally confirmed by the English
parliament and next year ratified by the estates of the French
provinces.[2] Not until July 1496[3] did Henry enter the Holy

---

[1] *Venetian Calendar*, i. 227, 241–5.

[2] *Statutes of the Realm*, ii. 635; *Foedera*, xii. 592–633. Ratification was given by
more than twenty 'estates'. The procedure occupied several months; whatever its
object it must have had the effect of making very obvious to Europe the understanding between England and France. Ferdinand and Venice united to convince
Maximilian that Henry would not break with France.   [3] Pollard, *Henry VII*, iii. 33.

League, and then he made it quite plain that he had no intention of taking up arms, and though he received with solemn
deference the sword and the cap of maintenance sent to him
by the pope in the following October,[1] he made no effort to
attack the old enemy of England. On the contrary he made a
commercial treaty with France in May 1497. When Charles
VIII died in April 1498 he celebrated the obsequies of his royal
cousin with splendour, and he declined to embarrass the new
king, Louis XII. Henry, in short, had used the preoccupation
of Europe in the affairs of Italy to his own advantage. He had
used the breathing space provided to crush disaffection in
England, to settle the affairs of Ireland and Scotland, and to
conclude favourable commercial treaties, especially with the
Netherlands.

His manifold activities were drawn together in a singular
fashion. Through a web of negotiation, conspiracy, and occasional war ran like a restless shuttle the figure of Perkin Warbeck, who claimed to be Richard of York, the younger of the
princes in the Tower.[2] A drawing in the public gallery at Arras[3]
shows him as a good-looking young man, whose eyes suggest
a dreamer rather than a man of action, and whose resolute jaw
accords ill with the weakness of a very pretty mouth. That he
was handsome, well-mannered, and able to wear his clothes is
obvious from all accounts. His history from the autumn of 1491
when he appeared at Cork[4] to that day in November 1499 when
he was hanged at Tyburn is well known, though there are a few
doubtful passages; but about his origin there is some mystery.
He himself gave two very different accounts. In the autumn of
1493 when things went well with him he wrote to his 'cousins'

---

[1] Kingsford, p. 211.

[2] There is a considerable literature on Perkin. In 1838 Sir Frederick Madden
published a well-documented account in *Archaeologia*, xxvii. 153 ff., and in 1898
James Gairdner, who had published many of the relevant documents in the
volumes edited for the Rolls Series (*Memorials of Henry VII* and *Letters and Papers of
Richard III and Henry VII*), appended 'The Story of Perkin Warbeck', illustrated by
some new evidence, to his *Life of Richard III*. His conclusions are usually regarded
as convincing. In the *Transactions of the Jewish Historical Society*, ix. 143-62, Cecil
Roth wrote on 'Perkin Warbeck and his Jewish Master'.

[3] The drawing reproduced in Gairdner's *Richard III* shows an irregularity in
Perkin's eyes. Soncino reported to the duke of Milan 'the young man is not handsome, indeed his left eye rather lacks lustre, but he is intelligent and well-spoken'.

[4] John Lewellyn, mayor of Cork, before whom he says he swore that he was not
the son of Clarence, did not assume office till October 1491. Gairdner, *Richard III*,
p. 272.

Ferdinand and Isabella,[1] telling them that he was indeed Richard Plantagenet, secretly spared by the murderers of his brother.[2] This story, lacking personal names and incorrect in the one point where the date can be checked, is plainly a tissue of improbabilities and, as the endorsement to the letter shows, the Spanish monarchs did not believe it. Yet the dignity of Perkin's bearing impressed his contemporaries;[3] he was accepted by several great princes[4] and there was plainly a notion that, if he were not what he claimed to be, he was none the less a scion of the White Rose. Maximilian is alleged to have said that he was really the son of Margaret of Burgundy and the bishop of Cambrai,[5] and it has been noted as odd that when the bishop was in London in 1498 he asked to see Perkin. Burgundy was certainly slow to abandon its belief in the 'duke of York', and even though Molinet refers to him as *Pierrequin Wezebecque*[6] after his capture, he continues to refer to him and writes his epitaph in an equivocal piece of *prose rythmée*:

'Ainsy fut la blance rose, qui tant avoit fait de traveil tant a la rouge rose, comme au bouton vermeil, pendue au vent et seschie au soleil.'

Bacon himself seems to have had some doubts, though formally he showed Perkin as an impostor, chance-found by Margaret of Burgundy, always on the look-out for likely young men. He represented that Henry's investigation into the murder of the princes had been perfunctory, noted that the pretender was 'never wanting to himself either in gracious or in princely

---

[1] Madden, p. 199. Also *Spanish Calendar*, i. 50, and Pollard, *Henry VII*, i. 95.

[2] The Milanese ambassador seems to have thought that Clifford was 'the first man who had this son of King Edward when he was in England'. Maximilian told the Milanese ambassador that Clifford had divulged the secret. *Milanese Calendar*, i. 292.

[3] Soncino, in describing how Perkin was made a spectacle in the streets of London, remarked: 'In my opinion he bears his fortune bravely.' *Milanese Calendar*, i. 335.

[4] At one time or the other by Charles VIII, by Maximilian and his son, by the kings of Denmark and Scotland, by the duke of Saxony, and of course by the dowager duchess of Burgundy.

[5] *Spanish Calendar*, i. 185. According to Puebla, Perkin swore that Madame Margaret knew as well as himself that he was not the son of King Edward; and Margaret, at this time, wrote asking pardon. The bishop of Cambrai was negotiating the whole affair, which was partly political, with great secrecy. Ibid. i. 196.

[6] Anxious to exculpate Margaret, Molinet says that Atwater, the mayor of Cork who suffered along with Perkin, was the *maistre* of the impostor, but it is noticeable that several of his few references to England deal with Perkin.

behaviour', hinted that some scandal lay behind the fact that
Edward IV had stood godfather to the son of a converted Jew,
and, when he came to tell of the deaths of Warbeck and Warwick, wrote 'it was ordained that the winding ivy of a Plantagenet should kill the true tree itself'. The parasite, it is true, is
no kinsman of the oak; but the phrase may reveal some inward
uncertainty on the part of the great biographer.

The lure of romantic mystery is always compelling, but there
is little of substance in all these arguments that Perkin was really
a Yorkist prince. Maximilian's story is reported only at third
hand, and even if the flighty emperor repeated it, it may have
been the self-justification of one who was 'ever the principal in
deceiving himself'; the king of the Romans had gone so far in
his support of Perkin that he was almost bound to claim a noble
origin for his protégé. It would be ignominious to admit that the
'duke of York', to whom he had given a seat of honour at his
father's funeral in Vienna, was a mere nobody.[1] His people,
too, were loath to admit that they had been hoodwinked, and
Molinet's account wavers between a determination to show
that Margaret was not the author of the imposture and a
lingering hope that Perkin was really an aristocrat after all. As
for Bacon, some of his doubt was founded upon a complete
misconception. Bernard André had said that Perkin was the
servant of a Jew whom Edward IV had sponsored when he
received Christian baptism; Speed, mistakenly, had made him
the son of this Jew; Bacon, carrying the error further, supposed
that it was the son of the Jew to whom Edward had stood godfather, and that there was something suspicious in the royal
condescension. In fact Edward's action was entirely regular;
the king of England was the official sponsor of all converted
Jews—there were not many—and it can, moreover, be related
to a real person. There is every reason to believe that the Jew
in question was Sir Edward Brampton[2] who had been employed
by Edward IV and Richard III and who, having been absent
from England at the time of his forfeiture, had reinstated himself by the good service he gave to Henry VII's embassy to
Portugal in 1489. This identification helps to explain both
Perkin's knowledge of the Yorkist's court and Henry's know-

---

[1] See p. 121 (infra).
[2] He was the first Jew to be admitted in England to the dignity of knighthood.

ledge of Perkin;[1] it even helps to explain Perkin's ambition, for Brampton had set a fine example of unscrupulous success.[2]

While the story of Perkin's royal birth dissolves under examination, the account which he gave of himself after his capture in 1497 bears the stamp of reality. According to his open confession he was the son of John Osbeck, a decent bourgeois of Tournai.[3] Having passed some time in the Netherlands in order to learn the language and the ways of trade, he had sailed to Portugal with the wife of Sir Edward Brampton and had entered the service of a Breton merchant named Pregent Meno[4] with whom he came to Cork. There the citizens, seeing him in his fine clothes—it is suggested that he was a 'mannequin' advertising his master's wares—had insisted that he must be a Yorkist prince—Clarence's son Warwick, or a bastard of Richard III, or, finally, Richard of York. In spite of his protests he had been made a prince, assured of the support of Desmond and Kildare, and taught to speak English. So, at first all reluctant, he had begun his career of imposture. The narrative gives the names of Perkin's friends and kinsfolk—some of which can be checked in the municipal archives of Tournai—of the houses in which he stayed, of the ship in which he sailed, and the details given tally with those of the desperate letter which he wrote to his mother from Exeter begging her to save him, presumably by confirming the account of himself which he had given.

Certainly the confession was given under duress and was used for the purpose of propaganda. Some things may have been omitted. It is hard to believe that the imposture began quite fortuitously in Cork.[5] Henry himself, André, and Polydore

---

[1] Henry's letters to Talbot, 20 July 1493 (Pollard, *Henry VII*, i. 93); cf. his Instructions for Richmond, king of arms, going to the king of France, 10 August 1494 (Gairdner, *Letters and Papers*, ii. 294, where he discounts the pretensions of the 'garson' against whom France had promised aid).

[2] John Ford's *Chronicle Historie of Perkin Warbeck*, produced a dozen years after the appearance of Bacon's *History of King Henry VII*, repeats the story:

> ... your father was a Jew
> Turned Christian merely to relieve his miseries. (Roth, op. cit.)

[3] Kingsford, pp. 219–21; Hall, p. 488; cf. Gairdner, *Richard III*, pp. 266, 329.

[4] Pryngent or Prijan Menon, or Prigent Meno, appears later, 1496, in the service of Henry VII. Gairdner, *Letters and Papers*, ii. 375; Conway, p. 167.

[5] Gairdner, *Richard III*, p. 269, inclines to the view that it was in Ireland that Perkin's adventure began. The fact that Kildare called him the 'French lad' may support Perkin's statement that he learnt English only after he came to Cork.

Vergil all assert that the conspiracy was engineered from Burgundy, and in 1490 James IV received a herald from Ireland whom he sent on to Margaret; it is quite possible that the king of England, now on good terms with the Netherlands, wished to suppress inconvenient references to the dowager duchess. Even if it be admitted that Margaret did not know of Perkin till 1492—and there is no evidence that she did—it is still unlikely that Perkin's recognition at Cork was the chance affair which he himself describes. It is plain that in 1490–1 a conspiracy was afoot. The Yorkists in England and France were planning to set up another Warwick,[1] and it was as Warwick, not York, that Perkin was first saluted in Ireland. He seems to have had Yorkist support, though not necessarily that of Margaret, from the very start.

It is just conceivable that Perkin was an illegitimate prince, fostered with decent folk in Tournai; but the great probability is that he was a conceited, ambitious youth with an engaging address, who became the tool of Yorkist malcontents and gained a European importance, because great princes sought eagerly for an instrument which would harass the Tudor king. His romance is an essay upon the uncertainties of Renaissance society, when new men were supplanting the old aristocracy, and when personal gifts would carry a bold adventurer very far.

Whatever his origin, Perkin was a thorn in the side of Henry for six years, and the ignominy of his end should not be allowed to veil the truth that for some time he was a figure of importance. It was in Ireland that he first attained notoriety, and Ireland seemed to provide a suitable scene for the manifestation of an 'apparition'—the word is Bacon's. In an inflammable atmosphere[2] a very small spark might well have sufficed to produce the attendant thunder and lightning, but in fact the squib lit by Perkin's supporters fizzled out with the merest sputter. The native chiefs, many of whom had believed in Simnel, were in no mind to be fooled a second time; Kildare showed no more than a passing interest; Desmond alone gave active support and this availed little against a very small English army under James

---

[1] See the correspondence of John Taylor. *Rotuli Parliamentorum*, vi. 454; usually dated September 1490, but may be 1491. Taylor is mentioned by Perkin as one of his supporters in Ireland. Hall, p. 489.

[2] See André's unflattering account of Ireland in *Les Douze Triomphes*. Gairdner, *Memorials*, pp. 147, 321.

Crmond and Captain Thomas Garth. Before long the pretender sailed off to France on the invitation of Charles VIII, brought to him by 'Loyte Lucas' and by Stephen Frion who had been French secretary to King Henry but had deserted his master. As the object of France was to secure a bargaining counter against England, Perkin was at first accorded the honours proper to a 'duke of York', and given a guard under the Sieur de Concressault; later, when France and England had made peace, he was dropped altogether. Charles declined on point of honour to surrender him to Henry, but expelled him from his dominions in accordance with the terms arranged at Étaples. The adventurer then passed on to the Netherlands, where Margaret, after a strict examination, she said, acknowledged his title, and continued his political education.[1] He commended himself to Maximilian, who invited him to attend the funeral of Frederick III at Vienna, rode with him back to the Netherlands and installed him in state at Antwerp. He had a guard of twenty archers, wearing as their badges the white rose, and on the house of the Merchant Adventurers, which he occupied, he displayed the arms of England in a manner which provoked the anger of certain English visitors.[2] How much actual help he received is uncertain, but just before Christmas 1494 he executed documents recouping Margaret for losses in her income consequent upon the fall of the house of York and acknowledging a debt of 800,000 crowns in respect of money spent upon the promotion of his cause. A month later, on 24 January 1495, at Mechlin he formally conveyed to Maximilian and Philip, in the event of his own death without issue, his rights in the succession to the kingdoms of England and France, the lordship of Ireland, and the principality of Wales.[3] On 8 May Margaret wrote to the pope on his behalf,[4] and in July Maximilian was gleefully assuring the Venetian ambassadors[5] that he hoped that the duke of York, who was in his pocket, would soon regain the throne of England, and, from that point of vantage, would launch an attack upon France.

Henry for his part affected to regard with complete contempt the *garson qui se fait renommer Plantagenet*[6] and rejected the notion

[1] In April there was a rumour that he might marry Maximilian's daughter Margaret. *Milanese Calendar*, i. 291.
[2] Molinet, ii. 398–9.
[3] Gairdner, *Richard III*, p. 290.
[4] Gairdner, *Memorials*, p. 393.
[5] *Venetian Calendar*, i. 221.
[6] Instructions to Richmond king of arms. Gairdner, *Letters and Papers*, ii. 294, 296.

that the discountenancing of the pretender by France was a good
office which demanded a *quid pro quo*. Perkin, none the less, was,
if not a danger to him, a nuisance, and that in two ways; he
gave a head to English malcontents and he interrupted the good
relations between England and the Netherlands. Although he
attracted no recruit of the first importance, no one knew how
deep the conspiracy went and there was a steady trickle of men
to his side.[1] Among them, as was confidently asserted at the
time, were 'espials' sent by the cautious Henry who, according
to Bacon, covered his agents' activities by having them 'cursed
at Paul's, by name, amongst the bead-roll of his enemies'.
Certainly the king's intelligence was good. In February 1494
he had a few traitors arraigned at the Guildhall, and later in
the year he arrested a group of more important conspirators
including the dean of St. Paul's, two Dominican friars, and two
priests. The churchmen were spared, but Sir William Mountford, Robert Ratcliffe, and William Daubeney were beheaded;
others were hanged, and Lord Fitzwalter, who had been taken
in custody to Calais, soon afterwards paid with his head for an
abortive attempt at escape. In December 1494 a free pardon
was granted to the most distinguished of Perkin's recruits, Sir
Robert Clifford,[2] who, in the following January, received a gift
of £500, and his return to England was followed by the dramatic
arrest of Sir William Stanley, the lord chamberlain, the man
who had set the battered crown on Henry's head at Bosworth.
It was rumoured that he had enriched himself with the spoils
of the dead Richard;[3] he had been given lands and offices by
the new king; and he passed for being the richest commoner in
England. Bacon, who says that he had an income of £3,000 per
annum and that in Holt castle in Denbighshire he had 40,000[4]
marks in money and plate besides other wealth, mentions the
'common opinion' that Clifford had been all along a spy, and
hints that Stanley had hard measure in being condemned upon
words uttered about a hypothetical case. It appears, however,
that Clifford went abroad in June 1493 as the result of a talk

---

[1] Cf. Molinet, ii. 398–9.
[2] Gairdner, *Letters and Papers*, ii. 374; Bentley, *Excerpta Historica*, p. 100.
[3] Bacon, *Henry VII*; Gairdner, *Letters and Papers*, ii. 374.
[4] Wyatt, as receiver, had in hand £9,062. 9s. 8d. of the ready money from the
Holt in May 1495. Conway, p. 75. Whether this was the whole sum obtained is
uncertain. Bacon may have been reckoning according to the value of money in
his own day.

which he had with Stanley in the preceding March, and although the purpose of the conspirators may not have extended to the clear treason set forth in the arraignment, there can be little doubt that there was conspiracy of a most serious kind. The probability is that Stanley, who may have been disappointed at not being made earl of Chester, was impressed by the reception given to Perkin by Margaret and determined at least to trim his sails. The best explanation of Clifford's conduct is that, when he went abroad, he believed in the pretender's claims, that his mind was disabused by his experiences in Flanders, and that in these circumstances he became susceptible to rewards offered by Henry VII. Whether the king was entirely surprised by the news of Stanley's disloyalty may be doubted. The truth may be that he had already some suspicion but hesitated to proceed against so great a subject without real proof, and that when he was provided with definite evidence he seized the opportunity to rid himself of a dangerous man whose wealth would furnish the royal coffers. It must be assumed that the evidence was good or the king would not have dared to destroy the brother of the earl of Derby and the brother-in-law of his own mother. Qualms he felt, for he paid the expenses of the 'traitor's' burial.

With the fall of Stanley the conspirators became in Bacon's words 'like sand without lime, ill bound together'; they 'were at a gaze . . . not knowing who was faithful'. Henry, however, was not yet quite secure; it is possible that already his avarice made him unpopular, and there was some risk that Yorkist conspirators might turn local unrest to their own advantage.[1] The acts of the parliament[2] which met in October 1495 are illuminating, especially the act often described as 'for security under the king *de facto*'. This provided that no person who assisted the king for the time being should be liable to conviction or attainder in respect of such assistance, and that if any such attainder were passed it should be null and void. Obviously such an act was of doubtful validity—no parliament could bind

---

[1] A Milanese report alleged that Henry's avarice had made him unpopular. *Milanese Calendar* (3 July 1496), i. 299.

[2] *Statutes of the Realm*, ii. 568–79. Many of the acts were for the maintenance of good order. Riots and unlawful assemblies were still causing concern and already vagabondage was presenting a problem; it is significant that the king checked the dishonesties of sheriffs by increasing the authority of the J.P.s, who were under the immediate control of the council.

the hands of its successor—and the principle which it established was open to serious criticism. It was a direct contradiction to that asserted by Henry himself, who had attainted the men who fought for Richard on Bosworth Field, and plainly it would legalize any successful usurpation. Henry, however, in accordance with his usual practice, valued exact logic less than real advantage. He was applying to the Crown the principle he had already asserted with regard to lands in the Statute of Fines. The past was past, and no useful purpose would be served by grubbing into old titles and reviving old animosities. He was vindicating the *status quo*. He must have seen very well that his act would not preserve his supporters if he himself were overthrown, but meanwhile it might serve to 'confirm the feeble knees'. If he were beaten no act of parliament would make much difference, but he did not intend to be beaten. Even continental observers, some of whom would have been glad to see the king of England in difficulties, began to realize that he was secure enough in his own land.

Perkin's activities, it seems fair to conclude, had little effect on the domestic affairs of England, but in the field of economics they had repercussions which seriously embarrassed Henry. In the summer of 1493 he had sent Poynings and Warham to protest to the government of the Archduke Philip, which replied that it could not interfere with the action of the dowager duchess in her own lands. So enraged was the king by this answer that he prohibited trade between England and the Netherlands and recalled the Merchant Adventurers from Antwerp to Calais. The Netherlanders were hard hit, but the king's subjects suffered too, and the only gainers were the merchants of the Hansa whose prosperity, at a time when the native-born were suffering, excited great animosity. On 8 October 1493 the Steelyard was attacked by a crowd of 500 apprentices and servants, and the tumult was not quelled until the mayor arrived at the head of an armed force. Eighty of the rioters were arrested, and, though some were released on bail, others were sent to the Tower or to the Counter. Among the prisoners there was 'not one householder'; but it was obvious that the masters shared the sentiments of their servants, and matters were not pushed to extremes.[1] On 1 November Henry notified Charles VIII[2] of his desire to be comprehended in the

---

[1] Pollard, *Henry VII*, i. 96, 98; Hall, p. 468.　　　[2] *Foedera*, xii. 550.

treaty which that monarch had made with Maximilian and Philip, but nothing seems to have come of his overture. Henry's prohibition remained in force and in May 1494 Maximilian and Philip replied by laying an embargo[1] upon English wool and iron; English manufacturers, no less than exporters and importers, suffered in consequence. The Netherlanders, however, were suffering too and their desire to be rid of Warbeck accorded well with Maximilian's belief that the 'duke of York' could easily establish himself in England. A considerable expedition was prepared and on 3 July 1495 Perkin appeared off Deal with fourteen ships. His attempt miscarried altogether. He himself did not dare to land, and the polyglot band of adventurers—200 or 300 in all—who did come ashore met an evil fate. None joined them; the county rose against them; many were killed and before long the sheriff of Kent, Sir John Peachey, was dragging some eighty prisoners, 'railed in ropes like horses drawing in a cart', to London, where most of them were hanged. Thus assured that England would have none of him, Perkin sailed off to try his fortune in Ireland, but by this time Ireland had felt the firm hand of the Tudor, and he got a cold welcome there.

In spite of Edgecombe's success the hold he had gained on Ireland was very slight. The famous Irish princes, O'Neill in Ulster, O'Brien in North Munster, MacMurrough in Leinster, kept the state of their kingly ancestors; O'Donnell in Tyrconnel and Burke in Connaught, albeit a de Burgh by descent, were among the many 'native' lords who paid little heed to the authority of a 'Saxon' king, even although, by right of his Yorkist wife, he was titular earl of Ulster, and lord of Connaught and Trim. Of the Anglo-Norman barons, the Fitzgeralds, much intermarried with Irish houses, were the leaders. Desmond, supreme in Kerry and in parts of Cork, Waterford, and Tipperary, nursed the old hate born of Tiptoft's executions in 1468; Kildare, despite his support of Simnel, was still deputy. The rival house of Butler, though Lancastrian in sentiment, was divided within itself. The absentee seventh earl of Ormond had used as his lieutenant Sir James Butler of Callan,[2] who when he died in 1487 had bequeathed his office to his son Piers; the

---

[1] Pollard, *Henry VII*, i. 99.
[2] Sir James of Callan was first cousin once removed to the seventh earl; his son Piers became the ninth earl in 1538.

Ireland, showing the English Pale and the 'countries' of the great chiefs.

earl, however, had appointed his base-born nephew James Ormond, who had been educated in England, and was a bencher of Lincoln's Inn. This James, a 'deep and far-reaching man', had designs upon the title, and the result of his advent to Ireland was a bitter feud which ended only with his murder in 1497.[1] The situation was dangerous and the forces which Henry sent to Ireland, though they were competent troops and well led, were surprisingly small.[2] Happily for him his opponents were divided and there was, at least among the inhabitants of the Pale, a feeling that English rule would be better than constant disorder.

Kildare was under a cloud for his alleged help to Perkin in 1491; in 1492 the ambitious James Ormond succeeded in ousting him from his post of deputy, and his brother from the post of treasurer which he had held, apparently without producing any accounts, for many years. An adherent of the Butlers, Archbishop Fitzsimons of Dublin, became deputy; Ormond contented himself with the post of treasurer and, after some sporadic fighting in which Kildare gave as good as he got (notably at Oxmantown Green near Dublin on 10 June 1493), the great Geraldine and his friends were attacked in a parliament which met in the capital. Before any active measures against him were taken the situation suddenly changed, perhaps because Henry himself suspected the motives of Kildare's enemies.[3] On 6 September Fitzsimons was replaced by Lord Gormanston, and though the authority of the new deputy was afterwards challenged on the ground that his principal, the duke of Bedford, had resigned the lieutenancy, it was for the moment sufficient to produce a surprisingly good order. Before the year was done Fitzsimons, Kildare, Gormanston, and Ormond had all set off to Henry's court. Kildare had a good

---

[1] Gairdner, *Letters and Papers*, ii. xxxvii ff.

[2] Conway gives the figures from the accounts. Garth and Sir James Ormond seem to have had 200 soldiers and 30 sailors when they arrived in 1491 (p. 51). In March 1493 after Kildare captured Garth, Sir Roger Cotton brought 200 more and Henry Mountford 100 (p. 54). In June Henry Wyatt arrived with a retinue which cost five-sixths of Mountford's and may be estimated at about 80 (p. 56). In the autumn of 1493 some of the troops were sent home, and Garth had pay for only 100 men (p. 59) till Poynings arrived on 13 October 1494. According to the usual account he brought 1,000 men, but he drew pay for only 427 (p. 78), and the summary of wages (app. x) shows for the first half of 1495 a grand total of 653 of whom 90 were gunners. No doubt Irish troops were used too, and if paid from Irish sources, or serving Irish lords, would not appear on the accounts.

[3] Conway, pp. 55-9.

reception from Henry, and in May 1494 entered into an indenture with him to secure Desmond's son as a hostage. Desmond
had already made his submission and all promised peace; but
in June 1494 Sir James Ormond gained the constabulary of
Limerick from Desmond, who had held it for years, and the
infuriated earl once more flew to arms. At this moment
O'Donnell, in the far north, was engaged in some suspicious
dealings with the Scots, and Perkin was threatening invasion.
Henry, aware that he was vulnerable in Ireland, decided
to use new men and new methods. Unable to trust either
Geraldine or Butler, he appointed his baby son Henry
lieutenant and gave him as deputy Sir Edward Poynings,
son of a Kentish squire, whose fidelity had been proved by
a record of good service begun before ever he fought on
Bosworth Field.

The new deputy at once set off to Ulster, then the most
'Irish' of the provinces, to chastise the abettors of Perkin and
the traffickers with the Scots; but before he reached his objective
he was attacked by O'Hanlon of Armagh, and it became plain
that the great Geraldine was in communication with his
enemies. He, therefore, abandoned his expedition and summoned to Drogheda the famous parliament which at first pardoned Kildare, but later attainted him. The earl was arrested
on 27 February 1495 and sent to England. His brother rose at
once and captured Carlow castle early in March; but the
advantage lay with Poynings, whose hands were strengthened
in June by the arrival of more men and fresh supplies of money
from England. Under his firm rule Ireland stood ready to greet
the 'duke of York'.

The parliament which met on 1 December 1494 has long
been regarded as a landmark in Irish history. How long it sat
is uncertain since the official record in the Rolls of Parliament
was destroyed in Dublin in 1922, but with various prorogations
and adjournments it must have remained for a considerable
time,[1] and, though only twenty-three of its acts are printed in
the Irish Statutes at Large, twenty-six other acts can be supplied

---

[1] Kildare was both pardoned and attainted; for his successful petition against
the attainder in the English parliament see *Rotuli Parliamentorum*, vi. 481-2. In a
paper in *Irish Historical Studies*, vol. vii (1950), Professor G. B. Sayles has produced
evidence to show that Kildare, in the course of his dealings with O'Hanlon and
Magennis, had really urged these chiefs to make peace with the deputy. If their
oaths (on the Host) were true, Kildare was not guilty of treason.

from other sources.[1] Of the measures passed a few were formal, and a few were of temporary importance only, but the general effect of its legislation was to provide a complete scheme for the reform of the English administration in Ireland. The best known of its acts are the two which placed the legislature of Ireland completely under English control. By chapter 9 no parliament could meet in Ireland unless the lieutenant and his council asked leave of the king and in so doing submitted, under the great seal of Ireland, a list of the acts it was proposed to pass; this list would be considered by the king in council who would then issue, under the great seal of England, a licence to hold a parliament for the ends therein set forth. Chapter 39 provided that all laws made in England should apply also to Ireland. The significance of these acts is obvious, and the phrase 'Poynings Laws',[2] in common parlance, is often used of them alone; but they would have failed of their purpose if the English control had not been made real by practical measures, and some of the other acts of the parliament at Drogheda are scarcely of less importance.

Provision was made for the inevitable hiatus which must occur with the comings and goings of the deputy, or the transferences of the office of deputy—or lieutenant—from one person to another. In the absence of the deputy the treasurer was to act as governor, and the chancellor was to preside over parliament. Moreover, treasurer, chancellor, the two chief justices, the first and second barons of the exchequer, the master of the rolls, and all officers of account were to hold their charges at the king's pleasure and not for life. No longer were the 'mighty men of Ireland' to find ready opportunity to seize control; no longer were the great officers, by long continuance of power, to behave as if they were independent of England and identify themselves with Irish families or parties. Provision was made, too, for the control of finance. The treasurer was himself to appoint all his subordinate officers as did the treasurer in England, and was

[1] Some of the acts are printed in Pollard, *Henry VII*, iii. 294, 310. Chapter 35 on 'coigne and livery' refers to the ineffectiveness of an earlier act passed by the same parliament.

[2] For 'Poynings Laws' see chapter vii by E. Curtis in Conway, *Henry VII's Relations with Scotland and Ireland*; see also the *Proceedings of the Irish Historical Society*, vol. i, D. B. Quinn, 'Anglo-Irish Local Government', and vol. ii for 'The Early Interpretation of Poynings Law' by the same author, and for a 'Historical Revision' on 'The History of Poynings Law', which was evidently somewhat flexible in application.

annually to make a declaration of his account before the barons
of the exchequer, who must certify it to the king's exchequer in
England. The country was to be self-supporting. In exchange
for his surrender of 'coigne and livery' the king was granted
for five years a subsidy of two marks (£1. 6s. 8d.) on every six
score acres of arable land actually under cultivation, and a
customs duty of one shilling in the pound was placed on all
exports and imports except those handled by the freemen of
Dublin, Waterford, and Drogheda. An act of resumption,
limited, however, by many exceptions, recalled all land held by
the Crown in 1327; the earldom of March, the lordships of
Ulster and Connaught, were declared to be crown-property,
and orders were given for the collection of the scattered records.
It was hinted that all revenues granted away since the days of
Edward III would be resumed, and inquiry was to be made into
alienations, grants, leases, and annuities in favour of spiritual
persons. For the first necessity of all, the re-establishment of
good order, many salutary measures were passed. The great
castles, Dublin, Trim, Wicklow, Carlingford, Carrickfergus
among them, were to have Englishmen born as their constables;
the 'pretensed prescription'[1] that rebels and traitors should have
immunity in Ireland was not only cancelled but flatly contradicted; the statute of Kilkenny (1366) was confirmed; the
making of private war, the hiring of private armies, the levying
of coigne and livery, the exaction of black rent, and the crying of
'Cromabo!' and 'Butlerabo'[2] were prohibited; the exaction of
private vengeance was forbidden and henceforth murder was to
rate as treason. By several acts a solid system of defence was
established. Every man was to have arms suitable to his rank,
though, without special licence from the lieutenant or his
deputy, none was to have 'great gunne' or 'hand gunne';
holders of land adjoining the Pale were made responsible for
the digging of a great ditch, and precautions were taken to see
that 'retinues' engaged for the defence of the frontier were not
used for other purposes. When the work of the parliament of

---

[1] See note on p. 71.

[2] *Crom* means monkey. *Cromabo* is still the motto of the duke of Leinster, whose
arms display, as crest, a monkey *statant* proper environed about the middle with a
plain collar and chained *or*. It is said that the first earl of Kildare was rescued from
the burning castle of Woodstock near Athy by a monkey. The same story is told,
with great circumstance, about Thomas Fitz Maurice, a 'nappagh' (An Appagh,
simiacus, 'of the ape'), father of the first earl of Desmond.

Drogheda is regarded as a whole its meaning is clear. This was not an attempt to reconquer all Ireland or to enslave the native Irish. It was an endeavour to secure the Pale, to put down the over-mighty subjects within and about the Pale, and to produce good order by the means already tried and found successful in England. The royal claim to Ulster and Connaught had the same sort of importance as the claim to Normandy and Aquitaine. It was reasserted to keep it alive—it might have its uses—but the Tudor government, despite high-sounding pretensions, knew the length of its arm. It limited its objectives and tried to secure its ends by practical measures.

With the new measures came new men and new methods. Poynings was accompanied by a few capable administrators; Henry Deane, bishop-elect of Bangor, was chancellor and Sir Hugh Conway treasurer. Perhaps because Conway, though not inefficient, was not efficient enough to please Henry, his authority was short-circuited, early in 1495, by the appointment of two officers trained in the king's own household, William Hatcliff, who was made under-treasurer, and Henry Wyatt, who, along with Hatcliff, was to act as auditor.[1] These two men, in accordance with Henry VII's new practice, dealt mainly with the chamber on the one hand, and with John Pympe, appointed to the new office of treasurer of the wars in Ireland, on the other. They avoided the formal exchequer in England, and after June 1495 relatively little money passed through the hands of the Irish treasurer.[2] Exact accounts were kept and attempts were made to reassess in carucates the land liable to subsidy, but the result of all effort was to show that Ireland could not pay its own way. Next year Wyatt explained the breakdown by complaining of the needless wars, the excessive support given to Sir James Ormond and others, and the decline of trade, but the figures produced in 1495 cannot have encouraged optimism. If the revenue expected, £2,691, had been received there would have remained, after paying fixed salaries, only £475 for general expenses including war; in fact only

---

[1] Hatcliff had been clerk of the Marshalsea of the household; his wife, a Paston, was Poynings's aunt. Wyatt had been clerk of the jewels and clerk of the mint, and later became first keeper and then treasurer of the chamber.

[2] From June to December 1495 Hatcliff received in all £4,035 of which £2,608 came from the king; he paid £3,349 straight to Pympe for the wars. Conway, who as treasurer of Ireland produced £900 for Hatcliff, dealt in far smaller sums. (Conway, app. xiv.)

£1,283 was received and it seems that no one received a salary
except Hatcliff and Sir James Ormond. Conway left Ireland in
November, and when in mid-December Poynings sailed home
in triumph, his place was taken by Deane, the chancellor, and
not, as the law provided, by the treasurer. Already it was clear
the new system was breaking down. Yet if it did not secure the
future of Ireland it had served her well in a present crisis. Whilst
Poynings was at the height of his power, Perkin came; and the
completeness of his failure is a measure of the deputy's success.

After the fiasco at Deal Perkin sailed to south-western Ireland,
where he made contact with Desmond and concerted with him
an attack on loyal Waterford. On 23 July the pretender
appeared off the city with eleven ships, and soon the earl
brought 2,400 men by land to his support. The citizens,
heartened by the presence of Poynings and of the mayor and
bailiffs of Dublin, were ready to meet the assault, and though
the siege was pressed with some vigour for eleven days the
defence, thanks to the competent gunners, prevailed with ease.
Three of the invading ships were taken and on 3 August Perkin
drew off in flight. Pursued to Cork he went on to Kinsale, and,
with the help of Desmond, Burke of Galway, O'Donnell and
O'Neill of Clandeboy,[1] managed to make his way to Scotland,
where he was well received.

Ireland had given little help to the alleged duke of York,
but the country was still restless, and when, after Poynings's
departure, the English army was reduced to a few hundred
men[2] there was a fresh rising under James Fitzgerald, brother of
Kildare. On 9 February 1496 the new deputy thought it necessary to order the preparation of bonfires against a sudden assault
on the Pale; but on 15 March Desmond made his peace with
the king and it is probable that the Geraldine activity was
merely meant to show that without Kildare there would be no
peace in Ireland. If this were so the device succeeded.

Things had gone well for Kildare in England. According to
the Book of Howth,[3] whose chronology is confused and whose

---

[1] For Perkin's supporters see the letter of Sir Ralph Verney to Sir Reginald
Bray, October 1496 (Conway, app. xliii). The adventures described in the Book
of Howth as occurring in 1497 probably belong to 1495.

[2] 330 Englishmen, 100 Kerns, and a trumpeter costing £294 a month. Conway,
p. 87.

[3] The Book of Howth is a sixteenth-century manuscript preserved among the
Carew manuscripts at Lambeth. It is early called 'The Book of Howth' and it is

spirit is light-hearted, the earl endeared himself to Henry by his naïve common sense. When John Payne, bishop of Meath, charged him with great offences he regaled the king with scandalous stories about his accuser (happily not recorded), and on being told that he must choose a counsel to defend him, chose the king himself on the ground that 'I can see no better man than you'. 'All Ireland cannot rule yonder gentleman' said the bishop. 'No?' answered the king. 'Then he is mete to rule all Ireland.' Whatever the truth of this narrative it is plain that the earl commended himself to Henry, always ready to let the theory go for the sake of the fact. His attainder was annulled in the English parliament, and once more he became deputy; on 6 August, in the presence of the king's council at Salisbury, he and the Butlers swore to end their feud, and about the same time he undertook to execute some at least of Poynings's laws. His allegiance was confirmed by a marriage with Elizabeth St. John, a second cousin of the king, and his relations with the people he was to rule were sweetened by the grant of a general pardon to all Perkin's supporters except Lord Barry and John Atwater. On 28 August he set sail and after a dreadful voyage of 21 days landed at Howth on 17 September. On 21 September he took the oath in a council held at Drogheda and by making Fitzsimons chancellor set his seal upon his pact with the Butlers. Very soon most of the great men of Ireland came in, and though his success was due in part to internal feuds in the houses of O'Neill and Butler, it was due also to his own popularity and to the desire of the Irish to be ruled by a man of their own blood.

The history of Ireland during the remainder of the king's reign reveals the working of an effective compromise. Kildare did much as he pleased; but in the main it pleased him to do what Henry wanted. There was more than a suspicion that he used his power to promote the interests of his own kin, and in 1503 he was summoned to England to give an account of himself; but, even though Henry may have doubted the motives which led him to rout the chiefs of the west at Knocktue (Cnoctvach) in 1504, his conduct was overlooked and he was made a knight of the Garter. Of his success there is no doubt.

certainly generous to members of the Howth family. In spite of obvious inaccuracies its narratives seem to bear the stamp of truth. It is in thirteen different hands but seems to be in the main contemporaneous. *Carew MSS.* (1871), 179 ff.

Henry's policy in Ireland was realist. Determined to stamp out the last remains of Yorkist discontent he tried to apply to Ireland the principles of the new monarchy, and he was so far successful that, after the day of Poynings, Yorkist conspiracies no longer throve. Soon, however, he realized that Poynings's reforms could be operative only in the Pale, and that even there they would arouse great opposition unless supported by a strong family interest. He realized too that they must needs cost money; his competent servants managed their business with very few troops but, even for these, Irish resources could not pay. Kildare, by whom he was probably impressed, offered an easy way out; he could produce order at small expense and the fact that he did so is a justification of the king's common sense. Yet Henry's policy, though it served the day, was fraught with danger for the future. It was a dangerous thing to place the powers of government in the hands of a great tribal chief who would inevitably, in the end, confound his two authorities and use the power delegated by the king for the pursuit of his family ends. Henry realized the risk; but the gentleman had proved that he could 'rule all Ireland', and while he did so the king was able to turn his mind and his hand to other things. That was a great matter. If troubles arose in the future no doubt fresh expedients could be found; and if, at the time when these troubles came, he was free from other cares these expedients might shape themselves into a definite policy. For the present he had good peace in Ireland and that sufficed him.

Rebuffed in Ireland, the indefatigable Perkin had gone on to Scotland. When he arrived there in November 1495 he found a prosperous country where the monarchy was well established, though the constitution was not yet fully developed. The advance of the central power had been slow owing to physical difficulties, to the racial distinction between Highlander and Lowlander, and to the constant necessity for resisting an aggressive neighbour who persistently advanced a claim of suzerainty. The attempt of James I to introduce the English system of government had been only partially successful, but during the fifteenth century, in spite of baronial wars, which vexed Scotland as they vexed most of the countries of west Europe, the monarchy had increased its strength and improved its machinery. Some have seen in James III, who preferred 'new men', who loved money, and who was interested in the arts, a prince of the Renaissance born before his time, but the general opinion, supported by all sub-contemporary writers, is that he was a *roi fainéant* who lacked the force necessary to rule wild Scotland. His overthrow at Sauchieburn in 1488 was not the expression of any constitutional or political principle but of baronial turbulence, and the government of his successor revealed no features essentially new. What it did reveal was a new spirit. James IV was only fifteen years and three months old when he became king but he developed quickly and soon showed all the qualities of a Renaissance prince, except that of caution. Restless, brave, curious about all things, and determined to rule himself, he made his own person the mainspring of the government. He was anxious to shine in the eyes of Europe and this was the easier because Europe was beginning to take an interest in the remote kingdom emerging from the northern mists. Scotland was beginning to feel her own importance; it is significant that the rebels of Sauchieburn decided to send a justification of their action to various princes and to seek an 'honorabill prencess' in France, Spain, and other needful places.

Though its horizon was thus widening, Scottish diplomacy was still dominated by the conflicting claims of England and France. The necessity of resisting English aggression had been the secular task of the Scottish monarchy, and for two centuries the situation had been complicated by the existence of the 'Auld Alliance'[1] with France. Born in the thirteenth century of a common fear of the mighty Plantagenet empire, this alliance had been renewed again and again. It had been signed by every Scottish king since the days of John Balliol—by James III as late as 1484. It was primarily defensive, and 'existing' truces with England were to be broken by some act of the English before a *casus foederis* arose; but it could easily become offensive and it specifically envisaged England as the enemy. Only when

---

[1] William the Lion had helped Louis VII against Henry II. Alexander II and Alexander III had both married French brides *en secondes noces*. The first formal treaty was that between John Balliol and Philip IV. As this reflected an unusual situation and as Balliol was regarded as a usurper by his successors, it was renewed in somewhat different terms by Robert Bruce and Charles IV in 1326, and these terms were virtually unaltered in the renewal of the treaty by every Scottish king till the days of James IV. In the renewal of 1512 France insisted that the clause relative to existing English treaties be excised as otiose since England had already committed acts of war. See Du Tillet, *Recueil des guerres et traictez d'entre les roys de France et d'Angleterre* (1558).

England and France were inclined to be friendly one with another could it be an instrument of peace for Scotland. Then it might serve as a basis for a triple *entente*.

Towards the end of the reign of James III the relations between England and Scotland had been good; indeed the 'inbringing of Englishmen' was one of the accusations levied against the king by the nobles who rose against him; but some of these nobles themselves had been intimate with England, and the battle of Sauchieburn did not greatly affect the diplomatic situation. It has been argued that the Tudor, true to his old ally, supported the partisans of the dead king, but an examination of the evidence does not support this view. Henry certainly gave some countenance to Buchan, Ross of Montgrenan, and John Ramsay, Lord Bothwell,[1] but he expected services in return and he extended his patronage to Scottish malcontents[2] who had been the opponents, not the friends, of the murdered James III. It is difficult to see in his suborning of Angus[3] and his scheme for kidnapping the young king and his brother[4] any purpose of delivering an innocent youth from the red hands of his father's murderers. Rather does it appear that Henry was pursuing the time-honoured English policy of keeping Scotland weak by the promotion of internal discord. Henry did not desire war with his northern neighbour; on the contrary he desired peace.

Fortunately for him the 'young adventurousness' of the Scottish king was curbed by the operation of diplomatic forces beyond his own control. Both France and Spain, though for very different reasons, were anxious that England and Scotland should be at one. Charles and Henry were in the main friendly and though the threat of war over the Breton question led to a renewal of the 'Auld Alliance' in 1491–2, the crisis passed with the settlement of Étaples, and thereafter it was the object of French diplomacy to keep things quiet in Britain; the Italian adventure depended in great measure upon the passivity of the English king. Spain, on the other hand, was anxious to embarrass France; it was to that end she sought to bring Scotland

---

[1] Conway, p. 25.

[2] Stephen Bull and the other English sea captains, who were defeated by Sir Andrew Wood, may have been engaged upon private enterprises, but Henry himself sent help to Lennox and Lyle who had been enemies to James III: Pitscottie, *Croniklis of Scotland* (Scottish Text Society), i. 226.

[3] Gairdner, *Letters and Papers*, i. 185.   [4] *Foedera*, xii. 440.

within the ambit of her diplomacy. She hoped not only to deprive France of a useful friend but also to secure the back of her Tudor ally if he could be induced to turn his face south and take an army across the Channel. As early as July 1489 Puebla, concealing the fact that the two Spanish princesses were already betrothed, proposed a marriage between James and another daughter of Ferdinand whose illegitimacy he concealed, and, though the catholic monarchs hastily broke off the unworthy negotiation, they retained an interest in Scotland and continued to urge a good understanding between James and Henry.

To the establishment of this understanding Henry himself worked unceasingly; not yet secure upon his throne, he was, on general principles, anxious to avoid war. When the three years' truce made at Coldstream in 1488[1] drew towards its end in 1491, he endeavoured to secure its renewal upon ampler terms and on 21 December his representatives and those of Scotland agreed to a new truce to last until December 1496. He himself was quick to ratify this arrangement (9 January 1492),[2] but James, then in active negotiation with France, apparently withheld his ratification, and all the Scots would agree to was a brief truce to expire in November 1492.[3] Undeterred by this failure Henry returned to the attack and in November 1492 procured a renewal until April 1494. In the following May he endeavoured to obtain a marriage alliance, offering to James Catherine, daughter of the earl of Wiltshire, who was, through her mother, a kinswoman of his own.[4] The proposal was not accepted but in June the truce was renewed to extend from April 1494 to April 1501 and this renewal was confirmed by both monarchs. In June 1495 the English king made another attempt to secure a family alliance, this time suggesting a match between the king of Scots and his own daughter Margaret.[5] It is patent that Henry really tried to establish good relations with Scotland and that it was from Scotland that the obstruction came.

Among the reasons for Henry's insistence and James's hesitation must be accounted the attitude of the Scots king to Perkin Warbeck. There had been much coming and going between

---

[1] *Rotuli Scotiae*, ii. 488.   [2] Ibid. 503, 504.   [3] *Foedera*, xii. 473, 494.
[4] Catherine's mother was Eleanor, daughter of Edmund, duke of Somerset, killed at St. Albans, 1455. *Foedera*, xii, 529, 534, 539.
[5] *Foedera*, xii. 572.

Scotland and the Netherlands[1] who were closely allied by the ties of commerce, and as early as 1488 there are signs of Yorkist machinations. In the spring of 1492 James received a letter from Ireland sent by 'King Edward's son' and Desmond.[2] Thereafter Henry's representatives in Ireland noted support given to the rebels from Scotland, and in the attainder of Kildare there was specific mention of the dealings which the earl had had with the king of Scots.[3] From the Scottish evidence it appears that in September 1495 James was already preparing to cross the border for the 'fortifieing and supleing of the prince of Ingland',[4] and a man of his nature was only confirmed in his purpose when in November Perkin arrived not as a conqueror but as a fugitive.

James honoured the 'duke of York' with a state welcome at Stirling in November 1495, introduced him to the nobility and gave him as wife Lady Catherine Gordon, daughter of the first earl of Huntly, who was, through her mother, a distant cousin of his own. If the love-letter printed in the *Spanish Calendar* was really written by Perkin, the adventurer knew how to plead his own cause. James spent much money in equipping his visitor with suitable clothes and seems to have given him a pension of £112 a month, though there is no evidence that payment was regularly made.

The attention paid to the pretender attracted the eyes of Europe towards Scotland. Ferdinand, disquieted perhaps by Maximilian's still optimistic vaunts, sent a Scottish ambassador home again with the offer, doubtless insincere, of a Spanish princess and in the summer of 1496 sent the able Pedro de Ayala,[5] who was in England, to persuade James to join the enemies of France. At the same time Concressault appeared in Scotland on behalf of France, and Henry's spy Bothwell suggested that he himself had countered a dangerous negotiation by disclosing the true story of Perkin's origin which he had had from Meautis, the English secretary.[6] About the same time Roderick de Lalaing appeared from Flanders. Henry, whose intelligence was good, was little moved by the news of this diplomatic activity. He understood that if the French offered

---

[1] Conway, pp. 39, 51. See Tytler, iv. 319–20, for evidence derived from the *Register of the Great Seal of Scotland* and the *Accounts of the Lord High Treasurer*.
[2] *Accounts of the Lord High Treasurer*, i. 199.
[3] Conway, app. xxxi.   [4] Gairdner, *Richard III*, p. 300.
[5] *Spanish Calendar*, i. 116.   [6] Pollard, *Henry VII*, i. 140.

100,000 crowns for the 'fenyt boy',[1] it was not in order to promote his cause but to gratify England by surrendering him, and he soon knew that de Lalaing had treated the pretender with scant courtesy. As both envoys had been closely associated with Warbeck in his earlier adventures the shrewd Tudor realized how the wind was blowing. He still hoped to bring Scotland into the ways of peace and in September sent Fox to renew the marriage negotiation.

How little the presence of Perkin had disturbed the equanimity of Henry appears from his dealings with the Netherlands. Unmoved by the threat of invasion, disregarding alike the boasts of Maximilian and the alleged interest of Ferdinand in Burgundian affairs, he had quietly settled his economic dispute with the Netherlands in a manner profitable to both countries. The *Magnus Intercursus* of February 1496[2] was a business-like arrangement which provided both for political peace and economic friendship. Among the usual provisions for the avoidance of all hostility, direct or indirect, were several clauses whereby each prince undertook not to assist the other's rebels, and it was specifically stated that Philip was to take action against Margaret if she persisted in her policy of maintaining Henry's enemies. The economic clauses are astonishing in their thoroughness; each prince was at liberty to exact proper dues and customs and to impose new if he wished, but everything possible was done to promote free commercial intercourse between the two countries, to prevent the common causes of dispute, and to assure that individual acts of wrong should not invalidate the peace itself. Fisheries, pirates, wrecks, contraband of war, the carrying of arms, scrutinies of cargoes, export of bullion, attachment of debtors—all these and many other matters were subjected to minute regulations. The event was to show that the causes of dissatisfaction, particularly among the English, still survived, and these were not allayed even by a fresh commercial treaty of 7 July 1497,[3] but the treaty of 1496 remained in force and in 1502 its general provisions were applied to the territories of Maximilian.[4] Well is it worthy of the name *Magnus Intercursus*.

---

[1] Ibid. 138.   [2] Ibid. ii. 285.

[3] It was stated that a reimposition by Philip would give Henry the option of cancelling the existing treaty. Philip agreed to withdraw a new imposition of a florin on English cloth, and England to discuss the removal of the mark on every sack of wool levied in Calais. *Foedera*, xii. 654.   [4] *Foedera*, xiii. 6.

How right was Henry's estimate of the situation was manifested by the event of a Scottish invasion in the autumn of 1496. Perkin had neither friends in England nor supporters overseas, and when James, disregarding his treaty with Henry, crossed the border on 20 September he met with no success whatsoever. The 'duke of York', who had generously promised to restore Berwick and recoup James for his expenses, issued a magniloquent proclamation offering handsome money rewards and the assurance of divine approval to anyone who would slay or take his 'mortal enemy'.[1] His proclamation is of interest to historians as supplying a list of the new men, 'caitiffs and villains of birth', on whom the king relied, but it brought forth not a single recruit and, according to Hall, when Perkin protested against the Scottish devastations James made the acid reply that he concerned himself overmuch about a land which showed very little interest in him. Without waiting to encounter an English force under Sir George Neville the invaders recrossed the Tweed on 21 September and the only effect of the inroad was to give Henry a pretext for raising more money, which he did with some difficulty.

His first device was to summon a great council to which he called besides the Lords 'certain Burgesses and Merchants'[2] from the cities and towns of England. This body, which sat from 24 October to 5 November, promised that parliament would grant £120,000 and the king endeavoured at once to raise loans on the security of its pledge. His efforts were only partially successful. London, when asked for £10,000, granted only £4,000 and the total yield throughout the country was some £58,000. The reluctance to pay must have been due, in part at least, to an instinctive constitutional sense, for the parliament which met in 1497 showed itself generous enough—England would pay for the king's necessities provided that the money was voted in a regular way. Stimulated by an address from Morton on the civic virtue of Romans and Maccabees and on the perfidy of the Scots, who had broken the seven years' truce,[3] the commons voted two fifteenths and tenths at once and promised to add a subsidy of equal amount if necessity arose. Henry's pre-

---

[1] Pollard, *Henry VII*, i. 150; where wrongly dated; also in Bacon, *Henry VII*. The money rewards offered were £1,000 down and 100 marks a year.

[2] Kingsford, p. 211.

[3] *Foedera*, xii. 534. The truce which was to last from 30 April 1494 to 30 April 1501. Dietz, p. 58. The clergy voted £40,000 in addition.

parations, which were on an adequate scale, can have been little delayed by a small inroad made by Lord Home's riders in February, but before Lord Daubeney was able to launch the projected attack the royal purpose was frustrated by a sudden rebellion in Cornwall. In making its grant parliament had endeavoured to protect the interests of the poorer subjects; it had stipulated that none was to pay the second tax unless he had either an income of 20s. a year from land or personal property to the value of 10 marks.[1] Even so the news of the subsidy gave great dissatisfaction. Men still felt that direct taxation was something of an extortion to be levied only under exceptional circumstances, and they may well have thought that the commons had already been generous to the king. It must have been common knowledge, moreover, that Henry had profited from the spoils of the attainted Yorkists and from the fines which he levied so readily after the failure of each successive rising. In Cornwall, a county 'sterile and without all fecunditee',[2] dissatisfaction produced an open explosion. Why, it was asked, should the poor Cornish miners be taxed for 'a smal commocion made of ye Scottes, which was asswaged and ended in a moment'? The discontent was general and leaders were found in a resolute blacksmith, Michael Joseph, and in Thomas Flamank, a lawyer, who seems to have argued that the defence of England against the Scots was a liability resting upon the tenures of the northern barons. Founding on this argument, and asserting that it was not Henry but his evil counsellors who were to blame, he succeeded, about the end of May, in persuading the populace to march upon London—not, as they said, in rebellion but to petition the king. They met with no opposition and they used no slaughter or violence. At Wells they found a new recruit in James Touchet, seventh Baron Audley, whose family had shed its blood for Lancaster during the civil war and who had himself accompanied Henry to France. Bacon asserts that Audley was by nature 'unquiet and popular'; it has been suggested, however, that he had been impoverished by the disappointing expedition to Boulogne, and probably was one of those Lancastrians who felt that past loyalty to Henry's ancestors had received little reward. Hall may be wrong in his assumption that Audley took command, for the city chronicler twice describes the smith as 'capitayne', but the adhesion of an

---

[1] *Rotuli Parliamentorum*, vi. 515.   [2] Hall, p. 477.

aristocrat certainly strengthened the insurgents who continued to advance with great rapidity. Bristol denied them entrance but they marched through Salisbury and Winchester without hindrance. Success seemed to be within their grasp, but, short as was the time allowed him, Henry had already made ample provision for the defence of London. Fortunately for him Daubeney had ready to hand the army prepared for the Scottish expedition, and by 13 June the trusty chamberlain took post at Hounslow Heath with a force of 8,000 or 10,000 mounted men. At the same time the king collected his own army about Henley, and the Cornishmen, after a brief encounter with 500 of Daubeney's spears near Guildford, decided to march for Kent, the traditional home of freedom, the breeding-ground of Wat Tyler and Jack Cade. Kent, however, was not disposed to welcome them, for the country-people were prosperous, and the king's competent servants, the earl of Kent, Lord Abergavenny, and Lord Cobham, had gathered force enough to ensure order. Sorely dismayed by the lack of recruits, the insurgents, who mustered 15,000 men, began to waver; some of them made a secret offer to Daubeney that they would surrender Audley and Flamank in return for a general pardon, and it was in 'greate agony and variaunce' that they encamped at Blackheath on the afternoon of Friday, 16 June. On the same day Henry had united his army to that of the chamberlain, and from his headquarters at Lambeth barred the way to the capital with a force of 25,000 men; London in arms stood firmly behind him; the queen and Prince Henry were safely in the Tower. Early in the morning of the 17th the royal army launched a sudden attack. At first the bowmen about Deptford Strand gave a good account of themselves—Daubeney, over-hasty in attack, was actually taken prisoner for a while, and according to Hall the royal army suffered a loss of 300 men killed. It was not long, however, before numbers and experience prevailed, and, once broken, the insurgent army collapsed in a dismal rout. Many were killed, many more, including the three leaders, were taken prisoners, and by two in the afternoon Henry crossed London Bridge in triumph to receive a hearty welcome from the mayor and his 'brethren in scarlet'. After knighting the mayor and two of his officers, the king, devout as always, gave thanks at St. Paul's and then rode on to join his wife in the Tower. He did not misuse his victory. On the 27th Flamank and Joseph were drawn from the Tower to be hanged at Tyburn and on the following day Audley, 'with a cote armour upon hym of papir, all to torne', was drawn from Newgate to Tower-hill and there beheaded. The three heads were set upon London Bridge but though the quarters of Flamank were exposed on the four gates of the City and the four quarters of the smith sent down to the west country, the victims had been allowed to die before the hangman began his grisly work. There were no other executions, and though Henry was at pains to get all the prisoners into his own hands he confined his vengeance to the exaction of fines over a wide area. Before he had got these in there was fresh trouble in the west country where the optimistic Perkin had appeared again.

Tenacious of his purpose Henry resumed his negotiation with Scotland at the earliest opportunity; it almost seems that for him the Warbeck episode and the Cornish rising were but temporary interruptions of a settled design. On 5 July 1497,[1] only a few days after the defeat of the Cornish rising, he issued a fresh set of instructions to trusty bishop Fox. The ambassador was to demand first of all the surrender of Perkin and, if James showed himself reluctant, was to argue that any engagement made with the 'duke of York' could be of no avail to a mere impostor. He was to be told that hostilities could be avoided provided that Scotland admitted the error of her ways according to diplomatic practice. James was to send the first embassy and this was to be followed by a personal meeting between the kings, which was to take place in England. Compensation was to be made for injuries done, hostages were to be given, James was to bind himself under ecclesiastical censures to maintain the peace. Considering that Scotland had broken the truce, Henry's attitude was surprisingly friendly and seemed to presage a renewal of the marriage negotiation.

James was in a quandary. Ferdinand and Isabella had been urging him to abandon the pretender and he had replied by inquiring about the possibility of a Spanish marriage. Ferdinand, though he did not immediately disclose the fact, had no daughter available,[2] but he was anxious to consolidate the alliance against France and believed that the best way to do this would be to arrange a marriage between James and an

---

[1] Gairdner, *Letters and Papers*, i. 104.
[2] The Infanta Maria was designed for Emmanuel of Portugal.

English princess. Accordingly he sent Pedro de Ayala to Scotland with instructions to promote the match. The ambassador, who arrived in June or July 1496, was favourably impressed both by the country and by its ruler, and before long he gained the complete confidence of the Scottish king.

James found a way out of his dilemma in characteristic fashion; he dismissed Perkin but at the same time showed his intention of continuing the war with England. He may have believed, as has been suggested, that Perkin was still a force in the political world and could give a lead to malcontents in Ireland and in Cornwall, but the circumstances of the pretender's departure[1] make it appear rather that he merely wished to be honourably quit of an unwelcome guest. The Spanish monarchs, apparently upon information sent by Puebla, asserted[2] that for some time the king of Scots had looked upon Perkin as a prisoner, but he was in fact paid his pension[3] in May and June 1497, and the king was plainly at pains to get him out of Scotland before Fox could press for his surrender. In the second week of July he sailed from Ayr in a single ship called, not inappropriately, *Cuckoo*,[4] but he was attended until his departure by Andrew Forman, later archbishop of St. Andrews; he took with him his wife and some thirty followers; his ship was well provided; his lady was given $3\frac{1}{2}$ ells of tawny Rouen cloth to make her a sea-gown;[5] and he was escorted by Robert Barton, one of the king's most trusted captains.[6] On 26 July he arrived at Cork to receive a poor welcome. Only his old ally John Atwater befriended him, and a month later he sailed off to Cornwall in the hope of profiting from the unrest there. His forces were negligible—120 men in two small ships and a Breton pinnace according to Henry VII[7]—and he was nearly taken at sea by four ships from Waterford. On 7 September, however, he landed at Whitesand Bay near Land's End, and,

---

[1] Henry wrote to Gilbert Talbot that Perkin had been 'set full poorly to the sea by the King of Scots'. Pollard, *Henry VII*, i. 162.

[2] *Spanish Calendar*, i. 140.

[3] *Accounts of the Lord High Treasurer*, i. 335, 340, 342.

[4] Several Flemish ships are called the *Kyckuit* or *Uitkyk* (Conway, p. 84) but the *Kekeout* which formed part of Perkin's expedition in 1495 was one of those taken by the English at Waterford and was bought by Sir Henry Wyatt.

[5] *Accounts of the Lord High Treasurer*, i. 343.

[6] Ibid. 344. Andrew Barton may have gone also, though this is not certain.

[7] Ellis, *Original Letters*, i. 32. Other evidence suggests that the ships were Spanish. Cf. *Spanish Calendar*, i. 186.

leaving his wife at St. Michael's Mount, set himself to fan the
smouldering embers of revolt. By the time he reached Bodmin his
following was reckoned at between 3,000 and 4,000 men, and
heartened by this rapid increase in his forces he proclaimed himself King Richard IV[1] and set off to besiege Exeter. His army now
mustered some 8,000 men but as they lacked armour, weapons,
artillery, and apparatus of any kind their attempt was foredoomed to failure. The city was resolutely defended by the earl
of Devonshire and the gentry of Cornwall and Devon, and
although the insurgents attacked the north and east gates with
the greatest courage on 17 and 18 September, their naked valour
was of no avail. Some accounts, indeed, say that Perkin burnt
the north gate and broke into the east gate, but it is evident that
the assault was repelled with great slaughter. Hardly a man
in the garrison was slain and the besiegers, leaving 400 dead
upon the ground, drew off and made their way to Taunton.
Their position was now desperate. Daubeney was hurrying
against them[2] with the levies of South Wales, Gloucester, Wiltshire, Hampshire, Somerset, and Dorset, and not far away a
royal army was forming under Henry himself. Behind them lay
uncaptured Exeter and a fleet under Lord Willoughby de Broke
cut off all hope of escape by sea. Towards midnight on the 21st,
knowing that Daubeney had reached Glastonbury, Perkin stole
secretly from his camp with sixty followers[3]—possibly all the
mounted men he had—and galloped for his life towards
Southampton Water in the hope of finding a ship. Advised of
his flight Daubeney sent in pursuit 500 spears who captured
most of the fugitives, but Perkin himself, along with John
Heron, a bankrupt merchant of London, Edward Skelton, and
Nicolas Ashley, a scrivener, who were his chief counsellors,
succeeded in obtaining sanctuary at Beaulieu. Aware, however,
that there was no hope of escape they decided to throw themselves upon the king's mercy and surrender at once. Perkin was
taken to Taunton and there on 5 October he made a full confession of his imposture. Henry then went on to Exeter where

---

[1] Soncino reported that he displayed standards showing the escape of a little boy
from a tomb, and from a wolf's mouth. *Milanese Calendar*, i. 326. The Milanese
account says he had 80 savage Irishmen—and later gives him 300 persons of
various nationalities. Ibid. 327.

[2] Ellis, i. 35.

[3] The Milanese account says that he fled with ten men only. *Milanese Calendar*,
i. 328.

the commons of Devon flocked in to make submission, and from that city[1] the pretender wrote to his mother that remarkable letter which goes so far to prove that he was an impostor indeed. Meanwhile, his wife, who had remained at St. Michael's Mount, was brought in all honour to the city and when she arrived Perkin was compelled to repeat in her presence the whole story of his deceits. Thereafter the luckless lady, who seems to have won golden opinions everywhere, was sent 'with a goodly sorte of sad matrones and gentlewomen'[2] to the court of Henry's queen where, according to Bacon, her charm gained for her 'the name of the White-rose, which had been given to her husband's false title'. She resumed her maiden name and after a space of eleven years made three marriages in quick succession —to James Strangeways, to Sir Matthew Cradock, and to Christopher Ashton. The first and third of these were gentlemen of the king's chamber, the second was a Glamorganshire squire from whom the noble house of Pembroke springs by a female descent. Content to have exposed the deceiver for what he was, Henry showed his usual clemency. A few of the ringleaders were executed at Exeter 'in sacrifice to the citizens whom they had put in fear and trouble',[3] but for the rest the king was content with fines imposed not only on the insurgents but upon those who had given them countenance. Bacon asserts that the commission under Lord Darcy appointed at Exeter proceeded with great strictness, and the extant figures suggest that this was so.[4] From the counties of Somerset, Dorset, Wiltshire, and Hampshire £8,810. 16s. 8d. was collected according to one account and an additional £4,629. 8s. 8d. according to another, and the total reached £14,699 when the last payment was made. Henry was thorough. Great persons, lay and clerical, are mentioned separately upon the roll, and even for boroughs, hundreds, and tithings, which are treated collectively, individual names are given; successive commissions were appointed lest anything be lost and the last payment was made in 1507; the king entrusted the actual collection, or part of it, to the reliable Hatcliff, clerk of accounts of the king's household; he scrutinized the accounts and added observations in his own hand.

Having set the machinery of punishment in train Henry set

[1] Gairdner, *Richard III*, p. 329.   [2] Hall, p. 485.
[3] Gairdner, *Memorials* (Bernard André), p. 75; Bacon, *Henry VII*.
[4] *Foedera*, xii. 696; Gairdner, *Letters and Papers*, ii. 335. Dietz, p. 58.

off for Westminster, which he reached on 27 November. Perkin, whom, as a monster, people had flocked to see along the route, was made to repeat his confession and then led through the City to the Tower of London. Behind him went in fetters one of his followers, who had been serjeant farrier to the king; and on 4 December this man and another deserter were hanged at Tyburn. Yet the pretender himself was soon released to an easy captivity in the hands of appointed custodians. So easy, indeed, was his durance that on 9 June 1498 he escaped and after hiding on the banks of the Thames took refuge in the Charter-house at Sheen, from which he emerged to repeat his confession from the stocks at Westminster and again in Cheapside. After that he was sent to a gloomy dungeon in the Tower.

While the dream of the adventurer had thus dissolved in ruin, the affairs of Henry had prospered in all directions. A renewed attack by the king of Scots, who assaulted Norham at the end of July 1497, was easily repulsed and Surrey, advancing with 20,000 men,[1] crossed the Tweed and threw down the castle of Coldstream. He destroyed a few peels, including Ayton, but could not lure the Scots to action. Hall represents the flight of the Scots as ignominious, but the romantic king, who did not summon the sheriffdoms till 9 August[2] and who challenged the English commander to single combat, evidently regarded his adventure as a knightly enterprise rather than a real campaign —a great raid.

The truth is that neither James nor Henry really wanted serious war and that Ayala had been working busily for peace. The patient Spaniard pursued his advantage and his efforts were rewarded when, at Ayton, on 30 September there was signed a truce between England and Scotland which was to last for seven years.[3] An understanding once achieved, good relations were established with surprising ease, and these were not seriously disturbed even when, in 1498, the garrison of Norham made an assault upon a few Scots whom they thought too

---

[1] Hall says 20,000 men. Surrey probably had only 8,000 or 9,000. He was supported by a fleet. *Navy Records Society*, viii, xlv–li. *Naval Accounts and Inventories of the Reign of Henry VII.*

[2] James had had to coin the great chain, worth perhaps £1,500, in order to equip his expedition, which included his guns, among them 'Mons Meg'. His offer was to fight Surrey 'puissaunce against puissaunce' or 'hand to hand' (Hall, p. 481) for the gage of Berwick; Surrey replied that Berwick belonged to his master and not to himself.

[3] *Foedera*, xii. 673.

inquisitive. Henry was quick to soothe the feelings of both James and Bishop Fox and the outcome of the incident was that in July 1499 the truce was prolonged to last during the lifetimes of both kings and until one year after the death of him who should survive the longer.[1] In the following September the marriage negotiation was resumed. Everything pointed to the establishment of a real peace between the two countries.

Important in itself, the treaty of Ayton was yet only one of the fruits of Anglo-Spanish friendship. The main outcome was the treaty for a marriage between Prince Arthur and the Infanta Catherine, which was concluded on 1 October 1496 and confirmed on 18 July 1497. The terms of the settlement followed those of the original treaty of Medina del Campo, though certain amendments suggested by Henry were embodied. The princess was to come to England when Arthur attained his fourteenth year; the marriage portion of 200,000 crowns was to be paid in money—half within ten days of the celebration of the marriage, the other half, in two equal instalments, in the two succeeding years; Catherine was to keep her rights to the succession of Castile and Aragon in the event of the other children of Ferdinand and Isabella dying without heirs.[2] Spain, for her part, secured her own interests with great care. The original jointure promised to Catherine was secured, and her rights to inherit in Spain were curtailed within the limit already mentioned. There was no question, for example, of Catherine's obtaining Aragon though her sister, the bride of Philip of Austria, gained Castile, nor was she to have any rights in Italy or in the rapidly expanding colonies as long as her sister lived. The Spanish inheritance, in fact, was to be kept intact and England was to have no share in it. Elaborate arrangements were made to ensure that the marriage should take place.[3] In August 1497 the parties were solemnly betrothed at Woodstock. Next year, in February of 1498, the Spanish sovereigns ratified the treaty at Alcalá. Each child made formal appeal to the pope for a dispensation to marry by proxy before the attainment of legal age; in May 1499 the marriage was celebrated by proxy at Bewdley, and a formal treaty was made in July.[4] *Ex abundante*

---

[1] *Foedera*, xii. 722, 729.
[2] *Spanish Calendar*, i. 130; *Foedera*, xii. 658.
[3] *Spanish Calendar*, i. 241.
[4] See pp. 158, 171-2.

cautela the Spanish monarchs insisted on another proxy marriage in the autumn of 1500 after Arthur had attained his fourteenth year. In both of these ceremonies the Infanta was represented by the somewhat unworthy ambassador, Puebla.[1]

The alliance of England with Spain and the establishment of peace between England and Scotland were rightly regarded by Ferdinand as triumphs of Spanish diplomacy, and he may have hoped to follow up his success by embroiling Henry in a war with France whilst he himself remained at peace.[2] A private report, indeed, expressed the opinion that, in view of Henry's difficulties, he could have secured any terms that he pleased, and hinted that Puebla had behaved as a servant rather of England than of Spain. The critic entirely misconceived the situation. Henry was never moved by fear of defeat and even the approach of the Cornish rebels to London did not disturb his equanimity; he knew his own strength and it was not the fault of Puebla that Spain could not secure better terms. What is true is that Henry emerged from the whole transaction a clear gainer. He had no intention of renewing the quarrel with the old enemy of England. When Maximilian suggested that the premature death of Charles VIII presented a splendid opportunity for an attack on his successor, Louis XII, he replied, not without irony, that he would like to see Maximilian at war with France but 'only by way of witnessing his wonderful feats'; he himself could not take part in the enterprise[3]. In July 1498 he renewed the treaty of Étaples. How wise was his decision appears from the fact that in the following month Ferdinand himself signed the treaty of Marcoussis with France. The real basis of this Franco-Spanish *entente* was a

[1] Pollard, *Henry VII*, i. 194; cf. Busch, p. 133. Puebla was in England temporarily in 1488–9 and permanently from 1494 to 1509. He was a difficult and undignified ambassador, and his government did not always trust him. Ayala was sent on a special errand about October 1496 and remained till 1502; Puebla traduced him to the king and queen and tried to get rid of him. In July 1498 Londõno and the Sub-Prior of Santa Cruz were in London, evidently to report upon him. He was accused of venality but, as his salary was not paid and he was given no benefice, he was often in difficulties. Henry spoke lightly of him and may have used him, but he seems to have managed Spanish affairs not too badly. In June 1500 Fuensalida was sent over to make sure that Henry had not pledged Catherine elsewhere.

[2] In January 1497 Ferdinand supplied Puebla with reasons why Henry should make war on France; on 8 April he announced that he had made a treaty with France in which he was prepared to include Henry as a relative, friend, and ally provided Henry wished to be included. *Spanish Calendar*, i. 133, 142.

[3] *Spanish Calendar*, i. 157.

design to exploit the situation in Italy;[1] from this Henry was excluded, but if he was aware of his exclusion he showed little concern. He professed to be disappointed that a 'League formed with so many ties should have been thus dissolved'[2] but he had, except on economic questions, no great interest in Italian affairs. If France or Spain should gain anything by adventures south of the Alps it would not be at his expense. From the intrigue and diplomacy of half a dozen years he had emerged a clear winner. Perkin was in the Tower and those who had thought to make him a stalking-horse had shared in his discomfiture. The conspirators were ruined or dead. Restless Ireland was in good order, warlike Scotland had made peace, the valuable trade of the Netherlands had been secured by the *Intercursus Magnus* and useful economic treaties had been made with other countries, a marriage alliance had been concluded with Spain, mighty France was friendly, and Henry, though the honoured son of the pope, was clear of all commitments. He stood before the eyes of Europe as a king whose title was firmly established, whose power was considerable, and whose wisdom made him impervious to diplomatic trickery. The ambassadors of all countries spoke of him with respect.[3]

[1] Louis XII, unlike his predecessor, had a claim to Milan. If he asserted this he must antagonize Maximilian. It seemed prudent to him to gain the support of Ferdinand by offering to share with Spain the kingdom of Naples to which both he and Ferdinand had claims.

[2] Pollard, *Henry VII*, i. 203.

[3] Puebla reassured his master and mistress as to the stability of Henry's power and his estimate was confirmed by Londõno, a special ambassador sent in the summer of 1498. The Milanese Soncino and the Venetian Trevisano were both impressed by Henry's state and riches, and they praised his clemency, his wisdom, and power of maintaining good order. Soncino, who remarked that the nobles of England either feared or loved him, was astonished by the discovery that Henry was extremely well informed on Italian affairs by his protégés in Italy insomuch 'that we have told him nothing new'. Pollard, *Henry VII*, i. 158, 161, 164, 178, 187, 196, 202.

VI

FOREIGN AFFAIRS

SECURE upon his throne, Henry was able, during the last decade of his life, to reap the fruits of the foreign policy he had pursued with such care. Latterly, when the deaths of his wife in 1503, of Isabella of Spain in 1504, and of Philip of Burgundy in 1506 seemed to open new vistas of matrimonial alliance and dynastic acquisition, he indulged in fancies, which were no less repulsive than impracticable; but he made no serious attempt to carry out these unworthy projects and guided his action by the sound principles which he had steadily evolved. Above all, in a world full of contendings and intrigue he refused to go to war. 'In the midst of this, his Majesty can stand like one at the top of a tower looking on at what is passing in the plain.' So wrote Soncino to the duke of Milan, giving his master to understand that there was no chance of Henry's intervening in Italy.

Looming behind the European unrest was the ever-present threat of the Turks. Fortunately for Christendom the forces of Islam were at this time divided by the rise in Persia of the power of the Shah, whose followers claimed to be more orthodox than those of the Sultan—they accepted as creed only the Koran and as caliphs only the son-in-law of Mohammed and his descendants. The might of Tabriz could not seriously challenge that of Constantinople but, with unrest in Egypt and the disastrous earthquake of 1509, it served to restrain the Sultan from any great adventure to the west. Even so the menace to Europe was serious. Between 1499 and 1503 Venice was engaged in a war in the course of which she lost some places in the Morea and suffered an invasion through Friuli which reached Vicenza. The Republic warded off the attack with relatively little loss, but the disunity of Christendom was made apparent. Seizing the opportunity of the jubilee of 1500 the pope preached a crusade but, as Polydore Vergil remarks, he considered not only the good of the world but his own advantage, and the princes of Europe followed his example. France sent some aid by sea and on one occasion the Spaniards gave a little help, but for both France and Spain the crusade was mainly an opportunity for the seizure of Naples on the ground that her king Federigo was bargaining with the Infidel, or at least was incapable of resisting him.[1]

It was upon Italy, not upon the Levant, that the eyes of Europe were turned. Despite their failure in Naples in 1495 the French had managed to keep a foothold south of the Alps and with the accession of Louis XII in 1498 their ambitions were re-awakened. Louis, unlike his predecessor, embodied the claim of Orleans to Milan and he set himself at once to make his pretensions good. For his adventure the situation, both in Europe and Italy, was favourable. He had renewed the treaty of Étaples in July 1498 and in the following August had come to terms with Spain at Marcoussis; in July 1499 he made peace with the Archduke Philip at Arras. Maximilian remained hostile but, embroiled with the Swiss, he could not seriously prevent the French design. In Italy the threatened duke of Milan found no support. Alexander VI came into the French camp when, in 1498, he granted Louis a divorce; he obtained for his son Caesar the dukedom of Valentinois, the hand of Charlotte d'Albret, and the patronage which would assist his design of making the papal states the patrimony of the house of Borgia. Venice, which had distrusted Milan since the day of Fornovo, signed a treaty with Louis in April 1499. Naples, although as a victim of French aggression she may have felt sympathy for Ludovico, could give no help, and the luckless duke was driven to rely on the imprudent king of the Romans, the discreditable Turks, and uncertain mercenaries, whom his great treasure could buy.

In September 1499 Louis occupied the Milanese without difficulty. The French abused their victory and when, in January, Ludovico returned from Innsbruck with a hired army he regained his capital with ease. His mercenaries, however, made no resistance to the returning French; in April they surrendered abjectly at Novara, and Ludovico, disguised in vain as a Swiss soldier, was taken prisoner. In less than a year Louis was supreme in Lombardy, but soon he committed what Machiavelli considered a capital error—he allowed another power to share in his success. Anxious to reassert the Angevin claim to Naples he took Spain into partnership and, on 11 November 1500, concluded with Ferdinand the treaty of

---

[1] Bridge, *History of France*, iii. 137.

Granada whereby the two powers, professing a zeal for a crusade, decided to share the kingdom between them, France taking the north (the Abruzzi, Gaeta, Naples, and the Terra di Lavoro) and Spain the south (Apulia and Calabria). In June 1501 the pope ratified the nefarious pact; in July Capua was bloodily sacked and Naples capitulated. Next month Federigo yielded to Louis who compensated him with the duchy of Anjou, and in March 1502 the son of the conquered king surrendered to the Spaniards at Taranto.

Almost at once the unworthy allies quarrelled; the treaty of Granada had not disposed of the whole of the Neopolitan territory and each side claimed as its own the provinces of which no mention had been made.[1] At first the French carried all before them and only in Barletta did the Spaniards maintain themselves under the redoubtable Gonsalvo de Cordoba, but in 1503 they lost everything. They were far from their base, the climate was against them, they were badly led by Nemours, and, though Bayard performed prodigies and d'Aubigny fought well, the hot valour of the *ordonnances* did not prevail against the hard courage of the experienced Spanish infantry. On 21 April an expedition from Sicily beat d'Aubigny at Seminara; a week later Gonsalvo, issuing from Barletta, beat and killed Nemours at Cerignola. In vain Louis tried to distract Ferdinand's attention by an attack on his northern frontier. The Spanish tide steadily advanced. In August Alexander VI died, Caesar was prevented by illness from exploiting the opportunity, and the cardinals elected to the papal chair not Georges d'Amboise, the trusty servant of Louis, but Giulio della Rovere who, as Julius II, was to show himself resolute to increase the papal states. The French party fell to pieces and when, in December, a new army was routed on the muddy banks of the Garigliano, Louis was beaten out of Naples altogether. Next year he sought to revenge himself on the deceitful Ferdinand by diplomatic means. Already in his attempt to keep his hands clear for Italy he had come to terms with the Habsburgs. At Lyons on 10 August 1501 he had arranged that his daughter Claude should marry Charles, son of Philip of Burgundy, the future Charles V,

---

[1] These were Capitanata, between Abruzzi and Apulia, the Principato, between Naples and Calabria, and the Basilicata, between Apulia and Calabria. For the omissions of these great territories mere haste cannot account; presumably each party hoped to profit from the negligence of the other.

and this pact had been renewed at Lyons again in April 1503;
at Trent on 13 October 1501 Maximilian had entered not very
cordially into the arrangement, and now at Blois on 22 September 1504 the three princes, 'one soul in three bodies', bound
themselves in an alliance which promised to dominate western
Europe. The betrothal of Charles and Claude was again ratified
and Louis promised that if he died without heir male his
daughter should have as her portion not only Burgundy and
Milan but several French provinces as well. All that Louis
gained was belated investiture of the Milanese (Hagenau, April
1505), and it was certain that a treaty which proposed to
dismember France would be hated by his own subjects—but
he had isolated the treacherous king of Spain.

In Spain itself the position of Ferdinand was shaken by the
death of his consort in November 1504. True to the last, Isabella
by her will appointed her husband as regent, but the Castilians,
who disliked the Aragonese, considered that Juana was their
queen and welcomed the action of Philip of Burgundy who
claimed the kingdom in her name. The plight of Ferdinand
seemed desperate, but he extricated himself with his usual
address. Aware that France was now regretting the heavy price
promised for the Austrian alliance, he proposed that he and
Louis should sink their differences over Naples and made overtures for a French bride, who should bring as her dowry the
old Angevin claim. Louis consented and, at Blois in October
1505, gave him the hand of his beautiful niece, Germaine de
Foix, he promising to pay a million gold crowns in respect of
the French expenditure in Naples. In May 1506, in response to
representations by the French Estates at Tours, Claude was
betrothed to Francis of Angoulême, heir apparent of France,
and the understanding between Louis and Ferdinand was sealed
by a personal meeting between the monarchs at Savona.

Dishonesty and war had kissed one another and the abandoned dignity of the Habsburgs was left to obtain what satisfaction it could. Philip and Maximilian moved closer to Henry
and, without any reference to Ferdinand, promised the hand of
Charles to Mary, daughter of the English king. At the same
time Maximilian announced his intention of going to Italy to
assume the imperial crown, but his advance was resisted by
the Venetians, who were helped by France, and in June 1508 he
was compelled to accept a three years' truce whereby he

abandoned territories to the Signory. The acquisitive Republic had overreached itself. On each change of fortune in Italy she had seized the opportunity to profit from the weakness of the loser. The misfortunes of Naples, France, Caesar, and the king of the Romans had contributed to her aggrandisement. Now she angered France by making peace without consulting her ally and she crowned her folly by quarrelling with the pope over the appointment to prelacies. Ferdinand, anxious to divert attention from himself, readily joined with the pope to devise a scheme whereby the great powers should forget their differences in the spoliation of Venice, and on 10 December 1508, under cover of a treaty between Habsburg and Valois, there was formed at Cambrai a league of the princes who, anxious as always to attack the Turks, proposed to begin their pious operations by 'recovering' the territories occupied by the greedy queen of the sea. In April 1509 a Bull of excommunication was launched; France advanced to the attack; and on 14 May at Agnadello Venice lost, in the words of Machiavelli, all that she had gained in 800 years. From the worst consequences of her overthrow she was saved by the fidelity of her subject populations, but before the drama was played out the estate of England was altered. Henry VII had died on 21 April 1509. Alone of the great princes he had had no share in the league—indeed his proposal that Ferdinand should be excluded seems to show that, as late as November 1508, he had no idea of the purpose of the congress of Cambrai. These facts are eloquent of his real greatness. The vicissitudes of European affairs during his last decade had provided many opportunities for interference; more than once his active intervention had been sought; he had not hesitated to seek his profit from the turning wheel of chance; but he had committed himself to no rash venture and had refused to be diverted from his essential aims.

Nowhere does his attitude appear more clearly than in his conduct with regard to the proposed crusade. He must have known very well that the money to be obtained from the great indulgence of the jubilee would not all be devoted to an attack upon the Turks; but he neither withstood the publication of the Bull nor endeavoured to profit by it, though he was pleased when the pope departed from his intention of sending a special legate.[1] He made no demur about the imposition of special

---

[1] The Articles of the Bull of the Holy Jubilee, 1501. Pollard, *Henry VII*, iii. 173.

papal taxation; he even contributed 20,000 crowns—about
£4,000—of his own money, and, unlike every other prince in
Europe, he allowed the whole yield of the levy to go to Rome.[1]
To the suggestion, however, made by the pope in November
1501, that he, 'with navy convenient', should himself take part
in the crusade he gave a calculated refusal veiled in polite
terms. He commended his two cousins of France and Spain for
their alleged willingness to go on crusade and he applauded the
preparations made by Germans, Hungarians, Bohemians, and
Poles, who were 'best acquainted and expert in faicts of war
and frauds of the said Turks and also nigh unto them'; but he
pointed out that his own dominions were so remote from the
scene of action that a great part of his cost and effort would be
wasted in mere transport. It was most desirable, he said, that
there should be a single leader to whose armament other
princes could contribute. If neither France nor Spain would act
he would, since Alexander was going himself against the Turks,
'give in proper person assistance to the Pope's Holiness',[2] but
of course he must have ample time to prepare for so distant an
expedition. Nothing could be plainer. Henry was a good churchman whose orthodoxy had earned the commendation of three
successive popes; his relations with the papacy had been uniformly good and he had established a 'sound business connection' profitable alike to himself and to the Holy See; but he
would neither embark on an impossible enterprise nor use the
pretext of that enterprise for making gain out of the cheated
hopes of Europe.[3] He did, in June 1502, give £10,000 to Maximilian for the Turkish war, but his generosity was not inspired
entirely by a crusading zeal; he wanted to detach the king of
the Romans from his Yorkist protégés and to bring him into the
good understanding which he had already established with the

---

[1] His agent in Rome, Adriano Castelli (bishop of Hereford 1502-4, of Bath
1504-18), later pointed out to the pope that, though most of the princes of Europe
had permitted the imposition of 'crusades' and 'tenths', they had kept the money
for themselves and *pontifex ne obolum quidem accepit*. (Letter to Henry, 4 January 1504
—Gairdner, *Letters and Papers*, ii. 116). The *cruzada* came to be a recognized part of
the income of the catholic kings.

[2] Pollard, *Henry VII*, iii. 166 ff.

[3] Henry received the sword and cap of maintenance from Innocent VIII in
1489, from Alexander VI in 1496, and from Julius II in 1505 (Gairdner, *Memorials*,
p. 46; Kingsford, pp. 211, 261). He had reached a good understanding with the
Curia as regards appointment to benefices (Busch, p. 230); he had used the pope's
aid for the improvement of the clergy, for checking the abuses of clerical privilege,
for the condemnation of enemies, and for the dispensations for marriages.

Archduke Philip. His procedure was characteristic. For him the high action of European politics was secondary to the immediate interest of his own country, and against the changing diplomatic background he can be discerned steadily at work upon his own designs. Four great things he accomplished. He bound Scotland to him in a firm peace, he crushed the last of the Yorkist pretenders, he made an alliance with mighty Spain upon equal terms, and, in all his dealings, especially his dealings with the Netherlands, he promoted the commercial prosperity of his country. A fifth achievement might be added—perhaps the most significant of all—he refused to go to war with France.

Secure in his *entente* with France and favoured by the absorption of the European powers in the affairs of Italy, Henry was able to bring his negotiations with Scotland to a satisfactory conclusion. The treaty of July 1499[1] with its guarantee of a truce which should survive the death of both kings seemed in itself to establish good relations on a permanent basis. Yet it was a truce, not a peace—there had been no official peace between England and Scotland since the treaty of Edinburgh—Northampton 1328 —and, to achieve a peace, a marriage alliance was, in the custom of the day, an almost necessary adjunct. As early as 1493 Henry had suggested a marriage between his kinswoman Catherine and the Scottish king; in June 1495,[2] when the danger from Warbeck seemed imminent, he offered the hand of his own daughter, Margaret, and this offer he renewed on the eve of James's invasion in September 1496.[3] The persistence of James in his support of the pretender, and that during the period of truce, might well have justified Henry in a refusal to continue the marriage negotiation, but when, by the mediation of Spain, Scotland and England were brought once more to friendship, the Tudor did not allow his sense of personal injury to interfere with the chance of political advantage. He wanted a complete settlement with Scotland; his dignity, he knew, could look after itself. On 11 September 1499 Henry again empowered Fox to negotiate for the marriage[4] and with his usual foresight procured, on 28 July 1500, a Bull of dispensation[5] which was necessary because the contracting parties were cousins in the fourth degree, both being descended from John Beaufort, earl

---

[1] *Foedera*, xii. 722.   [2] Ibid. 572.   [3] Ibid. 635.
[4] Ibid. 729.   [5] Ibid. 765.

of Somerset, the son of John of Gaunt. There seems to have been some delay upon the Scottish side, perhaps because James took too seriously the insincere Spanish offers made at an earlier date, and, although Henry issued a safe conduct in May 1501,[1] it was not until 20 November that James's ambassadors arrived in London. Robert Blacader, archbishop of Glasgow, the earl of Bothwell,[2] and Andrew Forman, postulate of Moray, were present at the rejoicings which marked the reception of Catherine of Aragon,[3] and one whom the *London Chronicle* calls a 'prothonotary of Scotland and seruaunt of the said Bisshopp', in all probability William Dunbar, produced the famous ballad in praise of London town.[4] In the midst of the celebrations the work went speedily on and, on 24 January 1502,[5] the negotiation was concluded in three separate treaties. One of these was for the regulation of the eternal Border disputes; another dealt with perpetual peace, league, and confederation between the two monarchs and their legitimate heirs and successors; the third was for the marriage.

The political treaty conforms to the diplomatic usage of the time and resembles in general outline the Anglo-Spanish treaty of 1499. Both kings pledge themselves to a good, real, sincere, true, entire, and firm peace by land, by sea, by fresh water, and everywhere else for ever, and by their promise their vassals of all degrees are bound. Each king promises not to make war upon the other directly or indirectly, not to receive the rebels of the other, and on demand to hand over to their proper lord any such rebels who have sought refuge in his territories. Existing letters of safe conduct are cancelled; in the future safe conducts are to be granted only with the consent of both parties and they are not to be valid for more than a year. If any other prince should wage war against one of the contracting parties the

---

[1] *Foedera*, xii. 772.

[2] This was Patrick Hepburn, third Baron Hailes, who was created earl of Bothwell in 1488 after the forfeiture of John Ramsay, Lord Bothwell, who acted as Henry's intelligencer in Scotland.

[3] Kingsford, pp. 252–3.

[4] It has been pointed out that the *Accounts of the Lord High Treasurer* read as if Dunbar was in Scotland in December 1501; he was not an official secretary and had no particular dependence on the bishop of Glasgow. The *Accounts*, however, are susceptible to another interpretation, and the Londoner may well have been misinformed as to the author's official position; it is difficult to believe that Scotland had two poets of the calibre of Dunbar.

[5] *Foedera*, xii. 787, 793, 800.

other, upon demand, shall aid him with all the power he asks, at the expense of the asker, to be reckoned according to the custom usual in the land of the asker, notwithstanding 'whatsoever contracts, pacts, conventions, friendships, leagues and confederations, with whatsoever prince or princes, in whatsoever form of words and however ratified already concluded contrary to this arrangement'. The allies of each party might be included in the treaty and a comparison of the names given shows that the friends of both countries were in the main the same; the king of the Romans, France, Spain, Denmark, and the Archduke Philip appear on both lists but England adds Portugal, Venice, Ferrara, Savoy, and the Hansa, whereas Scotland includes the dukes of Gelders, Holstein, and Cleves along with the margrave of Brandenburg. Any of these powers who did not ask for inclusion within eight months would be considered as excluded. If it should happen that either king should wage war against any of the princes so included, the other king might send to an ally aid for defence of his realm only, but he must abstain from direct invasion of the realm of the other contracting party. Arrangements were made for the mutual redress of grievances and it was expressly provided that, though the failure to obtain redress would justify the issue of what would later have been called letters of marque to an aggrieved subject, the issue of such letters and the action arising therefrom should not constitute a breach of the main treaty.

The terms of this far-reaching settlement undoubtedly exhibit a real political wisdom; the signatories recognized that 'incidents' on the Border and on the high seas were inevitable and endeavoured to discount their effect. Yet the treaty contained within itself the germ of its own destruction, for the obligation undertaken by the king of Scots was incompatible with the terms of the 'Auld Alliance' to which he, like so many of his predecessors, had already given his pledged word.

The details of the marriage treaty show the precision and the interest in money which mark every diplomatic instrument with which Henry was concerned. The princess was to be conducted to the Scottish border at her father's expense and at Lamberton Kirk, just outside the Berwick Pale, or some other convenient place was to be handed over to the representatives of the king of Scots; she was to be married to James *in facie Ecclesiae* within fifteen days of her handing over. She was to be given a jointure in real estate equivalent to £2,000 sterling or £6,000 Scots, and possession was to be given before 1 July 1503; if the lands handed over on that date were found to fall short of the stipulated annual yield other lands were to be added before 15 July, and Margaret's possession of the whole during her lifetime was to be guaranteed against any diminution or reduction by the Scottish parliament, or by any other authority. During his lifetime James was to enjoy the revenue arising from the dower lands, provided that he maintained his wife with her servants, of whom twenty-four were to be English, in the manner proper to a queen of Scots, and was to give her annually £1,000 Scots to spend as she pleased; after his death the queen was to enter into free possession of the whole income even if she elected to leave Scotland. Great care was taken to ensure that all the documents for the conveyance of the lands should be handed over in July but it was expressly provided that James should not demand his bride before 3 September 1503, when she would attain the mature age of thirteen years and ten months, though a proxy marriage was to take place before Candlemas 1503. For the marriage portion Henry promised only £10,000 sterling, and this was to be paid in three portions, the first at Edinburgh within six days of the marriage, the second at Coldingham one year later, and the third before the end of 1505. Receipts were to be demanded for each payment; if Margaret died without heir before the transaction was completed the English liability was to cease, and for the risk of transit of the money between Berwick and Coldingham James was to be responsible.

The marriage by proxy was celebrated on the day after the treaty was signed. In the course of the year ratifications were duly exchanged,[1] and on 12 July[2] James promised that he would make no renewal of the 'Auld Alliance' without first consulting the king of England. On 10 December,[3] however, possibly as the result of representations from Louis, he formally protested that he had in his original oath (22 February 1502)[4] given to Henry by inadvertence the title of king of France, and accordingly took his oath again in an amended form. Seven days later, this time presumably to gratify Henry, he undertook not to ask the pope for any relaxation of the oath or absolution from it.

[1] *Foedera*, xiii. 30–2, 43, 48–50. [2] Ibid. 12.
[3] Ibid. 46. [4] Ibid. xii. 804.

Despite a certain atmosphere of suspicion on both sides the arrangements for the marriage went on and Henry, who could be magnificent upon occasion, exerted himself to see that his daughter was properly equipped, insisting that where possible the Red Rose of Lancaster should be displayed. In the summer of 1503 he took her from Richmond to his mother's house at Collyweston in Northamptonshire, and there handed her over to the earl and countess of Surrey, who were to conduct her to her new home. Under their protection she set off on 8 July and in stately progress moved slowly through Newark, York, Durham, and Newcastle to Berwick. On 1 August she was met at Lamberton Kirk by the archbishop of Glasgow, who welcomed her in the name of the king, and thereafter the Scottish envoys took charge of the procession which advanced towards Edinburgh. At Dalkeith it was 'surprised' by a visit from James himself in the glory of fine raiment and a splendid horse and during the next few days the march was punctuated by an exchange of civilities. The little queen performed a basse dance with Lady Surrey in the presence of James who replied by singing to the lute—his well-instructed bride found the performance delightful—and, when he took his leave, leapt into the saddle without touching stirrup. On 8 August the marriage was celebrated with great splendour in the chapel of Holyroodhouse, and the following days were passed in church services, jousting, and banqueting.

A full account of the ceremonies survives in the manuscript of John Young, Somerset herald, who, while he rejoiced in the handsome appurtenances of the bridal train and of the English nobility and gentry who met it at the border of each county, was plainly surprised by the amount of civilization he found in Scotland. For James and for his clothes he had great admiration, his only criticism being that the king's beard was 'something long'. A commentary on his judgement appears in the *Accounts of the Lord High Treasurer of Scotland*: 'Item, the ix day of August, eftir the mariage, for xv elne claith of gold to the Countess of Surry of Ingland, quhen scho and hir dochtir Lady Gray clippit the Kingis berd, ilk elne xxij ħ.; summa iij^cxxx ħ.' The entry is eloquent of gaiety and good feeling; James, not presaging the day of Flodden, made much of Surrey, so much indeed that the poor bride felt herself neglected and wrote to her father a mournful letter in which she expressed the wish that

'I wer wyt your Grace now, and many tyms mor'. Truth to tell she found little joy in Scotland. The king's gallantries were known and the standards of his gay court, influenced by the French Renaissance, were not high, as may be gleaned from the pages of Dunbar. Yet he was not a bad husband, as his anxieties at the time of the births of his children attest. Much of Margaret's unhappiness was due to herself. As she grew up she showed herself headstrong and flighty; it almost seems as if her party at court was opposed to that of the king, and in after years her indiscretions earned her the rebuke of her correct brother, Henry VIII.

The marriage of 'the Thrissil and the Rois'[1] is justly regarded as a proof both of Henry's patience and of his wisdom; it was no small accomplishment to have united two countries divided by long years of war. His policy was not without risks, one of which may have been noted at the time. Bacon says that when the treaty was being discussed in council

'some of the table . . . did put the case; that if God should take the king's two sons without issue, that then the kingdom of England would fall to the king of Scotland, which might prejudice the monarchy of England. Whereunto the king himself replied; that if that should be, Scotland would be but an accession to England, and not England to Scotland, for that the greater would draw the less; and that it was a safer union for England than that of France. This passed as an oracle, and silenced those that moved the question.'

An eminent authority has pointed out that, when the treaty was made, Henry 'cannot have looked forward to the death without issue of two of his sons and of all his surviving son's children' and it may be added that the stem of Stuart did not appear to be very healthy at this time.[2] Yet the story is no *ex post facto* invention; it is found in the second edition of Polydore Vergil's *Anglica Historia* which was published at Basel in 1546.[3]

The fruits of the new relationship appeared almost at once in the joint attack upon the lawless borderers known as the Raid of

---

[1] The epithalamium of William Dunbar bears this title.

[2] James III had three sons but only one was alive after January 1504; of James IV's legitimate issue five children died in infancy and only one, James V, survived. James V left as legitimate issue only Mary, and Mary only one son.

[3] It does not occur in the 1534 edition nor in the manuscript version of 1513 recently edited by Mr. Denys Hay. It evidently attracted some attention, for it was quoted from Polydore by Bishop Leslie, whose book *De Origine, Moribus et Rebus Scotorum* was published in Rome in 1578.

Eskdale (1504), and, as long as Henry VII lived, England and Scotland remained at peace. Yet in spite of the new alliance the old animosities lingered on. James was building a fleet and the activities of his captains annoyed the English, whose trade with the Netherlands must have been interrupted. There was constant bickering on the Border and at length Sir Robert Ker, warden of the Middle March, was murdered by three Englishmen at a day of law. One of the culprits was surrendered but the other two flaunted their impunity until one was slain by the vengeful Scots. Towards the end of the reign it seemed that the smouldering embers of hate might be kindled by a wind from France. In 1506 Charles of Gelders, usually ranked as an ally of Louis, appealed to James for aid against an attack made upon him by Philip at the instigation, probably, of Maximilian. James, who was bound to the duke by kinship as well as by treaty, was temperate in his reply (July 1506),[1] observing that both he and Henry were now allies of Philip and recommending arbitration. Early in 1507, however, hearing that the Burgundians refused arbitration and persisted in their attack, he wrote to Henry VII a letter[2] which, though it still breathed peace, ended in a threat of war; if England joined in the attack upon James's ally he must regard her act as unfriendly. He was entitled, under the treaty of 1502, to aid an ally in his own country without a rupture with England, and in any case he must have known very well that Henry was not likely to send an expedition overseas for the purpose of promoting Burgundian ambitions in the Netherlands. He was really giving diplomatic support to his kinsman without running any risk of war. His action was correct and it was correct also in his dealings with France. In 1507 Louis, apprehensive no doubt as to the outcome of the Anglo-Burgundian *rapprochement* of the previous year, sent to inquire about the aid which might be given by Scotland for the defence of the Milanese against Maximilian. Probably his purpose was to ascertain Scotland's readiness for war, and James, who really believed in the much advertised crusade, treated his overture at its face-value.[3] There was, however, so

---

[1] Gairdner, *Letters and Papers*, ii. 206. In March 1506 James had written to Philip felicitating him on his alliance with Henry. Ibid. 211.   [2] Ibid. 225.

[3] *Flodden Papers*, Scottish History Society, 1933, pp. 1, 4, 6. James naïvely hoped to unite all Christian princes in the crusade, in which he wished to serve as admiral. His reply to Louis's overture about the Milanese was politely discouraging. *Epistolae Regum Scotorum* (1722), i. 83.

much coming and going between Scotland and France that
Henry's suspicions were aroused and early in 1508 he detained
the earl of Arran and his brother, Sir Patrick Hamilton, who
were returning surreptitiously from France without letters of
safe conduct. The Hamiltons, who were notable jousters, were
well treated; they were feasted both by the lord mayor in
London and by Henry at Richmond. Yet they were in fact
under open arrest and their master felt a grievance. Later in
the year James received a visit from Bernard Stuart d'Aubigny,
whom he welcomed both as a kinsman and as the 'father of war'
whose reputation in arms was international, but it must be
noted that d'Aubigny came openly by way of London with a
train of eighty horsemen and there is no proof that his mission
was directed against England. Whatever his errand he did not
discharge it himself. The long journey proved to be too much for
the ageing warrior, who died soon after his arrival; Dunbar,
who had hailed him with a triumphant 'Welcum', wrote also
his elegy. Plainly the old intimacy between Scotland and France
continued; but their intimacy was not dangerous to England as
long as England remained on good terms with France, and it
was a cardinal point of Henry's diplomacy to refuse to quarrel
with the French king.

The settlement of Scotland, though very important in Henry's
main design, was yet only a side-issue. Other things were more
essential. Among these was the necessity of dealing with rebels
and pretenders. Although, after 1497, Henry was in control of
the situation, the ghost of York was not yet laid. Henry's throne
was not seriously threatened, but his policy was embarrassed
by malcontents who justified their ambitions by a display of
the White Rose. Whether he himself felt any serious alarm may
be doubted, but he was aware of the effect produced upon
European opinion by the frequent 'apparitions'. Early in 1499
he was disturbed by yet another bogus Warwick. This was
Ralph Wilford, the son of a London cordwainer, who had been
seduced by an Augustinian friar named Patrick, and, though
he speedily detected and hanged this impostor[1] (12 February),
he may have decided that he would have no real peace as long

---

[1] Patrick was committed to a close prison for life. In mentioning this Polydore
Vergil comments upon the benefit of clergy accorded by English law even to great
offenders.

as the real Warwick lived. There can be little doubt that Spain urged him to this conclusion; it seems certain that he himself was troubled in his mind. He was not bloodthirsty, but he had the ruthlessness of the Renaissance and possibly some of its superstition too. In March Ayala reported that he had consulted a soothsayer who warned him that his life would be in danger, that he had aged perceptibly, and that he had grown very devout.

It is impossible to dissociate his *malaise* from the proceedings taken against Warbeck and Warwick in the following winter. On 16 November[1] Perkin, along with the mayor of Cork (John Atwater) and his son and John Taylor, was arraigned in the Whitehall at Westminster and all four 'for certyn treasons' were adjudged to be drawn, hanged, and quartered. Two days later a court at the Guildhall tried eight persons found guilty of a plot to slay the marshal of the Tower and release Warwick and Perkin. In the course of the proceedings the jurors set on record an indictment against the earl of Warwick, and on the morrow the luckless boy was tried at Westminster Hall in a court presided over by the earl of Oxford, created lord high steward *pro hac vice*; he pleaded guilty and was sentenced to the cruel fate appointed for traitors. On the following Saturday Warbeck and Atwater were drawn from the Tower to Tyburn where they were hanged, and on 29 November Warwick was beheaded on Tower Hill. During the next few days various other conspirators, among them one Astwood, were executed. There is no record of the evidence given at the trials but its nature can be safely deduced from the indictment against Warwick, of which a copy survives.[2] From this it appears that a conspiracy was hatched on 2 August between Warwick, a certain Robert Cleymound, and one Thomas Astwood, who had been a follower of Perkin, had been pardoned in 1495, and was now employed as a jailer. This conference took place in the chamber of the captive earl; two days later contact was made with Warbeck who was confined in a cell immediately below. Messages and tokens were exchanged, and then Cleymound, alleging that Perkin had betrayed the conspirators to the king and council, withdrew from the conspiracy saying he must seek sanctuary. The events thus rehearsed inevitably suggest the use

---

[1] Pollard, *Henry VII*, i. 211.
[2] *Third Report of the Deputy Keeper of the Public Records* (app. ii), pp. 216–18.

of an *agent provocateur*, perhaps Cleymound himself. It is surely remarkable that two important state prisoners should have been confined in adjoining cells between which communication could be easily established, that among the jailers employed should have been an old supporter of Perkin, that the plan included at once an escape to Flanders and the levying of war upon the king, and that the conspirators should have gone out of their way to bring their plot so definitely into the category of high treason; it is remarkable, too, that no action was taken against Cleymound. Even on the evidence presented, the part played by Warwick was naïve to the point of stupidity. As Polydore Vergil relates, having been brought up in prison from his cradle out of sight of man and beast, he could not even distinguish a hen from a goose. That such an one should venture on a bold stroke for liberty exceeds the bounds of probability: he can at best have been but a passive conspirator.

As regards Warbeck the case is different. His confinement after his unsuccessful attempt to escape seems to have been severe,[1] and Puebla, who saw him when he was produced for the bishop of Cambrai in August 1498 only a couple of months after his recapture, remarked, 'he is so much changed that I, and all other persons here, believe his life will be very short'. His plight was desperate; he was by nature given to the taking of risks, and his relative immunity in failure, no less than his surprising successes, perhaps encouraged an inborn optimism. It is easy to believe that he would readily try to escape and not unreasonable to suppose that he was concerned in a plan to fire the powder in the Tower—four or five persons, including his former adherent Astwood, were executed for some plot of the kind;[2] but even so it is possible that the design was rather the suggestion of others than the fruit of his own inventive mind. It is difficult to resist the conclusion that certainly Warwick and possibly Warbeck fell into a carefully arranged trap and that Henry, whether he formally connived or no, eagerly seized an opportunity provided by his servants. Certain it is that Ferdinand had been urging him to remove a dangerous competitor[3] and that Catherine herself afterwards believed 'that her marriage had been made in blood'.[4]

---

[1] Pollard, *Henry VII*, i. 199.   [2] Ibid. 211, 213.   [3] Ibid. 214.
[4] *Venetian Calendar*, v. 257–8, 7 September 1549 (Cardinal Pole himself appears to be evidence for this, and there is other evidence besides), Pollard, *Henry VII*, p. 179.

In January 1500 Puebla was rubbing his hands over the fact that not 'a drop of doubtful royal blood' remained in England.[1] His rejoicing was premature; a stem of the White Rose still flourished in the person of Edmund de la Pole.[2] A brother of that John who was killed on the field of Stoke, and eldest surviving son of Elizabeth, sister of Edward IV, Edmund might rank as a dangerous claimant to the throne and his treatment by Henry had not been ungenerous. In spite of his brother's attainder he had received back a portion of the family possessions; he had appeared at court and had been given a place of distinction in formal ceremonies. On the ground, however, that his reduced revenue would not support the dignity of a dukedom he was allowed only the title of earl, and the grievance rankled. In 1498 he killed a man in a brawl and although he received the king's pardon he felt insulted because he had been indicted before a common court of justice. It seems possible that Henry intended well in the matter, for the royal pardon would obviously be easier to bestow if the proceedings were not too formal; but the hot-headed young man felt a sense of injury, heightened no doubt by his own straitened circumstances and the whisperings of malcontents. In the summer of 1499 he suddenly fled to Calais, where he stayed awhile with Sir James Tyrrell, the governor of Guisnes; later he went on to St. Omer. As the Netherlands were the nursery of Yorkist plots Henry might have been pardoned if he had adopted stern measures, but, aware that Philip no less than himself was anxious to maintain the existing good relations, he was confident that the matter could be amicably settled. He contented himself with instructing Sir Richard Guildford and Richard Hatton, whom he sent to the Archduke in September, to induce the fugitive to return, and before long the earl was back at the English court. There he enjoyed his old privileges and when, in May 1500, the English king crossed to Calais for an interview with the ruler of the Netherlands he was accompanied by Suffolk; plainly he wished to convince Burgundy that the White Rose was now closely entwined with the Red. His efforts were unavailing. In July or August 1501 Suffolk fled again; this time he was accompanied by his brother Richard, and he sought not Philip but Maximilian to whom he had been recommended by Sir Robert Curzon.

---

[1] *Spanish Calendar*, i. 213.   [2] Busch, pp. 165, 362.

The part played by Curzon is equivocal. Polydore mentions, and evidently shares, a general opinion, that from the first he acted as an agent of the English king, and Bacon, improving on the story, avers that it was upon information supplied by him that Suffolk's adherents were arrested in 1502. It is true that Curzon had no sort of grievance against the king, that he could hardly have quitted his post at Hammes without the knowledge of his royal master, and that when he returned to England in 1506 he was granted not only complete immunity but, eventually, a pension of £400 a year with other favours. On the other hand, he certainly did go to fight the infidels, and in view of Henry's professed enthusiasm for a crusade there is nothing odd in his having received the royal permission to leave Hammes. There is, moreover, evidence that he did complain to Maximilian of the 'murders and tyrannies' of Henry VII,[1] that Maximilian in reply promised to help one of King Edward's blood to recover the crown of England, and that it was in reliance on this promise, conveyed by Curzon, that Suffolk took his flight. Unless one is to suppose a valiant soldier to have been a despicable villain one must assume that Curzon, shocked by the judicial murder of Warwick, did, for a time at least, play with the idea of supplanting the Tudor by a scion of the house of York. How long it was before he returned to his allegiance is hard to tell. That he was twice cursed at St. Paul's along with Suffolk might be part of the 'cover' normally supplied to a secret agent, but it is not easy to explain on this supposition the fact that in 1505 one of his sureties for the keeping of Hammes was made to forfeit a recognizance of £500. He himself did not return to England till 1506 when Suffolk was in Henry's hands. The immunity he then received may have been only an example of Henry's usual politic clemency, but in view of the confident statement of Polydore one cannot dismiss the possibility that even the stout-hearted crusader may have acted as Henry's spy; whether or not he did, the king was certainly well informed of the pretender's doings.

In May 1502 Sir James Tyrrell and several other persons including 'a yoman of the Croun, and a pursevaunt' were arrested and executed for certain treasons.[2] At the same time Lord William de la Pole, brother of Suffolk, and Lord William Courtenay, son of the earl of Devon, were taken into a confine-

---

[1] Gairdner, *Letters and Papers*, i. 134.   [2] Pollard, *Henry VII*, i. 223.

ment from which they did not emerge till after Henry's death. Courtenay's wife was Suffolk's cousin but she was also sister to Henry's queen, and unless we are to attribute the king's action to his settled 'aversion towards the house of York' it must be supposed that he suspected a far-reaching conspiracy. Next year there was some mysterious intrigue involving a secret mission to the bastard Sir George Neville at Aachen and the carrying off of an infant called James Ormond 'whiche shuldbe a great Inheretor & nexte unto the Crowne'.[1] A great authority has argued that the whole business was an invention of Henry's agents, but it is hard to avoid the conclusion that there was abroad a spirit of disaffection among the old families—among the Nevilles and the Montacutes, who were Yorkist in sentiment, and even among the Staffords and the Butlers, who were traditionally Lancastrian. No doubt the ambitions of the great families were aroused by the deaths of the king's sons, Edmund on 12 June 1500 and Arthur on 2 April 1502, but even without the disturbing influence of dynastic interests there was wavering in the ranks of the king's servants.

Of this there is evidence in a curious conversation which took place at Calais about the year 1504, between the deputy, Sir Richard Nanfan, the master porter, Sir Sampson Norton, and the treasurer, Sir Hugh Conway.[2] Conway, after casting doubts on the loyalty of Daubeney, the lieutenant, said that there had been a good deal of talk about what would occur 'yf hys grace hapned to depart', and that he had heard speech of Buckingham and of Edmund de la Pole but nothing of my lord prince. He alleged that Sir Nicholas Vaux, who had succeeded Tyrrell at Guisnes, and Sir Anthony Browne, lieutenant of the castle of Calais, had boasted that they could regard the matter with equanimity as they would be secure in their holds 'howsoever the world turned'. They themselves, however, who had no such safeguards, were in grave danger; Browne's wife was Lady Lucy Neville, niece of the king-maker and cousin of de la Pole; Calais was very close to Kent where they (he did not say who) had great alliance including Poynings, Guildford, and Bouchier; and even among the garrison were men who never 'loved the

[1] I. S. Leadam, An 'Unknown Conspiracy against Henry VII', in *Transactions of the Royal Historical Society*, N.S., xvi (1902). James Gairdner argued against the reality of the plot, and Leadam published a rejoinder in ibid., N.S., xviii (1904).
[2] Pollard, *Henry VII*, i. 240-50; Gairdner, *Letters and Papers*, i. 231-40.

king's grace and never would do'. When told that he should
have informed the king of all this Conway excused himself on
the ground that, when he had attempted to give information
about Lovell, the king was displeased with him, and Nanfan
corroborated that his highness was loth to believe evil of Sir
James Tyrrell or Sir Robert Clifford; but when the treasurer
went on to speak of a prediction that Henry would reign no
longer than Edward IV the others silenced him, professing
their complete loyalty to their master, and Nanfan made no
doubt that he could hold the town and marches for the king
and for the prince after him. The exact meaning of this communication is hard to discern; it may have been dictated by
Conway's dislike of Daubeney or by the informer's dislike of
Conway, but it certainly shows there were dissensions amongst
the 'new men' who served the king, and doubts as to whether
the house of Tudor would endure.

The king appears as one who was slow to take action against
men whom he suspected and who was prepared to take risks,
but it is easy to understand that he regarded the presence of
Suffolk in the Netherlands as a real menace to his throne. From
this menace he was delivered by accident. Maximilian, even
after he had come to terms with Henry in 1503, made no real
attempt to expel Suffolk from Aachen, but the earl, harassed by
his creditors, stole away without the knowledge of his august
patron in the hope of doing better elsewhere. Louis of France,
George of Saxony, new-made duke of Friesland, and Duke
Charles of Gelders were all interested in the fugitive, but the
event showed that they were merely seeking a diplomatic
counter of value. For some fourteen months Suffolk was virtually
a prisoner in Gelderland, and when, in July 1505, Duke Charles
was compelled to submit to Philip of Burgundy he changed his
protector, or jailer, once more. The archduke, no doubt, would
have surrendered the refugee to make a good market, but as he
had now received payment of a useful loan from England he
saw no need to throw away a valuable asset,[1] and in January
1506 Suffolk felt secure enough to instruct two of his servants,
Killingworth and Griffith, to open formal negotiations with
Henry; in return for his submission he demanded that his
dukedom and other lands should be restored to him, that various
prisoners should be set at liberty and that the agreement to be

[1] Pollard, *Henry VII*, i. 258.

reached should be confirmed by parliament. His confidence may have been feigned[1] and it was, in fact, ill founded. Even before he wrote Philip had fallen into the hands of the English king. Where his diplomacy had failed Henry was saved by his good fortune. On his way to Spain Philip encountered a storm in the Channel and was driven ashore at Melcombe Regis in Dorset on 15 January.[2] Before he was allowed to proceed the archduke agreed to surrender Suffolk, though under the condition that his life should be spared. On 16 March Edmund de la Pole was handed over to the English at Calais; on the 24th he was paraded through London and imprisoned in the Tower. There he remained till he was executed by Henry VIII in 1513. With his capture the hopes of the Yorkists were blasted. The colourful adventures of his brother Richard, a gallant flower of the White Rose, did not seriously affect the history of England.

At last the Tudor felt himself secure. He and his servants always spoke of the 'runagate' and some have supposed that the king magnified his anxieties in order to have a pretext for exacting fines and ridding himself of inconvenient subjects. The suggestion is unfair. In the Netherlands, so vital to English trade, Henry was engaged in a commercial struggle so even that a relatively small accession of strength might give the advantage to either party. The presence there of a Yorkist pretender was a serious embarrassment to his policy for, as the Venetian ambassador at Antwerp observed, Philip hoped 'by means of this individual to keep the bit in the mouth of the king of England'. The relentless pursuit of Suffolk was dictated not by greed or by mere vindictiveness but by the necessities of his diplomacy.

The claims of the White Rose which had so inconvenienced Henry in his dealings with Burgundy offered, after 1499, no impediment to his negotiations with Spain. Puebla's rejoicings over the deaths of Warwick and Warbeck were premature. Rid of his rivals the Tudor could bargain with Ferdinand upon equal terms and in a series of diplomatic exchanges the advantage passed steadily to England. The treaty of alliance, 10 July 1499,

---

[1] His correspondence with his brother shows that their plight was desperate. Gairdner, *Letters and Papers*, i. 273–6. Cf. i. 303, 312, 315.
[2] See *infra*, 184.

which flanked the marriage-compact was realist in tone and in its provisions absolutely equal between the contracting parties. The guarantees of territory, the conditions for mutual aid, the promises of free access for merchants, the undertaking to suppress rebels were all reciprocal, as were also the very practical measures for preventing 'incidents' at sea. The territorial guarantee differed from that of Medina del Campo in that it covered only the dominions in the possession of each power at the date of the treaty. Spain, no doubt, thought herself the gainer by the change; having secured Cerdagne and Roussillon for herself she would trouble no more about English claims to Normandy and Guienne. If such were her calculations the event proved them wrong; within a few years she realized with a shock that the new conquests she had made in Italy were in no way guaranteed by her dear brother in England. For the moment, however, the weakness of the Spanish position did not appear. In the course of the year 1500 the treaty was duly confirmed by both countries; yet Spain was in no haste to send her daughter. Puebla evidently expected her in June[1] but Ferdinand began to suspect that Henry's journey to Calais might presage the offer of his elder son to Burgundy and instructed a special envoy, Fuensalida, to discover what he could.[2] The Spanish suspicions were quite unfounded—it was the marriage of Henry's second son and of his daughter which were discussed—but they were strong enough to make Ferdinand hesitate and, in spite of English impatience, the coming of Catherine was postponed.

In April 1501 Isabella, satisfied at last by the proxy marriages and all the other precautions, announced that the princess of Wales was ready to take ship, but it was not until 27 September, after a false start occasioned by the weather, that she actually set sail from Laredo. On 2 October she arrived at Plymouth where 'she could not have been received with greater rejoicings if she had been the Saviour of the world'.[3] She was brought towards London in easy stages by the royal officers; before she reached the capital Henry and his son went forth to meet her, and, as the king assured her father, greatly admired her beauty and her manners. On 9 November the prince rode into London by Fleet Street and took up his quarters in the king's wardrobe

---

[1] *Spanish Calendar*, i. 226, 229.   [2] Ibid. 235.
[3] Ibid. 262.

near St. Paul's, and the princess turned aside to Lambeth where she rested for a few days. When on the 12th she made her formal entry she was given a reception of the utmost cordiality. Perhaps because the long delay had given more time than usual for preparation, the City, always hospitable, outdid itself in magnificence. As she crossed London Bridge she was greeted by a goodly pageant from which St. Catherine and St. Ursula made speeches of welcome; in the widest part of Gracechurch Street stood a splendid castle decked with red and white roses and other ornaments from which Policy, Noblesse, and Virtue delivered their compliments in appropriate verse. At Leadenhall Corner she turned left into Cornhill where there was revealed a wonderful 'volvell'—a sort of orrery—'wherein the twelve signs moved about the Zodiac and the moon showed her course of light and darkness'. Raphael the Archangel presided over that pageant supported by Alphonso X of Castile, Job, and Boethius, noted astronomers, and each recited his dutiful verses. As she went down Cheapside she saw opposite Soapers Lane (now Queen Street) 'the Sphere of the Sun' where Arthur sat in his golden chair and farther on, at the Standard in Chepe, was a fifth pageant called the Temple of God, decked with a great red rose, in which not only the prophets but the 'Ffader of Heven' himself appeared and said a few words; close at hand in the house of William Geffrey, haberdasher, stood the king, the queen, and the court, present unofficially that their presence might not stand between the City and its royal guest. For now it was the turn of the mayor, who with his brethren and citizens behind him advanced to greet the princess, and led her upon her way through the ranks of the mounted aldermen who lined the street. Her triumph was not yet done; at the Little Conduit, at the end of the Chepe, the mayor made her a present of plate and in the last of the 'goodly pageants' the Seven Virtues with a host of virgins in white stood below three seats, of which the middle was occupied by Honour, while the others carried the sceptres and coronets of the prince and princess. It must have been a tired, though gratified, princess who rode at last to the bishop of London's palace where she and her household were lodged.[1]

Two days later Arthur and Catherine were married in St. Paul's with great solemnity and the following fortnight was

[1] Leland, v. 356, Kingsford, p. 234.

given over to jousting and feasting in the midst of which ambassadors arrived from Scotland for the betrothal of their king to the Princess Margaret. On 28 November the Spaniards handed over the 100,000 crowns; two days later Arthur wrote to his bride's parents declaring his happiness and promising to be a good husband. Not without justice did Henry assure Ferdinand and Isabella that they need have no anxiety about their daughter 'who has been welcomed by the whole people'.[1]

In April the Spanish monarchs wrote to express their satisfaction with everything including the Scottish marriage which was partly of their making, but before their letter arrived the scene had changed. On 2 April 1502 Prince Arthur died at Ludlow where he had been sent to keep court. For his parents this was a heavy stroke and in a fragment preserved by Leland there appears a Henry very different from the thrifty politician of the history books. With the aid of his confessor the bereft father broke the news to his wife; she received it gallantly—'God is where He was and we are young', but shortly afterwards the king returned to find the queen in a state of collapse and it was his turn to play the comforter.[2] They took their 'painefull sorrowes' together.

The blow which smote the heart of Henry shattered the fine-spun schemes of Ferdinand; for him the severance of his close tie with England could not have come at a worse time. This was the moment when he was quarrelling with France over the spoils of Sicily, and d'Aubigny, who had the stronger force, was soon to carry all before him. He could not afford to let the English alliance go and though, on 10 May,[3] he instructed his English ambassador, Estrada, to demand the immediate return of Catherine along with the 100,000 crowns already paid and the portion due to her on her marriage, he empowered his representative, by an instrument of the same date, to arrange a match between his bereaved daughter and the new prince of Wales. The demand for the portion as well as the dowry plainly proceeded on the assumption that the marriage had been completed, but in July[4] Isabella claimed that Catherine was still a maid, in a letter urging Estrada to hasten on the new marriage. Her urgency was due largely to political causes; Spain had

---

[1] *Spanish Calendar*, i. 269.   [2] Leland, v. 373.   [3] *Spanish Calendar*, i. 267.
[4] *Spanish Calendar*, i. 272. Hall, p. 494, definitely states that the marriage was consummated.

realized that by the existing political treaty Henry was not bound to defend their acquisitions in the south of Italy and she was anxious to use a new marriage-alliance as an occasion for the renewal of the political treaties. The terms might remain unchanged; if only the date were altered Henry would be committed to the defence of Calabria and Apulia. Meanwhile, it must be represented to him[1] that he was bound to aid Spain in the defence of her possessions and he might be lured to attack France by promises of help in the recovery of Normandy and Guienne. By September[2] Ferdinand was more insistent than ever in his proposals for a grand confederation against France, but though in that month the privy council seems to have drafted a treaty for the new marriage, the cautious Tudor was slow to move. He valued his friendship with Spain, but he was not to be fooled again with the spurious offer of Normandy and Guienne; he probably realized clearly enough the limits of his obligations under the existing treaty, and he saw no need rashly to increase them. The death of Henry's queen in February 1503 doubtless delayed the negotiations, and in spite of the reproaches and persuasions of Isabella, who offered to Henry the hand of the young queen of Naples,[3] it was not until 23 June that the new marriage treaty was arranged. Two days later the young pair were solemnly betrothed *per verba de praesenti* in the bishop of Salisbury's house.

Even so the difficulties were not at an end. In the treaty it was explicitly stated that a papal dispensation would be necessary because Catherine had contracted affinity with Henry by her marriage with Arthur which had been consummated.[4] In August Ferdinand asserted that everyone in England knew that his daughter was still virgin;[5] but it appears that the treaty was ratified in its original form by Isabella on 30 September and by Henry in March 1504,[6] and a serious complication arose. Pope Alexander VI had died without giving the necessary dispensation and Julius II, who succeeded after the brief pontificate of Pius III, doubted, or affected to doubt, his authority in such a

---

[1] *Spanish Calendar*, i. 275.
[2] Ibid. 286.
[3] Ibid. 303, 11 and 12 April 1503. The young queen of Naples was Juana, widow of Ferdinand II of Naples; her mother was another Juana, sister of Ferdinand the Catholic, who had married Ferdinand I of Naples. There were thus two widowed queens of Naples who lived together near Valencia.
[4] Ibid. 306.   [5] Ibid. 309.   [6] Ibid. 317 and 325.

case.[1] There is extant a copy of a Bull dated 26 December 1503, but if this was authentic it cannot have been forwarded, for next month Henry's agents in Rome were still endeavouring to obtain the concession and in July[2] the pope himself wrote to Henry explaining the delay. He had wished, he said, to examine the matter more maturely, but he would send the necessary Bull by Robert Sherborne, dean of St. Paul's, who was in Rome. Next month Estrada, grumbling at the delay and lamenting the decline in Catherine's health and looks, announced that a dispensation had come; but, whatever this was, the discontented ambassador complained that he might have to await the coming of the Bulls by the hand of Sherborne, who could not arrive before mid-October.

Not till November, however, did the dean appear and even then he came without the promised documents. Henry wrote testily to the pope,[3] and in March 1505[4] Silvestro de' Gigli, bishop of Worcester, announced in his excellent Latin that he had been instructed by his Holiness to bring to England *bullas originales dispensationis matrimonialis*. He was also to explain the long delay and to express the regret of Julius that *copiae dictarum bullarum*, which had been sent, under pledge of secrecy, to comfort the dying Isabella, should have been sent on from Spain to England. Now on 24 November, two days before the death of his queen, Ferdinand had written to Henry telling him that he was sending to Puebla *dispensationis bullam . . . quam Sanctissimus Papa noster concessit*;[5] but if the documents so dispatched were copies of 'Bulls' they cannot be identified with the 'brief' for the comfort of Isabella produced by the Spaniards in 1528. Whatever was the tenor of the documents they were of little effect for, on 27 June 1505, the young Henry on the eve of his fourteenth birthday made secret but formal protest against the validity of his marriage to Catherine.[6] It is possible that Henry VII really was, as a Portuguese observer said, troubled in his conscience after the match;[7] it is certain that he was beginning to doubt the wisdom of tying his fortunes to those of Ferdinand, whose hold on Castile was shaken by the death of Isabella. He had been, for some time, discussing the

---

[1] Pollard, *Henry VII*, iii. 78 n.
[2] *Spanish Calendar*, i. 328.
[3] Ibid. 341.
[4] Gairdner, *Letters and Papers*, i. 243.
[5] Ibid. 241.
[6] *Spanish Calendar*, i. 358.
[7] Gairdner, *Letters and Papers*, ii. 147.

possibility of his own marriage with Margaret of Angoulême,[1] and now he began to think of withdrawing in favour of his son. To this proposal the pope, somewhat surprisingly, made no objection,[2] alleging that Ferdinand gave his consent. It seemed that the king of Spain would be abandoned by all.

The arch-intriguer, however, saved the situation by a characteristic volte-face. Forestalling Henry, he entered into alliance with France himself, suggesting that if he and Louis sank their differences they could rule all Italy, and in October 1505 he made his position secure by his marriage with Louis's beautiful niece, Germaine de Foix. The new orientation of forces caused some disquiet to the pope, who told Corneto 'those two kings have parted my raiment amongst them but we shall let them see something about that'. It may have surprised Henry, but if it did he was by no means put out of countenance. He still hoped to fish in the troubled waters of Spain, exploiting the angry breezes which blew between Castile and Aragon, and he had already begun to bind himself closer to Philip of Burgundy, who claimed Castile by the right of his wife. Whether so wise a prince really built castles in Spain after the death of his wife is uncertain, but it cannot be denied that he pursued the hope of a Spanish bride with the utmost diligence. The offer of the young queen of Naples had been renewed by Spain at intervals throughout the year 1504, and in November Ferdinand wrote as if the match were as good as concluded.[3] Henry, though he spoke much of the widowed duchess of Savoy, was plainly interested, and in June 1505 received from his envoys a report[4] upon the princess, framed to answer inquiries of his own so detailed that as Bacon says if Henry 'had been young, a man would have judged him to be amorous'. Personally, it appeared, the young queen was all that is desirable but her possessions and revenues were neither great nor certain, depending on the goodwill of the king of Aragon, and before long Henry was informing himself with great care as to the real strength of Ferdinand in Spain and particularly in Castile.[5]

The reply he received probably convinced him that though

---

[1] Gairdner, *Letters and Papers*, ii. 133–4; Bridge, *History of France*, iii. 224–5. One of Henry's proposals was that he should marry Louise of Savoy and his son Margaret of Angoulême.

[2] Gairdner, *Letters and Papers*, i. 247.

[3] *Spanish Calendar*, i. 324, 327, 333, and 338.

[4] Gairdner, *Memorials*, p. 223. [5] Ibid., p. 240.

Juana might be popular in Castile the Burgundians were not, and that Ferdinand, who, though disliked for his secrecy and his heavy taxation, commanded the support of many of his subjects, especially of 'the men of warre', was quite likely to maintain his authority throughout Spain. It was plain, however, that Juana, if she came to Spain, would have a considerable following; hence Henry, when Ferdinand allied with France in October 1505, abandoned all thought of the queen of Naples and gave his support to the Burgundian claimants to Castile. Before long fate presented him with another chance of a marriage alliance. In September 1506, soon after his enforced visit to England, the archduke Philip died, and Henry promptly made overtures for the hand of the widowed Juana. The proposal does him little credit, for the lady's husband was but recently dead[1] and she herself was soon patently mad. Yet by a masterpiece of bad taste the hapless Catherine—betrothed but not married to Henry—was made to write in commendation of the match,[2] descanting on the great admiration which the king had felt for Juana during her visit to England. Henry himself represented that the match would promote the weal of Christendom since, once established in Spain, he could conveniently attack the infidels. Catherine, no doubt at Henry's dictation, emphasized this point,[3] and it was believed in England that the bowmen could conquer the whole of Africa in a few years.[4] So said the sneering Puebla who gave his hearty support to Henry's design, remarking that the English thought little of a malady which did not prevent the bearing of children,[5] and that if the queen remained mad it might be well for her to be in England whilst Ferdinand administered Castile in her name. On the evidence it seems fair to conclude that Henry really did contemplate this repulsive alliance. Perhaps he suspected that Ferdinand was deliberately exaggerating his daughter's illness with the intention of keeping control of Castile, but it is well for his reputation that Juana's condition became so bad that marriage was obviously impossible.

---

[1] Henry's negotiations were well under way by March 1507.
[2] *Spanish Calendar*, i. 435, 439. Catherine professed that she was using the hope of marriage as bait for Henry. From a Venetian account it appears that Juana's capacity was doubted as early as April 1506. Pollard, *Henry VII*, i. 284.
[3] *Spanish Calendar*, i. 440.   [4] Ibid. 438.
[5] Ibid. 409, and Pollard, *Henry VII*, iii. 92 note. It is noted by Professor Pollard that Juana had, in fact, six children, all of whom grew up healthy.

Yet, though his matrimonial design was frustrated, Henry was still little the loser as the result of Ferdinand's sudden change of front. He was able to knit more closely the alliance with the Netherlands which was always a great object of his diplomacy. The real sufferer from the diplomatic revolution was the unlucky Catherine. She was neither married nor unmarried. The ceremony of 25 June 1503 was held to be invalid because the bridegroom, being under age, could not contract *per verba de praesenti*. The English king let it be understood that things could go no farther until the remainder of the marriage portion was produced[1] and, as Ferdinand, either because he could not find the money or because he distrusted Henry, was very slow to pay, the unfortunate princess remained in England in a very bad case. She wrote desperate letters to her father begging him to remember that she was his daughter; denouncing Puebla; complaining of lack of clothes, lack of money, lack of a good confessor. In April 1507[2] she asked that either Ayala, who knew England well, or the knight-commander of Membrilla, who as Fuensalida had been in England, might be sent as ambassador, but although Membrilla appeared in the early summer of 1508[3] and, according to a later statement by Ferdinand, was provided with money for the dowry, the negotiation made but slow progress. Henry had succeeded in betrothing his daughter Mary to the young Archduke Charles, and, secure in the Burgundian alliance, valued that of Spain less. He demanded that Ferdinand and Juana should both ratify the Anglo-Burgundian marriage treaty;[4] that the whole of Catherine's dowry should be paid in coin and that Spain should renounce any right to recover under any circumstances the 100,000 crowns already paid. Ferdinand, who had instructed his ambassador to demand the return of Catherine if the marriage could not be completed, went very far in concession, promising to pay all in coin and agreeing that Catherine might settle the whole of her 200,000 crowns on the king of England if she chose;[5] but the ambassador himself came to doubt the possibility of bringing the affair to a successful conclusion. He found little opportunity of seeing Henry personally and early in 1509 he began to smuggle out of the country the money he had collected towards the second

[1] *Spanish Calendar*, i. 376, 386; cf. Pollard, *Henry VII*, iii. 79 note.
[2] *Spanish Calendar*, i. 411.   [3] Ibid. 457.
[4] Ibid. 462 and ii. 3.   [5] Ibid. 462.

instalment of the dowry; by 20 March he had already disposed of more than 30,000 crowns.[1] His pessimism was not unjustifiable for, as appears from a report sent to Henry from Spain,[2] Ferdinand, for his part, proposed to make his consent to the Anglo-Burgundian match dependent upon the completion of Catherine's marriage. The negotiation had reached an impasse.

The plight of the unfortunate Catherine during these long-drawn-out discussions excites sympathy, the more that Henry, who still called her princess of Wales, was inclined to shed crocodile tears over her poverty at a time when his own coffers were abundantly full.[3] Yet there is another side to the matter. It is clear from the records that Catherine exaggerated her destitution[4] and when Membrilla, whom she herself had suggested as ambassador, appeared on the scene he took the view that the princess's misfortunes were largely of her own making. He reported to Ferdinand that her household was ill managed and that she herself was too much under the control of a young friar, who was 'light and haughty and scandalous in an extreme manner' and one whose conduct had been remarked by the English king. It must be added too that Catherine was her father's daughter. From her correspondence it is clear that she was quite willing to hoodwink Henry in the interests of Ferdinand, and as late as April 1506 she was, according to herself, still unable either to speak or to understand English.[5] She behaved rather as a Spanish infanta than as a future queen of England. Yet the queen of England she became. The death of Henry VII solved the dilemma which had seemed insoluble. The young king's councillors, who presumably knew the mind of their dead master, advised him to complete the match and on 3 June 1509 Catherine was at last wedded to Henry.

The long delays in the accomplishment of the marriage are eloquent of the changed relations between England and Spain. The alliance with Ferdinand, once so important to the newly established Tudor, had become a secondary matter. To Henry the gaining of a new ally was less important than the refusal to fight with an old enemy, and paramount in his policy was the desire to develop the traditional friendship between England

[1] Pollard, *Henry VII*, i. 323–6.
[2] Gairdner, *Memorials*, p. 431.
[3] *Spanish Calendar*, i. 432.
[4] Pollard, *Henry VII*, iii. 79, n.
[5] Ibid. 287.

and the Netherlands. The fact that Philip of Burgundy had, through his wife Juana, a great interest in Spain obscures the issue, but a study of Henry's diplomacy during his last decade shows a steady advance of the Netherlandish 'interest' at the expense of that of Spain. The promotion of English commerce was a cardinal point of Henry's foreign policy and the trade with the Low Countries was vital to England. Trade, however, breeds rivalry as well as friendship and though peace with the Netherlands was desirable it was not easy to maintain. There were endless controversies, arising sometimes on matters of principle and sometimes from personal disputes between the English traders and the merchants of Antwerp,[1] and the *Magnus Intercursus*, though it provided a code at once fair and practical, did not avail to end the disputes. In the very year in which it was signed quarrels arose about an import duty on English cloth, which Philip imposed in disregard of the treaty; Henry wrote a sharp letter to the archduke and, when he failed to gain satisfaction, removed the English cloth staple from Antwerp to Calais.[2] A partial reconciliation was reached on 7 July 1497,[3] but a conference planned for Bruges in April 1498 accomplished nothing,[4] and although the bishop of Cambrai, who was in London in the summer, secured the return of the English merchants to Antwerp[5] no final settlement was reached until the following year.

In May 1499, after another conference at Calais, there was concluded a treaty which, while it settled practical difficulties in accordance with the principles laid down in 1496, yet served to establish the position of the English traders.[6] The Netherlanders received a slight reduction in the price of wool sold by the staplers at Calais (except when some great *ovium mortalitas* made wool very scarce) and very exact arrangements were made to prevent fraud by the English vendors. Packers were to be sworn before the mayor and constables of the staplers at Westminster or Boston; they were to mark each sack according to the nature

[1] See, for example, the complaints of the Antwerp merchants in 1485 in Schanz, *Englische Handelspolitik gegen Ende des Mittelalters*, ii. 178, and notes in Busch, p. 357.
[2] *Spanish Calendar*, i. 112, for Puebla's report of the new levy (July 1496); Gairdner, *Letters and Papers*, ii. 69 for Henry's firm letter (June) and ibid. i. 327 for a letter to Margaret recounting the story of his controversy with Philip.
[3] *Foedera*, xii. 654.
[4] Ibid. 695. Henry's powers are dated 25 August.
[5] Hall, p. 483.   [6] *Foedera*, xii. 713.

and quality of the wool; they were to append their names to the certificate and were liable to a fine of £20 (£15 to the king and £5 to the informer) if fraud were detected. Practical arrangements were made for dealing with claims that wool was damaged while still in the hands of the vendors. On the other hand, English merchants were protected against unjust claims and guaranteed against improper obstruction of their suits in the archduke's courts. They were to be paid in good money and to be allowed to take out of the country plate and jewels as well as bullion. Their cloth was to have free access to every place in Philip's dominions, except Flanders, though they must limit themselves to wholesale trade. Henry had meanwhile obtained an apology from Margaret for her dealings with Perkin[1] and in 1500 he endeavoured to cement his friendship with Philip in a personal meeting.

Choosing characteristically a season when an outbreak of plague made his absence from London desirable,[2] Henry, accompanied by his queen and by a gallant train,[3] crossed to Calais early in May 1500. Philip, though he was all friendliness, declined to enter an alien fortress lest he should establish a precedent which might be awkward if the king of France gave him a similar invitation, but the meeting took place on English territory at St. Peter's church outside Calais on 9 June 1500. The scene was one of great magnificence. The church was divided into various chambers with rich hangings and the feast was sumptuous—young kids, an English fat ox 'poudred', venison baked into cold pastries, spiced cakes and wafers, cream, strawberries and sugar, 'vij horselode of cherys and abundant drink'. 'The plente was so moche that the peple could not spende hit that day' and on the morrow the king had it divided among the peasants. After the banquet 'the Duke of Bourgoyne dauncyd with the ladyes of England', but behind the gaiety lay serious purpose. Access to the royal party was closely guarded and amid the gay arrases was a council chamber. The secret discussions which took place there aroused much speculation. Ferdinand suspected that the design might be to marry Arthur to a daughter of Philip,[4] and instructed a special

---

[1] 1498. *Spanish Calendar*, i. 196.

[2] Polydore Vergil (edition 1570), p. 609.

[3] Gairdner, *Letters and Papers*, ii. 87–92; *Chronicle of Calais* (Camden Society, 1846), pp. 4 and 49.

[4] *Spanish Calendar*, i. 234 ff.

envoy to find out what he could about the matter but, in fact, nothing was definitely settled. There was a general assertion of friendship and some talk of a double marriage between Henry of York and his sister Mary with Philip's daughter, Eleanora,[1] and his infant son, Charles; but that the arrangements were only tentative appears from the fact that next year on 10 August Charles was betrothed to Claude, daughter of the king of France.

Thereafter for some time the relations between England and the Netherlands hinged round the person of Suffolk, with whose affairs Maximilian became involved, and the object of the Habsburg princes was to get money from Henry, if possible without surrendering the fugitive. After a long negotiation Maximilian signed a general commercial treaty in June 1502,[2] and was promised £10,000 for the Turkish war in return for an undertaking not to countenance Henry's rebels. In October the money was paid but the light-hearted king of the Romans was at no pains to implement his bargain. Whilst the luckless White Rose was still in the hands of his patrons or jailers the situation was complicated by the outbreak of a new tariff war in the autumn of 1504. This was waged with some acrimony; Henry showed favour to the Hansa[3] and again recalled the English merchants to Calais.[4] The situation was complicated by England's dealings with the duke of Saxony, who had acquired the lordship of Friesland, and by her alleged help to Gelders, but while the economic strife continued occurred the event which drove the king of England into a political alliance with the archduke. This was the death of Isabella on 26 November 1504. Henry, at this time not unwilling to embarrass Ferdinand, supported Philip's project of sailing to Spain; in April 1505 he lent him £108,000 towards his expenses and in the following September made a second loan of £30,000. Meanwhile the ingenious Maximilian had thought of another plan for strengthening the Anglo-Burgundian alliance and in the autumn of 1504[5] had proposed a match between his daughter Margaret and the Tudor king. By March 1505 the project was under active discussion[6] and in November the king of the

---

[1] Hall, p. 492; Busch, p. 167.

[2] 19 June. Busch, pp. 174, 366; *Foedera*, xiii. 3, 6–10, 12–27.

[3] *Statutes of the Realm*, ii. 665; Busch, p. 179 note.

[4] Busch, p. 368; Gairdner, *Letters and Papers*, ii. 379.

[5] Busch, p. 212.   [6] *Spanish Calendar*, i. 351.

Romans issued formal powers for its conclusion though his able daughter, who had a mind of her own, had not consented to the arrangement. While the diplomatists hesitated accident brought the matter to a head. On 10 January 1506 the archduke and his wife set sail for their new monarchy. Two days later there sprang up a tremendous storm which swept the Channel;[1] the fleet was scattered in all directions and Philip's ship, which had been in danger from fire as well as from tempest, was glad to make shore at Melcombe Regis. Philip and Juana, who had both behaved with great courage though they had given themselves up for lost, were anxious to continue their journey as soon as possible but Henry refused to let the occasion slip. The unexpected visitors were conducted in all honour to Windsor where they were met first by the prince of Wales and later by the king himself. On 9 February Philip was installed as a knight of the garter and replied by investing Prince Henry with the Golden Fleece. There was a great exchange of civilities in which the magnificence of the English train contrasted with the simple raiment of the Burgundian; but though each monarch deferred to the other with the utmost politeness the advantage was plainly with Henry.

In the height of the storm the vane of St. Paul's, a brazen eagle, was carried away and in its fall destroyed another eagle which served as a sign for a tavern in Cheapside. According to Hall this was held to presage evil for the house of Habsburg and the death of Philip in the following September was regarded as a fulfilment of the omen. Others have thought that the omen began to fulfil itself before ever the prince left England in that Philip was compelled to make great concessions to his cordial host. That Henry used the occasion to establish a firm alliance with the Burgundian is not to be denied and certainly he consulted his own advantage; but the actual terms of the treaties made were not in themselves unequal. In a treaty of friendship and alliance signed on the day of the investiture each party undertook to maintain the other in all the dominions which he possessed 'or which he, his heirs and successors, in future, should have a right to possess';[2] each party undertook to discounten-

---

[1] On the storm and Philip's reception in England see Pollard, *Henry VII*, i. 263–85; the Venetian evidence is particularly valuable. See also Polydore Vergil (edition 1570), pp. 613–14; Hall, pp. 501–2; Kingsford, p. 261; Gairdner, *Letters and Papers*, ii. 364; *Spanish Calendar*, i. 379.

[2] *Foedera*, xiii. 123; *Spanish Calendar*, i. 380; Pollard, *Henry VII*, iii. 94.

ance, to arrest, and, if required, to surrender the rebels of the other, and if both should have begun a common war neither was to make peace without the other's consent. The one place in which the text departed from absolute parity was in the promise made by Philip that Maximilian would formally adhere to the pact within four months, and even so it was stipulated that his failure to do this would not invalidate the treaty itself.

Verbally the advantage might seem to lie with Philip. He was given the title of king of Castile throughout and was promised, if he asked it, the help of an English army to make good his claim; in return he bound himself only to give military support to protect Henry in his British dominions, which were in no danger, and to vindicate claims in France, which Henry was at that time not in the least inclined to assert. As so often happened, however, the real gain was with the Tudor. He was pledged to help Philip only with 'such an army as he may be able to spare and as the circumstances may demand'; in exchange for a promise not to support Philip's rebels, to whom he had never given any real help, he closed the Netherlands to Yorkist malcontents and was able to demand the surrender of Suffolk.

A few weeks later the political alliance was flanked by a marriage contract to which Juana, who had accompanied her husband to England, gave her formal assent. By the treaty signed on 20 March[1] Henry, in accordance with the proposals already made by Maximilian, was to obtain the hand of Margaret of Savoy; the lady was to have a dowry of 300,000 crowns, and in respect of her Spanish jointure and her revenue as dowager-duchess of Savoy an annual income of 30,850 crowns, of which Henry was to have the full use as long as he lived. In return the new queen of England was to have a jointure of 20,000 gold nobles a year, which was to continue after Henry's death and was to be at her free disposal provided that she did not spend the money in the country of a declared enemy of England. Careful provision was made for the payment of the sums promised and the details of the marriage which was to follow hard on the transmission of the first instalment of the portion were carefully arranged. To the children of the marriage were guaranteed any inheritances which might fall to the

[1] Pollard, *Henry VII*, iii. 96; *Foedera*, xiii. 127; *Spanish Calendar*, i. 382.

archduchess in Spain, Flanders, or elsewhere. Philip was bound to ask the pope to excommunicate him if he did not produce the money punctually or otherwise failed to fulfil the treaty. The king of the Romans, who had already given his son power to conclude the treaty, was to give his formal ratification by 1 August, and both Habsburg princes were to employ all their influence to persuade Margaret, whose formal consent was to be delivered by the same date.

If Philip found the bargain hard he may have consoled himself with the reflection that marriages arranged did not always 'take place'. In July Maximilian was complaining that Henry had not sent a promised embassy to Mechlin to perfect the arrangements,[1] but the real delay was due to Margaret herself who made it quite clear that she was not going to marry the English king.

A few days after the treaty was signed Philip left Windsor, but he was compelled to remain (at his own costs) at Falmouth whilst his fleet was reassembled and it was not until 23 April that he at last set sail. During the expensive and tedious period of waiting he was brought to sign, on 4 April, a power for the final settlement of the trade question, and after his departure, on 30 April, a commercial treaty was concluded. To this Bacon gave the title *Intercursus Malus* and the suggestion plainly is that the archduke was coerced into making very great concessions.[2] He, however, was safely out of England before the transaction was completed, and the power which he left behind him authorized only the adjustment of the treaty of 1496 and the correction of outstanding abuses. The English certainly made gains, particularly in being allowed to sell their cloth freely throughout the Netherlands, apparently even retail, except in Flanders; but they gained no monopoly in the sale of their own cloth; and the privileges given them were valid only in Philip's Netherlandish territories. To the Netherlanders were promised, in a vaguely worded clause, reciprocal privileges in England, and they received an additional guarantee against English fraud in that the wool sold in the staple at Calais was in twenty-nine categories. In any case the treaty was never operative, for

---

[1] Gairdner, *Letters and Papers*, ii. 153. Henry's reply of 12 August is very outspoken (Gairdner, *Letters and Papers*, ii. 155); the marriage was already concluded; the diet at Mechlin was for certain secret matters for which Philip had made overtures when he was in England.

[2] *Foedera*, xiii. 132; Pollard, *Henry VII*, ii. 322, for the text of the treaty.

Philip died before he ratified it, and when his sister Margaret renewed the *Magnus Intercursus* in 1507 the concession regarding the sale of English cloth was withdrawn.

Thus the matrimonial and the commercial settlements, made by Philip when he was in England, alike came to nothing. Yet the Anglo-Burgundian alliance had been strengthened and after the death of the archduke in 1506 it was confirmed in spectacular fashion. The proposal of a match between Henry and the mad Juana, as already noted, was futile, but it may have served to stimulate the desire of Maximilian to secure the English alliance for his grandson through his own mediation. Soon after Philip's death he revived the old project of a match between Henry's daughter Mary and the young Archduke Charles, and at Calais on 21 December 1507 a treaty for the marriage was concluded. As usual this was accompanied by a treaty of alliance, and in 1508 Henry stood before the world as the most successful diplomatist of his time.

The year 1508 was filled with active negotiations and ratifications.[1] It became plain that Margaret had no intention of marrying Henry and it is doubtful whether Maximilian's persuasions were as sincere as he alleged. None the less in May the English king tried to arrange a meeting with the reluctant lady at Calais, and the preparations for the wedding of Charles and Mary were pushed on with great activity.[2] At the same time Henry renewed his proposals for a match between the prince of Wales and Margaret of Angoulême; plainly he hoped to detach France from Aragon and commend himself to his Habsburg allies by isolating Ferdinand, who was the great obstacle to the realization of their ambitions in Spain. All his fine schemes ended in smoke. None of the projected marriages took place; he had made himself somewhat ridiculous in proposing for the hands of several girls young enough to be his daughters and in being rejected by them all; he was unable to keep all the gain of the *Intercursus Malus*; he was excluded from the inner secrets of his professed friends. How little he appreciated the situation in Europe appears from his letter to Margaret of

---

[1] Pollard, *Henry VII*, iii. 124 and 128.

[2] Margaret alleged that she had been married thrice, including her betrothal in infancy to Charles VIII, and was not minded to try matrimony again. Gairdner, *Letters and Papers*, i. 323. For the proposed meeting at Calais and the very detailed provisions for Mary's wedding see the *Chronicle of Calais*, pp. 52–66 (Camden Society, 1846).

November 1508;[1] he assumed that the conference, which was arranged for Cambrai, was essentially a meeting between Burgundy and France, and that Ferdinand should not be invited. He did not realize that Louis, valuing his position in Italy above all things, would never part from Ferdinand; that Ferdinand was the architect of a grand design of aggrandisement, and that the real object of the conference at Cambrai was to perfect the plan for the attack on Venice, which took place the following year.

Yet behind the appearance of failure lay the reality of success. The new dynasty had established itself in the eyes of the world, the parvenu was accepted in the courts of the great. Unlike most of the powers of Europe England had not wasted her strength in war and if her allies hoodwinked her as to their designs in Italy their friendship was valuable in north-west Europe. Thanks to their aid Henry had been able to overthrow his rivals, to maintain peace, and to promote the commerce of his country. Against his failures in the marriage market must be set his victory in markets of a more prosaic kind.

Not only did he promote English trade in the Netherlands, but he challenged the monopolies of the Hansa and of the Venetians, and though his efforts brought no immediate gain, his encouragement of the Bristol merchants in their western ventures set England upon the path of ocean navigation. The rounding of the Cape and the crossing of the Atlantic had changed the geographical importance of England; no longer was she on the outer edge of world commerce. That Henry perceived the full implications of the change is unlikely; but it is significant that with a kind of instinctive skill he directed the economic energies of England to meet the new conditions.

What is true of his commercial policy is true of his foreign policy as a whole. He seemed to depart little from the practice of his predecessors—except that he would not go to war with France; yet he succeeded in winning for England a part of importance on the new-set diplomatic stage, and in preparing her to gain fresh triumphs as the political drama unfolded.

[1] Gairdner, *Letters and Papers*, ii. 365.

VII

THE ACHIEVEMENT OF HENRY VII

THE achievement of Henry VII was that he rescued England from the disorder of civil war and set her upon the way of prosperity in a difficult world where the new jostled with the old in a strange confusion. The disorder of the fifteenth century was great, but it should not be exaggerated. For many Englishmen life must have gone on in the ordinary way. The machinery of government remained intact. The crown was in dispute between rival parties, but it was still the crown, and though it had passed strangely from head to head, always it had carried with it a royal authority; all men knew that England must have a king, that the king should govern through his servants, that he was entitled to hold land and privileges, and that when his 'own' did not suffice for the work of government, he should receive aid from the contributions of his subjects. Yet, as Fortescue had so stoutly asserted, the king of England was not absolute; he must govern according to the law. Whence the law came is not entirely clear. Fortescue had derived it conveniently from the contract made by 'Brute and his fellowship' when the realm of Britain was founded, but for most men the argument *ab origine* was not necessary. Their feeling was expressed in the words of Augustine, quoted more than once by the clerical orator who opened parliament, 'Sublata justicia quid aliud sunt regna quam magna latrocinia'.[1] Law was made by God; the law of nature was an expression of the divine law; and the civil law must be derived from the divine law either directly or through the law of nature. Neither king nor parliament could 'make' law; they might 'declare' it and appoint penalties for its breach, but, in theory at least, their pronouncements were invalid unless in accord with divine justice. There was a body of law, long tried and found good, administered after set forms and generally accepted. Its machinery, not distinguished clearly from administration or government generally,

---

[1] *Rot. Parl.* vi. 520, 409. But see *Statutes and their Interpretation in the Fourteenth Century* (T. F. T. Plucknett), pp. 26–31. A statute might make 'new' law over existing common law.

might be defined as a 'constitution'. In this the royal judges, the council, and the council in parliament occupied established places, but no one supposed that the recognized machinery of government represented the whole of the authority of the king. In him there was a reserve of power which might be used either through the recognized channels of authority or outside them. The king could make a new statute in parliament or he could issue a proclamation with the advice of his council; he could do justice through the great law-courts or he could exercise a special jurisdiction outside that of the common law. Yet in all that he did, he must not transgress the divine law as it was understood by the long tradition of his people. The royal authority, in fact, was checked by the constitutional sense of England.

When, therefore, he set about his task of restoring order to England, Henry had as assets both the established machinery of government and his personal reserve of power. He used both, but in the main he employed his reserve to supplement and strengthen the existing institutions. Some of his acts appear high-handed to modern eyes, but in the main he did not outrage the English sense of what was fitting. He had the support of the part of England which mattered and which was to matter more and more as the Tudor monarchy developed. If he was an autocrat, he was, for the most part, an autocrat by consent. Henry did not, as a reader of Hallam might suppose, break a constitution; nor did he make a constitution. His reforms were not effected simultaneously; they cannot be said to represent the working out of a definite plan; they were, however, the expression of definite spirit, the spirit of efficiency. Nothing that he did can have seemed very new to the men of his own day. Except in the realm of finance he did not create a new machinery; he made the old machinery work.

To the Crown, which was the mainspring of the state, he restored the prestige and the power which had been sadly diminished during the troubles of the fifteenth century. For him it was a prime necessity to restore to the minds of his subjects the old reverence. He himself was well aware of this. His attitude has been misunderstood. Because the Lancastrians were believed to be more 'liberal' than their Yorkist rivals, and because the Tudors made an alliance with the third estate, it has been supposed that Henry was *bourgeois*; because he was a practical monarch who sought practical ends and was extremely careful of his money, it has been imagined that he was indifferent to display. These are misconceptions. The Lancastrian tradition was aristocratic—it was Edward IV who tended to be hail-fellow-well-met with the Londoners; even feeble Henry VI was very conscious of his royal office, and Margaret of Anjou was as authoritarian as any champion of the White Rose. Henry, as has been remarked, was 'a medieval man'; he was not without a touch of superstition and by no means inclined to neglect the tradition of awe and sancrosanctity which should hedge about a king.

He believed that his royalty came to him by descent, though, illogically, he dated his reign only from the day of Bosworth or perhaps from its eve, and began to exercise his kingly power empirically as soon as he was able to do so. Yet he believed that the unction conferred a special grace. His coronation was performed with the utmost ceremony; of the three swords borne before him, one was understood to signify his spiritual jurisdiction, and he did not hesitate, upon occasion, to attack the privileges of the clergy. He had no mind, however, to dispute the authority of Rome, and showed himself friendly to successive popes. He professed himself interested in the crusade against the Turks and in the conversion of the heathen; his every victory was celebrated by an act of public devotion; the list of his foundations,[1] though not magnificent, is respectable, and it is significant that he favoured the Observants (reformed Franciscans) whose reputation for piety stood high. He performed exactly the duties of a good churchman, and gave money unostentatiously to priests who sang masses for him. In all that he did he behaved as a Christian prince.

Along with this medieval piety went a medieval splendour. The great feasts of Easter, Pentecost, and Christmas began with religious exercises, but were continued with mirth and gaiety, and when the feast of Saint George was celebrated along with that of Easter, the magnificence was doubled. In 1488, for example,[2] there was a superb display at Windsor in which not only the king, but the queen and 'my lady the king's mother' wore the robes of the Garter, and ambassadors from Maximilian, Philip, James IV, and the duke of Brittany were present.

[1] Polydore Vergil (edition 1570), p. 617.
[2] Leland, *Collectanea*, iv. 238.

Family events were made the occasions of colourful ceremony. The preparations for a royal birth were attended by the performance of exact rites; the coronation of the queen, the christening of the royal children, the knighting of Arthur, the betrothal of Margaret, all were celebrated with great circumstance. Henry's meetings with Philip near Calais in 1500 and at Windsor in 1506 each had a background of lavish dignity, and the reception of Catherine at Westminster was marked by a noise, a brightness, and an opulence which made it remarkable even in that colourful age.[1] There was tilting in New Palace Yard, which had been sanded 'for the ease of the horses' and surrounded with stages, for the king and his court on the south, for the mayor and citizens on the north. The spectacle, which was repeated on several days, began with displays of fine clothes and good horse-flesh, but it was a display of manhood as well. Many spears were broken—on one occasion they were sharp spears—and on another occasion swords were broken as well. The harness was good, and no serious casualties are reported but there must have been sore bones amongst the competitors, who numbered in their ranks the first nobles of the land. The athletic exercises of the crowded days, interrupted on one occasion by rain, were accompanied by banqueting and revelry in the evenings. Westminster Hall was the scene of wonderful 'disguisings' in which were introduced those mechanical 'pageants' so dear to men of those times. A castle, a ship, an arbour, a lantern, and various mountains were at diverse times presented, and from each there issued a troop of reluctant ladies or admiring lords, who in the end danced gaily together. Catherine and one of her ladies 'in apparel after the Spanish guise' danced two 'bass daunces'—dances, that is, where the feet did not leave the ground—and the English princes and princesses displayed their accomplishments too. It is significant that Prince Henry, 'perceiving himself to be accombred with his clothes, suddenly cast off his gown, and danced in his jacket'. The hall was hung with arras and made splendid with a display of the royal plate; there was abundance of music and 'voydes' or collations of wines and spices; and at the end was a 'distribution and delivery' of prizes—to the duke of Buckingham a great diamond, to the marquis of Dorset a ruby, to the other lords and knights precious stones and rings of gold. The grand

---

[1] Leland, *Collectanea*, v. 356.

banquet was given in the Parliament Hall since the great hall was needed for the pageants and, perhaps because it was necessary to give places of honour to the ladies and gentlemen of Spain, the mayor and citizens were not invited. When, however, feasts were given in Westminster Hall itself, as at the coronation of Queen Elizabeth,[1] and at the Christmas feast of 1493,[2] not only the mayor and the aldermen, but other merchants and citizens were present. It is true that for the jousting in the Palace Yard 'the common people' hired seats or places at 'great price and cost',[3] but even so it is plain that Henry's hospitality and display must have been expensive. Though he was by nature parsimonious, he knew what was due to the dignity of a king, and he knew that his outlay paid a dividend in the gratification of his loving subjects, of the nobles who shared in the festivities, and of the people who beheld them from afar.[4]

Behind the outward splendour lay the inward power. This radiated from the royal person and informed all the machinery of state. It expressed itself most of all in the council, which, like the king himself, exercised an authority in administration, justice, and legislation; but inasmuch as its activities impinged upon those of the more formal institutions of government, it will be convenient to consider first the law courts and the parliament, which were at once expressions of the royal power and checks upon it.

The three great courts which sat in Westminster Hall—king's bench, common pleas, and exchequer—administered the common law and were manned by justices who, though they were the king's nominees, had had a long professional training and were conscious of the majesty of the law. They tried cases between party and party; they got more from fees than from salaries, and were therefore willing to widen their jurisdiction; but, theoretically, they sat in the king's presence and, though they were far from being royal hirelings, it was a great part of their business to assist the king in his affairs. Sometimes they sat together in the exchequer chamber to discuss and advise upon

---

[1] Leland, *Collectanea*, iv. 216.
[2] Kingsford, *Chronicles of London*, p. 323.
[3] Leland, *Collectanea*, v. 357.
[4] The information regarding Henry's feasts is taken from Kingsford's edition of the *Chronicles of London*, but chiefly from Leland, *Collectanea*, vols. iv and v.

difficult cases. Before parliament assembled they met to discuss
the efficacy of old statutes and the probable effect of new ones in
contemplation. They helped to draft bills and they advised the
council on points of law. Henry made full use of their authority,
as when he reinforced a judicial committee of council by the
presence of the two chief justices in the famous star chamber
act, and of their expert knowledge in arranging his parliamentary business. Sometimes he met with opposition as when,
in 1489, the judges insisted that the commons must share in an
act of attainder,[1] but, in the main, he profited by their sound
legal advice. It was the judges who showed him how to deal
with the attainders pronounced against himself and his supporters before he gained the throne.

Almost equal in authority to the courts of common law was
the court of chancery. This was really an equity court—a court
that is, to remedy injustice which might ensue from a strict interpretation of the common law— but by this time the chancellor,
in his judicial capacity, was something more than a keeper of
the king's conscience. His court dealt with cases which fell
outside the somewhat cumbrous processes of the common law—
for instance, it recognized uses—and, partly because it sat out
of term time, it gradually assumed a jurisdiction comparable
with that of other central courts.

The justice administered at Westminster was applied to the
whole of the kingdom. The great palatinates, save Durham,
had now fallen into the hands of the Crown, and the royal
judges often sat in them as in other parts of England. Theoretically, these judges were appointed by special commission, but
in fact it was the king's justices who executed the criminal
commissions of oyer and terminer and jail-delivery, by this
time virtually the same, and the commissions of assize[2] which
could deal with almost every form of civil business. Lower down
in the scale of authority were the sheriffs who, since they were
often allied with noble families, had come much under suspicion
during the civil wars. Already under Henry VI and Edward IV
attempts had been made to curb the sheriff's power.[3] His
tenure of office had been limited to a single year, and many of

---

[1] Pollard, *Henry VII*, ii. 19.
[2] Pickthorn, *Henry VII*, pp. 57–58, relying largely on Holdsworth.
[3] 4 Henry VI, c. 1, 9 Henry VI, c. 7, 23 Henry VI, c. 9 (renewing the limitation
of his tenure of office to one year), and 1 Edward IV, c. 2 (checking the abuses of
the sheriffs' tour).

his duties were being transferred to the justices of the peace, who at the beginning of Henry's reign were already charged with the administration of many statutes. When, before the meeting of Henry's first parliament, 'all the justices were at Blackfriars to discuss the king's business', it was recalled that in the days of Edward IV there had been compiled a list of useful statutes which was to be sent to the justices of the peace in each county, 'viz. Winchester and Westminster for robberies and felonies, the statute of riots, routs and forcible entry, the statute of labourers and vagabonds, of tokens, and liveries, maintenance and embracery'.[1] The powers enjoyed by the justices of the peace under these various statutes were considerable; as individuals, they were specially bound to arrest criminals and suspects; in their petty sessions when two or three sat together, they could deal with various misdemeanours; and in their general sessions, which they must hold at least four times a year, they could try any indictable offence except treason. Their authority was limited to their own counties. The commission issued by the chancellor always included a councillor or two, but the real work was done by the gentry of the shire, of whom some, specially mentioned in the commission, were to form a quorum in matters of importance. The nature of their duty appears from the titles of the acts which they administered. They dealt with agriculture, industry, vagabondage, and economic matters generally; but their main business was to preserve good order and to recover for the Crown the authority lost to the over-mighty subject, who retained men to his own service by the giving of liveries and signs, who maintained his dependants in the face of the law and who tried to corrupt or intimidate the officers of the Crown, especially the local juries. Local disorder was a dire legacy of the civil war and it continued long after the battle for the crown had been lost and won.

The uneasy society of the day, disturbed by changes of ownership, the presence of unemployed soldiers and incipient vagabondage, was susceptible to acts of mere power, and if a cunning man of law joined hands with the unscrupulous scion of some established house, there was almost no limit to his power of illegal action. How varied might be his misdemeanours appears from a complaint by the inhabitants of the Isle of Purbeck against Harry Uvedale of Corfe castle and his jackal,

[1] Williams, *England under the Early Tudors*, p. 169.

William Bayle of Wareham. Harry gave 'hy countenans, beryngowte and mayntenauns', William supplied 'lernyd councell and sotell practyse', and between them they took sheep, avoided liability to pay fines and to supply men to the king's army, annexed treasure trove, smuggled wool, despoiled wrecks, released felons for bribes, poached gulls' eggs, took bribes for permitting hunting on ground which was not theirs, embezzled clients' money, and terrorized complainers into flight.[1]

It was against disorders of this kind that Henry had to struggle, and the chief agents of his authority were the justices of the peace whose powers were extended in each successive parliament. An act of 1485 enabled them to arrest and examine persons suspected of hunting under cover of darkness in disguise; the famous act 'Pro Camera Stellata', which dealt mainly with the suppression of great offenders by high authority, also empowered justices of the peace to hold inquests into concealments by other inquests. The parliament of 1495 actually empowered them to hear all statutory offences short of felony without indictment, empowered them to punish sheriffs and bailiffs found guilty of extortion, gave them extended authority in economic matters and in the suppression of riots, and placed local juries beneath their supervision.[2] The legislation of 1504 added to their competence in all directions, especially in giving to them and the sheriffs the right to certify who were retainers and embracers when an intimidated or corrupted jury failed to find a riot, and in emphasizing their powers to deal with livery and retaining. It was not only directly that their authority was increased, it was, all the while, being magnified because the king enlarged by successive statutes the offences with which they might deal. An act of 1489, for example, dealt only with unlawful retaining of the king's tenants; the act of 1504 dealt with retaining generally.

While their powers were thus lavishly increased, the justices were brought more and more under the control of their royal master. The council was vigilant in its surveillance, and as councillors were named in each commission, it was well informed. The courts of common law could control the quarter sessions by writs of *mandamus*, for omissions, and *certiorari*, for

---

[1] Campbell, *Materials*, ii. 255–6; Gairdner, *Letters and Papers*, ii. 75.
[2] *Statutes of the Realm*, ii. 589.

things done; the Crown was at pains to inform the public of the means of remedy against justices who exceeded their powers. No doubt it feared lest its own servant, the justice of the peace, might himself become the tool of some local aristocrat, and the provisions of an act of 1489[1] exhibit an almost excessive distrust of the agents on whom it needs must in great measure rely. The justices were commanded, under penalty, to promulgate four times a year, in four principal sessions, a proclamation which, after setting forth the disorders of the realm and the royal efforts to cure these, bluntly asserted that the laws had not been put into execution on account of the failure of the justices of the peace. The justices were therefore commanded to do their duty and at once to report all persons who put obstacles in their way. The public were informed that if any man were abused by a justice he could seek remedy from a neighbouring justice, then from a justice of assize, and, in the last instance, from the king or his chancellor.[2] The expressed intention of the statute was 'that the king's subjects might live in surety under his peace in their bodies and goods'; the promotion of good order would certainly benefit the public, but it would also strengthen the royal authority. In making his laws effective both in the central courts and in the provinces, Henry, without departing from established precedent, materially increased the power of the Crown.

Henry's dealings with parliament exhibit the working of the same principle—the old institution was turned to a new end. Parliament, though it did not bulk so largely in the machinery of government as it does now, was already an important institution in the days of Henry VII; it was still a function of the royal council, it was still a high court; but it was already taking on the status of a grand jury of all England, and though it did not meet regularly or sit very long, its voice was one that must be heard. In the parliament-chamber sat only the lords spiritual and temporal. The two archbishops, the nineteen bishops, and twenty-eight mitred abbots were summoned,

---

[1] 4 Henry VII, c. 12, *Statutes of the Realm*, ii. 536.
[2] The retainder of the king's servants was condemned by an act of 1487; retainder generally by an act of 1504 (ibid. 522, 658), and the evils of maintenance and embracery are mentioned in the acts against riots of 1495 and 1504 (ibid. 573, 657), as well as in the act of 1495 against perjury (ibid. 589), and the famous act 'Pro Camera Stellata' of 1487 (ibid. 509).

though few of the abbots came,[1] and an act of 1513 declared that their presence was not necessary.[2] The attendance of the temporal lords is less capable of exact definition; indeed, a great authority has described the English peerage as a fiction.[3] Since Richard II had begun the practice of creating peerages by letters patent which conveyed heritable estates,[4] the idea had inevitably grown that a place in parliament depended upon tenure by barony; but, in fact, the lay lords attended because the king summoned them, although the right to summons had come to be attributed to the holders of certain titles. Death and attainder had thinned their ranks, and to the first parliament of Henry VII only twenty-nine were summoned; but the decline in number was not permanent, for though the first Tudor made only five new creations, his later parliaments were attended by some forty lay peers.

The commons, though theoretically outside parliament, were by this time an integral part of it.[5] From each of thirty-seven shires came two knights, elected, since 1430, by the forty-shilling freeholders; these were not always *gladio cincti*, but they were substantial men, sometimes connected with the great families, and it was they who took the lead in the *domus communis*. Along with them sat 224 representatives of the boroughs. They were chosen by an uncertain franchise, but this was usually in the hands of an oligarchy of traders and wealthy craftsmen[6] in close relation with, if not identical with, the administrative authority of the town. Their influence in parliament grew with the growth of the economic factor in politics and, in 1533, a burgess, Humphrey Wingfield of Yarmouth, was chosen speaker. The knights and burgesses in the common house represented some of the most solid elements in English society. Although they obtained access to the parliament-chamber only in the person of their speaker, except on special occasions, they had already gained considerable power. With them lay the initiation of money bills and of many other bills too, though a bill *formam actus in se continens* was, strictly speaking, less proper than a petition. Procedure,[7] including that for the election of

---

[1] Tanner, *Tudor Constitutional Documents*, p. 514.
[2] Pollard, *Parliament*, p. 207.   [3] Ibid., chapter 5.
[4] Professor Pollard pointed out that the letter patent was not truly the instrument of the creation but only evidence of it.
[5] Pickthorn, *Henry VII*, pp. 96, 111.   [6] Ibid., p. 100.
[7] A transcript of the *Modus* is prefixed to the *Lords' Journals*, which began in

a speaker, was definitely fixed; it has been shown that the rule laid down in the *Modus Tenendi Parliamentum* was carefully preserved by the clerk of the parliaments, and a contemporary report by a Colchester burgess shows that this rule was substantially observed.[1]

Parliament met when the king summoned it, and usually he summoned it when he wanted to increase his ordinary income by a special grant; normally, it sat only for a few weeks, though Edward IV had kept his parliament of 1472 in being until January 1475 by means of six successive prorogations.

It is characteristic of Henry that he realized the importance of parliament and, without seeming to make any innovation, used its strength for his own ends. The machine lay ready to his hand. The clerical peers inclined towards him as the upholder of law and order and over them he had a real measure of control by the operation of his business-like arrangement with the *curia*. He saw that Rome got her dues and Rome appointed the men whom he preferred—the statutes of 'Provisors' and 'Praemunire' were conveniently forgotten. The lay peers who survived to attend his parliament were not unnaturally his partisans, and so secure was he that he did not hesitate soon to summon some of his erstwhile enemies, the powerful Surrey among them. There is no evidence that he packed or tried to pack the commons, though he did keep an eye on local elections. On the ground that the elections at Leicester had been disorderly, he directed by his own writ that henceforth the choice should be in the hands of the bailiffs and forty-eight of the 'moost wise and sadde commons',[2] and he saw to it that the burgesses of Reading were duly paid for their attendance at the customary rate of 2s. a day.[3] There was little need for him to interfere with the popular choice; once parliament was assembled, he could get his own way. With the aid of the judges, he had already prepared his programme; the lords were his own supporters, sitting under the presidency of his officer, the

---

1510, and an English version was in a 'little old parchment book', kept by the clerk of the parliaments. *Transactions of the Royal Historical Society*, 3rd series, viii (1914), 36, and *English Historical Review*, xxiv. 213. For formal procedure see *Rotuli Parliamentorum*, *passim*.

[1] Williams, *England under the Early Tudors*, p. 147 (from the *Red Paper Book of Colchester*).

[2] Campbell, *Materials*, ii. 456.

[3] Williams, *England under the Early Tudors*, p. 133, from the *Reading Records*.

chancellor; the speaker was, if not a royal nominee, a person entirely agreeable to the king. Among the speakers of Henry's reign were Thomas Lovell, Richard Empson, and Edmund Dudley; of the other four, three had careers in the royal service, and Robert Drury, not otherwise distinguished, had among his sponsors Sir Reginald Bray, Sir Richard Guildford, and Richard Empson. The speaker received a 'reward' of £100 for his services, and the records show that every man who held the office tasted in other ways of the royal bounty. The real reason, however, of the harmony between king and parliament was that the king, in the main, did what the commons wished. They did not like granting money; but they were not asked to make excessive grants; special subsidies were always justified by special occasions, and the commons, who in the end must pay, were often given a *douceur* in the redress of economic grievances. Because he and his servants conducted their business with skill, and because his programme was in general accord with the wishes of his people, Henry had immense power in parliament. Much of his legislation came from his own initiative; some of the statutes begin with 'Prayen the Commons', but others begin 'The king oure sovereign lorde calling to his remembraunce', without any hint that the royal memory had been jogged by the faithful subjects.

He could, as in the attainder of 1485, bear down opposition by sheer authority. He could, and throughout his reign he did, unchallenged, add 'provisos' of his own to measures which parliament had already passed; on one occasion, at least, his action was so arbitrary that the unfortunate clerk of the parliaments did not know to which act a proviso should be attached. So complete was his hold that in 1504[1] parliament authorized him to repeal acts of attainder at his pleasure, forestalling in spirit the famous act whereby Henry VIII was enabled to dispose of the Crown at will.

An analysis of the statutes of his reign shows that his control of parliament was exercised towards the achievement of definite ends. Apart from various acts which replenished the royal coffers—acts of resumption, of attainder, and for occasional subsidies—most of his measures fall into two categories, acts for the establishment of good order, and acts for the control of trade and industry. The general effect of the acts for the pro-

[1] Pollard, *Henry VII*, ii. 17.

motion of good order was to strengthen the hands of the council at headquarters and of the justices of the peace in the provinces. Not unnaturally they commended themselves to the knights of the shire who represented the class from which the justices of the peace were drawn. In like manner, his economic legislation commended itself to the merchants and wealthy traders represented by the burgesses. Henry believed in governmental control. As regards trade, he was a mercantilist; he set the interest of the merchant company and the Staple before that of the individual trader, and he set the interest of the manufacturer before that of the consumer. Yet he defended the consumer by passing many regulations to ensure that the articles supplied to him were of good quality, he endeavoured to confine each industry to its own field— tanners were not to be curriers nor curriers cordwainers. Finally, he was at pains to see that if profit were to be made from the English consumer it should be made by Englishmen and not by aliens. He kept a close eye upon bullion and upon currency, and was the first to bring the actual coins used into relation with the denominations employed in accountancy. He was the first to coin the sovereign of 20s. (1490) and the shilling of 12d. (1504).

Parliament, which in the days of the 'constitutional experiment' had served as an 'opposition', became now the ally and partner of the Crown. While the royal authority was thus increased in the established machinery of government, the great feature of the reign was the expansion of the uncanalized power of the Crown. Part of it was diverted into new channels which began to attain for themselves recognition alongside the older institutions, but part of it still remained in the hands of the king himself who could use it in emergency. Its chief instrument was the council, which reflected every facet of the royal might and which served less to advise the king than to ensure that his wishes were carried out. This council is hard to define. It sometimes took the form of a great council which was summoned on special occasions as in 1485, 1487, 1488, 1491, and 1496.[1] Normally, however, it was something smaller and more permanent than the great council, yet far bigger than the little group of trusted advisers to whom, somewhat later, the name

---

[1] Steele, *Tudor and Stuart Proclamations*, i. LXXVI, LXXVII.

of privy council was given. The Renaissance prince had few real confidants—'the king's mule carried king and council both'; on the other hand, as the work of government became more centralized and more comprehensive, he needed an increasing number of administrative officers.

The effective core in some ways resembled a cabinet. Its members had, most of them, shared in the king's exile. There were three or four churchmen and a few territorial magnates, there were some plain knights and some men without title; but amongst them they held, or came to hold, the great offices of state and household which controlled the government of England. The king used them to lead his armies, to build and sail his ships, to look after his ordnance, to supervise his administration, to rule his finances, to conduct his embassies abroad, and to do justice at home. Associated with these men in their various activities were other councillors who, taken collectively, might be described as an administration. The general body of councillors, so far as is known, never sat together and its outer fringes were indistinct.[1] Technically, the taking of the councillor's oath should be the test of membership and, under Henry VII, the receipt of a fee of £100 a year; but the king was not compelled either to rely solely on sworn councillors or summon to any particular meeting all who were sworn. Now and again, as in 1504 when 41 members attended,[2] the council must have resembled a great council, but the average attendance as computed from the surviving *sederunts* was usually between 6 and 10. From all sources of evidence, the names of 172 councillors appear, of whom 12 are not known to have sat elsewhere than in the court of requests, while others are of relatively small importance. Polydore Vergil, from whom Hall borrowed, gives a list of 17 councillors (including 3 bishops), but adds the names of 37 *sapientes* (including 6 churchmen) whom the king *continenter* summoned to his councils and remarks that there were others besides. The proclamation of Perkin Warbeck recites the names of 19 councillors, 11 of whom do not occur in Vergil's list, and mentions others, 'caitiffs and villains of simple birth'. It is

---

[1] Tanner, *Tudor Constitutional Documents*, p. 213; Pickthorn, *Henry VII*, p. 28; Pollard, *Henry VII*, iii. 315, and *Wolsey*, p. 105; *English Historical Review*, xxxvii. 337, 516; Gladish, *The Privy Council under the Tudors*. Conway, *Henry's Relations with Scotland and Ireland*, app. xli, notes the occurrence of 107 council meetings and the survival of 46 *sederunts*.

[2] *Calendar of Patent Rolls, 1495-1509*, p. 388.

plain that the lists include councillors who attended at different dates, and none of them can be reckoned complete: that of Vergil, for instance, omits the chancellor (Alcock) and the keeper of the privy seal (John Gunthorpe), and the offices held by the councillors are sometimes mentioned and sometimes not.

Vague as it was in composition, the council was none the less a definite entity. It had a clerk who was nominally the *secundarius* in the privy seal office, and towards the end of the reign a president sometimes appears, though his office was the less necessary because the king presided himself when he attended. The chancellor, treasurer, lord privy seal (who could issue writs and warrants), and latterly the president, acted in a general way as an executive committee; but the council was a unity and its true function was to advise. Even when, as will be shown, the work of the council was divided up for greater efficiency, the *disjecta membra* were still understood to be royal councils, and in every one of them the king in his own person was supposed to be. The various bodies to whom special duties were entrusted gradually took on the appearance of separate institutions, but definition was slow of growth and never absolute. The new developments were dictated by convenience, by the sheer necessity of governing at the same time different parts of the realm and of administering different kinds of justice. None of them was, in fact, quite new; several had Yorkist precedents and others had precedents which dated to an earlier age than that of York. They represent the gradual hardening into a system of expedients adopted to meet particular ends.

There had long been a tendency to distinguish between the councillors who remained about the king *ubicunque fuerit* and those who remained in his capital, and in Henry's reign there is a clear differentiation between the council attendant and the council remaining at Westminster. This differentiation no doubt helped to develop the body which was later known as the privy council, but it must not be exaggerated; the council attendant was not always composed of the same persons, and business begun before it might well be continued before the council at Westminster.

The provision that the king should always have about him reliable councillors did not suffice for the administration of remote parts of the realm. His progresses seldom took him north of the midlands, and latterly he was much at Woodstock and in Windsor when he was not in London. For dealing with Wales and the Marches, he adopted a plan tried by the Yorkists as early as 1471, namely, the establishment at Ludlow of a council nominally under the presidency of the prince of Wales. Early in his reign, Henry issued commissions to hear and determine causes in Wales and the Marches thereof, and in March 1493 he included in such a commission the name of prince Arthur who was then six years of age. It is probable[1] that his representatives were at this time appointed a council, and certain that a council was in being by 1494. A definite organization is evident in 1501, when Arthur, soon after his marriage, went down to Ludlow to preside over a council of ten, some of whom are elsewhere mentioned as councillors; but the death of the prince in April 1502 seems to have interrupted the development, and though the council may well have continued, little is known of its activity. A Welshman himself and husband to a descendant of the Mortimers, Henry encountered little difficulty in Wales and the Marches.

The uncertain North gave more trouble. Here the over-mighty subjects, far removed from the royal control, were able to unite feudal privileges with an influence which rested on 'the olde good wyll of the people, deepe-grafted in their harts to their nobles and gentlemen'.[2] As their land offered neither much agriculture nor much industry to absorb the surplus population, their authority readily expressed itself in terms of man-power. Obviously they were a danger to the king, but the king could not dispense with them, since it was they who defended the frontier against the Scots. Rather was he compelled to entrust to them the machinery of government, making them wardens of the Marches, justices in eyre of the forests, justices of oyer and terminer, and justices of the peace. The king-maker, who inherited from the Percies after their defeat at Towton, collected a huge complex of offices, but before long many were concentrated in the hands of Richard of Gloucester who, from his castle at Middleham, wielded an immense power. When war broke out with Scotland in 1482, Gloucester was made sole king's lieutenant in the North, and as his capture of Berwick increased his reputation and his popularity, he established a clear precedence over Northumberland. He had had a council

---

[1] Skeel, *The Council of the Marches*, p. 30.
[2] Reid, *The King's Council of the North*, p. 21.

to help him to manage his estates, and now the council began to resemble that of a royal officer rather than that of a private man. When Richard, having gained the crown, left it still to administer the North, it took on the appearance of a royal council, and when Richard died at Bosworth his conqueror had a useful instrument ready to his hand. He could not at first use it; he was compelled to trust Northumberland who had been his ally. When, however, Northumberland was killed in 1489 Henry was quick to seize the opportunity. He appointed the infant Arthur as warden, giving him as lieutenant the competent Surrey, newly released from the Tower; Surrey had a council, but the king was careful to afforce it with official members, and before long it was not only administering the law but acting as a sort of star chamber and court of requests.

While the royal council was thus expanding in terms of geography, it was expanding too in its activities. It helped the king in every department of his government, it issued proclamations on all sorts of matters, saw to the defence of Calais, made appointments, received ambassadors, discussed armaments, superintended local administration, and saw to the suppression of offenders against the king's peace. Government being closely allied to justice, much of its work was necessarily judicial, and a machinery already operative in the days of the Yorkists was given a new precision and a new force. Its judicial capacity was most obvious when it sat in the star chamber, a building near the Receipt of the Exchequer at Westminster, which had been used for council business ever since the days of Edward III. The powers which it exercised were not new, but in Henry VII's time it sat regularly, at least four times a week during the legal terms, and it dealt with all kinds of cases, particularly with those which concerned good order. Theoretically, it should abstain from matters involving life and limb and freehold and from appeals in error from the courts of common law; but as it did not hesitate to instruct judges of other courts to interfere with actions and control the findings of juries, its arm was very long. Much of its business came to it by way of petition, but much was initiated by the Crown itself, by writs under the privy seal, and as petitions and 'privy seals' were not hampered by the formalities which surrounded the processes of common law, its procedure was rapid and effective. Often it remitted the cases of individuals to other courts with instructions as to how they should be decided, but on matters affecting the king's interest it kept a jealous eye. Some of its judicial powers seemed oppressive; it might imprison both parties pending a decision, or restrain them by bonds; it might examine parties as well as witnesses upon oath, and use the admissions of one co-defendant against others; it might refuse to disclose an accusation before a case began and it might refuse counsel for the accused. If it left life and limb intact, it could imprison, dismiss from office, and impose fines beyond a man's capacity to pay. Summary as its procedure was, it was scarcely harder than that of the common law, and the fact that it dealt with strong injustice and the great offender made it popular.

Although the star chamber was virtually identical with the council—its clerk was the clerk of the council—it did not execute the whole of the council's judicial power. Always there remained that reserve of authority which in later ages was called prerogative, and the king could invest his council, or part of it, with special powers to deal with particular matters. The famous act of 1487, to which some scribe of Elizabeth's day appended the name *Pro Camera Stellata*, is a case in point. Convinced that the attack on the over-mighty subject was not pressed home, and possibly suspicious that among great offenders were those of his own household,[1] Henry gave to a powerful committee of council special powers against 'unlawful maintenance, giving of liveries signs and tokens and retainders by indenture promises oaths writing and otherwise, embraceries of his subjects, untrue demeaning of sheriffs in making of panels and other untrue returns, taking of money by juries, great riots and unlawful assemblies', which things, he said, were virtually unpunished to the prejudice of all the subjects and the great displeasure of almighty God. The chancellor, the treasurer, and the keeper of the privy seal, or two of them, along with a bishop and a temporal lord of the council and the two chief justices, or failing them, other two justices, were authorized upon bill or information on behalf of the king or of a private person to call the misdoers before them by a writ of privy seal, to examine them and punish those found guilty according to the existing statutes. In 1493 it was held that the only judges appointed by the act

---

[1] An act of 1487 had already authorized a special inquiry into conspiracies by the king's servants to murder the king or his councillors or great officers. *Statutes of the Realm*, ii. 521.

were the chancellor, treasurer, privy seal, or two of them, the other persons being merely assessors,[1] but whether this opinion were adopted or no, it is patent that the court so constituted, sitting in private, with a definite personnel in an indefinite place and at an indefinite time, was entirely different from the court of star chamber which sat in public, in Westminster during term time and which dealt with all kinds of matters without respect to the limitations imposed by the act.

How long the committee maintained a separate jurisdiction is uncertain. There is no particular reference to it in an act of 1504[2] which mentions star chamber, king's bench, and council attending, and the fact that the lord president of the council attending was added to the other three great officers by an act of 1529[2] may suggest that the powers given by the act of 1487 had been associated with the council attending rather than with the star chamber. This act of 1529, however, may reflect the antipathy felt against Wolsey, who had tended to ignore the council attending and do all in the star chamber, whose powers he extended so much that later ages regarded him as its creator. There can be little doubt that the councillors, whether sitting in the star chamber or elsewhere, availed themselves of the act of 1487, and the mistake of the Elizabethan clerk and of the Long Parliament in 1641 is readily comprehensible.

The jurisdiction exercised in the star chamber or in the powerful committee did not exhaust the powers of the council to administer justice. A delegation of the council, named the court (at first the 'council') of requests, was a poor man's court of equity. This institution had taken definite form under Richard III when a special clerk was appointed; in the reign of Henry VII, under the supervision of the privy seal, it became increasingly active, and about the middle of the reign, special masters of requests were appointed[3] to ensure that suitors were not neglected while the attention of councillors was directed to other affairs.

Informed by the power of the king, manned by his servants, with its branches established in Wales and in the North, with its jurisdiction expanded into new forms but never exhausted,

---

[1] By judge's interpretation. Pollard, *Henry VII*, ii. 57 (quoting *Year Books*, 8 Henry VII, p. 13).

[2] *Statutes of the Realm*, ii. 659, iii. 304.

[3] Selden Society, *Select Cases in the Court of Requests, 1497–1569*, xii and xvi.

the council was the main instrument of the royal power. Yet, as in his dealings with law courts and parliament, Henry introduced nothing that was absolutely new. All he did was to endow existing organizations with more precision, and above all with more force.

Whence came the royal power? As already shown, it came partly from the exhaustion of competitors, partly from the general desire for peace, and partly from the fact that the king carried his people with him in his attempts to restore order and develop husbandry, industry, and trade. Yet it is obvious that the king had at his disposal some military power. This will be found to be, in appearance at least, surprisingly small. In England the 'new monarchy' cannot be said to rest upon a 'standing army'.

The yeomen of the guard, who made their appearance in 1485, formed a small personal guard for the king. It is said that they were formed upon a French model, but, if so, it must be upon *la petite garde*,[1] for they were yeomen, not gentlemen, 'good Archers, as of diverse other Persons, being hardy, stronge, and of Agilitie'. It has been suggested that the scarlet of their uniforms symbolized the dragon of Cadwaladyr; certainly, it set them apart from ordinary soldiers, who were generally 'white coats', though in Tudor days special troops sometimes wore white and green. There is no record of their having taken any part in the warfare of the reign. Empson was accused of having used a number of them to carry out an unwarranted act of force against Sir Robert Plumpton in 1501,[2] and in 1509 they conducted the body of their dead master to its tomb.

Of more practical importance was the artillery. The Tower had long been an arsenal, and the office of 'master-general' of the ordnance was created in 1483. But although the king had gunners with him at Kenilworth whilst he awaited Perkin's attack,[3] it does not appear that cannon played any part in Henry's English warfare. Guns were freely used in Ireland,[4] but there is no evidence that they played any part in Morley's adventure at Dixmude, and although Henry had artillery at the siege of Boulogne its success was not conspicuous. The

---

[1] Pegge, *Curialia*, p. 4.   [2] *Plumpton Correspondence*, cvi–cix.
[3] Conway, *Henry's Relations with Scotland and Ireland*, p. 56.
[4] Ibid., pp. 8, 78, 85.

ultima ratio regis was little used by Henry who, like other English soldiers, thought the old ways best. It may be remarked that he legislated against the use of cross-bows,[1] and kept down the price of yew wood,[2] much of which was imported from Italy. His trust was in the English archers, not 'regular soldiers' but strong-armed ploughmen who could be called out by commissions of array when necessity arose. As soon as the Breton question seemed likely to lead to war, he arranged for the calling up of his bowmen in the old way. On 23 December 1488 he issued commissions[3] for musters in every shire. The sheriff was on each commission, but at its head was always a trusty nobleman, and, indeed, reliable soldiers like Oxford and Welles served in more than one county.

But it was only for war on a grand scale that the king depended on shire-levies, and even when he did so he used also the method of indenture which had been introduced by Edward II. He bargained, that is to say, with nobles and gentlemen to provide him with troops who should be paid at a fixed rate—18*d.* a day for a lance with his attendant 'custrell'[4] and page, 12*d.* for a spear on horseback with his custrell and for a master gunner who wore half armour, 9*d.* for a demi-lance, 8*d.* for a mounted archer or bill, 6*d.* for an archer or bill upon foot, a gunner, and a drummer, and 10*d.* for a trumpet. Of the trusted Gilbert Talbot, for example, Henry demanded 80 mounted men in 1493 and 120 in 1497, and on the eve of his expedition to France he made a whole series of indentures with his nobles and gentlemen which would produce for his army a hard core of well-armed men.[5] An exact date was fixed for the muster, and the leaders were promised conduct money at the rate of 6*d.* for every twenty miles between the men's homes and Portsmouth; exact arrangements were made for monthly pay, and the king appointed a treasurer of the war to superintend the whole finance. For the maintenance of discipline he appointed Sir Robert Willoughby de Broke marshal, strengthening his hand by a statute which condemned alike captains who defrauded their men of pay and soldiers who deserted their captains; he undertook to embody his regulations in a book of Statutes and

---

[1] *S.R.* ii. 649, 19 Henry VII, c. 2.
[2] *S.R.* ii. 521, 3 Henry VII, c. 13.   [3] *Foedera*, xii. 355.
[4] A custrell was, by derivation, a dagger-man, but the word had come to mean an armour-bearer, a general attendant upon the man-at-arms.
[5] *Foedera*, xii. 477.

Ordinances of the War. For the conduct of the army in Ireland, the same system was adopted. The king sent a relatively small force of all arms and supplemented it by enlisting kerns and gallowglasses locally. Troops, in garrison or otherwise, on a semi-permanent footing, were distinguished as a 'retinue', but their number was small.

Henry's naval strength was organized in much the same fashion as was his land army; he had as a nucleus a small permanent force, and this he expanded when occasion arose, by the hiring of merchant-ships and the enlistment of crews. The permanent nucleus for sea-service was greater than that for the land. This was natural since, for an island power, control of the sea was of the first importance; indeed, the keeping of the sea was always mentioned as among the reasons for a grant of tunnage and poundage. The considerable navy collected by Henry V had almost disappeared in the troublous reign of his son, but Edward IV had laid the foundations of a new fleet, part of which descended to the Tudor king. Of the four ships which came down to him, all disappeared or were reconstructed within a few years and Henry added only six vessels in all.[1]

He built, however, for himself, and the *Regent* and the *Sovereign* were big ships for their day. Theoretically, the work was under the control of the clerk of the ships, but he was responsible to the council, and in fact it was two councillors, Guildford and Bray, who exercised the authority and possibly obtained some financial benefit. The *Regent* was modelled on a French ship, seen by Henry in his exile, and was designed to be of 600 tons; she was built on the Rother under the superintendence of Guildford and was ready for sea by 1490; she was a four-masted vessel; the fore and the main were equipped with topmasts and the main carried a topgallant as well. In manœuvre she must have been awkward for she boasted a summer-castle as well as a poop and a forecastle with two decks, and it was in her upper works that many of her guns were carried. Of these, she had 225 in all, a good proportion being Serpentines, mostly of iron, though a few were of brass. The Serpentine was the biggest naval gun

---

[1] *English Historical Review*, xxxiii. 472, C. S. Goldingham, *The Navy under Henry VII*, and *Accounts and Inventories of Henry VII*, edited by M. Oppenheim, Navy Records Society, 1896. The king's barque mentioned in the Breton war may be another small vessel built by Henry.

then in use, and it weighed probably about 300 pounds; it was a breech-loader, but the ball that it fired must have weighed only about a quarter of a pound, and it was discharged with perhaps a quarter of a pound of powder.[1] The purpose of the guns, therefore, was less to sink an enemy than to sweep his decks and destroy his rigging, and it is significant that the *Regent*, like all other warships, carried bows and sheaves of arrows as part of her armament. The *Sovereign*, constructed about the same time at Southampton under the vigilant eye of Bray, was a little smaller. She carried only 141 guns and had no topgallant mast. For each vessel three boats were provided. The great boat with masts and sails was usually towed astern; the cock-boat and the jolly-boat were hoisted inboard, though not by the davits which were used for the numerous anchors. It was probably that he might maintain these sea-monsters in good order that Henry built the first dry dock ever made for the royal navy. This was constructed, not without some difficulty, at Portsmouth between 1495 and 1497. In the latter year the king added two new vessels to his little fleet, the *Sweepstake* and the *Mary Fortune*. These, however, were small, and after they were completed he built no more. Possibly he found that construction was expensive; possibly he felt that he could always increase his strength, when he wished to do so, by recourse to the merchant fleet with which his relations were intimate—he used to hire out even his great vessels to merchants.

He never fought a naval war as we understand it, but when, as for the Breton war and the Boulogne expedition, he needed ships to control the seas, he hired merchant vessels at the rate of a shilling a ton a month. These he equipped with a few guns and many arrows from his store at Greenwich, some with gunners and with a body of soldiers under a captain. It was the master and his own crew who sailed the ship, and they were paid at regular rates. These were high for the times. The master had 3*s*. 4*d*. a week, the bosun 1*s*. 8*d*. to 2*s*., the gunner 1*s*. 3*d*. to 1*s*. 10½*d*., and the cook 1*s*. 3*d*. to 1*s*. 6*d*.; the seaman had 1*s*. 3*d*. a week at sea and in harbour; boys got from 6*d*. to 9*d*. Generally speaking, the soldiers aboard in time of war outnumbered the sailors in a proportion of five to three. Relying, as he did, upon

[1] Ibid., p. 19, n. 2. The Elizabethan Serpentine weighed about 400 pounds, fired a 5⅜-ounce ball with about 5¼ pounds of powder.

hired or impressed ships, Henry was at pains to develop the merchant navy; he gave a bounty for the building, or the purchase from abroad, of ships fit for the Crown's service—ships, that is, of a minimum of 80 to 100 tons. So doing, he followed the example of Ferdinand, and it has been observed that a good proportion of the ships which he hired came from Spain. It does not follow either that the Spanish ships were better or that they were cheaper than English vessels; Henry may have felt that the ships of his own land were best employed on their proper business of trade. At sea, as on land, Henry had no great 'regular' force; his strength depended, in the end, upon the goodwill of his people.

Obviously it was not by military power that Henry gave to his government its compelling force. His strongest arm was money. Like Cosimo dei Medici, whose taxes were his daggers, he used the monetary weapon not only to ruin his enemies, but to increase his own strength, and it is in the field of finance that his achievement is most remarkable.[1] An English king normally drew his revenue from his lands; from the customs, which since the middle of the fourteenth century had always been supplemented by the parliamentary grants of tunnage and poundage and the wool-subsidy; from the farms of shires and towns; from the dues which came to him as the head of the feudal system, and from money which by arrangement with the papal curia he obtained from the church—he took the temporalities of a dead bishop into his hands *sede vacante* and made his successor pay for their 'restitution'.

Only occasionally did he obtain from parliament power to levy direct taxation, and since this was computed upon an old and unreal assessment, the revenue obtained, even when supported, as often it was, by a clerical subsidy, was small. The 'fifteenth and tenth', supposedly a fifteenth of movables in shires and a tenth in towns, had since 1334 settled down to a levy on real estate which produced always the same gross figure, about £39,000; allowance, however, was made for

---

[1] See Frederick C. Dietz, *English Government Finance, 1485–1558* (University of Illinois Studies in the Social Sciences, vol. ix, September 1920, no. 3); A. P. Newton, 'The King's Chamber under the Early Tudors' (*English Historical Review*, xxxii (1917), 351), and in *Tudor Studies*. Tables for the customs' returns under Henry VII and Henry VIII are in Schanz, *Englische Handelspolitik gegen Ende des Mittelalters*, ii. 37 ff.

decayed towns, and the yield of the tax steadily fell. Peers did not pay on their demesne land, and the landless did not contribute. When a fifteenth and tenth was voted the king knew that he would get about £30,000; every area knew exactly how much it was expected to produce and divided out its liability amongst individual tax-payers, each of whom knew what he was expected to pay. If more than one fifteenth and tenth were granted, every contributor still paid *pro rata*. Such a system had the benefit of simplicity, though there were quarrels about the rebates, but it took no account of fluctuations in the possession of wealth, and the forced loans or benevolences exacted by Edward IV represented an attempt, not universally unpopular, to utilize fairly the taxable capacity of the kingdom. Certainly the money which, from all sources, reached the exchequer of receipt was well guarded, but the machinery employed, tellers and tallies, pells and pipes, was cumbrous and expensive; there was no sure means of controlling the revenue from the provinces (except in the case of direct taxation), and no means of discovering and exploiting new sources of supply.

When Henry came to the throne, the royal finances were at a very low ebb. The revenue from the crown lands had fallen, for Edward IV had alienated much of what he confiscated, and what remained was charged with pensions and leases; the customs, which had averaged £47,000 under Henry IV, produced only about £20,000 under Richard III, and the farms of towns and shires yielded only £2,500[1] instead of the £17,000 of Henry VI's day. The fifteenth and tenth still produced something like its constant figure, but benevolences had been condemned by an act of Richard III. From the very moment that he came to the throne, Henry began to set things right. He used no wonderful alchemy; he merely increased revenue when he could and scrutinized expense with an unremitting care. Only towards the end of his reign, when money had become something of a god to him, did he use the machinations associated with the names of Empson and Dudley.

The figures speak for themselves. Thanks to his great act of resumption, to the cancellation of grants and leases made by Richard III, and to the strict economy of Bray, who was made chancellor of the duchy of Lancaster, the crown lands, which in his very first year yielded £13,633, rose steadily

---

[1] Dietz, p. 12.

in value; in the year 1504–5 the receipts from the crown lands and wards' lands in the king's hands reached the total of £32,630.[1] The rise of the customs receipts was less spectacular. Henry had the means of increasing revenue by the conclusion of trade-treaties, by the laying of special imposts upon the commodities of opponents (as when he enlarged the duty on the Levantine wines imported by Venetians), and by altering the relation of the quantity to the value of any commodity by the issue of a book of rates. None the less, the customs revenue, though it rose sharply after his accession and averaged £32,950 during his first ten years, did not thereafter expand rapidly. For the last ten years of the reign it averaged over £41,000.[2] These figures, however, include the returns from wool paid by the Staplers at Calais and locally spent; they include also the returns from Newcastle and Hull which were spent on Berwick. For the year 1505–6—rather a lean year—the exchequer received only £27,597 plus £893 for goods confiscated at ports, and against that must be set the expense of collection shown as over £2,700.

The figures are not easy to reconcile, but it is at least obvious that the mercantilist policy did not bring in great revenue by indirect taxation; the use of the economic weapon in politics is not necessarily profitable in finance, and Henry's quarrel with the Netherlands was probably disadvantageous to both sides. Yet it must be noted that although the king issued a new book of rates in 1507, he does not seem to have made any startling profit, and it may not be unfair to suppose that he was more interested in the expansion of trade than in the collection of customs. As Bacon says, 'being a king that loved wealth and treasure, he could not endure to have trade sick'. He was something of a trader himself; he hired out his ships; lent great sums to the merchants—£87,000 between 1505 and 1509; he sold alum—in one single transaction he sold alum to the value of £15,166. 13s. 4d. to Lewis de la Fava. Whilst he kept abreast of new ideas, he expanded his old revenues.[3] He kept a strict eye upon his feudal dues, not only levying distraint of knighthood on all who possessed £40 a year in land in 1486, 1500, and 1503, but exacting every year after 1494[4] penalties

[1] Dietz, p. 82.
[2] Schanz, *Englische Handelspolitik gegen Ende des Mittelalters*, ii. 46; Dietz, p. 80.
[3] Dietz, pp. 24, 45.   [4] Ibid., p. 27.

from those who did not become knights of the Bath, and levying in 1504 two 'aids', one for the marriage of his daughter, the other for the knighting of Arthur, who was already dead. There can be no doubt that he employed in England the system which he certainly used in Ireland[1] of scrutinizing closely all transferences of land, following up the writ *diem clausit extremum* with a writ *ad melius inquirendum* in cases of doubt. From the bishops he gained more than his predecessors by the simple device of elevating even his favoured ecclesiastics from benefice to benefice, gaining his profit on each promotion.

All these ways of raising revenue were merely adaptations of old methods,[2] but to them Henry added a system which, if it was not in itself new, was rather a perversion than an improvement, and became in the end an abuse. He used the royal power of punishing or of remitting punishment to fill his coffers, and some of the means he adopted are little short of disgraceful. From the very first he had punished by fine, and the opportunities provided by Perkin's conspiracies and by the Cornish risings were turned to good account. Towards the end of his reign he was employing fines ruthlessly against nobles and against wealthy citizens. Sometimes he had justification, as when he demanded £2,000 from the faithful Daubeney[3] in respect of money for which he had failed to account as captain of Calais; the famous fine of £10,000 imposed upon Oxford for keeping too many retainers, if it was ever imposed, was legally at least justifiable, though Oxford had served the king so well in arms that something might have been overlooked to him. There was justification in law for the fines imposed on Northumberland (£10,000), Fitzwalter (£6,000), Burgavenny (£5,000), and Conyers (£1,000); but the sums demanded were excessive even when, as in the case of Northumberland, only half was demanded at once. With regard to the fines imposed on lord mayors and aldermen[4] 'for their offences in office' the chroniclers are very guarded, but it is plain that they thought the king's action high-handed,[5] and indeed Henry seems to have been exacting benevolences under colour of law.

His exactions have been regarded as part of his attack on the

---

[1] Gairdner, *Letters and Papers*, ii. 66.
[2] Dietz, p. 31.
[3] Lansdowne MSS. 127, f. 34 (Edmund Dudley's Account Book).
[4] Dietz, p. 43.
[5] Kingsford, *Chronicles*, pp. 205, 261, 262.

over-mighty subject. It is pointed out that Empson and Dudley were not specifically denounced by contemporary writers; that their ill fame rests mainly on the account of Polydore, which was embroidered by Bacon, and that they were executed by Henry VIII not for extortion, but upon a charge of constructive treason, trumped up no doubt by their influential victims.[1]

That Henry used his exactions for political ends is certain. The story that, because Thomas More had opposed the aids of 1504, his father was sent to the Tower until he paid a fine of £100 may not be exactly true;[2] but it is not improbable, and certainly foreign observers[3] believed that the king had 'an intention to keep his subjects low, because riches would only make them haughty'. It is beyond doubt, however, that the king loved money for its own sake, and beyond doubt that his agents, even if they did not, as has been alleged, call old statutes out of oblivion, exercised their master's power for unworthy ends and in dishonourable ways. They sold pardons for murder; they sold the king's favour, not only for promotions but even in judicial suits; they used fraud and chicanery and they feathered their own nests. Bacon tells a circumstantial story of Empson's Book of Account with its discreditable entry, and 'the king's hand almost to every leaf', and there is extant an account book of Dudley[4] for 1504–8, signed daily by the king, which shows receipts of large sums for the purchase of the king's favour, for the royal interference in suits, as well as for pardons and fines. For one year, that which includes £5,000 each from Northumberland and Burgavenny, £18,483 was received in fines alone. It is impossible to deny that, even if his motives were sometimes political, Henry exploited not only his special powers but even his justice for a financial end. It seems that he himself was sometimes troubled in spirit by his conduct.[5] In 1504, before Dudley entered his service as a financial agent, he issued a proclamation for the redress of grievances about 'any loan or prest or injury

---

[1] Dietz, p. 48.

[2] It rests on the authority of Roper who wrote perhaps twenty years after More's death. A letter from More to Ruthall of 1506 speaks in high terms of Henry. A letter of Erasmus of the same date seems to show More as a successful barrister, whereas Roper's story represents him as in disgrace and even in danger as long as Henry lived: F. M. Nichols, *The Epistles of Erasmus from his Earliest Letters to his Fifty-first Year* (hereafter Nichols), i. 405–7. On the other hand, More's Latin poem on the accession of Henry VIII denounces the rapacity and the delation which had marked the reign of Henry VII.   [3] Pickthorn, *Henry VII*, p. 70.

[4] Dietz, p. 37.   [5] Ibid., p. 47.

done'; in his will he made provision for a like proclamation, and it was as a consequence of the resultant clamour that Empson and Dudley lost their lives.

Enriched from his own resources, Henry made no particular demands upon the generosity of parliament. The fifteenth and tenth was not, as shown, very remunerative, and when, in 1489, an attempt was made to raise money by a combination of income-tax and a mild 'capital levy' on movables, the result was so poor that the deficiency must be made good by recourse to the old system.[1] Forced loans, which sometimes supplemented parliamentary grants, were a little more fruitful, but even when their yield is included, the king's revenue from direct taxation was not great. Yet must it be noted that these special grants were for special purposes upon which the king always contrived to spend less than he collected. For the French war of 1492 he got over £100,000 in fifteenths and tenths and benevolence, and a clerical grant besides; his outlay was less than £49,000 apart from his expenditure in Brittany, for which he was recouped at his own estimation with a French pension (£5,000 a year). For his war against Scotland and the Cornish rebels, he obtained about £160,000 of which he spent less than £60,000; he gained, that is, a clear £100,000, besides the fines imposed on the Cornishmen and Perkin's supporters, which brought in almost £15,000.

What he gained he kept, and fruitfully employed. In handling the money that was so dear to him, he developed what was, in effect, a new system of accountancy, though, like his other innovations, it was not entirely without a Yorkist ancestry. While he retained the old exchequer of receipt, he gradually supplanted it by the chamber over which he maintained, through the treasurer of the chamber, a direct and personal control. At first the chamber was supplied by grants from the exchequer, but as early as 1487 it began to get directly the surplus of some receivers of lands. By the end of the reign the exchequer received only the profits of the customs, of some of the old crown lands, and the sums paid by sheriffs and boroughs; from these resources it met old-established outlays on the household, the wardrobe, and various wages, remitting the balance direct to the chamber. The treasurer of the chamber,[2]

[1] Pickthorn, *Henry VII*, p. 21.
[2] Thomas Lovell, who was chancellor of the exchequer, was made treasurer of

who accounted not to the exchequer but to the king direct, handled almost the whole of the royal income except for the revenues of Calais and Berwick, which were locally collected and locally spent. The results of the royal economy may be summarized thus. During the first five years of the reign the king's income averaged about £52,000. This was not enough to meet new expenses and old debts, and the king borrowed from the bishops, from the Italian merchants, from the Staplers, and from the City of London—which did not always give as much as was asked. The loans were repaid, and by 1492 the royal accounts showed a surplus, part of which was spent on buildings and jewels. By 1497 the king began to save on a large scale. For his last five years his income averaged about £142,000 and though his expenditure was shown at about £138,000, much of this took the form of loans and much was spent on jewels. The balance which he left behind him when he died was not by any means all in currency. It consisted of jewels and plate, bonds, and obligations, but it was a real balance which was estimated at £1,300,000 by the Venetian ambassador, and by Bacon, relying only on tradition, at £1,800,000.

Intent upon making himself rich, Henry enriched his country. Well did he understand the importance of commerce. The economic aspect of his dealings with the Netherlands has already been emphasized, and it must be remarked that in his 'political' treaties with all countries the welfare of the English merchant was carefully consulted, either in special clauses or in supplementary instruments. He pursued a settled design for capturing trade.

The lurid picture of decay presented in the *Libelle of Englyshe Polycye* (1436) must not be taken too literally. Though English trade had certainly suffered from the civil war, the occurrence of depressions was less frequent than might have been expected, and from 1471 on there was a period of prosperity under Yorkist

the chamber, probably in September 1485. In 1492 he was succeeded by John Heron, who had assisted him in the chamber since 1487. Heron was plainly a member of the 'inner ring' from 1496. He held office until 1521 and occupied other important financial posts. He was supervisor of the customs in the port of London, clerk of the hanaper of the chancery (profits of the great seal) and, significantly, clerk of the jewel house. His successor, Wyatt, had been keeper of the jewels to both Henry VII and Henry VIII. A. P. Newton in *English Historical Review*, xxxii (1917), 357.

rule. England was well supplied with natural harbours and their development reflects the history of English trade. The total number of official 'ports' was small—about twenty; but that is because each 'port', which was a unit of customs collection, had dependent upon it a number of smaller harbours known as 'members' or 'creeks'.

Broadly speaking ports owed their existence to three circumstances, a good harbour, a situation convenient for foreign markets, and a connexion with a hinterland producing exportable commodities. London, with its broad estuary, its connexions by road and river with a rich inland area, its own manufactures, and its accessibility from the busy Netherlands, early obtained a prominence which soon grew to supremacy. It is not too much to say that as the strength of the English state was organized its economic structure became rather metropolitan than national. The trade of London, however, was at first largely in the hands of aliens, and in the development of English shipping the 'outports' played a great part. As long as commerce confined itself to the narrow seas the Cinque Ports and those of East Anglia throve best. Some of them developed close relations not only with 'opposite numbers' on the Continent but with other towns upon the English coast. Sandwich, for example, dealt largely with Caen and London; Lynn traded much with Cologne and got from Newcastle, probably in exchange for corn, the coal which she sent abroad. With the development of ocean navigation other areas began to flourish. Southampton, which had good road connexions with London, became the depot for the Mediterranean trade which was long in the hands of the Italians; it may be noted that, after a riot in 1457, the Italian merchants[1] for a time abandoned London in favour of Winchester. Bristol, which had a close connexion with Waterford, developed an active commerce with Iceland and the Iberian peninsula as well as with Ireland.[2] The ships used varied in size and it is not always easy to discover who owned them. It is clear that as the fifteenth century developed the relative importance of alien shipping declined and the number of vessels owned by Englishmen, either in partnership or alone, increased.[3] Romney paid £73 for a ship; at Hull John Taverner, 'with the Divine assistance and the help of divers of the king's

---

[1] Lipson, i. 542.
[2] Power and Postan, p. 240; Gras, p. 114.   [3] Lipson, i. 569.

subjects', built a great carrack which could compare with those
of Genoa and Venice. About 1461 William Canynges of Bristol
had a fleet of ten ships with a total tonnage of nearly 3,000, and
the great merchant John Tanne of Lynn used the ships of his
own port for eighteen of the twenty ventures which he made in
1503–4.[1]

With the desire for 'self-sufficiency', *autarkeia*, which was part
of the make-up of the Renaissance prince, Henry set himself to
make England independent upon the seas. Besides encouraging
his merchants on every occasion he struck a shrewd blow for
their shipping by the passing of two acts which a later age
would have called 'Navigation Acts'. An act of 1485, lamenting
'the grete mynishyng and decaye' of English shipping in recent
times, ordained[2] that for the future no wines from Guienne or
Gascony should be sold in the dominions of the English king
save those brought in English, Irish, or Welsh ships manned by
crews of whom the majority must be English, Irish, Welsh, or
men of Calais; in 1489 the prohibition was extended to cover
Toulouse 'woad' as well as wine and it was provided that both
master and mariners should be subjects of the king of England.
By another clause the king's subjects (as opposed to aliens in
England) were forbidden to use foreign shipping for export or
import trade if sufficient English shipping was available in the
port which they used. Even though the royal navy remained
small, Henry was always adequate at sea; his fleets encountered
little opposition in the Channel—France was busy in the
Mediterranean—and, if the complaints of the Antwerp merchants may be taken as evidence, it was English ships which
took the initiative in individual acts of hostility.

It was not only by his encouragement of English shipping
that Henry helped his merchants. They were in need of help.
Two great groups of alien traders, by reason of their strength,
experience, and organization, had obtained the commanding
position in English commerce—the merchants of Venice and
the Hansa, which by the treaty of Utrecht in 1474 had obtained
great privileges in England.

During the fifteenth century the English merchants, organized
somewhat loosely as Merchant Adventurers, endeavoured to
secure reciprocity of trade with the powerful Easterlings, but,
as the English monarchy was often weak and as the League

---

[1] Gras, p. 114.   [2] *Statutes of the Realm*, ii. 502, 534.

behaved like a sovereign power in its own right, they had the worse of the encounter. After a struggle, which culminated in open war and a complete breach of intercourse, the Hansa had regained all its privileges in England on terms of complete reciprocity. The Easterlings had recovered their famous Steel-yard in London, and though they did not return to their house in Boston they acquired land at King's Lynn; as compensation for the damages suffered by them they had been given £10,000, to be deducted from dues payable in the following year. Secure in England by the goodwill of Edward IV they had made no effort to perform their part of the bargain. The little groups of Merchant Adventurers, who had established themselves in Prussia and Scandinavia, were ruined; their posts in Danzig and Bergen were lost and they were excluded from the Baltic trade.

Henry, when first he came to the throne, was in no case to quarrel with the powerful corporation, and in March 1486 he confirmed the Utrecht treaty. The inequality of this settlement was obvious; the English towns protested against it and the king, who can have had no predilection in favour of allies of the house of York, soon set himself to curtail the monopoly of the aliens. This he did at first by indirect means, complaining of the piracies of the Hansards, permitting them to export only completely finished cloth, limiting the privileges which their 'own commodities' enjoyed to the products of the Hanseatic towns alone, and broadly hinting that if the Germans expected to enjoy special rights in England they must accord similar rights to English merchants in the areas under their control. In 1487 the German merchants in England were driven to propose a conference where differences might be settled, but the diet of the Hansa, which met at Lübeck in February 1488, rejected the plan. They argued that as the treaty was already renewed there was no occasion for debate and asserted that the English would demand new concessions, whereas they merely wanted to preserve the *status quo*; that they were not, in fact, giving the reciprocity promised at Utrecht did not concern them since they supposed that England had no means of enforcing it. They reckoned without Henry who continued to make the position of the Easterlings in England uncomfortable and, after he had come to terms with Spain at Medina del Campo, advanced boldly to the attack by making a commercial treaty with Denmark in January 1490.[1] This established England as a most-favoured nation in the area from which the Hansa had endeavoured to extrude her, and gave to her merchants the privilege of trading freely in Denmark and Iceland. Impressed by these vigorous measures the League agreed to the holding of a conference and in May 1491 sent a powerful delegation to Antwerp. Henry's representatives appeared late, and by an agreement made in June the Germans granted very few concessions in return for a confirmation of their privileges.

Henry, who was moving towards a war with France, probably felt that he could not have another quarrel upon his hands at this time, but he had no intention of allowing the matter to rest; he continued his negotiations with Denmark and when the unpopularity of the Germans was made obvious in the attack on the Steelyard in 1493, he used the opportunity to oust them from their traffic between England and the Netherlands and to subject the Steelyard to the supervision of English customs officers. After his settlement with Philip by the *Magnus Intercursus* Henry pushed his attack farther. He brought to nothing a conference at Antwerp in 1497 and at Westminster in 1498 compelled the acceptance of the English interpretation of the Utrecht treaty. In 1498–9 he renewed his attack upon the Baltic trade by making a treaty with Riga.[2] His attempt was unsuccessful because Riga, in her isolation, could not resist the pressure of the Hanse towns, and later, when England's relations with the Netherlands were embarrassed by the presence there of Suffolk, the cautious Tudor thought it well to come to terms with the powerful Hansa. In 1504 an act of parliament formally guaranteed the privileges which the Germans enjoyed in England, providing only that the freedom and privileges of London should remain intact. These had been confirmed by the king in 1498 in response to an offer of £5,000, and although the Hansa had been excepted from the resultant curtailment of retail trade the resolute attitude of the English king had had its effect and the Almains showed themselves more accommodating than they had been.

No less resolute did Henry show himself in his conduct towards the powerful republic of Venice, whose merchants bought much cloth and wool and in exchange brought to England not only the essential spices from the east but also

---

[1] *Foedera*, xii. 373, 381. See Busch, 75, 179.   [2] *Foedera*, xii. 700–9.

various Mediterranean commodities, among them books, glass-ware, yew for bows, and the heavy sweet wines of the Levant. Unlike the Hanse merchants the Italians possessed no special privileges in England, which indeed they regarded somewhat as a stopping-place on the way to Antwerp, but their skill and knowledge had given them a great hold upon English trade and they were in consequence unpopular. Richard III had tried to gain the favour of his people by imposing heavy customs upon the aliens, and the Venetians at first hoped that a change of dynasty would mean a change of policy. Henry revoked Richard's act,[1] but he maintained the customs and in his first parliament stopped an obvious loophole by decreeing that a naturalized Englishman must pay the same dues as a foreigner. About this time the capture, by French pirates, of four of the 'Flanders galleys' laden with English wool brought unemployment to the Venetian weavers, and the incident served to show Henry his strength. For the Greek wines which they brought to England the Venetians charged a freight of 7 ducats a butt; English ships charged only 4 ducats and in 1488 the signory imposed an additional duty of 4 ducats per butt upon wine carried by foreign ships. The effect of this measure would have been to deprive England of her carrying trade in wine, and Henry was quick to accept the challenge. Well informed as to Italian affairs,[2] he was aware of the rivalry between Venice and Florence. He had already, in 1486, established a consul at the Florentine port of Pisa[3] and in 1490 he concluded a six years' treaty with Florence,[4] establishing free intercourse of trade, setting up an English wool staple at Pisa, and limiting the supply of wool to Venice to 600 sacks per annum which must be carried in English ships. Conscious that she controlled a great part of the Mediterranean trade, the proud republic sought to deprive the English vessels of their return cargoes by forbidding her ships to carry wine to Pisa; but Henry pursued his attack and in 1492 not only imposed an additional duty of 18s. a butt upon Malmsey wine but cast the burden upon the importer by fixing £4 as the maximum price of the butt and insisting that the butt should contain 126 gallons. The tariff war continued long until it faded away in the complications of Italian politics; each of the parties was too useful to the other to make a

---

[1] *Statutes of the Realm*, ii. 507. [2] *Venetian Calendar*, i. 274.
[3] Campbell, *Materials*, i. 543. [4] *Foedera*, xii. 389.

complete breach desirable, and Henry, in the end, relaxed the severity of his demands; even so the result of the struggle was to enhance the power of England and limit that of the signory, already shaken by the Turkish advance.

The great trade-routes from the east were not effectively cut until Selim conquered Egypt in 1517, but the commerce of Venice was injured by the growing Turkish control over the Levant, and it received a serious blow when India was reached by sea. In the development of ocean navigation it was the Portuguese who led the way. All through the fifteenth century their seamen had worked down the African coast; in 1486 Bartholomew Diaz rounded the Cape, and in 1497-9 Vasco da Gama sailed to Calicut and back. His venture is said to have paid 'sixtyfold'; for the voyage from India, though long and hazardous, was cheaper and safer than the old channels of commerce, complicated by trans-shipment, menaced by pirates, Semitic robbers, and the greedy European barons whose castles commanded the rivers and the roads. Not only by way of Africa did the Portuguese seek the treasures of the East. In the course of their voyages they had discovered the Atlantic islands from the Azores to the Cape Verde group; remembering the old tradition, they kept a sharp look-out for the great lost island— Atlantis? Brasil? the island of St. Brandan? the island of the Seven Cities? at all events some great island to which the name Antilha was given. When men began to realize that the world was a sphere, they began to wonder whether this island might not be 'Cipangu', reported, though not visited by Marco Polo, as fabulously rich; they consulted the geographers, and in 1474 obtained from the Florentine Toscanelli his famous map; ignorant of America, and underrating the breadth of Europe and Asia, they believed that it would be possible to pass to the north of Cipangu, which was believed to lie in the tropics, and reach by a relatively short voyage the splendours of Cathay and the wealth of the spice islands.

In these golden dreams England had her share. Bristol, which traded with Madeira as well as with Iceland,[1] was the place where the sailors from north and south met one another; from her harbour in 1480 John Lloyd set off to find 'Brasil'; according to Ayala her merchants, during the nineties, sent out every year a fleet of three or four ships to look for the wondrous

---

[1] Williamson, *Voyages of the Cabots and the Discovery of North America*, p. 18.

isle, and according to Robert Thorne, writing in 1527,[1] his father and Hugh Elliot, Bristol merchants both, actually discovered the 'Newfound Landes' in 1494. Christopher Columbus, who had lived for some years in the Madeiras, seems to have visited Iceland and known of its connexion with Bristol. When, finding himself cheated by the Portuguese, he sought a new patron, he dispatched his brother, Bartholomew, to engage the interest of the king of England. Bartholomew was taken by pirates,[2] but it is said that when at last he gained audience with Henry, the English king was so impressed with the map that he presented that he invited the bold adventurer to his court. But then it was too late. 'God had reserved the offer for Castille.' Ferdinand and Isabella, grateful for their triumph at Granada, had yielded to the importunity of the Genoese. The three small ships which sailed from Palos on 3 August 1492 carried the flag of Spain and their crews, afforced by jail-delivery, contained only one chance Englishman.

It was not long, however, before England found her venturer in John Cabot, generally believed to be Genoese, though it has been suggested that he may have been born in England.[3] In 1476 he had become a citizen of Venice. On a visit to Mecca he had discovered that the spices came from very much farther east, and had convinced himself that they could be obtained easily by a ship which sailed due west from Europe. Having vainly sought support in Lisbon and Seville he seems to have come to England about 1490, and it was from Bristol that early in 1496 he applied for a patent[4] of privileges with a view to a voyage of discovery.[5] The commission granted to him in 1495 bespeaks not only optimism but caution on the part of the English king. Cabot and his three sons, their heirs, and deputies were empowered to sail under the English flag with five ships to all parts of the eastern, western, and northern seas to discover and annex all lands hitherto unknown to Christians. They were to bear the whole expense of the venture themselves, and were to pay the king one-fifth of the profits accruing from every voyage; on the other hand, they were to be exempt from customs

---

[1] Williamson, op. cit., p. 24. The date is supplied by John Dee.
[2] Pollard, *Henry VII*, ii. 325.
[3] Williamson, op. cit., p. 21.
[4] Ibid., p. 25. *The Voyages of John and Sebastian Cabot* (Historical Association Pamphlet, 1937).
[5] *Foedera*, xii. 595.

on goods which they brought back, and no other subject of the king was to trade in the newly discovered lands without their licence. The grant in effect repudiated the Bull of Alexander VI which divided the unknown world between Spain and Portugal; it recognized the rights of Spain and Portugal in the southern seas, but it plainly supposed that the adventurers could reach the Indies by following a route to the north of that taken by Columbus. Cabot, in short, seems to have guessed, as did others of his contemporaries, that Columbus had merely found an island, and that the coast of Asia was still to be discovered. On 2 May 1497 he set out in *The Matthew* with a crew of eighteen men, and on 24 June made a landfall on a coast which he supposed to be that of Asia, though it was probably in Newfoundland or Nova Scotia. He saw no people, though he found signs of habitation, and he was back in Bristol on 6 August. He had certainly found land far to the north of that discovered by Columbus, and entertained no doubt that this was indeed the realm of the great Khan. That it was unfertile did not discourage him overmuch, for he believed that by following the coast south-west he must inevitably reach the splendours described by Marco Polo, and he probably thought it wise to return with a stronger expedition before he encountered the might of Cathay.

He seemed to have opened up for England a new and wealthy trade, and on his return he was greeted with the utmost enthusiasm. The king gave him a gratuity of £10, and later an annual pension of £20; he appeared at court dressed in silks; the people of London ran like mad after 'The great Admiral', and he had no difficulty in enlisting recruits for another venture. This was made in 1498. The king contributed a large ship and empowered him to impress other ships, paying the standard rate of hire,[1] and there was talk of giving him a gang of prisoners to do the rough work of a new settlement—a stopping place, no doubt, on the road to golden Cathay. The merchants of London and Bristol, some of them with the aid of the king, found the cargoes, and in May 1498 he left Bristol with the five ships. One of these was damaged in a storm and put back to Ireland, but the fate of the others is unknown, though there is some reason to suppose that they reached America.

Yet neither the king nor the Bristol merchants lost hope, and

---

[1] This was normally, about the year 1490, 1*s.* a ton per month. *Naval Accounts and Inventories* (Navy Records Society), pp. viii, xxxi.

their interest was probably stimulated by the competition of the Portuguese (one of whose explorers was João Fernandez, a squire or *labrador*, whose title has been mistakenly transferred from Greenland, which he rediscovered, to a part of Canada). To him and to five others, of whom two were Portuguese, Henry issued, in March 1501, letters patent empowering them to annex all lands hitherto unknown to Christians; next year he authorized three members of the syndicate and a new associate, Hugh Elyot, not only to discover, but to recover, heathen lands, and though he forbade them to interfere with the possessions of other Christian princes, he stipulated that such possessions must be in the actual occupation of the power which claimed them. If the wording of the second patent may be relied on, Henry was not only establishing a new conception of possession, but also envisaging a new sort of settlement. It is plain that the patentees were to establish a permanent colony even although the lands which they occupied were not already the seat of a rich civilization—they were to erect a depot or depots along the route to the Indies.

Several voyages were made, and by 1506 the Bristol merchants were organized as 'the Company Adventurers into the New Found Lands'. The king's interest appears from his privy purse expenses; rewards are given to merchants who had been in the new land, to a priest who was going there, and to explorers who brought home interesting trophies—an eagle, hawks, and popinjays. According to the London chroniclers, there were brought in 1502 three savage men who dressed in skins, ate raw flesh, spoke unintelligibly, and generally behaved like brutes; two years later two of them were seen about the court of Westminster wearing civilized clothes, and indistinguishable from Englishmen.

Henry's interest in the new world was not confined to the satisfaction of curiosity, nor to the hope of making converts; he still hoped to find a short way to the Indies. In 1505 he granted a pension to Sebastian Cabot who, some time before the king's death, undertook a voyage in which, in all probability, he penetrated into Hudson Bay, confident that at last he had found the passage to Cathay. When, alarmed by the ice, his men refused to go on, he determined to circumnavigate the obstacle from the south and, regaining the Atlantic, sailed past Newfoundland and explored the coast as far south as latitude

38—almost to the modern Delaware Bay. He examined several promising inlets, and returned home still hopeful that a channel might be found; when he arrived in England, however, Henry VII was dead and the new king, set upon winning glory in Europe, had no mind for unremunerative adventures in distant seas. Sebastian left England and entered the service of Spain, but there were still a few Englishmen who interested themselves in the new discoveries. In 1511 some of the *Quatuor Americi Vesputii Navigationes* were translated into English. Plainly they were known to Thomas More, and early in 1517 More's brother-in-law, John Rastell, set off on a voyage to the New Found Land with the aim of founding a colony. His expedition was soon broken up by mutiny, fomented, it has been suspected, by the lord high admiral, Surrey, who thought English ships were best employed against France, and the French war of 1522 stopped a project on which the king had seemed to smile in 1521. Small expeditions set out in 1527 and 1536, but the main discovery they made was that Portuguese, Bretons, and Normans had forestalled them. For the moment England missed her destiny in the New World.

The foundation of a maritime empire resulted from the Tudor achievement, but long before the dream was realized the first of the Tudors was dead. For some years foreign observers had noted his failing health, and on 21 April he died at the age of fifty-two in his palace of Richmond. A few days later, after magnificent obsequies, his body was buried in the great chapel which he had begun to build in 1503 and for which he had designed the tomb which to this day attests his good taste and the skill of Torrigiano. The chapel was not completed until Henry VIII had sat upon the throne for some years, and it is significant that the son altered his father's plan as it is set forth in the remarkable will which he executed about a month before his death.

Chapel, tomb, and will are eloquent of Henry's mind. It was essentially a medieval mind. Henry was determined to make his title clear to future generations. Rightly he ordained that the effigy of his Yorkist queen should be beside his own upon his tomb. Yet, not without purpose did he choose to lie near Henry V and Catherine of France, and propose to translate from Windsor the body of his sainted predecessor, Henry VI; not without purpose did he provide for the display of the

Red Rose and the portcullis, and for the making of a statue of
himself to be set upon St. Edward's shrine kneeling in his
armour and holding 'betwixt his hands the Crowne which it
pleased God to geve us, with the victorie of our Ennemye at
our furst felde'.[1] He bade his executors in arranging for his
'Funerallis' to have a 'special respect and consideracion to the
laude and praising of God, the welth of our Soule, and somewhat to our dignitie Roial; eviting alwaies dampnable pompe
and oteragious superfluities'.[2] It is significant that in providing
for 10,000 masses to be said for his soul, he set down the sum to
be paid for each mass and named, with the precision of an
account-book, the number of masses which should be said in
the honour of various celestial powers. His alms were generous,
and they were appointed for definite purposes, such as the
relief of poor prisoners and the founding of hospitals. He provided for the finishing of King's College, Cambridge, for the
church at Westminster, and for the houses of the Observants
which he had established at Greenwich, Richmond, Canterbury, Southampton, and Newcastle. Exact arrangements were
made for the furnishing of the money, but its expenditure was
carefully governed by indentures made with the abbot of
Westminster and the heads of other religious foundations. In
instructing his executors to pay his debts he ordered them to
offer redress to those who had been injured; yet he stipulated
that a close inspection should be made of grievances and compensation given when 'the complaint be made of a grounded
cause in conscience, other then mater doon by the course and
ordre of our lawes';[3] as Empson and Dudley as well as Lovell
were named amongst the scrutineers it is obvious that some
great miscarriages of justice might be found to be within the
law. Even when he was clearing his conscience and casting
himself upon the divine mercy, Henry was still the heir of
Lancaster, the exact accountant, the attorney-at-law. His works
speak for him.

'Upon the whole, we are indebted to him for many excellent laws
and regulations in favour of the people; and to his wisdom, as well
as to the great events which happened in the age in which he lived,
we owe the foundation of our present constitution. But as he appears
even in these to have acted rather upon principles of policy than

---

[1] Astle, *The will of King Henry VII* (*1775*), p. 36.
[2] Ibid., p. 8.   [3] Ibid., p. 12.

those of justice or humanity, we are perhaps indebted more to the effects than to the intentions of his conduct; and if we cannot refuse him in many instances the character of a wise statesman, it may however be doubted whether he deserved the applause and gratitude due to a good king.'[1]

This is a hard judgement. True it is that there were flaws in Henry's character. His face as portrayed in a portrait by an unknown Flemish artist shows cunning eyes, wide-set and small, and a mouth that is somewhat mean. After the death of his wife the evil that was in him was accentuated; he contemplated curious marriages, one of them revolting, and he gave way to the sin of avarice. Yet he had great qualities, courage, especially in the time of danger, coolness, resolution, industry, and a great power of organization. Bacon, whose biography, as the writer hoped, has remained a monument worthy to stand beside that of Torrigiano, is more just and more generous than the eighteenth-century scholar.

'Yet take him with all his defects, if a man should compare him with the kings his concurrents in France, and Spain, he shall find him more politic than Lewis the Twelfth of France, and more entire and sincere than Ferdinando of Spain. But if you shall change Lewis the Twelfth for Lewis the Eleventh, who lived a little before, then the consort is more perfect. For that Lewis the Eleventh, Ferdinando, and Henry, may be esteemed for the *tres magi* of kings of those ages. To conclude, if this king did no greater matters, it was long of himself: for what he minded he compassed.'[2]

It may be true that Henry lacked imagination and that his attitude to the changing world of his day was guided rather by instinct than by the computation which was so dear to him. He burned heretics, he consulted astrologers; yet he revived the ancient strength of the English monarchy, turned it into new channels, inspired it with fresh energy, and sent it forth upon the path of future greatness. He has some claim to be regarded as the greatest of Tudors.

[1] Astle, Preface, pp. xiv–xv.   [2] Bacon, *Henry VII.*

VIII

SPLENDOUR OF YOUTH

Meta haec servitii est, haec libertatis origo, Tristitiae finis, laetitaeque caput.[1]

HENRY VIII succeeded to the throne on 22 April 1509, at the age of seventeen years and ten months. It is a tribute to his father's achievement that he succeeded without incident. Ferdinand anxious to emphasize his own importance, to win the gratitude of the young king, and to establish the position of his daughter, busily proffered help,[2] whilst at the same time he warned his ambassador in England and Catherine herself that they must conclude the marriage as soon as possible. He might have spared his pains. Henry encountered no opposition to a marriage with his brother's widow. On 11 June, scarce a month after the splendid obsequies of his father had been completed, and only a fortnight before his own coronation on 24 June, Henry VIII was married to Catherine of Aragon by Archbishop Warham. The ceremony took place 'in the Queen's closet at Greenwich'. Its privacy was doubtless justified on the ground that the new king was still in mourning, but it is not without significance. That his son should complete the marriage had been the expressed wish of the dying monarch; the councillors hurried on the match because they wanted the new king to produce an heir as soon as possible, because they wanted to secure the Spanish money, and because they sought above all things to carry out the policy of their dead master.

The advent of the new king made little alteration in the conduct of affairs. What change there was was brought about by the attempt to execute Henry's will. The public promise of redress of grievances produced a storm of complaints, and the council saved itself on the backs of Empson and Dudley, who

[1] 'In suscepti diadematis diem Henrici octavi', &c. More's poem, published amongst his *Epigrammata* in 1520 (p. 17), draws a sharp contrast between the unhappiness of Henry VII's reign and the new day believed to be dawning with the accession of Henry VIII. Delation, exaction, sale of offices, neglect of laws, advancement of the ignorant—all these things were to end. There is a reference to the imprisonment of Empson and Dudley. Cf. Alexander Barclay's glowing tributes to Henry VIII in *The Ship of Fools*, 1509 (2 vols. ed. Jamieson, 1874), i. 39, ii. 16, 205–8, and in the *Eclogues* (1514), i and iv where Henry VII is praised also, though there is some animadversion upon his ministers.

[2] *Spanish Calendar*, ii. 4.

were sent to the Tower. Some of the 'obligations' extorted from the nobility were cancelled, but others were left standing, and when the late king's general pardon was formally confirmed on 23 April the royal grace was withheld from some eighty offenders including, besides Empson and Dudley, the three brothers de la Pole.[1] The government was carried on in the established way and by the established councillors. Among these, the chancellor, William Warham, as archbishop of Canterbury, took official precedence. His importance has been dimmed to the eyes of history by the greater abilities of Fox and by the bright splendour of Wolsey, but he was a competent man; his portrait is conspicuous among those of the archbishops at Lambeth and it does not owe all of its pre-eminence to the genius of Holbein. Yet it was Richard Fox, keeper of the privy seal since 1487, and since 1501, bishop of Winchester, who was the most important member of the council. He had had a long training in affairs in the realist school of Henry VII and had been credited with the invention of 'Morton's Fork'. His young master, with characteristic Tudor gratitude, warned the Spanish ambassador of his vulpine quality,[2] and his conduct may sometimes have lacked the scrupulosity proper to a bishop. Yet the main feature of his character was the resolute common sense which he had inherited from a yeoman stock of Lincolnshire. He had never resided either at Exeter or at Wells, and in 1516, 'thinking of all the souls whereof I never see the bodies', he resigned his office and went to reside in his see. Even after his retirement he seems to have been from time to time consulted as an elder statesman, but he devoted the last twelve years of his life to the promotion of piety and good learning. His interest in letters was not feigned, for he had already shown himself a good friend to both the universities, but if his conscience was troubling him as regards his see as early as 1509, he gave no outward sign of anxiety; for the first few years of the new reign he was the mainstay of the government and was regarded by the Venetian ambassador, Badoer, as *alter rex*.[3] Another clerical councillor was Thomas Ruthall, who was made bishop of Durham in June 1509; he had served Henry VII as

---

[1] *Letters and Papers of Henry VIII* (henceforth *Letters and Papers*), i. i. 8.

[2] *Spanish Calendar*, ii. 42; *Letters of Richard Fox* (edited P. S. and H. M. Allen, 1929); E. C. Batten, *Fox's Register for Bath and Wells*, 1889.

[3] *Venetian Calendar*, ii. 30.

secretary, and he performed the same office for Henry VIII until 1516 when he succeeded Fox as privy seal. To these churchmen the continuation of a policy of peace seemed all-important, and it may be assumed that Lovell, the treasurer of the chamber, Poynings, the controller—though Poynings was a man of his hands—and Marney, chancellor of the duchy of Lancaster, were mainly concerned to preserve the system of Henry VII.

Thomas Howard, earl of Surrey, may have represented a 'war party', but aristocratic hotheads had been excluded from Henry VII's counsels and he had no supporters of great weight. He seems to have used his position as treasurer to promote a lavish expenditure, and though the machinery of the chamber still functioned, the old economy was gone. During the first three years of the new reign only £156,000 was spent through the chamber, and in that sum are included the expenses of a royal funeral, a royal coronation, and £37,000 paid to the executors of the dead king. The appearance of thrift, however, is misleading. Although the outlays do not appear in the chamber accounts, Henry VIII's treasure was rapidly disappearing, and from the chamber accounts themselves it is plain that nothing was expended now on jewels and loans, while the wages of musicians, minstrels, falconers, and yeomen of the chamber rapidly increased. While the councillors, not without dissensions among themselves, were in the main following the system of Henry VII at home and accepting the direction of Ferdinand from abroad, another minister, who became a councillor in 1509, was rising to power with giant strides. Thomas Wolsey was to play so great a part in English history that his career is worthy of special study. Here it suffices to say that he was probably pushed forward by Fox to counterbalance the influence of Surrey, and that he soon used his own influence to further a policy of war. Already he saw that the quickest road to promotion was to gain the favour of the king.

The king, though at first he was content to allow his council to act for him, was from the moment of his accession a commanding figure. 'His goodly personage, his amiable visage, his princely countenance, with all the noble qualities of his royal estate', not to mention his clothes, in which he was vastly interested, proclaimed him a true king. History, which regards him as a masterful man, but one as ruthless as he was capable, and as corpulent as he was cruel, must in honesty remember that when he mounted the throne he had all the qualities of a prince of romance. His comely red and white suggested the face of a pretty woman, but his frame was that of a strong man. He rode and tilted with great skill, he drew the bow as well as any man in England, he played on the lute and the harpsichord, and 'sang from the book at sight'. He was learned too; Paolo Sarpi says that he 'not being borne the king's eldest son had been destinated by his father to be Archbishop of Canterbury, and therefore in his youth was made to study',[1] but the Venetian wrote a hundred years after the event and his assertion may be no more than a surmise founded upon Henry VII's reputation for thrift. All the Tudors were well educated. The new king's grandmother, the Lady Margaret, had established professorships and founded colleges; his father, if not much given to letters, respected learning and hated idleness; his mother was pious and amiable, and the account of the nursery at Eltham, given by Erasmus,[2] suggests that the royal children were carefully instructed. The young Henry was early put to his books; the poet Skelton supervised his education; Bernard André no doubt saw to his Latin; Giles d'Ewes taught him French; later he learned some Spanish, presumably from Catherine, and he understood Italian too; besides his gift for tongues he also had an aptitude for mathematics. With the ceremonies of public life he was not unacquainted, for he had been given titles and offices when he was a mere infant. He was made warden of the Cinque Ports and earl marshal when he was about a year old; when he was three he became lord-lieutenant of Ireland, duke of York, and a knight of the Garter.

His father's object in thus heaping honours upon him had been to direct to the royal treasury money which would otherwise have gone in fees, and to concentrate in his own hands power which might have strengthened the over-mighty subject. Obviously the young king performed his duties by deputies. He had no political education; if he learned anything, it was that trusted servants would do the work whilst he remained a figurehead, and though all along he was determined to be popular,

---

[1] *History of the Council of Trent,* in the translation from the Italian (1619), published by Nathaniel Brent in 1629, p. 16.

[2] Nichols, *The Epistles of Erasmus,* i. 201, translating the Preface to the *Catalogue of Lucubrations,* presented in the *Life of Erasmus* (by John Jortin, 2 vols., 1758), ii. 419.

he was at first content a figure-head to be. Always he was masterful, always he knew himself to be a king. At times his arrogant royalty flashed into view, as when, soon after his accession, he saw a letter from Louis XII replying to one in which the new king of England asked for friendship and peace; 'Who wrote this letter?' he demanded. 'I ask peace of the king of France who dare not look me in the face, still less make war on me?'.[1] In the main, however, he relied on his father's old servants at home and accepted the guidance of Ferdinand in foreign affairs; dazzling himself with his prowess in the lists and his magnificence at court, he gave the impression that he was interested, as the Spaniard thought, 'only in girls and hunting'.

The first years of his reign, as recorded by the admiring Hall, were given over to sport and gaiety; masques and jousts and spectacles followed one another in endless pageantry. Yet the English court was free from the licence which disfigured that of France; the athletic contests were serious, and although 'the kynges garde came sodenly, and put the people backe' when they exceeded the bounds of decorum, the commons had their share in the festivities of the king. It was as if an incubus had been exorcised from the spirit of England with the death of that *avarus* who had been pilloried by More, and the country with a sigh of relief greeted the dawn of a gentler day. The life-giving airs of the Renaissance seemed to be blowing upon England.

So thought the humanists.

'Oh, my Erasmus, if you could see how all the world here is rejoicing in the possession of so great a prince, how his life is all their desire, you could not contain your tears for joy. The heavens laugh, the earth exults, all things are full of milk, of honey and of nectar! Avarice is expelled the country. Liberality scatters wealth with bounteous hand. Our king does not desire gold or gems or precious metals, but virtue, glory, immortality . . . . The other day he wished he was more learned. I said, that is not what we expect of your Grace, but that you will foster and encourage learned men. Yea, surely, said he, for indeed without them we should scarcely exist at all.'[2]

Humanism was not unknown in England, but it was as yet known only to a few. It has been defined as an attitude of mind

[1] *Venetian Calendar*, ii. 5.
[2] Lord Mountjoy to Erasmus, 27 May, 1509, *Erasmi Epistolae*, ed. P. S. Allen, i. 450.

which regards the classics not as a repository of facts and quotations, but as an inspiration powerful to improve both the content
and the style of literature. During the fifteenth century this way
of thought had appeared in England, but its development had
been slow.[1] England, unlike Italy, did not preserve eternal
monuments of the classical age; she was far removed from
Greece whence, long before the fall of Constantinople, fertile
Greek influences spread westwards; her political framework did
not embody city-states administered by men who knew Roman
law; English society did not produce a wealthy leisured class
which delighted in literature; and the English climate was not
conducive to all-night discussions in the company of the roses
and the stars.

There had, indeed, been some slight infiltration due to the
presence of papal collectors in England and to the desire of men
like Duke Humphrey, John Tiptoft, and George Neville to play
the part of Maecenas, but for long the main service of the men
of letters had been to acquire books, especially Latin translations
of Greek books, and to cultivate a latinity better than that of
the middle ages. Some of those who acquired the new learning
were moved, it has been thought, less by the love of truth than
by the knowledge that good scholarship was a recommendation
for employment in diplomacy. The allegation that early English
humanism was utilitarian in motive is partly true; it does not,
however, contain the whole truth; there were men in England
interested in knowledge for its own sake. The great school of
Canterbury, in close touch with Rome, produced some men of
real learning and its connexion with Canterbury College spread
the new light to Oxford. Winchester College, too, bred good
scholars like Thomas Chaundler and John Farley who went on
to New College, and the university as a whole must have been
aware of the new movement. About the middle of the century a
few Englishmen visited Italy and at the famous school of
Guarino da Verona in Ferrara learned of the new approach to
the classics. One of these, John Free, actually mastered Greek
and though he never returned to England, a younger contemporary, William Sellyng, who became prior of Christ Church,
Canterbury, not only acquired the treasure in Italy but brought
it home to his own land. Sellyng it was who first revived in

---

[1] See Dr. R. Weiss, *Humanism in England in the Fifteenth Century*, for an account
of the whole matter.

England the teaching of Greek; between 1470 and 1472 he renewed in the school of Canterbury a tradition which went back to Theodore of Tarsus.

Meanwhile, foreign scholars had begun to visit England. Polydore Vergil explains that *perfectae literae*, both Latin and Greek, being ejected from Italy by the dreadful wars,[1] crossed the Alps and spread through Germany, France, England, and Scotland. Germany was affected first and her inhabitants once *'minime omnium literati'* *'nunc maxime docti sunt'*. He gives to Cornelio Vitelli of Corneto in Tuscany the credit of having been the first to teach the youth of Oxford Greek letters, but it seems probable that he exaggerates the services of his patron, who did not arrive in England till about 1490,[2] and certainly he ignores the work of Sellyng and his pupils.

Vitelli, moreover, had predecessors. A Milanese, Stefano Surigone, had been able to make a living by teaching Latin eloquence in Oxford at some period between 1454 and 1471, and during the Yorkist period several learned Greeks tried their fortunes in the unknown island. Andronicus Callistus and George Hermonymos had little success; but Emanuel of Constantinople found a useful patron in Neville, archbishop of York, a brother of the king-maker, Demetrius Cantacuzenus copied some excerpts from Herodotus, and John Serbopoulos made various transcripts including, very significantly, several copies of the Greek Grammar of Theodore of Gaza.

Along with the Greeks came Italians who brought with them not only the improved latinity, but also the rationalistic method of the Renaissance. In 1482–3 Dominic Mancini, a humanist who managed to get a place in the *Catalogus Scriptorum Ecclesiasticorum* of Trithemius (Basle, 1494),[3] paid a visit to England and witnessed the usurpation of Richard III. This he describes in *De Occupatione Regni Anglie*, a work important in that, omitting the fictitious speeches, it presented an objective piece of history.[4] Pietro Carmeliano from Brescia, who seems to have

---

[1] Polydore Vergil, *Anglicae Historiae Libri Viginti Septem, XXVI, ad finem* (Basle edition, 1570, p. 617). See the Introduction to Mr. Denys Hay's edition in the Royal Historical Society's Camden Series (1950) for a brief account of Vergil's life and works. For a fuller account see the same writer's 'Life of Polydore Vergil of Urbino' (*Journal of the Warburg and Courtauld Institute*, 1949).

[2] Weiss, p. 173.

[3] For Trithemius see R. W. Seton-Watson in *Tudor Studies*.

[4] C. A. J. Armstrong, *The Usurpation of Richard III*, 1936.

gained some minor administrative post, taught at Oxford and helped to edit the books produced by the printing-press there established by a German, Theodoric Rood.

With the dawn of the sixteenth century appeared in England a number of Italians who, though mainly concerned to make careers, helped the cause of learning by maintaining relations with the home of the Renaissance and sometimes by the efforts of their own pens. Adriano Castelli de Corneto, who first came to Britain as nuncio to Scotland in 1488 and remained in England as collector of Peter's pence, was given in succession the bishoprics of Hereford and Bath and Wells before he returned to Rome in 1511 to become a cardinal, and in some sense a representative of English interests. Silvestro de' Gigli, who obtained the bishopric of Worcester, acted as Henry VII's ambassador at Rome, and later as Wolsey's confidential agent; but he came to England in 1504 as the envoy of Julius II and was for some time master of ceremonies at court. More important from the point of view of letters are Polidoro Vergilio of Urbino, Andrea Ammonio of Lucca (a *protégé* of Gigli), and his kinsman Pietro, who is generally known by his English name of Peter Vannes. Polydore, who arrived in England in 1501 or 1502, as sub-collector under Castelli, remained long in the kingdom and obtained several benefices, notably the archdeaconry of Wells; before he arrived he had won renown by writing a book of proverbs and *De Inventoribus Rerum* (1499), and he has the distinction of having produced a history of England—*Anglica Historia*—instinct with the Renaissance spirit. Ammonio, who also began his career in England as papal collector, rose to be Latin secretary to Henry VIII, and his kinsman, Peter Vannes, was secretary in turn to Wolsey, Henry VIII, and Edward VI. Another humanist worthy of mention is Pietro Griffo, who was in England only between 1506 and 1512, but who signalized his presence by describing in *De Officio Collectoris in Regno Angliae* the office which had brought so many of his countrymen to England.[1]

Native genius responded to foreign inspiration, and by the time that Henry VII ascended the throne there was already a tradition of good letters in England. Humanist versions of

---

[1] There are many references to these Italians in the *Letters and Papers* and in the *Erasmi Epistolae* of P. S. Allen. Ammonio was intimate with Erasmus; for Vannes, see Allen, iii. 76, 77. For Griffo, article by Denys Hay in *Italian Studies*, 1939.

classical texts and even humanist books of grammar were not uncommon.[1] It is significant that amongst the books which issued from Rood's press there appeared, in 1483, the *Compendium Totius Grammaticae* of John Anwykyll, master of Magdalen College School; this was a not unsuccessful attempt to unite the old grammatical system of Alexander of Villedieu with the humanistic methods of Valla's *Elegantiae* and Perotti's *Rudimenta Gramaticae*, both of which had been known in England for some time.

Yet the influence of humanism in England must not be exaggerated. It is worthy of note that Caxton, who began to print in England in 1476 or early 1477, produced very little that was tinged by the new thought; his first book, and this is characteristic, was the *Dictes and notable wise Sayings of the Philosophers*. Although in Jesus College, founded in 1497 by John Alcock, bishop of Ely, with the revenues of a dissolved nunnery, there was an attempt to unite the cultivation of piety with the pursuit of good letters, the new learning was little known at Cambridge, and in Oxford, in spite of the advent of new ideas, truth was still approached, mostly, through the door of scholasticism. The fate of Peacock[2] was a warning against innovation, and even where the new tools of scholarship were known, they were used mainly to strengthen the old building. The scriptures, *ex hypothesi* verbally inspired, were regarded, after the doctrine of Aquinas, as containing infinite truth. Accordingly they might be interpreted in many senses—'their literal sense is manifold; their spiritual sense threefold—viz., allegorical, moral, anagogical [mystical]'. A subtle logic was used to overcome difficulties and to veil discrepancies. Certainly the use of Bruni's Latin version of the *Ethics*, of Decembrio's Latin version of the *Republic*, and of more scholarly texts of other established classics increased the apparatus at the disposal of the expositors; but the method of exposition was unchanged. Aristotle was still 'the philosopher'. Plato was venerated because his system approximated to Christian doctrine, but there is little evidence that Canterbury and Oxford were affected by the heady neoplatonism which had made so deep an impression

---

[1] Weiss, p. 175.

[2] Reginald Peacock, bishop of Chichester, deprived and confined for life, in 1457, because in his *Repressor of over-much weeting of the Clergy* and in other works he used rationalistic arguments against rationalistic criticism of the church.

on the thought of Italy, at once the repository of a great tradition and the home of a degenerate papacy. Among the early humanists, much of the old scholasticism remained, and even when active minds gave themselves to the pursuit of the new learning, it must not be supposed that they had necessarily enlisted themselves under the banner of ecclesiastical reform. Amongst the factors which produced the reformation the new scholarship was a powerful force, but along with it worked other forces with which humanism had little to do.

While care must be taken to ante-date neither the disappearance of the old scholasticism nor the union of the new learning with the cry for ecclesiastical reform, it is none the less true that towards the end of the sixteenth century English thought began to feel the influence of the fresh spirit which, vivified from Greece, enlivened old institutions and old forms. This fresh spirit was closely concerned with religion. In England there was outside the church little learning; there was hardly any secular literature and almost no real poetry; as the careers of the Italian visitors show, it was next to impossible to make a living out of letters alone. Architecture and the decorative arts, in which Englishmen had great skill, were usually employed upon ecclesiastical buildings. Everywhere the renaissance was closely connected with the 'reformation', which was at once a continuation of it and a revulsion from it, but in England the connexion was more than usually close. Here the renaissance began in the church and it was mainly upon ecclesiastical institutions that its criticism played.

Five men—William Grocin, Thomas Linacre, John Colet, Thomas More, and the Dutchman Erasmus—have long been regarded as the fathers of the revival of learning in England. Others must be remembered too, but these five are worthy of their reputation; yet in each of the five the new spirit manifested itself in a different way. All save More were in orders; it is significant both that More should have meditated a religious life and that Erasmus chafed against his fetters until he was relieved of them by papal dispensation in 1517. Grocin was a scholar; he discovered that the writings attributed to Dionysius were not really by the Areopagite, and until his death in 1519 he was the true friend of good learning. A scholar he remained; he held the living of St. Lawrence Jewry from 1496 to 1517 and he does not seem to have felt, or at least to have expressed emphatically, any desire for ecclesiastical reform. Linacre, who was in London from 1501, was also a friend to good learning; his study of Greek medicine must have shown him some of the follies of medieval empiricism, but he too took priest's orders in 1520, and received ecclesiastical preferment. Apparently his rationalizing spirit confined itself to definite fields. With Colet, scholarship informed life, and a great authority has urged, not without reason, that it was with him that the true spirit of the renaissance first showed in England both its destructive criticism and its creative power. What he saw of truth, he tried to realize with all the power of his being. More saw farther than Colet. He understood that life was greater than scholarship, married, and followed a political career; but he tried to reconcile the things of the world with the things of the spirit by the practice of asceticism, and when reconciliation was no longer possible, he died a martyr. For Erasmus, scholarship was all, or almost all; to do his best work the scholar must have comfortable surroundings which it was the duty of society to provide. Never did he macerate the flesh by self-imposed discipline; he found the natural evils of head-ache and gravel more than enough, and loudly complained of his sufferings. Intellectually he had a passion for truth and to the service of truth he brought great learning, a heroic industry, and a brilliant pen, but though he grieved or laughed at human folly and ignorance he felt no overwhelming urge himself to put things right, and his own life was not always a mirror of the truth he taught.[1]

William Grocin certainly, and Thomas Linacre probably, knew some Greek before they set off to Italy, and both pro-moted the study of Greek when they came home. Grocin, who had received his early education at Winchester and New Col-lege, and had later become reader in divinity at Magdalen, was a man of forty when he went abroad in 1488; having sat at the feet of Politian and assisted Aldus to produce his famous text of

---

[1] The *Encomium Moriae* indicates a personal acquaintance with certain kinds of folly (cf. Allen, ii. 66, and cf. More's joke on clerical celibacy in his letter to Erasmus, February 1516—Allen, ii. 196). Erasmus alleged that in his youth More 'non abhorruit a puellarum amoribus, sed citra infamiam' (Allen, iv. 17); More told Erasmus that he would not shrink from a fib occasionally (Allen, ii. 193) and gave with relish an account of his own sharp practice on behalf of Erasmus with the financier Maruffo (Allen, ii. 259). Upon occasion their correspondence was that of men of the world. Erasmus's attitude to servants in his correspondence is not that of the New Testament.

Aristotle, he came home a finished scholar and between 1491 and 1493 lectured on Greek at Oxford. Thomas Linacre of All Souls, who went to Italy in 1485–6 and was for some time a fellow student of Grocin, was mainly interested in medicine. Later, he became physician to Henry VIII, and besides founding the College of Physicians in 1518 made provision, shortly before his death in 1524, for the establishment of two lectureships in medicine, one at Merton College, Oxford, and one at St. John's College, Cambridge. His services to classical study must not, however, be overlooked; he taught some Greek after his return to Oxford about 1492 and in 1501 was appointed tutor to Prince Arthur; he translated Galen into Latin, wrote an elementary Latin grammar which Colet found too hard for little boys, and later acted as tutor to the Princess Mary.

John Colet, son of Sir Henry Colet, who was more than once lord mayor of London, went up to Oxford in 1483 at the age of about sixteen and there, according to Erasmus, read the works of Plato and of Plotinus as well as those of Cicero. These, presumably, he read in Latin translations, for there is no evidence that he sat at the feet of Grocin or, indeed, that he ever acquired more than a smattering of Greek.[1] In 1493 he went abroad, and though little is known of his travels, it seems likely that he studied theology and philosophy in Paris and then proceeded to read law at Orleans and Bologna. It is probable that he stayed for a while at Florence; certainly he felt the influence of Savonarola's reforming zeal, of the spirit of Pico della Mirandola who offered his brilliant youth upon the altar of asceticism, and of the learning of Marsilio Ficino, whose book *De Religione Christiana*, simplifying the Christian doctrine into a gospel of love, came near to asserting the doctrine of justification by faith.

It was under the inspiration of these great spirits and as a disciple rather of Origen than of Augustine, that Colet, soon after his return to England in 1496 or 1497, began at Oxford[2] his famous expositions of the Epistles to the Romans and to the Corinthians. He continued his teaching for some years and, after 1505 when he was appointed dean of St. Paul's, he preached his new ideas in London both from his pulpit and in his life. Associated with Grocin, who had come to reside in his living of St. Lawrence Jewry, near the Guildhall, in 1499, with

[1] J. H. Lupton, *Life of Dean Colet*, London, 1887, p. 67.
[2] Traditionally at Magdalen.

Thomas More, who was now practising law in London, and later with Erasmus and Linacre, he set himself to foster the growth of true piety based upon careful study of the word of God.

He founded upon the Vulgate text; for long he accepted the writings of the pseudo-Dionysius as the works of the Areopagite converted by St. Paul; he used, especially when he dealt with the Old Testament, some of the mysticism which had become traditional in exegesis, but for all that he struck a new note in the interpretation of the Scriptures. He took the Epistles to be the 'real letters of a real man'; rejecting the idea that every phrase of holy writ should be understood in 'manifold senses', he strove to get the plain meaning of the Apostle's words; he did not hesitate to quote from the pagan writers as well as from the Fathers; above all, he developed a theory that the Apostle, great teacher that he was, in expounding sublime truths to uninstructed hearers 'accommodated'[1] his teaching to the capacities of the taught. Greatly daring, he applied this principle to the study of the Old Testament, and expressed the opinion that the story of the creation as told in Genesis was a 'poetic figment'[2] used by Moses to explain the divine purpose to a primitive people: 'Thus he led forth those uninstructed Hebrews like boys to school.'[3] Following Pico and through Pico Dionysius, he concluded that 'knowledge leads not to eternal life, but love. Whoso loveth God is known of Him. Ignorant love has a thousand times more power than cold wisdom.'[4] Even the discovery, made by Grocin in 1498, that the so-called works of Dionysius were forgeries did not shake the belief of Colet in the simplicity of the Christian faith. 'He set a very high value on the Apostolic Epistles, but he had such a reverence for the wonderful majesty of Christ that the writings of the Apostles seemed to grow poor by the side of it.' So wrote Erasmus who, in his Colloquy on *Pietas Puerilis*, was at pains to emphasize the depth and the simplicity of Colet's piety. He makes his Neophite say that 'I believe firmly what I read in the holy Scriptures, and the Creed, called the Apostles', and I don't trouble my head any farther: I leave the rest to be disputed and defined by the clergy, if they please; and if any Thing is in common use with Christians that is not repugnant to the holy Scriptures,

[1] Frederic Seebohm, *The Oxford Reformers*, 1887, p. 83.
[2] Ibid., p. 54.   [3] Ibid., p. 58 n.   [4] Ibid., p. 73.

I observe it for this Reason, that I may not offend other people'. Asked 'What Thales taught you that Philosophy?', the lad replies, 'When I was a Boy and very young, I happen'd to live in the house with that honestest of Men, John Colet'.[1]

When, pursuing this idea of simplicity, Colet denounced the abuses which disfigured the church, notably in his convocation sermon of 1512, he laid himself open to the accusation that he was a Wyclifite. It has been argued that his profound belief in the love of God made him tend to disregard the doctrine of the atonement, and that for him the Eucharist was not a sacrifice; but while his thought certainly approximated in some ways to the doctrines of the later protestants, his sympathy with an attitude such as theirs can easily be exaggerated. There is no evidence that he proposed to attack the institution of the papacy; misled, no doubt, as to the value of the Dionysian writings, he accepted without question the idea of a hierarchy; he regarded celibacy as the ideal and marriage as a concession to human frailty; the *Cathechyzon* which he provided for his school, although in the Creed it defined 'the Chyrche of Christ' as the 'clene congregacyon of faythfull people in grace', contained a simple exposition of all seven sacraments, saying that in the gracious Eucharist 'is the very presence of the person of Chryst under form of brede'. In his old age he went into retirement with the Carthusians at Sheen.

While Colet's thought thus differed from that of the later reformers, his attitude to the new learning differed from that of the true humanists. He felt that the pagan classics were useful only as a means towards the better understanding of the Holy Scriptures. 'Do not become', he wrote, 'readers of philosophers, companions of devils. In the choice and well-stored table of Holy Scripture all things are contained that belong to the truth.'[2] In his exposition of the Scriptures, it must be added, he sometimes blended in a singular mixture[3] 'Dionysian mysticism, scholastic terminology and grammatical interpretations' which were erroneous.

None the less, when every reservation has been made, Colet stands forth as the champion of the new ideas. Although he held that the Scriptures were the treasury of all truth, he realized that the door must be unlocked with the key of learning.

---

[1] *The Colloquies of Erasmus* (trans. N. Bailey), London, 1878, i. 98.
[2] Lupton, p. 76. [3] Ibid., p. 86.

His knowledge of Greek was limited and some of the Latin
grammarians whom he trusted were unreliable; but he aimed
at, and achieved, a command of Latin as it was understood by
the new scholarship. Convinced that the cause of Christianity
must be defended by true knowledge, he used his wealth to
establish a school which should teach not only piety but good
letters. There was already a school connected with the cathedral,
but this was under the control of the chancellor, Dr. William
Lichfield, who, on his own profession, was quite unable to
lecture *continue*. The dean therefore established, about the year
1509, in a house to the east of St. Paul's churchyard, an entirely
new foundation of three masters and 153 boys. The High
Master he selected was William Lily, of all Englishmen the one
most fitted for the post. Lily, who was a godson of Grocin,
entered Magdalen College in 1486, and so must have been a
younger contemporary of Colet. After taking a degree in arts, he
went on a pilgrimage to Jerusalem and seized the opportunity
to gain a complete mastery of the classical tongues. On his way
home he made a long stay at Rhodes, where he learned Greek
from men who spoke Greek, and he gave a final polish to his
scholarship in Italy under the tuition of great teachers. He
returned to England an accomplished scholar, but, as he
preferred marriage to the priesthood, he cut himself off from the
ecclesiastical promotion which was the usual reward of good
scholarship.

The very bar to a career in the church made him the better
qualified to instruct youth; he was teaching in London when
Colet, in 1512, invited him to be High Master of his new
school, and so doing, set that school upon the path of good
learning from which it has never diverged. St. Paul's was the
first school 'in which Greek was publicly taught in England
after the revival of letters', and it taught good Latin too. Colet
himself, determined that the boys should have the chance to
'come at the last to be gret clarkes', prepared a small accidence
fitted to the 'tendernes and small capacity of lytel myndes', and
when he discovered that the grammar prepared by Linacre was
not entirely suitable, he employed Lily, to whom Erasmus gave
some assistance, to produce a book which soon became a
standard. John Rightwise who succeeded Lily in 1523 was a
good grammarian too, and his *Tragedy of Dido*, played before
King Henry VIII at Greenwich in 1527, had considerable

success. Along with Lily and Rightwise must be reckoned William Horman, a scholar of Winchester, later fellow of New College and vice-provost of Eton, who issued in 1519 his *Vulgaria*,[1] a collection of sentences on various topics calculated to instil not only sound latinity but good morality, general knowledge, and common sense. Another Oxford doctor, Robert Whittinton, ventured to attack the new latinity and drew down upon his head the vigorous *Antibossicon* (1521), wherein Lily and Horman denounced his old-fashioned pedagogy. It is not clear that Greek was taught at Eton, though some of Horman's sentences suggest that it was; but it is plain that Colet's desire for sound latinity commended itself at least to some school-masters outside the precincts of St. Paul's.

It was not scholarship alone that Colet sought. No one can read the rules which he made for his new school without being impressed by their author's passion for truth and by his deter-mination to translate the passion into action. What he did in his school he tried to do in public life. For him it was not enough that a man should realize the wisdom and the love of God for the salvation of his own soul. He felt it part of his duty to con-vince others, and he attacked abuses with whole-hearted zeal. Although his learning was clogged by some of the errors of medievalism, he was a true son of the renaissance in his indomit-able conviction that truth must prevail, and that it was man's duty to obtain the truth for himself and to share his gain with all his fellows.

Associated with Colet in his quest for truth was Thomas More, who was born in 1478. The son of a good lawyer, John More, who eventually became a judge, Thomas was fortunate in having the best education the age could provide. He began at St. Anthony's School in Threadneedle Street, considered the best of the London schools, where he was well grounded in Latin. When, at the age of twelve, in accordance with the practice of the day, he left home to learn manners (and much else), under an alien roof, he was happy enough to be admitted into the household of Archbishop Morton, and there he was taught nothing but good. As a boy of fourteen, he went up to Oxford, possibly to Canterbury College, which had long been a home of sound learning, and although some modern scholars

---

[1] First printed by Richard Pynson in 1519, reprinted by the Roxburghe Club, 1926, see *infra* 577.

are sceptical about the story that he was sufficiently instructed in 'Greake' as well as in Latin, the scepticism is not altogether justified. Roper's story is supported by Harpsfield, who adds that More's father was so much alarmed at his interest in the 'liberal sciences and divinity' that he withdrew him from the university and put him to the law. Both these authors, it is true, wrote long after the event and their testimony is not entirely independent; but Erasmus, whose letter to Ulrich von Hutten in 1519 is the earliest biography of More, tells the same story as Harpsfield. After all, there was no reason why More should not have learned some Greek as an undergraduate—the opportunity was there, and the men who studied Greek at Oxford in his time were the men who in after life became his friends. At all events, either because his father became alarmed at his absorption in the humanities, or because, as with other young men, his years at the university were regarded only as the preparation for a legal career, he left Oxford after two years and entered one of the Inns of Chancery, known as New Inn. In 1496 he went on to Lincoln's Inn, of which his father was a member, and was 'pardoned four vacations'. About 1501 he became utter barrister and before long was made reader in Furnivall's Inn, where he taught for three years. His interests, however, were not all in the law, for soon after he was admitted barrister he lectured publicly on the *De Civitate Dei* in the church of St. Lawrence Jewry, no doubt upon the invitation of Grocin. About this time he thought of entering the priesthood, lived much with the London Carthusians, and fell under the influence of Colet, whom he chose as his spiritual director.

A letter which he sent to Colet in October 1504 shows him as very conscious of the world's evils, apprehensive of his own frailty, and anxious for his confessor's return; meanwhile, he says: 'I pass my time with Grocin, Linacre and our dear friend Lily. The first as you know is the guide of my conduct, while you are absent, the second my master in letters, the third my confidant and most intimate friend.'[1] Obviously, he was interested in learning as well as in piety. He had met Erasmus; he continued his Greek studies; along with Lily he was translating Greek epigrams into Latin, and it was perhaps under Lily's influence that he decided in the end to lead a spiritual life in a secular world. In 1504 he was elected to parliament as a burgess

[1] Stapleton, *Life of Thomas More*, translated Hallett, p. 13.

and, according to one account, imperilled his career by his bold resistance to the king's financial exactions. Next year he married Jane Colt and settled down in Bucklersbury in London. The tradition in his family was that when he decided to marry, he resolved to take as his example of lay perfection Pico della Mirandola, but it seems more likely that his *Life of John Picus, Earl of Mirandola* was written when he himself was contemplating an entry into religion; it was dedicated to a nun, and Pico, who himself did not marry, plainly regarded life as an ante-chamber to death. The other works of More's youth, his epigrams and English poems, are not lacking in humanity though the humour is stilted, and even though he found it necessary always to be instructing his wife, he lived with her in great content until her death in the autumn of 1511. Children were born to him; he was happy in his friends, including Erasmus; the accession of Henry VIII improved his professional prospects, and, in 1510, he was appointed one of the under-sheriffs of the City. His office was important—that of the permanent official who advised a principal, chosen for his citizenship rather than for his knowledge of law—and he executed it with integrity to the satisfaction of all. He became so popular with the Londoners that, at the instance of the merchants, he was allowed to execute his office by deputy and joined Cuthbert Tunstall in an embassy which Henry VIII was sending to the Low Countries in order to settle some commercial difficulties. The mission was successful and, on his return, More was invited, and almost compelled, to enter into public life.

Meanwhile, there had been changes in his home. A few months after the death of his wife he had married, a lady, according to his own ungallant saying, *nec bellam admodum nec puellam*, because, apparently, he wanted a mother for his children. The lady, Alice Middleton, was a widow with one child; she was a notable housekeeper and after her arrival the house at Bucklersbury was less attractive to visitors. Erasmus, cloaking his criticism in Greek, wrote to his friend Ammonius of the 'hooked beak of the harpy'. Yet both of them, it must be observed, made use of a hospitality which they affected to despise.

It is time to say something of Erasmus. An illegitimate child born in 1466, Desiderius Erasmus had been forced by the chicanery of his father's executors into the order of the Augustin-

ian canons. Soon he showed a talent for letters as marked as his
unsuitability for a monastic life, and after some unhappy years
in the priory of Steyn, near Gouda, he gained an uncertain
freedom by becoming secretary to the bishop of Cambrai, under
whose patronage he was able to go to the university of Paris.
The harsh life and the bad food of the College of Montaigu—
Mons Acetus, as he afterwards called it—injured his health, but
increased his learning; he gained a reputation in letters and
earned a precarious living as a teacher of rhetoric, but he was
extremely anxious to find a patron more generous than the
bishop of Cambrai. The fact that he dedicated his first book,
Carmen de casa natalitia Iesu, to Hector Boece, who had been a
student and a teacher at Montaigu, has been held to show that
he was hoping for employment in the new university of Aberdeen
which had been founded at the instance of Bishop Elphinstone
by a papal Bull of 1494. That Bull, however, was not published
until 1497; Boece does not seem to have been invited to become
principal of Aberdeen until 1498, and he did not enter upon
his duties until 1500. As Erasmus's book was published in 1496,
he can hardly have been hoping for a chair in Aberdeen, and
it is unlikely that he contemplated a journey to so remote a city.
The probability is that he had heard from Boece that Elphinstone was generous and hoped to receive a share of the good
bishop's patronage. Fate, however, found him another patron.

Amongst the pupils who were attracted by his learning was
the young William Blount, Lord Mountjoy, at whose invitation
he came to England in the summer of 1499. Probably he had
then no intention of making a career in England, but meant
merely to gain money towards the cost of the journey to Italy
which had long been the object of his hopes; but he was both
surprised and gratified with what he found in the unknown
island. He fell in love with the country life of the English
gentry, and he found himself impressed by the learning which
he found in London and Oxford. When he was staying near
Greenwich he was taken by More to visit the royal nursery at
Eltham and was chagrined because, not having been forewarned, he was unable to offer to the young Prince Henry the
literary tribute which was evidently expected of him. He proved
himself worthy of the fame which had plainly preceded him, by
producing in three days a poem *Prosopopœia Britanniae* with a
dedication to 'Duke Henry'. His departure to France was

delayed by the troubles consequent upon the flight of de la Pole, and he went to Oxford where he stayed in St. Mary's College, a house of his own Augustinian order, then ruled by prior Richard Charnock. Colet, who knew of his reputation, wrote him a friendly welcome; he made the acquaintance of Joannes Sixtinus, a learned Friesian, who was practising as a lawyer in the English ecclesiastical courts, with whom he afterwards maintained a correspondence. He was delighted with his surroundings—'the good cheer would have satisfied Epicurus, the table-talk would have pleased Pythagoras; the guests might have peopled an academy'. He praised Colet's battle for 'the restoration of genuine theology to its pristine brightness and dignity', but he declined himself to enter whole-heartedly into the conflict. He represented that though he did not propose to himself the profession of secular literature, he must improve his learning before he advanced his banners in defence of theology, and represented too that, having seen London, 'I do not now so much care for Italy. When I hear my Colet, I seem to be listening to Plato himself. In Grocin who does not marvel at such a perfect round of learning? What can be more acute, profound, and delicate than the judgment of Linacre? What has Nature ever created more gentle, more sweet, more happy than the genius of Thomas More?'[1] The truth is, however, that he was determined to pursue his secular studies and was determined also to visit Italy. Somehow or other he had collected twenty pounds, but when he left London in January 1500, all but two pounds were confiscated in accordance with the English law,[2] and it seemed that all the fruits of his venture were lost.

The publication of the *Adagia* in Paris in the following June brought him his first great literary success, and the appearance of the *Enchiridion Militis Christiani* (the Handbook or Dagger of the Christian soldier) at Antwerp in February 1504 established his reputation; but though he applied himself with vigour to his Greek studies and strove hard to gain patrons, he had no very certain income and in April 1505 he tried his fortune in England once more. Later he alleged that he was tempted by the letters of friends and by their promise of 'mountains of gold',[3] but though he received an invitation from Mountjoy, with whom he had

---

[1] Allen, i. 274.
[2] 4 Henry VII, c. 23, renewing statutes of Edward III and Edward IV.
[3] Nichols, i. 393; cf. Jortin, ii. 418.

kept in touch,[1] it does not appear that any definite offer was made. Probably he came to England because the promotion of his friend Servatius to be prior of Steyn increased the danger that he would be recalled to the convent, while the promotion of his friends in England increased their power to be of help. Mountjoy was become a councillor and could commend him at court—Grocin and Linacre, as well as Colet and More, were all in London. He wrote to Servatius as from the bishop's palace, saying that 'there are in London five or six men who are accurate scholars in both tongues, such as I think even Italy itself does not at present possess',[2] and representing that it was not the desire for money but the invitation of learned men which had recalled him to England. In fact, however, he sought advancement in his usual way, dedicating to Fox a translation of Lucian's *Toxaris* and to Warham a translation of the *Hecuba*. Warham, to whom he was presented by Grocin, gave him a disappointing reward, perhaps under the suspicion that the work had already been offered to some other patron, and later he took his revenge by adding to the *Hecuba* the *Iphigenia in Aulis* and dedicating both to the archbishop. This he did in June 1506, when he was in Paris, at last on his way to Italy.

Either because his friends' hopes had proved deceptive, or because the unexpected advent of Philip of Burgundy had made the king of England forget his promise of a benefice,[3] his hopes of promotion were not realized. He found consolation in the company of More whom he loved so much that 'if he bade me dance a hornpipe, I should do it at once just as he bade me';[4] he pursued his study of Greek, and, thanks to the good offices of Bishop Fisher of Rochester, was offered a facility of taking the degree of doctor of divinity at Cambridge. Probably he became a bachelor of that university,[5] but he gladly seized the opportunity of going abroad as tutor to the sons of John Baptist Boerio, a Genoese who was chief physician to the king. After an unpleasant journey across Europe he arrived at Turin, where he took the degree of doctor of divinity in the space of a few weeks or even a few days. He then conducted his charges to Bologna, where he superintended their education. He made himself known to the learned in the university; he paid a visit to Florence; and when, in December 1507, a quarrel with the boys' father

[1] Nichols, i. 386; Allen, i. 406.   [2] Ibid. i. 415.   [3] Ibid. i. 421.
[4] Ibid. i. 422.   [5] Nichols, i. 401.

ended his employment as tutor, he took the opportunity to see as much of Italy as he could. He went to Venice, where he produced a greatly enlarged edition of the *Adagia*, printed by Aldus Manucius, and towards the end of 1508 went on to Padua, where he gave lessons in rhetoric to Alexander Stewart, a natural son of James IV of Scotland, who had been made archbishop of St. Andrews, and to his brother James, already earl of Moray. With the young archbishop he visited Siena, Rome, and Naples, and when the Scots boys returned home in May 1509 he was so far advanced at the papal court that he seemed certain of promotion. Soon, however, he received an enthusiastic letter from Mountjoy[1] describing the new and happy day which had dawned with the accession of Henry VIII, promising a living in the name of Warham and enclosing ten pounds, half of it from the archbishop, for his travelling expenses. Erasmus seems to have supposed that Henry, who had written to him in January 1507, was himself a party to the invitation. He bade good-bye to the cardinals and set off hot-foot for England.

Anxious to remove his fastidious mind from the discomfort of his long journey, he gave his fancy free play over the whole field of human folly and when, on his arrival in London, he found himself confined to More's hospitable house by an attack of lumbago, he committed his random thoughts to paper. In a few days he produced the *Moriae Encomium*; the name bespeaks both his love for Greek and his friendship for More.

What Erasmus did was not entirely new. 'The ferment of new ideas' which preceded the reformation had produced what has been called the 'fool literature' and, in 1494, Sebastian Brandt (1458–1521) of Strassburg had produced the *Narrenschiff* in which he collected 112 specimens of different kinds for trans-shipment to their native 'land of fools', and hinted that other ships might easily be manned with crews of the same sort. His book, which was written in German, had immediate success. It was translated into Latin and French and, amplified by a Belgian addition of six shiploads of foolish virgins, it was printed again and again. At the very time when Erasmus was writing, two English versions appeared. The first, in prose, made by Henry Watson at the instance of Wynkyn de Worde and the countess of Richmond, had little success. The second, produced

---

[1] Allen, i. 450; quoted p. 235.

in verse by Alexander Barclay, chaplain to the college of St. Mary Ottery, was popular from the first; partly because the author, who used a bold freedom in translation, dressed his characters in English clothes and made the *Shyp of Folys* an English ship.

Erasmus then, had a precedent for his book on folly, but a comparison between his work and that of earlier writers shows great differences both in style and in content. The *Shyp of Folys* is a mere catalogue of offenders interspersed, in the case of Barclay's book, with 'Envoys' which point the morals; the fools whom it condemned are those who offended against the canons of righteousness and good conduct familiar to the middle ages, amongst them despisers of God, blasphemers of God, and those who keep not the holy day. There is animadversion of the clattering and vain language used by priests and clerics in the choir, and an oblique stroke at Rome in a lament for the state of Europe; but while the author deplores the 'ruyne, inclynacion and decay of the holy fayth catholyke', his object is plainly to re-call Christendom to its duty, especially in the face of the Turkish advance, and most of the criticism he expresses might have been urged by St. Bernard himself.

Erasmus, though he notices in passing many of the fools specifically denounced by Brandt—the fool, for example, who wastes his substance upon building he cannot complete—presents no mere catalogue of follies. He veils his criticism in irony and, so protected, delivers an attack far more trenchant than that of his predecessor. He puts Folly into the pulpit and makes her deliver in her own apology a rambling address which denounces the short-comings of the clergy, and even of the church, in an uncompromising way. Folly begins by explaining that she is the daughter of Wealth and Youth, nursed by Drunkenness and Ignorance, attended by Self-Love, Flattery, Laziness, Pleasure, Sensuality, and Madness, and goes on to claim not only that she is popular and amusing in herself, but that she is the mainspring of all human action and all human happiness. Birth, marriage, conviviality, friendship, games, field-sports, war, and patriotism all come from Folly; so do governments, inventions, and schools of learning. Nature, who made man her masterpiece, surely made him complete without all the trappings added to human life by human endeavour. Valour, industry, wisdom itself are all children of Folly, and 'those fall shortest of happiness that reach highest at wisdom'. The student who sells his youth and his health for knowledge is miserable; the drunkard pays for short happiness by a long *malaise*; but a fool is happy all the time.

Having indulged in a *tour de force* familiar enough to readers of modern plays, Erasmus goes on to attack institutions and creeds generally regarded as sacrosanct. Folly chastises in turn, the grammarians, choked with dust in their unswept schools or houses of correction, bullying the boys, glorying in the discovery of some new word or new usage, triumphing when they detect a colleague in error and unable in the end to produce a satisfactory grammar; the poets, vainly hoping to become immortal and rashly promising immortality to others; the rhetoricians, trained to elude an argument by a laugh; the authors whose few good works remain unread; the lawyers with their irrelevant precedents; the logicians who lose the truth in their quibbling; and the natural philosophers who measure the stars and the sky. The trusted servants of the church do not escape condemnation. First under the lash come the divines with their rival schools— Realists, Nominalists, Thomists, Albertists, Occamists, Scotists. Unlike the Apostles, who converted men by their lives and their miracles, knowing no difference between *gratia gratis data* and *gratia gratificans*, these men employ hair-splitting arguments to discuss truth and hell and purgatory and heaven in shocking Latin, ready to explain how the Virgin Mary was preserved immaculate from original sin, or to demonstrate in the consecrated wafer how accidents may subsist without a subject. The religious orders themselves are found unworthy; much of their time is spent in meticulous attention to rules, and their austerities, prayers, and fastings are a poor substitute for obedience to the rule of Christ, who would say on the Great Day, 'I left you but one precept, of loving one another, and none of you pleads that he has done that'. The artificialities of the preachers are exposed and then Folly goes on to attack the leaders of society. Courtiers and princes are found to be dolts; the king of actuality, breeding horses, selling offices, grabbing taxes, and making war, is very different from the ideal king. A king indeed, if he realized his responsibility, would hardly be able to eat or sleep; only because he gives himself up to Folly is he able to survive. Greatly daring, the preacher applies the same argument to bishops, cardinals, and even to the pope himself. These men, if they understood their callings, would either resign or would spend their time in poverty, watching, and preaching; they would not neglect their spiritual duties. In all this, says Folly, she is not attacking good princes; merely is she showing that those alone who take her for a friend can be happy in their high offices.

After buttressing her case with quotations from classical and scriptural authorities, presented in the dialectical method of the schools, the preacher concludes with the plea that the Christian religion bears some relation to Folly. St. Paul, in his epistle to the Corinthians, said that he spoke as a fool, and advised his hearers to become fools that they might be wise. Cloaking her diatribe by thus representing it as the familiar antithesis between divine simplicity and human artifice, Folly bows herself out of the pulpit.

Not until June 1511 was the *Praise of Folly* published, and it appeared not in London, but in Paris. Probably Erasmus felt that Louis, used to the freedoms of the *sottie* on the French stage, would not resent overmuch its outspokenness, and, indeed, might even welcome its criticism of the warlike pope against whom he was, along with Maximilian, concerting an attack. If Erasmus so reckoned, his calculation was correct. The French king took no umbrage and a later pope merely read the book and laughed. The work, as its author boasted, 'pleases the learned of the whole world; it pleases bishops, archbishops, kings, cardinals and Pope Leo himself who has read the whole of it from beginning to end'. Erasmus made all learned Europe laugh, and his book was frequently reprinted; but he was attacked by theologians, especially by Martinus Dorpius who issued a somewhat laboured remonstrance from Louvain in 1514, and it is possible that the author himself felt that he had been over-bold. When, in 1514, he resumed his assault on the pope in *Julius Exclusus*, he was most careful to preserve his anonymity and even tried to attribute the work to others,[1] although in fact, a draft in his own hand was in England in the custody of Thomas Lupset.[2]

It is possible that the plain speech of Folly injured Erasmus's prospects in England. From the king, who reverenced the papacy, he got nothing save fine words. At the persuasion of

[1] Chambers, *Thomas More*, pp. 114–15.
[2] Allen, ii. 420; see Nichols, ii. 448.

Fisher, he went to Cambridge to lecture upon Greek in August 1511, and next year the university asked Mountjoy's help to pay his *immensum stipendium* (twenty pounds a year). There was some talk of his becoming Lady Margaret Reader in divinity, too, with a salary of twenty marks but, according to himself, he actually lost money in Cambridge.[1] He had few pupils,[2] and in October 1511 he alleged that he had already spent nearly all the seventy-two nobles he had brought from London. In the spring of 1512 Warham gave him the rectory of Aldington in Kent and allowed him, in resigning a living whose cure he could not exercise, to charge it with a pension of twenty pounds a year, payable to himself. When he was anxious to impress possible patrons abroad, he laid stress upon the munificence of his English patrons, but his correspondence with his friend Ammonius shows him a disappointed man. He disliked Cambridge where he felt that Colet was leaving him to carry on unaided the struggle with the Scotists.[3] He gave up the idea of translating Basil on Isaiah because Fisher was not interested.[4] Using Greek letters as a cypher, he wrote that 'these people (apparently the court and the university) are nothing but Cyprian bulls and dung-eaters,—they think they are the only persons that feed on ambrosia and Jupiter's brain'.[5] 'What a university!' he wrote; 'no-one is to be found at any price who can even write tolerably.' The carriers stole the wine which Ammonius sent him and he had to drink beer which gave him the stone; he found the lodgings proposed for him in London unworthy of a civilized man; worst of all, he realized that as England drifted into war, the royal interest in the humanities would certainly wane. In the midst of his labours on Jerome and on the New Testament he found time to dedicate a translation from Plutarch to Henry, and to Colet he inscribed a rhetorical treatise, *De Copia Verborum et Rerum*. But his efforts produced no great reward, and possibly the promotion of Ammonius to be king's secretary in the autumn of 1513 heightened his feeling that he was living 'a snail's life at Cambridge'.[6] By November he was resolved to leave no stone unturned ($\pi\acute{\alpha}\nu\tau\alpha$ $\lambda\acute{\iota}\theta o\nu$ $\kappa\iota\nu\epsilon\hat{\iota}\nu$) to secure promotion in England during the winter months, and, if he failed, to seek his fortune elsewhere. He left Cambridge

---

[1] Allen, i. 484, 542. Nichols, ii. 33.   [2] Allen, i. 473.
[3] Allen, i. 467, 468.   [4] Ibid. i. 477.
[5] Ibid. i. 492.   [6] Ibid. i. 542.

unobtrusively in January 1514, and after a dedication to Wolsey had produced no good result, he quitted England in the following July.

He did not break off his relations with English scholars and English patrons. In dedicating to Warham his edition of Jerome (1516),[1] he gave a high-sounding compliment to the advance made of recent years in English letters, and certainly there was in London, in Oxford, and in Cambridge an interest in the new learning which obviously owed much to his inspiration. In 1516 Colet 'at his age' was learning Greek;[2] Bishop Fisher was studying too;[3] William Latimer of All Souls was a scholar whose help Erasmus sought in the revision of the New Testament[4] and Bullock wrote from Cambridge 'people here are hard at work upon Greek'.[5] Not all the English interest in scholarship was due to Erasmus. Some of the ornaments of the new learning had made their studies in Italy before ever he came to England. Richard Pace of the Queen's College, Oxford, who later gained eminence not only as a diplomatist but as reader of Greek at Cambridge and dean of St. Paul's, had visited the universities of Padua, Ferrara, and Bologna under the patronage of Thomas Langton, bishop of Winchester, before he went to Rome in the retinue of Cardinal Bainbridge in 1509. Cuthbert Tunstall of Balliol and Trinity College, Cambridge, who was also to distinguish himself in diplomacy and in politics, had visited the university of Padua before 1506. None the less it seems certain that Erasmus left behind him at Cambridge a tradition not only of good letters but of inquiry into the Scriptures. It has been claimed that 'the English reformation began at Cambridge, and the Cambridge movement began with Erasmus';[6] and it is true that Thomas Bilney, Robert Barnes, Richard Croke, and Thomas Cranmer himself were all Cambridge men. Yet it was not concern for English learning but consideration of his own interests which brought Erasmus back to England on short visits in 1515, 1516, and 1517. He still drew an English pension; he hoped to obtain a canonry,[7] and in the end secured a pension charged against a canonry from Wolsey as bishop of Tournai; above all, he was anxious to obtain, through the good offices of Ammonius, dispensations which

---

[1] Ibid. ii. 211, 221. [2] Ibid. ii. 347, 351. [3] Ibid. ii. 347, 371.
[4] Ibid. ii. 248; see Nichols, ii. 466. [5] Ibid. ii. 313.
[6] *Cambridge History of English Literature*, iii. 28. [7] Allen, ii. 194, 284.

would release him from his monastic vows and which would permit him to accept ecclesiastical promotion in spite of his illegitimacy. Perhaps because he had been at pains to commend himself to the cardinals and had dedicated to Leo X his great New Testament, dispensations were sent to him through Ammonius in 1517;[1] but it was from Basle and not from London that there issued in 1516 the three works which were the crown of his career, the *New Testament*, the *Jerome*, and the *Institutio Principis Christiani*, with its noble plea for peace, dedicated to the young prince, so soon to become the Emperor Charles V.

Yet in the same year, England produced a book of the first magnitude. The *Utopia* of Thomas More, though it was written in Latin and published in Louvain, was essentially the fruit of the English scholarship which had ripened in the early days of Henry VIII. In his history of Richard III, a fragment extant both in Latin and English of an uncompleted history, More had already launched an attack upon the *Realpolitik* practised by the princes of his day. In this work, although in the classical manner he made his actors explain themselves in imagined speeches, he was resting upon good information, supplied apparently by Morton, and his work stands forth as a monument both of English prose and historical interpretation. In the *Utopia* he gave freer play to his imagination; forestalling Defoe and Swift in his pointed ironies and in his skilful blending of fiction with fact, he produced a work which is at once a real picture of his own age and an imaginary scheme of reform. He combined the idealism of Plato with the speculation aroused by the discovery of strange lands beyond the seas, and professed to describe an unknown island where all things were organized for good. To this island he gave the name of 'Utopia', which certainly meant 'Nowhere', for he himself refers to it in his correspondence as 'Nusquamia', and Budé uses the name, 'Udepotia'; but the pun between οὐ and εὐ was readily made, and almost from its birth Utopia symbolized the country where everything went well. The other names used were all ridiculous derivations from Greek roots, but their meaning might well escape the notice of the uninstructed reader; and at a time when the wildest tales were, if not true, at least current, about mysterious islands, some readers may well have supposed that Utopia

---

[1] Nichols, ii. 461, 541.

was a real place, especially when it was introduced to the public in a narrative conspicuous for its verisimilitude.

The *Voyages* of Vespucci[1] had recently been translated into English. In one of these the explorer professed to have met a people who lived a simple life, having no government, holding all things in common and disregarding the gold and jewels which lay about them; in another, he told how twenty-four men had been left in Brazil when the fourth expedition returned home. More, beginning with an account of his mission along with Tunstall to the Low Countries, which was severely factual as regards time, place, and persons, alleged that Peter Gillis, the town clerk of Antwerp, to whom he had been commended by Erasmus, had introduced him to a man who turned out to be one of the twenty-four left by Vespucci in South America. This man had come home at last by way of Ceylon and Calicut, and had in the course of his travels passed through many countries never heard of before or since. Among these was Utopia, which he described in the course of a conversation which dealt with real things, and the pretence of reality was skilfully maintained because the reader was presented not only with a preface which referred to the conversation in a natural way, but also with a rough map of the island and—this was a brilliant thought of Gillis—with a reproduction of the Utopian alphabet and a quatrain of Utopian verse. The wanderer, Raphael Hythloday (the purveyor of babble), readily discussed his experiences, including an incident at Morton's table which had probably occurred to More himself, and in so doing animadverted upon the woes of contemporary England. An English lawyer had commended the severe laws against theft—the Italian Relator wrote that 'there is no country in the world where there are so many thieves and robbers as in England', and Sir John Fortescue in *The Governance of England* had almost boasted of the valour of English thieves. Hythloday argued that the presence of so many thieves was proof of weakness in the English system and

---

[1] Vespucci wrote an account of his four voyages in Italian to the Gonfaloniere of Florence in 1504. A French version of this letter was translated into Latin in the *Cosmographiae Introductio* published by Waldezeemüller at St. Dié in 1507 under the title of *Quatuor Americi Vesputii Navigationes*, and the third voyage found a place in the *Paesi Novamente Ritrovati* of the same year. In 1511 some of the voyages were printed in English by John of Doesborowe. The truth of the stories, especially of the first two voyages, has been questioned, but the second two certainly took place between 1501 and 1504.

went on to analyse the causes of the poverty which begat the thieves. The foolish wars were a source of wickedness and of expense and a breeding-ground of unemployables, but the main cause of the evil was the unreasonable covetousness of a few persons. In their quest for wealth the landowners, even 'certeyn Abbottes, holy men god wote', were inclosing their fields, depopulating the country-side so that the sheep might be said to devour men; the monopolies of wool and cloth were harmful, the luxury of all classes bred unproductive expenditure; the noblemen and their 'families' made no contribution to the national wealth. Through the mouth of a convenient fool the criticism was carried against the idle clergy, and when Hythloday returned to this topic in the second book, he included 'the idle company of priests and religious men, as they call them', along with the households of nobles and the women, as persons who failed to do their share in production. His remedy for the social ills was to end useless wars, to abolish the death penalty for theft, and to set offenders to useful labour. Kings should live of their own and abandon their practice of squeezing money out of their subjects on all sorts of pretexts; they should not fear at all lest their subjects should 'wax fat and kick', but should reflect that none are so prone to rebellion as the desperate who have nothing to lose. The prime cause of all evils was the possession of private property, and when More, in criticism, asked a sensible question as to how there could be plenty when there was no incentive to labour, Hythloday replied that he himself had seen in Utopia a commonwealth which was rich, prosperous, and contented simply because all things were held in common. In 'Book Two' he sets forth the ideal state.

He gives a picture of a completely planned community with fifty-four cities, each a garden-city with fine streets twenty feet broad. With each of the cities twenty miles of soil is associated and every town-bred child must take a turn of country life in an artificial family presided over by a *Hausvater* and a *Hausmutter*. Even in the towns the families are kept artificially to a level of ten to sixteen persons, though the core is a natural family following a hereditary trade. Each city contains 6,000 families and when it gets too big a new city is founded, perhaps upon land 'acquired' from persons who do not make good use of it. Magistrates are elected from mathematical constituencies; every year thirty families choose a magistrate, anciently called the 'Syphogrant', but now a 'phylarch', and over every ten syphogrants is a 'tranibor', now the 'chief phylarch'. The 200 syphogrants elect the prince out of four persons nominated by the four quarters of the city. The prince is chosen for life and the tranibors, though chosen annually, are often continued from year to year. There are, however, very few laws, for everyone lives in plenty, because everyone works for six hours a day and no wealth is wasted on unnecessary things. A national economy ensures that one area supplies the deficiencies of another and that there is always a surplus for export; of imports they have little need and as they have no regard whatever for bullion they establish credits abroad which procure them useful allies. Of their own abundant gold they make no use, for they neither buy nor sell. Each producer simply takes his finished wares to a depot from which he and other citizens take what they want as they want it. There are no fashions in clothes, men and women wear the same costume; meals are taken in common, each family cooking in turn for the whole group and obtaining the necessary victuals from a central store after the hospitals have drawn their daily ration. There are hospitals, though euthanasia is practised. In their marriages they follow the laws of eugenics.

From the universal system of labour there is no escape save for a very few who, after commendation by the priests, are secretly elected by the syphogrants to join the ranks of the learned. Those who fail to make progress are returned to their ordinary occupations; the others form a select group from whom the king, priests, ambassadors, and magistrates are chosen. The only other persons whose life lies outside the common lot are the bondsmen. These are offenders against the social order who are kept continually at work with a severity which varies according to their offences, some being distributed to do the rough work of the country families. Prisoners of war are not enslaved, but sometimes poor fellows from other countries prefer bondage in Utopia to liberty at home; such men serve voluntarily and if they want to go home are readily dismissed with a luck-penny.

The men are daily exercised in arms and the women too 'upon certayne appoynted dayes'. Their training is practical and includes swimming in armour; their armour is excellent, they have effective 'engines of war' which they keep very secret; but for all that they regard war as a monstrous thing and avoid it when they can. Even when they must fight they prefer to do so by indirect means;—they make use of their allies; they hire the convenient Zapoletes (obviously the Swiss) who will fight for the best payer; they employ an intense propaganda and do not hesitate to offer vast sums to anyone who will kill the prince and the leaders of their enemies.

In matters of religion they are very tolerant. Everyone must believe in an overruling providence and in the immortality of the soul; persons who cannot do so are excluded from office and, though allowed to hold their own views, are forbidden to dispute upon them with the common people. The overruling providence is worshipped under many forms; augurs and sooth-sayers are despised, but miracles are highly regarded as proofs of the present power of God. When the Utopians heard of Christ, some of them were readily baptized; these considered the possibility of obtaining a priest by election 'without the sendyng of a christian bysshoppe'. No convert suffered any sort of persecution save one, who spoke disparagingly of other religions; he, after he had continued his diatribes for a long time, was exiled, not as a heretic but as a seditious person and raiser up of dissension among the people. A few men show their religion not by study, but by undertaking good works of an arduous kind; these are of two sorts, one celibate and vegetarian, which is counted the holier, the other, abstaining from no pleasure that does not hinder their labour, is held the wiser. Their priests are of exceeding holiness and therefore few, and women are not excluded from the priesthood. Their public worship is cere-monial, but they regard the pursuit of virtue as the great end of life and, in the main, they take virtue to be the leading of a life in accordance with the rules of nature. In their leisure time, which is abundant, they cultivate virtue, and even their games teach either mathematics or morality.

Both in his diagnosis of England's ills and in his proposed remedies More lays himself open to some criticism. The economic survey of contemporary England is faulty in several respects. The statement that women contribute nothing to productive industry is absurd, and of an age which followed a 'domestic economy' it is more than usually absurd. Again, the denunciation of the men who raised rents takes no account of the rise in prices due to the development of the German silver-mines; the landlords, most of whose land was let out on fixed leases, suffered heavily from the increased cost of living and were bound to recoup themselves as they could. From a national standpoint an increase in the wool-clip was desirable; some inclosure was justified, and the evidence seems to be that when *Utopia* was written the process of inclosure had not gone very far. The gilds were undoubtedly monopolistic, but on the other hand they could not be expected willingly to share with outsiders privileges which they had gained by their own effort at their own cost. None the less the fact remains that inside the gilds full privilege was vesting itself more and more in the wealthy and, in spite of the statistics, the evidence of Starkey and Crowley,[1] as well as of the statute-book, makes it clear that inclosure was already inflicting harm upon many helpless sufferers. The labourer, excluded from his holding in the country and debarred from employment in the town, was truly in evil case and worthy of the championship of More. Whether he would have really been happier in Utopia is a matter of doubt.

In spite of More's seeming reasonableness his ideal state was highly artificial; in spite of his liberalism, it was really a managed state whose people were not free. To bring it into existence the author had to postulate an extraordinarily fortunate geography, a fruitful island rich in minerals, surrounding a large navigable bay which could be entered only by one narrow channel; even the Robinsons of Swiss origin who 'looked up and beheld the butter tree' had never such an island. He had to postulate, too, allies who never thought of repudiating their debts and the convenient Zapoletes who would fight as mercenaries without ever considering that it might be easier and simpler to take the Utopian gold by force of arms.

His planned state was a danger to world-peace—it resembled strangely the Germany of Hitler. It was an organized community wherein everyone had his place; where there was no unemployment; where the rough work was done by alien labourers or by the forced toil of persons who did not conform to the standards set by the state; where all citizens were trained to arms yet where few citizens lost their lives in wars which were conducted

---

[1] Thomas Starkey (?1499–1538) wrote *Dialogue between Pole and Lupset* (Early English Text Society, 1871–8) and attacked papal supremacy. Robert Crowley (?1518–1588) was a protestant reformer and a social reformer. See *The Way to Wealth* and other works (Early English Text Society, 1872).

by 'secret weapons', propaganda, and 'fifth column'; where aggression was justified whenever *Lebensraum* was needed on the ground that an intelligent people could use land better than their uninstructed neighbours.

This ideal state was a danger also to its own people—as always the planner considered his own theories rather than the true benefit of the planned-for. For the ordinary individual, life in Utopia must have been intolerably dull; it seems probable that in the well-ordered feasts conversation must have languished and the food must often have been cold. The fact that the citizens attended lectures before breakfast needs no comment. Life must also have been sterile; to what use could a man profitably turn his leisure if the capacities which he developed in himself could never be satisfied in action by individual enterprise?

Yet it remains a great thing that an Englishman had the courage to envisage a state which would be free of the social evils which afflicted the England of his day, and the genius to work out a system so complete in everything except in appreciation of the infinite variety of human personality. In designing his Utopia More was plainly influenced both by the old theories of the *Republic* and the *Timaeus* and by the new possibilities created by the great explorations. He lived in a world which was expanding in every direction and in which the fresh discoveries seemed to justify the wisdom of the ancients. In warfare, for example, the heavy masses of spears, developed as a counter to the charge of the men-at-arms, obviously recalled the Macedonian phalanx, and Aelian's *Tactica*, dedicated to the emperor Hadrian, became a text-book for the sixteenth-century soldier. Why should not Atlantis take on reality in some remote island? Yet there is evidence that More used authorities other than Plato and Vespucci. It seems likely that he had read of the travels of Marco Polo; the route adopted by Hythloday on his return to Europe is not unlike that taken by the bold Venetian; Amaurot in its plan and organization has some resemblance to the new-built city of Tai-du in Cathay, and the bridge over the Anyder recalls the wonderful bridge over the Pulisangan or Hoen-ho. More cannot have read *Il Principe* in print, for it was not published until 1532; but it was written, or at least begun, in 1513 and it must have circulated to some extent in manuscript, for it was plagiarized soon afterwards. In any case, *The*

Prince was only a fixation—perhaps even a criticism—of the amoral doctrine of politics which was current in Italy at the beginning of the sixteenth century, and to these doctrines the *Utopia* was certainly a reply.

No doubt More had in his mind the *Praise of Folly*, which was a skit upon contemporary morals and manners; but he was an Englishman, whereas Erasmus was a cosmopolitan, and in the end it will be found that his main inspiration lay in the 'condition of England'. Obsessed by the social distress which he saw around him, he devised a country in which want should be unknown, and his approach to the ideal state, unlike that of Plato, was primarily economic. For that reason it is very much in tune with modern thought. One great scholar has pointed out that it is almost hedonistic. Another, arguing that Utopia represented the ideal of a people with the pagan virtues of wisdom, fortitude, temperance, and justice, but without the faith, hope, and charity of christianity, concluded that 'religion is the basis of all' and that the phantasy of More's early life is in complete accord with the spirit which made him a martyr for the Roman catholic church.[1] Certainly More was gay as well as pious in his youth and to the scaffold he took his gaiety as well as his piety; but it is hard to reconcile the liberal tone of *Utopia* with the austerities practised by More at various periods of his life and the tolerance of *Utopia* with the attitude towards heretics which he adopted in his later years. The truth seems to be that at one time More, like Erasmus, hoped to see a liberal reformation inside the old church, and that experience convinced both men that reform as it gathered force would tear down many things which seemed to them holy and precious. Confronted with a choice, they decided for the old and the known; Erasmus was deaf to the plea of Dürer: 'Hear, thou knight of Christ! Ride forth by the side of the Lord; defend the truth, gain the martyr's crown!' More gained the martyr's crown, but not as a protestant.

In 1516, however, there was no thought of hard choice, and none of martyrdom. Forward-looking Englishmen still hoped for a glorious summer; others, not ill contented with their lot, were confident that the year would bring its increase in the ordinary way.

[1] Contrast the view of Sir Ernest Barker in *The Political Thought of Plato and Aristotle* with that of R. W. Chambers in *Thomas More*, p. 126.

To most Englishmen More's ideal state must have appeared as a mere phantasy; political thinkers for the most part contrived to fit the facts to the theories in the old-fashioned way. *The Tree of Commonwealth*, written by Dudley during his imprisonment, may claim some connexion with the new thought because the author attacked not abstract injustice but definite malpractices of government *quorum magna pars fuerat*, and because he was very frank about the worldliness of the clergy. But the book in its main outline is an allegory conceived and executed in the medieval manner, and its argument is a direct antithesis to the view that the science of politics lay outside the field of morality.

Even in philosophy tradition was strong; except in classical scholarship and in the philosophy and divinity which depended upon scholarship, English culture was little affected by the Renaissance in the early sixteenth century. The list of books produced by the first printers shows that the old standards held their place. English prose, though it gained nobility in the hands of More and Hall, was mainly applied to old themes; English poetry, which was at a low ebb, gained no real inspiration from without before the days of Wyatt and Surrey, and the English drama made no advance beyond the well-established interlude. Italian ornament was occasionally introduced into English architecture, but the building of the period, some of which was magnificent, was essentially English in tradition, and English music, though it may have borrowed from the French court something in the matter of light songs, preserved in its graver compositions the tradition of Dunstable. The English alabasters, though much sought after, reproduced the medieval forms, and English portrait-painting achieved nothing of note before the advent of Holbein. No doubt the spirit of the Renaissance stimulated existing traditions into new life; but it did not alter the traditions, and where, as in poetry and painting, no good tradition existed, it produced no marked effect at all.

In the realm of politics these truths hold good. Henry and his councillors were anxious, in the renaissance fashion, to produce a strong state and to cut a great figure upon the European stage; but their conduct of affairs both at home and abroad was in accord with the good old English way. Certainly there is evidence of the amoral ruthlessness associated with the

Renaissance in Italy—Empson and Dudley and the luckless Edmund de la Pole were executed for reasons of state. On the ground that they had—it was not unnatural—summoned their men about them when King Henry VII died, the 'fiscal judges' were condemned for treason, Dudley at the Guildhall in July 1509, Empson at Northampton in the following October. Both were sent to the Tower, and at one time it seemed likely that Dudley at least might be spared, but Henry, who was gaily spending the money they had helped his father to amass, allowed them to be beheaded on 18 October 1510—'a purchase price for his subjects' love'. Edmund de la Pole, despite the fact that his life had been guaranteed when he was handed over by Philip, was executed early in June 1513, simply because Henry, about to invade France, did not care to leave behind him a possible claimant to the throne. Political murders, however, had not been uncommon in medieval England and, as in the case of two dukes of Gloucester, the pretence of justice had sometimes been very slight. It may be said, perhaps, that Henry was less careful than his predecessors to conceal his summary proceeding beneath a cloak of legality, but this is all that can be said.

Henry's political executions represented no startling innovation in English government, and in English foreign policy his spectacular adventures were in the main only variations upon a time-worn theme. The ramifications of his policy, it is true, went wider than those of earlier kings, and his resolute championship of the papacy at the start of his reign was something of a novelty; yet behind the high-sounding professions and the imposing leagues may be discerned the workings of the familiar principles. A king of England must be a foe to France and to Scotland, the ally of France; he must, if possible, assert his own right to the crown of one country and the suzerainty of the other; he must, if he was to attain to full majesty, wage a successful war.

For war, as it was understood on the Continent, Henry was little prepared; he lacked the heavy cavalry and the mercenary infantry which played so great a part in the Italian wars, and though he had both guns and gunners, the English artillery was not trained to mobility. To the Yeomen of the Guard he added early in his reign a guard of gentlemen known as the King's Spears which sufficed only to provide him with a stout retinue and a 'pool' of reliable captains who could command either by land or by sea.[1]

His military power, however, was far greater than it seemed to the European experts who were applying, not without pedantry, the *Tactica* of Aelian to the warfare of their day. The English king could draw from his quite well organized shire musters a number of competent archers.[2] From the Borders especially the king could raise some very useful light horse; his guns, as was shown at Waterford, were not to be despised, and in the hands of men who had an arrogant tradition of victory, 'bows and bills' were to prove themselves surprisingly effective.[3] Moreover, Henry possessed in his fleet a weapon whose value became more apparent as the years passed. From his father he had inherited five good ships, including the *Regent* and the *Sovereign* of 1,000 tons apiece, and from the moment of his accession he devoted himself to the increase of his naval might. In the first year of his reign he built the *Mary Rose* of 600 tons and the *Peter Pomegranate* of 450; soon afterwards he began work upon the great *Henry Imperial*, or *Henry Grace à Dieu* of 1,000 tons, which was not 'hallowed' till 1514, though it was evidently hoped to send her to sea in 1513. At the same time he bought ships both in England and abroad, and he retained the old power of hiring merchantmen when he required them. He could thus commission a considerable fleet at short notice and, as the evidence shows, he could readily find mariners, gunners, and soldiers by the system of indenture. From the annotations in his own hand which appear on various lists, it is plain that his interest in the navy was no less than his interest in horses and in arms.[4]

He rejoiced in his manhood and was anxious to prove it in

---

[1] Samuel Pegge, *Curialia* (1791).

[2] The pay-rolls provide evidence that efforts were made to turn the levies into a competent fighting force. As the issue of coat-money shows, provision was made for some sort of uniform—the coat was usually white—and the payments made to petty captains argues an improvement in the subordinate command. There is some appearance, too, of a small tactical unit of a hundred men.

[3] In 1512 Henry had passed an act enforcing practice with the long-bows under supervision of the J.P.s, and in the same year had renewed an act of Henry VII restricting the use of the cross-bow to the nobles and the wealthy (3 Henry VIII, cc. 3 and 13). In 1515, after his successes at Flodden and Guinegate, he confirmed the act compelling practice with the long-bow (6 Henry VIII, c. 2).

[4] For the growth of the navy see M. Oppenheim, *History of the Administration of the Royal Navy and of Merchant Shipping in relation to the Navy from 1509 to 1660* (London, 1896), and particularly Navy Records Society, *The War with France, 1512–1513*. See also *Letters and Papers*, I. i. 750.

action. It was not enough to be Adonis and Maecenas; he must
be Alexander too. Erasmus, Colet, and Dudley taught that war
was the mother of all evil, but before long the banner of St.
George was in the air. From the statutes and ordinances of the
corps it is plain that Henry, in establishing a gentlemen's guard
on the French model, meant to turn the jousting prowess of his
young courtiers to a military end. Each spear was to have at
least two great horses upon which he and his page should ride
as men-at-arms, to be accompanied by a custrel, who should
rank as a demi-lance, and to have at call two mounted archers.
His pay was to be 3s. 4d. a day, but as the ordinary wages were,
for a man-at-arms 1s. 6d., for a demi-lance 9d., and for a
mounted archer 8d., it is plain that he would gain very little on
the transaction; when his archers were actually in service it
would seem that he would lose at the rate of 3d. a day even
though he did not draw a penny for himself. The ordinary
soldier, however, drew his wage for only a short period, and
the spear's retinue need not necessarily have been paid all the
year round. The corps, which survives today as the Gentlemen
Pensioners, could hardly bear the weight of its own magnificence, though it did not, as is often alleged, disappear soon after
birth to reappear only in 1539. When Howard took to sea in
1512, seven of his eighteen ships were commanded by 'spears'
who drew no other than the established wage, though one of
them, Stephen Bull, was afterwards treated as an ordinary
captain.

When the king came to the throne, all promised peace. His
father had managed to unite France and Scotland with England
in a kind of triple *entente* and had, almost alone amongst the
European princes, refused to embroil himself in the affairs of
Italy. Soon after his accession, on 29 June 1509, he renewed
the treaty of 1502 with Scotland,[1] and on 23 March 1510 he
renewed also his father's treaty with France;[2] both treaties were
signed *pro parte* of each of the contracting monarchs by most of
the West-European princes, and both were under the sanction
of papal excommunication. In diplomacy, as in other things,
Henry soon scattered his inheritance. Less able than his father
to distinguish between practical gains and lordly pretensions,
contemptuous of the Scots and jealous of the French, he fell
an easy victim to the wiles of Ferdinand who affected to treat

---

[1] Ibid. 45.   [2] Ibid. 186.

a callow youth as a brother in statesmanship. Eagerly he sought the path of adventure, and his progress was the more rapid because he was propelled, unwittingly, by the ambitions of a most able servant who, for his own purpose, sought at once to flatter his royal master and to gratify the spiritual head of Christendom.

The political stage soon resounded to the martial strut of the new *jeune premier*. The relations between England and Scotland became strained partly because of incessant 'incidents' on the Borders and at sea, partly because Henry refused to hand over to his sister Margaret the jewels bequeathed to her by her father, most of all because the haughty young Tudor flaunted his claim to suzerainty. Arrangements were made for a meeting with Scots commissioners to redress border grievances in the summer of 1511, but in the August of that year Sir Edward Howard, who was later appointed lord admiral (1512), killed one of James's favourite captains, Andrew Barton, in the Downs, and made prizes of his ships *The Lion* and the *Jenett of Purwyn*. The Bartons had received letters of reprisal against the Portuguese in respect of an injury suffered in the days of James III; it is possible that they interpreted their powers widely, and certain that they interrupted commerce in the North Sea—as appears from Howard's own career as admiral, the rights of neutrals were at that time little regarded.[1] There is, however, no clear evidence that the Bartons attacked English ships before the action in the Downs, and the Scottish story was that their captain had been treacherously surprised by the vessels of a friendly power.[2] Whatever was the truth of the matter, Henry was pledged by treaty to consider Scottish grievances, and when he replied to James's protest that 'kings did not concern themselves with the affairs of pirates', the Stewart was not unnaturally incensed. On 5 December 1511 he wrote to Julius that since the king of England had waged war upon him, both secretly and openly, he presumed that the pope had liberated both monarchs from their oaths and from liability to apostolic censure.[3] He was in fact denouncing his treaty, but though he seemed to threaten war, he was not inclined to fight. He really believed in

---

[1] For examples of Howard's brutalities see Navy Records Society, *The War with France, 1512–1513*, xvi.

[2] Leslie, *History of Scotland*, S.T.S. ii. 135.

[3] *Epistolae Regum Scotorum* (1722), i. 122.

the great crusade of which the other princes talked, and understood that his voyage depended upon the existence of peace in Europe; Louis XII, fully occupied in Italy and anxious to avoid a war in the north, was urging him not to break with Henry; Henry, after all, was his brother-in-law and he had at this time some hope of succeeding to the English crown;[1] lastly, as a good churchman, honoured by the pope in 1507 as Protector of the Christian Faith, he was most unwilling to quarrel with the Holy Father. In spite therefore, of the omens, peace might have been preserved had not the explosive atmosphere been kindled by a spark from the European bonfire.

The war-like breath of Julius II had roused to a new blaze the smouldering embers of Cambrai. From the point of view of the pope, the attack on Venice had succeeded only too well; the French, who had gained the great victory of Agnadello (14 May 1509), had advanced to the Mincio; their ally, the duke of Ferrara, was impiously producing salt at Comacchio to the detriment of the papal monopoly at Cervia. Plainly, a powerful monarchy was going to be a worse neighbour than a recalcitrant republic and, disregarding altogether his treaty with Louis, Julius determined to expel the ally whom he had himself invited. Speedily he gathered supporters. On 24 February 1510 he absolved the Venetians on heavy terms—which the Council of Ten secretly repudiated by a formal instrument; in March he made a treaty with the Swiss, who undertook to supply 6,000 men for his protection, and on 3 July he obtained the co-operation of Ferdinand by granting him the investiture of Naples. This he gained the more readily because the king of Aragon, secure now in the south, was secretly hoping to create another Spanish dominion out of the confusions of north Italy.

Military action followed hard on this diplomacy. The Venetian army took the field once more; in July the papal ships attacked Genoa—in vain; in August the Swiss invaded Lombardy—only to retire; nothing daunted, the martial pope, putting himself at the head of his own troops, began in the same month a campaign which eventually led to the capture of

---

[1] James took quite seriously his chance of succeeding to the English throne, and the bishop of Moray asserted to the French king that the English ambassadors had promised him parliamentary recognition as heir-apparent. *Flodden Papers*, Scottish History Society (1933), 38, 73, 74. There is no record in the English State Papers of any such pledge; from them it appears that the ambassadors made threats rather than promises. Cf. *Flodden Papers*, lxiv.

Concordia and Mirandola (20 January 1511). To these un-sacerdotal proceedings Louis found a ready reply. From his own clergy, gathered at Tours in September 1510, and at Lyons in April 1511, he procured a formal declaration that a general council was necessary. On 16 May half a dozen dissident cardinals, with the concurrence of the emperor and the king of France, issued from Milan a formal invitation to a general council, to meet at Pisa on 1 September 1511, which they requested the pope to attend. Julius, who, in 1503, had promised to hold a general council, replied by summoning a Lateran council for April 1512.[1] He condemned the Milan decree and deprived most of the cardinals who had issued it. Already he had sought to improve his position by creating in March 1511 eight new cardinals on whom he might rely. Among these was Matthew Schinner, bishop of Sion, who was able to influence the Swiss, Matthew Lang, bishop of Gurk, Maximilian's chief adviser (provisionally), and Christopher Bainbridge, arch-bishop of York, Henry's ambassador to the Holy See. These promotions were significant; the pope was widening his con-federation. He had hopes of gaining over the flighty emperor, who was disappointed with the fruits of his ill-conducted campaign, and still higher hopes of winning to his side the king of England, naïve, enthusiastic, and under the influence of his father-in-law. By the tactful dating of his instrument he had made it appear that his forgiveness of the Venetians was due to the intervention of Henry to whom, on 10 April, he had sent a golden rose. The news that England had renewed her treaty with France in March 1510 was unwelcome, but the pope was still confident that the hot-headed young monarch would con-form to the old English tradition and launch an attack upon France, whose *fleur-de-lis* occupied the place of honour on his royal shield.

His confidence was not ill founded. Some of the elder states-men of England hesitated, but their influence was less than that of Ferdinand who, after some hesitation, decided that, under cover of an English invasion, he himself might overrun Navarre, a small land-locked kingdom to the south of the Pyrenees.

---

[1] The council sat, with some interruptions, from May 1512 to March 1517; it endeavoured in vain to reform the Calendar and made no other great reforms. While it was in being Leo X issued a Bull in May 1514 reasserting the immunity of the clergy from lay jurisdiction. A. F. Pollard, *Wolsey*, p. 55 n. The *Conciliabulum* moved from Pisa to Milan in December 1511.

Throughout the year 1511 the balance swayed uncertainly. France showed herself punctilious in the payment of the pensions due under her treaty,[1] curbed the impatience of James of Scotland, and promoted the genuine efforts for peace made by Andrew Forman, bishop of Moray, whom James sent, via England and France, to the pope towards the end of 1510.[2] Her efforts were in vain. Henry showed his inclination by sending forth, in 1511, two expeditions each of about 1,500 men in support of the enemies of France. One of these, led by Sir Edward Poynings, aided Margaret of the Netherlands against the duke of Gelders and achieved some success in a brief campaign from July to September; the other, which sailed from Plymouth in May under Lord Darcy to join in Ferdinand's attack upon the Moors, found on its arrival at Cadiz that it was not wanted, gave a display of indiscipline and returned home after about a fortnight's stay. Ferdinand meant to avail himself of English help, but not against the Moors. On 4 October he joined the pope and Venice in a Holy League to defend the papacy from its enemies; on 13 November Henry VIII joined this Holy League, and four days later made with Aragon a definite treaty binding himself to attack France before the end of April 1512.

The die was cast, and though Henry still maintained the pretence of peace he was now resolute for war. In December 1511 he recalled the artillery which he had lent to Margaret on the ground that he needed it for his expedition to Scotland. The parliament which met on 4 February 1512 was a war parliament. Warham preached peace, but the pope's appeal for aid was read to the lords and communicated to the commons; some of Henry VII's legislation for the conduct of war was renewed; two fifteenths and tenths were granted, and the preamble of the act expatiated on the insatiable appetite of the king of France and the misconduct of the king of Scots 'very homager and obediencer of right to your highness'. Louis made a final effort by sending the bishop of Rieux to London about Easter time, but his envoy could not gain access to the king and understood from the councillors that war was intended. England

---

[1] *Letters and Papers*, I. i. 495. A payment was made as late as December 1511.
[2] For Moray's mission and France's endeavour to keep England at peace see *Flodden Papers*, and J. Herkless and R. K. Hannay, *Archbishops of St. Andrews*, ii. 35, 43.

made no declaration of war,[1] but on 7 April Sir Edward Howard was appointed admiral of the fleet and on the next day an indenture was made with him for the providing of the crews for eighteen ships. Before long he was at sea pursuing French fishing-boats and threatening the coasts of Normandy and Brittany. At first, the French believed that a landing would be made on their northern shores; but Henry, bribed by his father-in-law with empty promises of Guienne, had decided to send his army to Galicia to co-operate with the Spanish army. Whilst his host was gathering at Southampton, bad news came from Italy. For long the papal fortunes had flourished. The projected council at Pisa, when at length it assembled on 1 November, turned out to be a mere *conciliabulum* of French ecclesiastics; the Swiss invaded the Milanese, Julius laid siege to Bologna which had rebelled against his unpopular governor and fallen into the hands of the French. The pope, however, had alienated sympathy by his overweeningness, and in the spring of 1512 Louis seemed likely to recover all. Under Gaston de Foix, 'the thunderbolt of Italy', his armies swept across the northern plain, sacked Brescia, and in a brilliant action on Easter Sunday, 11 April 1512, routed their enemy at Ravenna. The papal defeat was less serious than it seemed at first, for Gaston was killed, and without his inspiring presence the French army soon fell to pieces; but it was bad enough, and for the moment it seemed overwhelming. It is characteristic of Henry that he pursued his way undaunted. On 7 June an army of 10,000 men, safely convoyed past Brest by Howard, landed near Fuenter-rabia. It accomplished nothing. Withholding the transport which he promised, Ferdinand kept the English inactive before Bayonne, where they masked a French army, while he himself overran Navarre. Without tents, ill fed, short of beer, and un-accustomed to wine, the English army lost both its health and its discipline; many of the men died of dysentery; the others bluntly told their inefficient commander that they would not abide after Michaelmas 'for no man', and in October made their way home without their commander's leave.

This utter failure was to some extent redeemed by the successes of Howard at sea. On the way home from the convoy, he burnt Le Conquet in Brittany; after refitting at Portsmouth,

[1] The English breach of treaties with both France and Scotland was emphasized by James in a letter of 7 August to John of Denmark. *Epistolae Regum*, i. 165–6.

where he waited in vain for the promised help from Spain, he reappeared off Brest on 10 August with twenty-five ships of war, among them the *Regent* and the *Sovereign*, and a large number of Flemish hulks and victuallers. Outside the harbour he surprised a French fleet of twenty-two vessels of which the biggest were the *Louise* of 790 tons and the *Cordelière* of about 700. He attacked at once, damaged the *Louise* by cannon-fire and drove her with most of the French vessels into headlong flight. The *Cordelière*, which stayed to fight, was grappled by the *Regent*; her magazine exploded and the two great ships with their captains, Sir Thomas Knyvet and Hervé de Porzmoguer, were utterly destroyed. Though his loss must have been rather the heavier, Howard had the victory in that he penned the French fleet in Brest. For a few days he remained off the port, putting ashore parties which ravaged and took prisoners, and before he returned to Southampton at the end of the month he had swept the coasts of Brittany, Normandy, and Picardy, taking many prizes. The English guns, perhaps, were not as good as the French, but in seamanship and in manhood the English sailors had proved themselves superior.

Attacked thus from the sea, hard-pressed in Italy, beaten in Navarre, and uncertain of Maximilian, who eventually joined the Holy League in November 1512, Louis, changing his policy, now tried to involve Scotland in a war with England. On 16 March the 'auld alliance' had been renewed at Edinburgh on the usual terms, but when, immediately afterwards, the failure of Rieux's mission convinced Louis that war was inevitable, La Motte was sent back to Scotland with a revised version of the treaty. In this, the clause which made the alliance operative only when Anglo-Scottish truces had been broken by the act of the English was omitted on the ground that English outrages, unredressed, were hostile acts. On 10 July James, with some reservations of his own, accepted the document in its amended form and thereafter the French king insisted that England and Scotland were definitely at war. The king of Scots did not accept this interpretation. As late as April 1512 he was negotiating for peace, rather naïvely through Ferdinand, the principal architect of the war, and throughout that year he still hoped to persuade the pope to gather the nations of Europe for the great crusade.[1]

[1] For James's dealings with Ferdinand, see references to Leonardo Lopez in

West and Dacre, whom Henry sent up in the early summer, reported favourably on James's attitude, though the Venetian ambassador observed that the Scots people wanted war though their king did not. There was much coming and going between Scotland and France, and between August and October the English government kept on foot 'an army in the North' under the command of Surrey, who was sent to York.[1] This was a mere skeleton force, which cost in all only £2,166. 11s., but Surrey took the opportunity to prepare for the rapid mobilization of all the strength of the North, and, as the sequel was to show, his work was well done. There was, however, no sign of Scottish hostility and during the winter of 1512–13 it seemed likely that peace would return to Europe. Ferdinand, possessed of Navarre and hopeful of gains in Italy, began secret negotiations with France which resulted in the arrangement of a year's truce at Orthez on 1 April 1513, and France, for her part, was privately negotiating with Venice, with whom she came to terms on 26 March.[2]

The pope, however, was resolute to continue the struggle. No less resolute was the young king of England whose simple piety urged him to assist the Holy Father and whose instinctive statesmanship told him that the military honour of England must be redeemed from the disaster of 1512. By a treaty of 5 April 1513 Henry bound himself along with Julius, Maximilian, and Ferdinand jointly to declare war on France within thirty days, and though the Aragonese, exposing his duplicity, hastily withdrew, he hurried on his martial preparations. Already Howard was at Plymouth with twenty-three king's ships and five hired vessels;[3] he was short of both beer and biscuit, but he had plenty of men and guns, and on 20 April he put to sea without waiting for his victuallers. Arrived off Brest he found himself worse placed than he had been in the previous year; the weather was bad, for it was early in the season, and to the defence of Brittany Louis had summoned the redoubtable Prégent de Bidoux whose galleys had worked wonders in the

---

Letters and Papers, i. i. 513, 516, 530, 531, 547, and *E.R.S.* i. 131, 133; for his hopes of a pope-made peace, see references to Octavian Olarius, *Letters and Papers*, i. i. 584, 691, 734, 736, 741, and *E.R.S.* i. 179, and to Andrew Forman, whom James hoped to send to France by way of England, *Letters and Papers*, i. i. 594, 599, 603, 623, 691.

[1] Ibid. 660.
[2] Bridge, pp. 197, 200.
[3] Navy Records Society, *The French War of 1512–13*, xxxiii.

Mediterranean. Light of draught and carrying very heavy guns
the galleys were most effective for in-shore fighting, and the
English admiral's attempt to come to grips with his foes ended
in disaster. In an endeavour to enter the harbour, Arthur
Plantagenet[1] lost his ship upon a hidden rock, and Howard,
realizing that assault was impossible, fell back on his tactics
of blockade which he maintained in sorry lack of food and
drink. On Sunday, 25 April, provoked, it seems, by an unworthy letter from his master, he made a foolhardy venture
on the galleys in four small boats. It was a forlorn attempt. His
archers did manfully and he himself succeeded in boarding
Prégent's own vessel; but the painter slipped, or was cut, and
before long he was thrust over the side by morris-pikes. Discouraged by this failure, the weather-beaten, under-fed fleet
returned to Plymouth on 30 April. There it endeavoured to
reorganize under a new admiral, Thomas Howard, elder
brother of the valiant Edward; but the captains were dashed
by Henry's undeserved rebukes, the men deserted, provisions
and casks were scarce, and it was bad weather rather than the
English fleet which held off the French counter-attack. None
the less, the English naval power recovered so quickly that
Henry was able to carry out his invasion of France without
interruption.

For this, great preparations had been made. In March
Nicholas West, dean of Windsor, was sent up to Scotland
equipped with a monitory brief from the pope and a copy of a
'Bull executorial', which had been confirmed by the new pope,
Leo X, to warn James of the dire consequences of an attack
upon England.[2] James, in irritation, said he would appeal to
Prester John, whom Henry in a letter to Bainbridge identified
with Prégent,[3] but he replied on 13 April[4] still urging peace and
evidently hopeful that he could deter Henry from invading
France. Henry was not to be deterred. Under the direction of
the panurgic Wolsey, preparations were carried on apace. In
the middle of May the earl of Shrewsbury crossed to Calais
with the van of 8,000 men. Soon afterwards he was joined
by the rearguard of 6,000 under Sir Charles Somerset, Lord

---

[1] He was an illegitimate son of Edward IV.
[2] *Letters and Papers*, I. i. 791 and 809.
[3] Ibid. 806. 12 April (Prégent was called Pregianni).
[4] Ibid. 810.

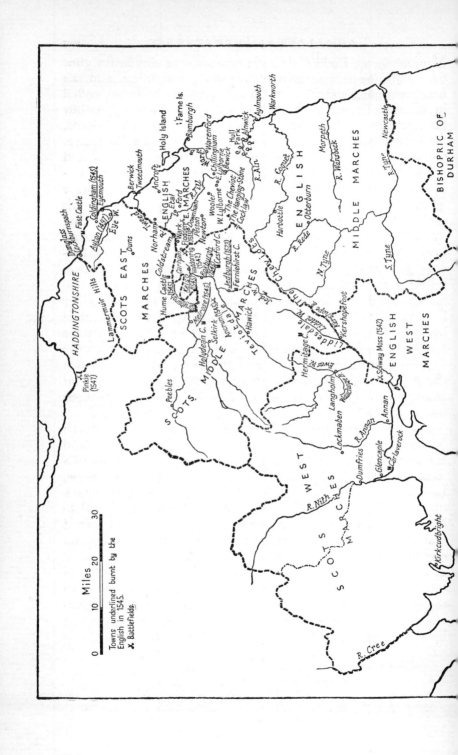

Miles
0   10   20   30

Towns underlined burnt by the
English in 1545.
✗ Battlefields.

BISHOPRIC OF DURHAM

HADDINGTONSHIRE

SCOTS EAST MARCHES

ENGLISH MIDDLE MARCHES

ENGLISH E. MARCHES

ENGLISH WEST MARCHES

SCOTS MIDDLE MARCHES

SCOTS WEST MARCHES

TEVIOTDALE

LIDDESDALE

Lammermuir Hills

Pinkie (1547)

Dunglass
Cockburnspath
Fast Castle
Ayton (1497)
Eye W.
Eyemouth
Coldingham (1545)
Berwick
Tweedmouth
Duns
Holy Island
Farne Is.
Ancroft
Bamburgh
Warenford
Chillingham
W.Lylborne
E.Lylborne
Hull
Newcastle
Warkworth
Aylmouth
Morpeth
R.Wansbeck
R.Ain
R.Coquet
Alnwick
Hull Park
Wooler
The Cheviot
The Hangingg-Stone
Cocklaw
Harbottle
Otterburn
R.Rede
N.Tyne
S.Tyne
R.Tyne
Norham
Ford
Etal
Tweed R.
Wark
Coldstream
Hume Castle (1545)
Roxburgh
Kelso (1545)
Maxton (1545)
Melrose (1545)
Ancrum (1545)
Ednam
Newton
Cessford C.
Jedburgh (1523)
Ferniehirst C.
Hawick
Selkirk (1545)
Holydean C.
Peebles
Jed
Ale W.
Teviot R.
Hermitage C.
Ewes W.
Wauchope W.
Langholm
Langholm
Kershopefoot
Liddel W.
Solway Moss (1542)
Lockmaben
Annan
R.Annan
Caerlaverock
Glencaple
Dumfries
R.Nith
Kirkcudbright
R.Cree

S C O T S
E N G L I S H

Herbert. In the middle of June these two commanders marched out, and after a feint on Boulogne sat down before the little town of Thérouanne. On the last day of June there arrived at Calais, thus conveniently cleared, in a fleet such as Neptune never saw before, the main battle of about 11,000 men (including carters and others) under Henry himself. The host was both magnificent in dress and well provided for war. The king had enlisted some German mercenaries, 800 of whom marched 'all in a plumpe' before the royal person; the 'Spears' were in attendance and so were the royal guns, amongst them twelve great pieces called the 'twelve apostles'. Though threatened from Boulogne, Henry reached the camp at Thérouanne without serious interference. Two of his guns, however, were upset on the way; one of these, 'St. John the Evangelist', fell into the hands of the enemy, but the other, 'the redde gonne', was gallantly brought off in the face of superior force.

On 12 August, outside Thérouanne, Henry was joined by Maximilian, who is said to have urged him to cross the Lys and invest the town completely.[1] This he did, and as the result of the new dispositions, a French attempt to re-provision Thérouanne was routed on 16 August at Guinegate in an engagement usually called the 'Battle of the Spurs', from the speed of the French retiral in which many noble prisoners, including Bayard himself, were taken. The town fell on 23 August, and early in September the English formed the siege of Tournai, which capitulated on the 21st. The French king, who had just lost Dijon to the Swiss, could make no riposte and Henry, after renewing his engagement with the Netherlands and preparing for his sister's marriage to the Archduke Charles, sailed home in triumph from Calais on 21 October.[2]

During his absence Surrey, who, sorely against his will, had been left behind in England, had gained a resounding victory over the Scots. After a vain attempt to reconcile his contradictory obligations by limiting his aid to France to a naval expedition which left Leith on 25 July, James had sent Lyon King with an ultimatum to Thérouanne. Henry must withdraw from France or he must expect an invasion from the North. His envoy

[1] Bridge, p. 228.
[2] John Taylor, clerk of the parliaments, left a diary of events from his arrival at Calais (25 June) to the king of England's embarkation for England (2 September). *Letters and Papers*, i. ii. 1057 ff.

received a contemptuous reply and, hastily assembling his forces, the king of Scots crossed the Border on 22 August. He brought with him the available strength of his country. His guns, some brought from France, some cast in Scotland, were good, and his own entourage was probably well armed, but the bulk of his army consisted of shire levies, summoned by letters issued late in July, and the followings of feudal lords and highland chiefs. Norham castle, to the disgust of the bishop of Durham, fell on the 28th to a violent bombardment and assault, and the Scottish army took, without undue difficulty, the smaller castles of Wark, Etal, and Ford. After the fall of Ford, James established himself about 4 September in a strongly fortified camp commanding the road which led north from Wooler down the valley of the Till. He has been blamed for his indecision, but he could not possibly have moved south leaving Berwick unmasked, and he was in no case to besiege Berwick with an undefeated English army hurrying up to offer battle.

The Scottish invasion was by no means unexpected. As early as 9 August, apostolic censures had been pronounced against James at Rome, and in England, as the financial records show, war-like preparations were being made soon after the middle of July. On the 16th day of that month Sir Philip Tilney, treasurer of the king's wars under Surrey, received £1,000 from the treasurer of the king's chamber, and from the subsequent account it is clear that the 'viage' was reckoned to begin on 21 July. Surrey, with a 'retinue' of 500 trusty men, left London on 22 July. On his way north he had dispatched to the assistance of the Border garrison 200 mounted archers whose discipline and valour made short work of a raid conducted by Lord Home—the 'Ill Raid', the Scots called it—and when he arrived at Pontefract about the beginning of August, he at once put into train the arrangements he had made in the previous year. The royal artillery, which—this is significant—had already arrived at Durham, was sent north to Newcastle, and when, on 24 August, Surrey heard of the invasion he was able to summon the whole strength of the North to be with him at Newcastle on 1 September. His arrangements worked well and, reinforced by 1,000 men from the fleet under his son, Thomas, the lord admiral, he took the field at Bolton in Glendale with an army of about 20,000 men.[1] These were stout fellows drawn from the

[1] On the opposing armies see *Scottish History Society, Misc. VIII*, 35.

northern counties according to a prearranged plan and it may
be concluded both from the speed of concentration and from
the rates of pay that many of them were mounted. They were
given conduct money from their own homes and taken into the
king's pay as from 1 September. Among them appeared members of those great fighting families of the North whose leaders
at the time of the Pilgrimage of Grace were able to gather in so
short a time so large an army—'34,000 or 35,000 men well-tried
on horseback'—and throughout the host was a spirit of confidence
which appears in the lord admiral's assertion that he and his
men would neither give nor take quarter.

The strength of James's army is hard to assess. As the total
population of Scotland at that time was only about 600,000 it
can never have approached the traditional 100,000, and as it
had been weakened by desertion it may not greatly have
exceeded that of the opposing force on the day of battle. But
it was well posted on Flodden Edge and when Surrey made
contact on 7 September he hesitated to attack a position that
resembled a fortress. He attempted to lure James from his stronghold with a knightly challenge only to be told that it was not
for an earl to dictate the terms of a combat to a king, and on the
8th crossed the Till and marched off north-east, apparently
with the intention of invading Scotland by way of Berwick. He
halted for the night under cover of Bar Moor and, perhaps as the
result of a reconnaissance which may have shown the Scottish
camp in some disorder, he decided upon a bold stroke. Breaking
camp before five in the morning he sent the admiral with the
vanguard and the artillery on a long march which brought them
across the Till at Twizel Bridge about six miles north of Flodden,
whilst he himself with the main battle worked across a ford or
fords near Etal. His intention was to unite the two divisions and
attack the Scottish position from the north, thus depriving his
enemies of the protection afforded by their strong camp which
was on the southern slope of Flodden Edge. His success was
complete; if he did not take James altogether by surprise, he
took him at a disadvantage.

James, who seems to have lost touch with his enemy, may
have been preparing to withdraw by way of Coldstream, though
according to the Scots account he felt himself compelled to stay
in position until noon in accordance with his pledge to Surrey.
He must have realized his danger soon after 11 a.m., when the

admiral crossed at Twizel Bridge, and either because he was already *en route* for home, or more probably in the hope of denying a good position to his enemy, he endeavoured to occupy Branxton Edge, a hill about a mile north of his original position. The manœuvre was evidently difficult for his clumsy host, and before it was completed his enemy was upon him. His left wing indeed, under Huntly and Home, gained the height and, descending, scattered the extreme right of the English under Surrey's younger son, Sir Edmund Howard, and for a moment the situation was so critical that the admiral tore the *Agnus Dei* from his breast and sent it to his father. The old 'white lion' was equal to the occasion. He brought the main battle upon the left of the van and advanced on the Scots in one grand line. On his own left, Sir Edward Stanley, no doubt taking advantage of the dead ground, gained the height of Branxton Edge before the Highlanders of Lennox and Argyll, whose headlong valour quailed before the English archery. In the centre of the field the Scots gained the height only to find themselves exposed to the English guns and hand-guns, against which their artillery, perhaps because it was hardly moved from its original position, made but an ineffective reply. Galled beyond endurance the Scots came down the hill in silence 'almain fashion'—like the landsknechts, that is, in heavy masses. The passage over rough country broke their formation and in the mêlée which followed, the English bills were far more effective than the long Scottish spears. Staking all on a last desperate throw, James hurled himself with his household upon the main battle and there, only a spear's length from Surrey's banner, he was slain. With him there died his son, the archbishop of St. Andrews, the youth of twenty who had been a pupil of Erasmus, a bishop, two abbots, twelve earls, fourteen lords, and representatives of most of the families of consequence in Scotland as well as a mass of lesser folk, reckoned by the triumphant English as over 10,000, perhaps even 12,000. Their own losses they put at 1,500 at most.

The battle was not joined until four in the afternoon and the increasing darkness may have concealed from the English the true extent of their gain; in any case, men who had marched so hard and fought so fiercely in a long day were in no case to pursue. Next day they secured the Scottish artillery, seventeen fine guns, and found the dead body of the king. They realized that the power of Scotland was broken, for the time at least, and Surrey did not hesitate to spare the royal exchequer by dismissing the great bulk of his army on 14 September.

The pious Catherine, knowing that James had died excommunicate, was uncertain what to do with his body, but Henry gained the pope's permission to have him buried in consecrated ground and visualized a funeral in St. Paul's. Characteristically, he asked that St. Andrews should be reduced to a bishopric and that he should be given control over appointments to Scottish sees. Characteristically, too, while he remarked that James had paid a greater price for his perfidy 'than he would have wished', he did not remember his own provocative actions or his own breach of a sworn treaty with France. The pope was with him, who could question his morality?

The king of England had shown himself the paladin of the church and his was the glory. Yet either he or Wolsey soon began to fear that while England had done the fighting, others were drawing the benefits. Before he left the Netherlands Henry had become party to a treaty (signed at Lille on 17 October) by which he, Ferdinand, and Maximilian bound themselves to a combined invasion of France to be begun before June 1514;[1] but in the following December the Aragonese, who had descanted largely upon the sacrifices he had made, told his ambassador in England not to ratify the treaty which he himself had signed.[2] He calculated that Louis, who had submitted to the pope in October 1513, would now be amenable to a scheme whereby his daughter Renée should be married to Charles's brother Ferdinand and given as a marriage portion all the French rights in Italy. Maximilian, who would in any case approve a design for his grandson's benefit, might be confirmed in his purpose by the offer of gains at the expense of Venice, but this part of the plan must be kept secret at first since France was the ally of the republic. The whole scheme could be justified on the ground that it was the necessary prelude to a crusade, and with the pope's approbation might obtain respectability in the eyes of the world and acquiescence at least from the enthusiastic Englishman. On 13 March 1514[3] Ferdinand renewed his truce with Louis and soon after Maximilian gave his adhesion; the marriage between Charles and English

---

[1] *Letters and Papers*, i. ii. 1052.   [2] *Spanish Calendar*, ii. 195.
[3] Ibid. 207.

Mary which, as arranged at Lille,[1] was to be completed before 15 May 1514 did not take place.

The king of England had as the result of his war and of his diplomacy only the towns of Tournai and Thérouanne, and the consciousness that he alone of the European princes had kept his word. It was a poor return for honest effort, but Henry was not his father's son for nothing. He realized that, Julius being dead, war was no longer commendable at Rome, and that Leo X wished to promote an understanding between England and France; he realized, too, that since it was he who was the real danger to Louis, he might fare better than any other prince if he came to terms with France. When, early in 1514, he discovered the dishonest practices of Maximilian and Ferdinand, he had his reply ready. At once he started secret negotiations with the French king, and as soon as the date fixed for the Netherlandish marriage had passed he pursued active negotiations for a match between his sister Mary and Louis XII who, at the age of fifty-two, was already reckoned an old man. By mid-July the Venetians were confident that an arrangement had been made,[2] and a marriage treaty was in fact completed on 7 August. On the 10th of that month a peace between England and France was formally proclaimed; Henry kept Tournai, and the pension promised under the peace of Étaples was almost doubled. Two days later the English king wrote to the pope explaining that as Charles had failed to keep his engagement Mary had been betrothed to the king of France; on 18 August a marriage by proxy took place in London and in the following October the princess, suitably equipped with train and trousseau, set out for her new realm.[3] The marriage was formally proclaimed on 9 October, but Mary was a most reluctant and discontented bride. She had given her consent only on condition that when her ageing husband died she should choose her second husband herself,[4] and already her mind was set on Charles Brandon, newly created duke of Suffolk who, it may be remarked, had already at least one wife living.[5]

---

[1] This marriage had been arranged in 1507. See p. 187.

[2] *Letters and Papers*, I. ii. 1317, 1336, 1338.

[3] Ibid. 1351, 1401-8.

[4] Ibid. II. i. 75.

[5] Brandon had married Anne, daughter of Sir Anthony Browne, to whom he had been contracted in youth, after he had married by dispensation and then repudiated Margaret Mortimer. To secure Mary he had his marriage to Anne

As an essay in *Realpolitik* the Anglo-French treaty is a master-piece. A reluctant princess contracted a loveless marriage; two treacherous monarchs, Maximilian and Ferdinand, were themselves deceived and their dreams of controlling Italy were brought to nothing; Charles of Castile was driven to accept comprehension in the Anglo-French treaty, and France, recently expelled from Italy, was able once more to threaten Milan. The king of England, though pledged with Louis to render mutual aid against common foes, withdrew from France with his conquered towns and his pension. He was resolved to set his kingdom in order, to be no more the catspaw of the princes of Europe, and to settle with the Scots who were to be 'comprehended' in terms they could not possibly keep. He was still at one with the pope, who gave his assistance and benediction to the Anglo-French marriage; but the paladin of the church, now an experienced man of twenty-three, had learnt a great deal in three years. He had learnt amongst other things how competent was his servant Thomas Wolsey.

Browne declared null and void on the ground that the dispensation obtained was invalid. A. F. Pollard, *Henry VIII*, p. 199.

IX

THE CARDINAL

FOR more than fifteen years the history of England was dominated by the great personality of Cardinal Wolsey. Seldom in England has a subject enjoyed such power. Thomas Becket, indeed, had been both chancellor and archbishop, but the powers which he exercised had been in his person mutually opposed, and if he served the king well as chancellor, he defied him to the death as archbishop. In the fifteenth century Henry Beaufort had been cardinal, chancellor, and legate *a latere*; but he was not an archbishop and his career was run at a time when neither the English monarchy nor the papacy was in full vigour. Thomas Wolsey held power at a time when the realm of England was flourishing and when the papacy, albeit in danger of becoming an Italian state, still exercised great influence on European affairs. He was at the same time chancellor, archbishop, cardinal, and legate *a latere*, and in his capable hands the powers belonging to each of his offices were welded into a single compelling force which defied all authority save that of the king of England, who might sometimes be hoodwinked, and that of the pope, who might sometimes be disregarded. Before the rulers of Europe he stood forth as the man who controlled England's destinies. 'This cardinal', said a Venetian ambassador who, by the year 1519, had been in England for four years, 'is the person who rules both the king and the entire kingdom.' The cardinal, he alleged in another letter, might be styled *ipse rex*.[1] Cardinal d'Amboise, who did everything for Louis XII—'leave it to Georges', they said—may be regarded as a prototype, but the real parallel to the greatness of Wolsey is that of Richelieu, who ruled France by his own personality and by the agency of commissioners dependent upon himself.

To the chariot of the state Wolsey harnessed mettlesome horses of different breeds and different tempers; guiding their course by the whip of authority and the rein of expediency, he seemed himself to be the true type of a Renaissance prince. Yet they were wrong who thought that the king was but his

[1] Pollard, *Wolsey* (1929), pp. 102–3, quoting *Venetian Calendar*.

shadow. Henry VIII remained, as he is shown in the famous play, the hidden power which lies behind the bright action, and in his brooding presence lies much of the significance of Wolsey's achievement. The cardinal lived for the hour. Of the vast movement which we know as the Reformation he understood very little, and to the solution of political questions which he understood well enough he made no real contribution; at home he did all by 'authority'; abroad he relied not on 'the balance of power', but upon opportunism tempered by a belief that the papacy must win in the end. When he fell his power vanished utterly, yet it left more than a wrack behind. Its legacy was not in the way of institutions or even in machinery, but in its record of things done and of authority realized. The great cardinal had shown how the strength of the law could be made the instrument of the Crown, and how the power of the church could be used not to check the royal authority but to enhance it. His master, never at any time a mere figure-head, understood very well what he saw. If his minister, at once chancellor, cardinal, and papal legate, could weld the forces of church and state into so powerful a dominion, what might not one do who was himself a king, the chosen of God, the champion of the papacy, the governor of England? The rule of Wolsey presages the appearance of Henry's 'empire' and island papacy. The cardinal's power was ephemeral, but while it lasted it was real indeed. Power was his God.

'Truth it is, Cardinal Wolsey, sometime Archbishop of York, was an honest poor man's son, born in Ipswich.' So begins the *Life* written by George Cavendish which, as is natural in the work of a gentleman usher, deals much with the outward splendour of the cardinal's *ménage*—his crosses, his pole-axes, his rich robes, and his handsome attendants. He was the son of a butcher and inn-keeper who was frequently presented at the court leet for offences against the rules of trade, but who was also a churchwarden,[1] and the lowliness of his birth is worthy of remark since it affected his mental outlook throughout his life. In his youth he was a rebel against authority;[2] when he was bursar of Magdalen he seems to have spent, though not to

[1] The malicious coat of arms printed in Roy's *Rede me and be not Wrothe* quarters the three bulls' heads with three bloody axes. From manuscripts in the Town Hall at Ipswich it is plain that Wolsey's father was indeed a butcher and apparently of some truculence. Letter in *The Times*, dated 24 June 1930.

[2] Pollard, *Wolsey*, p. 13.

have misspent, the college funds without a proper warrant; when he was rector of Limington he was put in the stocks by Sir Amyas Paulet for excessive gaiety after a fair. Later, however, when he had power in his own hands, he showed himself a vigorous disciplinarian and he delighted to exercise discipline against members of the aristocracy. He was born in 1472 or 1473, entered Magdalen College at a very tender age, and at the age of fifteen had already gained celebrity as 'the boy bachelor'. Before long he was fellow and bursar, but he had no intention of wasting in an academic life the vast powers of intellect and industry which he felt himself to possess. Through the favour of the marquess of Dorset, whose sons were at Magdalen College School, he gained the living of Limington in Somerset and on Dorset's death he found a new patron in Henry Deane, archbishop of Canterbury, whose chaplain he became. Characteristically, he had obtained in 1501 a dispensation to hold Limington along with two other benefices *in absentia*, and on Dean's early death he sought a new road to advancement in the service of Sir Richard Nanfan, deputy lieutenant of Calais. Here he found full scope for his abilities, for Nanfan, who was old and ailing, entrusted him with much business and, on his own retirement in 1506, recommended his energetic chaplain to the king. In 1507 Wolsey became a royal chaplain and though Cavendish's story of his phenomenal speed on an errand to Flanders is no longer believed, he certainly commended himself to his master by his thoroughness and dispatch. Once again fate robbed him of a patron, and on Henry's death he was in some danger of losing all, since the old king's favourites were not popular in the new reign. But his abilities, coupled with his indomitable desire for advancement, brought him into the royal service as almoner and councillor in November 1509, by which date he had somehow secured the deanery of Hereford. Probably he owed his appointment to the good offices of Fox (privy seal), who was in need of assistance in opposing Surrey (treasurer), but before long he had got so close to the king that he was able to dispense with the help of lesser men. As early as 1511 he presented to Warham a bill signed by the king which lacked the authentication of both signet and privy seal, and the chancellor accepted it,[1] apparently because 'Dominus Wulcy' gave him the letters by the king's direct

---

[1] Pollard, *Wolsey*, p. 15.

command. By this time he was dean of Lincoln and, as part of the reward for his vigorous conduct of the war of 1512–14, he was promoted to the bishopric of Lincoln, along with which he was soon allowed to hold that of Tournai *in commendam*. Other honours came to him, or, to speak more truly, were seized by his acquisitive hands. He had become dean of York in 1513 and when, in July 1514, Christopher Bainbridge died at Rome, not without a suspicion of poison, he was quick not only to secure the archbishopric for himself, but to pay the expenses of his promotion out of his predecessor's estate. As Warham showed no sign of dying—he lived until 1532—Wolsey had now reached the highest position open to him in the English church; yet still he was not satisfied and he sought to gain from Rome promotion which would make him not only supreme in England but renowned throughout Europe.

Clearly a king and his minister who had done so much for the Holy See could expect recognition of their service. Wolsey determined to become cardinal and legate *a latere*. The situation was favourable, for, Bainbridge being dead, there was now no English cardinal, and although his candidature was prejudiced by his own greed and by the ill conduct of his agent at Rome, Silvestro de' Gigli, bishop of Worcester, he obtained the coveted red hat in 1515; its arrival in November was made the occasion of a magnificent ceremony which emphasized the supremacy of the cardinal over archbishops and bishops, over dukes and temporal lords.[1] The legacy was a different matter. By this time the title *nuncio* was given to a special ambassador from the pope; that of legate was assigned to a permanent representative. Every metropolitan was a *legatus natus* and administered some portion of the papal authority in his province, but the *legatus a latere* had special power not exactly defined as of course, but far-reaching. He might even be empowered to release land from the dead hand of the church if the transaction could be shown to be *in evidentem utilitatem*, as easily it might in a day when customary arrangements accorded ill with the sharp fall in the value of money. Obviously, so wide a delegation of papal power was made only after much consideration and, as a rule, for a short period; Leo X was shaken when the ambitious Englishman demanded to be made legate *a latere* for life, but

---

[1] The archbishops of Armagh and Dublin as well as the archbishop of Canterbury assisted at the religious service.

Wolsey was not to be put aside. In May 1518 he secured the office he demanded by *force majeure*—he kept Campeggio waiting for three months outside the walls of Calais on the plea that it was not 'the rule of this realm' to admit a legate *a latere*. The legacy was given him under the decent pretext that he was to act with Campeggio who had come over on a special mission, and the pope intended that it should be temporary; but the resolute cardinal soon had his commission extended to three years, to five years, to ten years, and at last, when Clement VII was in great difficulty, he procured a grant for life. He did not at any time acquire the whole of the faculties which he demanded, but he obtained a power of 'reformation' which gave him a very free hand in regard to persons and benefices.

By this time, no doubt, he realized that the greatest ambition of all was not to be fulfilled. He had meant to be pope. His ambition was not unreasonable at a time when the affairs of the papacy were in an uncertain state and when he and his master had rendered signal services to the Holy See. As early as 1520,[1] the imperial ambassador in London, the bishop of Badajoz, reported that the French were promising to make Wolsey pope, and hinted that Charles himself might act more efficaciously than Francis in the matter. When Leo died in December 1521, Charles wrote promising his good offices; but in fact he promoted the election of his former tutor, the devout and narrow Adrian Dedel of Utrecht, and the cardinal of York obtained but seven votes out of thirty-nine in the only scrutiny in which he was considered at all. When Adrian's death in October 1523 presented a new opportunity, the emperor again played a double game and the English cardinal did not receive a single vote. Wolsey, who had been unreasonably optimistic, was furious and he did not forget the injury. His reputation in Europe remained high, however—if he was not loved, he was feared —and in England, at least, his power over ecclesiastical affairs was paramount.

This power he used primarily for his own aggrandisement. He was quick to assert that a legate *a latere* took precedence over an archbishop of Canterbury on formal occasions and that a cardinal should have more dishes at table than a lord of parliament.[2] He dressed magnificently and encouraged the

---

[1] Pollard, *Wolsey*, p. 126.
[2] Steele, *Tudor and Stuart Proclamations*, i, no. 75.

clergy to appear publicly in silk and velvet. Maintaining the power and dignity of the church at a time when the church was much criticized, he showed himself a champion of his order, but to the spiritual life of England he contributed nothing at all and it was upon the spiritual side that the English church, like other churches of the period, most needed help.

How general was the criticism of the established order appears from the situation which greeted Wolsey at the outset of his career as archbishop. In the parliament of 1512 there had been passed, not without difficulty, an act to deny benefit of clergy to men in minor orders who committed murder.[1] This act had been welcomed by some good churchmen; it was not severe, and it was to last only until it was reconsidered by a new parliament. In May 1514, however, Leo X had declared in his Lateran council that laymen had no jurisdiction over churchmen and the English clergy prepared to dispute the confirmation of the offending act in the parliament which had been summoned for 5 February 1515. They still held a small majority among the lords[2] and they endeavoured to sway public opinion by appointing the abbot of Wynchcombe, Richard Kidderminster, to preach a sermon at St. Paul's Cross just before the session opened. Founding upon *Nolite tangere Christos meos* Kidderminster made an uncompromising attack upon the act of 1512 which he represented as the outcome of heresy, but his eloquence was of little avail, for by this time had occurred an event which had stirred anti-clerical feeling to boiling-point.

On 14 December 1514, Richard Hunne, a well-to-do merchant-tailor of good repute, was found hanged in the bishop's prison at St. Paul's, called the Lollards' Tower, where he had been confined pending trial for heresy. His enemies, alleging that a guilty conscience had driven him to suicide, had him condemned by a clerical court over which the bishop of London presided, and his body was duly burnt as that of a heretic. The anger of the citizens was aroused. Hunne's friends asserted that his heresy consisted in a refusal to pay a mortuary due for the burial of an infant and in a threat to bring an action of praemunire in the king's bench. A coroner's jury, ignoring the

[1] 4 Henry VIII, c. 2.
[2] Pollard, *Wolsey*, p. 29. Summonses had been sent to 49 spiritual lords and 42 lay peers and to the prior of St. John, whose position was ambiguous; the actual attendance varied between 40 and 18 during the session of 1515; on 15 days the spiritual lords had a majority and on 12 the laymen.

finding of the clerical court, pursued its quest with vigour, effected the arrest of the jailer, Charles Joseph, who had absconded, and, largely on his evidence, cast for murder not only Joseph himself and the bell-ringer, but the bishop's chancellor, Dr. Horsey, who was in full orders. Horsey was arrested and placed in the custody of the archbishop of Canterbury.[1] Even amid these excitements the clerical influence was strong enough to reject two bills in favour of Hunne and to postpone a decision upon the act of 1512, but the prorogation of parliament on 5 April did not end the matter. The debate was continued elsewhere. The king had been persuaded by some of the commons, and notably by Sir Robert Sheffield, who had been speaker in 1512, that his royal rights were being infringed, and had presided over a debate at Blackfriars in which the case was argued by spiritual counsel. One of his champions, Dr. Henry Standish, had defended the act of 1512, mainly on the ground of English practice, and it was clear to the clergy that the validity of a papal decree was directly challenged. During the recess Standish developed his theories in public lectures, and as a result he found himself summoned before convocation when it reassembled in November 1515. There he was asked four pertinent questions: Could a secular court try clergy? Were minor orders holy? Did a decree of the pope and clergy avail against the established constitution of any country? Could a lay ruler restrain a bishop? Standish appealed to the king, and Henry, after consultation with the dean of his chapel, determined to defend Standish on whose behalf lords, judges, and commons had made representations. In a second assembly at Blackfriars,

---

[1] More, whose account of the affair is 'special pleading', alleges that Hunne's heresy was established by mere chance in a later investigation, and represents the evidence against Horsey as hearsay (a *Dialogue concerning Heresies*, iii. 15). That Hunne, like other citizens, attended private meetings to read the Bible is not improbable, but it is not clear that he was a suicide. The flight of the gaoler and the shameful alibi which he tried to set up impugn the character of the prison officials. It seems quite possible that they did Hunne to death, tried to make the death look like a suicide, and, failing to do this, pleaded instructions by Horsey, who may have suggested that they need not be too gentle with the recalcitrant prisoner. The bishop of London asserted that the gaoler's confession had been extorted by violence, and that public opinion in London would be quick to condemn any cleric though he were 'as innocent as Abel'; but the action of the clergy is suspect. Hunne had been accused, not found guilty, of heresy during his life, and as there was no proof of final contumacy, he should not have been condemned after he was dead. The speed and the irregularity of the proceedings against him suggest a desire to forestall the findings of a coroner's jury.

attended by the judges, the councillors, and others, the clergy
present at Standish's citation were declared guilty of praemunire; and in a meeting of council and both houses of parliament at Baynard's Castle Cardinal Wolsey, on his knees, made a
partial submission on behalf of the church. With the support of
Fox and Warham he pleaded that the matter should be remitted to Rome, though he denied completely any intention
to derogate from the privileges of the Crown, and the lame
compromise was ratified in effect by parliament where the bill
for renewing the act of 1512 was talked out in a very brief
session.[1] The victory, however, was with the king, for though
the charge against Horsey was dropped, he was compelled to
pay £600 and to leave London; convocation, declared by the
judges to have incurred the penalties of praemunire, withdrew
its action against Standish, who was soon promoted to the see of
St. Asaph, and begged Henry, very humbly, to suffer it to meet
and act as convocations had been wont to do.

Wolsey's part in these affairs is noteworthy. It was to him and
not to Warham that the clergy appealed in their desperation;
it was he who made their peace with the king, and he who, in
making it, was at pains not to alienate his royal master. The
conclusion he drew from the affair was that representative
bodies, as critics of authority, should be discouraged. He urged
the dissolution of parliament, and before long found means to
send Sheffield to the Tower, where he died; in the next fourteen
years parliament met only once (1523). To convocation he was
no more sympathetic. When he could, he made its authority
subservient to that of his legatine council; even in 1523, when
the summons of parliament made necessary the assembling of
the convocations for the voting of money, he seems to have
succeeded, in spite of opposition, in reducing their action to
formality, in keeping real power in his own hands, and in
exacting a heavy tax.[2] This he did by calling before him at
Westminster the convocation assembled by Warham to St.
Paul's.

> Gentle Paule laie downe thy swearde;
> For Peter of Westminster hath shaven thy beard

wrote Skelton. The great questions at issue, however, he did
not attempt to solve; the relations of universal papacy to

---

[1] From 12 to 22 November.
[2] Pollard, *Wolsey*, pp. 187 ff. for full discussion.

nation-state, of bishop to king, of spiritual to temporal juris-diction he left unsettled. He must have been aware of the weight of anti-clerical opinion and aware that this opinion might well be infected with heresy, but it does not seem to have occurred to him either to close the ranks of the church or to attack the evil at its roots. If, on the one hand, he was no persecutor he was, on the other hand, no real reformer. A few conspicuous abuses he was prepared to repress, not without benefit to his own pocket; but further he would not go. That hungry souls were unsatisfied with the spiritual nourishment given to them by the official church, did not seem important; all could be put right, he thought, if the church were strongly and wisely governed. Strength and wisdom, he felt, were in himself, and with the aid of his legacy he directed at his own will the fortunes of the church of England.

Over the bishoprics he asserted his power. One of the few reforms of the Lateran council had been to forbid the holding of bishoprics *in commendam*, but Wolsey unconcernedly held first Bath and Wells, then Durham, and finally Winchester, along with the archbishopric of York; he 'farmed' the sees of Salisbury, Worcester, and Llandaff, whose non-resident Italian bishops must content themselves with fixed stipends, and between 1514 and 1518 he held Tournai as well; he claimed as legate the spiritualities of all bishoprics during vacancy and, though a secular, he took to himself, *in commendam*, St. Albans, the richest abbey in England. The courts Christian became the courts of the legate and he used their authority to coerce the bishops, even the archbishop if need be, with threat of praemunire. He attacked their jurisdictions and particularly gathered to himself the profits arising from probates. He used his legatine powers of visitation to usurp ecclesiastical patronage, overriding the right of visitation *sede vacante* which belonged to the archbishop of Canterbury. He gradually took control over the smaller secular benefices; in every diocese he had his agents who advised him of vacancies and these he filled without consulting the bishops; he intruded his nominees everywhere and woe betide him who resisted. Over the monastic houses, too, he spread his acquisitive hand. Here his task was easier because his legatine power of reformation enabled him to suppress more than twenty small foundations and to threaten stronger houses which disputed his authority. His action has been justified on the ground that the houses which he suppressed were useless, and that he transferred their resources to the promotion of good learning in the colleges he founded at Oxford and Ipswich, but he cannot be acquitted of the charges of disregarding the rights of founders and of making personal gain. He deprived, or threatened, heads of houses who would not subscribe to his colleges. Having obtained control over conventual suffrages he used his power to promote his own friends, and some of his promotions were scandalous, especially that of the abbess of Wilton[1] in 1528. He thought no shame to provide for an illegitimate daughter in the nunnery of Shaftesbury, and on his son, Thomas Wynter, he showered small livings and considerable offices. To do him justice he seems to have had no mistress save their mother, 'the daughter of one Lark'; but it is significant that while the affairs of his family must be regarded, the affairs of the church of England might wait. He had at the back of his mind a plan to turn monasteries into churches and to create thirteen new sees, but this plan was never carried out. That he failed to execute it was due rather to lack of interest than lack of power. His strength may have been less than it seemed, for he had gradually alienated the goodwill of many of the bishops and clergy; but if, surrounded by sycophants and dependants, he realized the strength of the opposition, he disregarded it. *Oderint dum metuant.* If he forbore to interfere overmuch with the ecclesiastical polity, it was because his own interests were mainly personal and political; he was, in truth, supreme over the church of England.

No less commanding was his voice in civil matters. In 1515, having, according to Hall, pestered Warham into resignation by constant interference with his jurisdiction in the name of the king, he became lord chancellor. Of his conduct in that office a great authority has written: 'it has been said to be the mark of a good judge to amplify his jurisdiction. If this be a valid criterion, there can be no doubt that Wolsey was by far the greatest chancellor England ever had.' The chancery was, in essence, the writ-issuing department, and originally its head was, in the words of Stubbs, 'secretary of state for all departments'. He was reckoned the first councillor of the Crown.[2] His

---

[1] *Letters and Papers*, IV. ii. 1960.
[2] Even in the seventeenth century, though other officers were gaining in impor-

jurisdiction was only incidental to his office, yet it became in the end the greatest part of his business and it is as head of the legal system of England that the chancellor is important today. Early in the sixteenth century he administered a justice outside the common law in two distinct ways. In his own court wherein, as was not inappropriate for a churchman, he represented the king's conscience, he dealt with much business connected with trusts and uses, contracts and libel. His concern with these matters had begun before Wolsey's day, but in the hands of the tireless archbishop of York it exhibited itself with increasing activity. The eulogies of his friends, the accusations of his enemies, and the very bulk of the records are eloquent testimony to his all-pervading energy.

In the star chamber, which exercised the traditional jurisdiction of the council, particularly in the maintenance of good order, his dominating force was no less obvious. When he spoke of teaching recalcitrant gentry 'the new law of the star chamber', he did not mean that he proposed to introduce new legislation, or even to apply new methods. What he meant to do, and what he did, was to apply the old system with rigour. To his credit it may be said that he established order in a society where the over-mighty subject was still a danger, that he made the star chamber popular as the defender of the impotent, and that in the maintenance of order he punished with a heavy hand not only crimes of force, but crimes of fraud, forgery, perjury, contempt, and libel as well as all sorts of offences against the economic legislation of the period. His industry was such that the public came to believe that this majestic court, which sat only at Westminster and in public, was the new creation of the ever-vigilant cardinal.[1]

In the eighth year of Henry VIII the court sat on Monday, Tuesday, Thursday, and Saturday, Wednesday and Friday being reserved for chancery,[2] and even the endless business of star chamber and chancery did not satisfy Wolsey's unfailing

tance, the chancellor controlled an immense patronage and influenced administration in many fields. 'There is not in the whole Christian world so great and so powerful a magistrate as is the lord chancellor of England', wrote Bishop Goodman, *Court of King James I* (1839), i. 275. Clarendon was chancellor, but after his fall the chancellor was not what we should call 'prime minister' in England.

[1] Cf. Sir Thomas Smith, *De Republica Anglorum* (ed. Alston), p. 118.

[2] Dorothy M. Gladish, *The Tudor Privy Council* (1915), p. 122. After Wolsey's fall, star chamber became less assiduous. In 1529–30 it sat only on three days a week, and latterly on Wednesday and Friday only during term time.

lust for control. Over the court of requests, wherein the poor
men's complaints had obtained conciliar justice under the direction of the lord privy seal, he stretched forth his hand. When Fox
resigned in 1516 his successor Ruthall found that the privy
seal's jurisdiction was virtually absorbed by the star chamber,
whose business was so much swollen that poor men's suits were
referred to four committees of the council, one in the White
Hall, one in the Rolls, one in the treasurer's chamber, and one
under the presidency of the king's almoner. All these courts
were, in the long run, under the cardinal's hand and even they
did not suffice him. By means of injunctions he interfered with
the ordinary work of the judges; he issued commissions very
freely for all sorts of purposes,[1] and though many of these were
purely *ad hoc*, their cumulative effect was to exalt prerogative
justice at the expense of common law. Lastly, by the creation
of special courts, he confirmed his grip over whole areas of
jurisdiction. An admiralty court appeared after 1524; in 1525
a council in the Marches of Wales was established under the
nominal authority of the Princess Mary, and in the same year
the Council of the North, which had been dissolved in 1509, was
re-erected as the council of Mary's half-brother, the duke of
Richmond. In one way or another, the chancellor controlled
the whole judicial system of England; in chancery or in star
chamber he dealt with almost any matter he pleased, and when
his central courts rose or when they were overburdened with
business, he it was who decided to which court unsettled cases
were to go. He regarded the great seal almost as a personal
possession, took it abroad with him when he went on embassy
to Calais and Bruges in July 1521, and kept it with him in full
operation for months. He took it again to France in 1527 and
on the eve of his fall refused for a time to surrender it on the
ground that he held it for life.[2]

He was, in fact, making England a unitary state, at least as
regards law, and he was justified in his boast that he gave good
order. As Hall says, 'he punyshed also lordes, knyghtes and men
of all sortes for ryottes, beryng and maintenaunce in their
countreis, that the pore men lyved quietly'.[3] Until his financial
exactions drove England into discontent in 1525, the only

---

[1] For his proceedings under the act of apparel see Hall, *Henry VIII*, i. 148.
[2] Hall, i. 227, ii. 156; Pollard, *Wolsey*, pp. 92, 254.
[3] Hall, i. 152.

serious disturbance under his régime was that of the 'evil May-day' of 1517, and for that he was hardly responsible. During a period of economic depression, produced perhaps by the continental wars, the citizens of London convinced themselves that their plight was due to the incursions of foreign merchants and the dishonesties of foreign artisans to whom the government was unduly lenient. A few of the citizens, and especially a broker named John Lincoln, stirred up a Dr. Beal to give a sermon on the subject at St. Paul's Cross, and the upshot was an attack on the foreigners delivered mainly by a gang of irresponsible apprentices, though watermen, serving-men, strangers to London, and a few priests were also involved. The rising was not dangerous, but the civic authorities were taken by surprise; the lieutenant of the Tower, Sir Richard Cholmeley, fired a few rounds into the City and order was restored by harnessed men brought in by Shrewsbury and Surrey. Nearly 300 prisoners were taken at the time and in all over 400 persons including eleven women were arrested. Of these, Lincoln and thirteen others were executed on the absurd ground that an attack upon foreigners when the king was at peace with all princes was tantamount to treason. It was Norfolk and his sons who restored order and incidentally made themselves unpopular by their severity. The cardinal, who had fortified himself in his house during the crisis, had little to do with the matter, though he was at pains to take a prominent part in the dramatic scene in Westminster Hall when the king gave free pardon to 400 wretches brought before him in their shirts with halters round their necks.

In this instance he cannot be blamed except, perhaps, in that his foreign policy led him to favour strangers whilst he showed himself contemptuous of Englishmen, but an examination of his administrative work reveals much which calls for criticism. With all his doings he did nothing to reform the legal system. He may have meant to introduce more of the Roman law, which was already familiar to the ecclesiastical courts and to chancery, but all he did, as it seemed to his contemporaries, was to put too much power into the hands of spiritual men. They and his other servants administered a law which, though it might conceal itself beneath the veils of royal conscience or public good, was, in fact, merely the will of the chancellor. In the day of Wolsey's power, men murmured only under their breath, but it was plain enough to England that the cardinal used his justice for his own ends.[1] Members of the aristocracy were fined for keeping retainers, or were ejected from the council. After the cardinal's fall criticism was vocal, and John Palsgrave, tutor to Mary and Richmond, made the cardinal assert in an imaginary confession: 'we have Towered, Fleeted, and put to the walls at Calais a great number of the noblemen of England and many of them for light causes . . . we have hanged, pressed and banished more men since we were in authority than have suffered death by way of justice in all Christendom beside'. Had a strict suppression of disorder gone hand in hand with an alleviation of poor men's grievances, arbitrary action might have found some defence; but it is plain that as his power became established Wolsey began to care less for the generous impulse, or possibly the dislike of inefficiency, which had justified his seizure of power. The four committees of council to deal with poor men's complaints were of short duration. When Wolsey fell, only those of the Rolls and of the White Hall survived, and before long these, too, were merged in the court of requests which sat in the White Hall. The commissions to deal with inclosures, which Wolsey issued in 1517, 1518, 1526, and 1528, were, in the main, ineffective. The cardinal saw the grievance and may have hoped to heal it; he saw, too, that he might gain popularity in helping poor folk and might strike at his rivals, even at Fox who had been his friend. He made, however, no sustained effort to put things right; his mind was on higher things.

Wolsey was not only legate and chancellor; he was first minister too. To call him prime minister might be misleading, for he depended upon no political party, was not responsible to parliament, worked with no cabinet. He was, in theory, only one of the royal councillors, but, in fact, he dominated the council. As the Tudor council was a consultative and not an executive body, there was no need that its decisions should be unanimous, but before long the vigorous cardinal took control. Thomas Ruthall, bishop of Durham (May 1516), 'who sang treble to Wolsey's bass',[2] was made privy seal. Wolsey's own secretary, Richard Pace, was promoted to be king's secretary. No successor was appointed to Edmund Dudley as president.

[1] Pollard, *Wolsey*, pp. 71, 76.   [2] *Letters and Papers*, II. i. 672.

The ageing Norfolk, though treasurer, was thrust into the background, and his active son, Surrey, was employed in Ireland or in war. Suffolk, become suspect to the nobles owing to his hasty marriage, early in 1515, with the Princess Mary, lost influence for a while. The old trusties of Henry VII, men like Lovell and Poynings, left the stage, and Wolsey, rising above the rivalry of churchmen, nobles, and administrators, became supreme.

He used his power ruthlessly to maintain his party, punishing disaffection, quelling opposition, rewarding ready service. In 1514 he successfully resisted Leo X's proposal to make Bainbridge legate *a latere*, and on Bainbridge's sudden death stoutly upheld his own agent, Gigli, who had been accused of administering poison. In 1515 he imprisoned Polydore Vergil, who had been his confidential agent to Rome in the previous year, partly because Vergil had tried to secure the papal collectorship which his master wanted for Ammonio, partly because his own desire to become cardinal had not been at once realized, but chiefly perhaps because he had seen an intercepted letter in which his conduct had been frankly criticized.[1] About the same time he was at pains to have deprived of his bishopric of Bath and Wells, Castelli, who had represented English interests at Rome for twenty years and whom he had professed to support for the papacy. It is probable that he fomented the ill feeling against Suffolk, who survived because the king did not in his heart of hearts dislike the marriage of his companion in arms, but only the manner of it. He was instrumental in the fall of Buckingham in 1521, though the haughty duke was really the victim of his own high descent,[2] and his speculation as to his prospects if Henry should have no male heir. Buckingham had despised the parvenu and had spilt water over his shoes when he was serving the king from a basin; he had said that 'my lord cardinal was an idolator', and when his careless words touched the monarch as well as the minister, Wolsey was quick to direct the royal wrath against a high-born offender. As he grew less sure of his own position, he became more and more sensitive to criticism. When he felt himself attacked in a masque produced at Gray's Inn at Christmas, 1526, he sent to the Fleet the producer, a well-known sergeant-at-law named John Rowe, and one of

---

[1] *Letters and Papers*, II. i. 71.
[2] He was descended from Thomas of Woodstock, sixth son of Edward III, and his mother was Catherine Woodville, sister to the queen of Edward IV.

the actors. According to Fox another of the actors was Simon
Fish, who managed to escape abroad. Incensed by *Colyn Clout*
and *Why come ye nat to court?* he drove Skelton, the poet, into
sanctuary at Westminster. Even of his faithful servants he was
jealous; he would let no man save himself influence the king.
Suspecting even the competent Pace, Wolsey removed him from
his post as secretary, employed him abroad as much as possible,
and, in 1527, sent him to the Tower from which he was released,
with his mind affected, only after the cardinal's fall. Arrogant
towards Englishmen, he was no less overbearing towards the
ambassadors of foreign powers, to whose privileged position he
paid little heed. He rated the Venetian, Giustiniani, and ordered
him not to put anything in his dispatches without his consent
(1518); he actually laid hands on the papal nuncio, Chieregato,
and threatened him with the rack; in 1527 he opened the
correspondence of the imperial ambassador, de Praet, and
confined him to his house.

To foreign observers it seemed that he was the real ruler of
England. When he wrote, *ego et rex meus*,[1] he was only conforming to Latin usage, though a more modest official would have
avoided a sentence which necessitated such a construction; but
as early as 1519 the Venetian ambassador noted that he no
longer said 'his majesty will do so and so', or even 'we shall do
so and so', but simply, 'I shall do so and so', and in his private
correspondence he did not always maintain the fiction that
everything he did was done by royal command. According to
More, Wolsey conducted the negotiations for the peace of 1518
so autocratically that 'even the king hardly knows in what state
matters are'.[2] Sometimes he even deceived his master. When,
in 1514, he sent Polydore to Rome to urge his claim to be made
cardinal, he was careful to make it appear that the proposal
came from Leo himself; and it seems plain that Henry did not
know of the 100,000 crowns which his servant accepted from
Louise of Savoy when he came to terms with France in 1525.[3]
In 1527 he concealed from king and council the bellicose instructions which he sent to the English ambassadors in Spain.
Yet the foreign observers misconstrued the situation; in the end
the cardinal depended upon the king and his position resembled

---

[1] *Venetian Calendar*, iv. 43: 'Nec Ego, nec Rex meus dedit talem facultatem.'
[2] Pollard, *Henry VIII*, p. 110, quoting from *Letters and Papers*, II. ii. 1364.
[3] Pollard, *Wolsey*, p. 148.

that of grand vizier. It is part of his genius that he was able to bring Henry to his own way of thinking and so, whilst seeming to obey, to carry out his own designs. His task was the easier because he and his master were, in fact, like-minded. Each valued greatness and prestige and each identified his own glory with that of England. For long Wolsey was able to reconcile differences of opinion, usually by getting his own way, and though his position obviously became more difficult as his master grew slowly to his full stature, king and minister worked in harmony for more than a decade.

Secure in his master's approval, Wolsey established an absolutism, and under his rule the country had not only good order, but the appearance at least of great prosperity. The Venetian, Falier, writing in 1531,[1] was evidently impressed by the magnificence of the court, the active trade, and the general appearance of well-being. Yet, though the country was wealthy, and the royal income sufficed for ordinary needs, the Crown could not easily obtain the vast supplies of ready money demanded by an expensive foreign policy, and in spite of his high administrative skill Wolsey was unsuccessful in the matter of finance. It may be questioned whether any administrator would have been more successful—in every country in western Europe the increased cost of government was creating difficulties, and Englishmen as much as others disliked being taxed; but Wolsey's methods lay him open to criticism upon three grounds. He encouraged the king in extravagance and was extravagant himself.[2] He failed to convince his countrymen that there was any justification for his ambitious diplomacy. In pursuing his political designs he did not stop to count the cost; he decided what he and his master wanted to do and expected England to pay the bill. That there might not be enough ready cash in the country at once to meet his demands and to supply the needs of English economy never occurred to him, as it occurred to his successor Cromwell. His attitude was autocratic; the king wanted money, there was money in England, and the king should have it. It is fair to say that in 1515 he resumed some

---

[1] *Venetian Calendar*, iv. 292. Wolsey's wealth was tremendous—perhaps £35,000 a year when the normal 'ordinary' revenue of the Crown was about £100,000 a year and the revenue from all sources in 1529 less than £150,000. Dietz, pp. 138, 115.

[2] After his fall his debts were eight times as large as the sum shown in his statement to the king. For his resources and debts see *Letters and Papers*, iv. iii. 2763–70, 2783, 2869, 2945, &c.; he seems to have expected '4000l yeerly' in his retirement.

of the lost crown lands; that, repelling the efforts of the exchequer
to recover its lost authority, he took finance into his hands in
the manner of Henry VII and kept careful control;[1] that he
tried by the ordinances of Eltham (1526)[2] and on other occasions
to reform the royal household; that he had prepared new and
accurate assessments; that he endeavoured to make his foreign
policy pay by exacting pensions, or as Henry VIII preferred to
say, tributes from France. Yet a glance at the financial records
of the day makes it plain that the costs of the wars and the
diplomacy far exceeded the value of the territories and the
pensions gained. The ineffective campaigns of 1522-3, for
example, cost about £400,000; the French pension, irregularly
paid, brought in no more than about £20,000 a year at the very
best of times, and even between 1527 and 1529, when France
was increasing her promises, Henry actually advanced Francis
£50,000 in money and a jewel worth £10,000,[3] besides remitting £50,000 of pensions.

The necessity of providing funds drove Wolsey into courses
which made him unpopular, not so much because his demands
were irregularly made—though the irregularity aroused some
protest—as because they were very heavy. In 1522-3 he raised
by forced loans upon a strict assessment no less a sum than
£352,231, but as he wished to undertake fresh enterprises he
was driven to summon parliament in April 1523. The fact that
the commons chose as their speaker Sir Thomas More shows
that they were not ill disposed towards the king, but when the
cardinal in person demanded 4s. in the pound of every man's
lands and goods, he met with violent resistance. He dismissed
with contempt a deputation sent to beg a 'more easier sum',
but when, in the full panoply of his power, he came down to
coerce the commons, he was given a sharp lesson in procedure.
Members declined to answer his questions and More, who had
tried to get the subsidy voted, excused the 'marvellous silence'
by explaining, upon his knees, that though the commons would
hear a message from the distinguished stranger, they must

---

[1] Hall (i. 236) shows him admitting a new chief baron; his position as chancellor
gave him great opportunities to promote adherents. Palsgrave, writing after
Wolsey's fall, tried to blame Wolsey for the innovation made by Henry VII.
Letters and Papers, IV. iii. 2557.

[2] *Letters and Papers*, IV. i. 860. See also 'Tudor Reforms in the Royal Household'
by A. P. Newton in *Tudor Studies*.

[3] Dietz, p. 102 n.

debate among themselves. The feeling which prevailed was summarized in a speech prepared, though perhaps not delivered by young Thomas Cromwell, who said that the winning of Thérouanne had cost the king more than 'twenty such ungracious dog holes were worth' and that England, unsure of her allies abroad but certain of Scots hostility, would have difficulty in gaining and keeping anything of value. In the end parliament offered a subsidy to be paid in four instalments, which produced little more than £150,000 in all, and the failure of parliamentary supply led Wolsey into what has been described as 'the most violent financial transaction in English history'. The 'Amicable Grant' of 1525 was nothing less than a demand for one-sixth of the movables and incomes from the laity and twice as much from the clergy, with concessions to the poorer contributors both lay and clerical.[1] Opposition was general and resolute; even when, as at Norwich, men alleged a willingness to pay, they pleaded that they had no coin, and in most places there was not even a pretence at willingness. Wolsey's attempt to bully London was a complete failure though he threatened that resistance might 'fortune to cost some their heads', and, as the news of London's opposition spread, the provinces showed themselves more and more recalcitrant. Despite the efforts of Warham in Kent, of the dukes in Norfolk and Suffolk, and of other noblemen and prelates in their own shires, the revolt spread. Markets and industry came to a standstill. A general insurrection seemed possible, and in East Anglia the gentry, aware of the peasants' risings in Germany, cut the bridges to hinder the concentration of the malcontents, though they would give no help in the collection of the tax. There was, however, no disloyalty to the Crown—when Norfolk asked a band of insurgents who was their leader, he was told 'Poverty and his cousin Necessity'. Henry with a sure instinct abandoned the imposition, representing that he had not understood its severity, and exacted no penalty save the reprimand of a few ringleaders in the star chamber. Wolsey, after attempting to argue that the judges' opinion legalized the demand and that the whole council had supported it, realized that he must yield and sought popularity by asserting that the remissions had been granted as the result of his own instance with the king. England was not deceived; Wolsey was universally denounced. Yet,

[1] *Letters and Papers*, IV. iii. 3089; Hall, ii. 35–36.

though his position was shaken, he still kept the confidence of the king, and his fall was due not to the breakdown of his government at home, but to the failure of his policy abroad.

It is by his foreign policy that he is usually judged. He had brilliance and resolution and in his day England played a great part on the diplomatic stage. 'Nothing pleased him more', says Giustiniani in 1519,[1] 'than to be styled the arbitrator of the affairs of Christendom', and certainly in his day the English alliance was a decisive factor in the politics of western Europe. It is too much to say that he consciously developed the doctrine of the balance of power, since more than once he sided with the strong against the weak, but he did endeavour to play off the mutual antipathies of the European princes and to give his country an effective control in continental affairs. This control he meant to use for his own ends. He was a medievalist at heart. Possibly he thought he could make his master emperor; certainly he thought that he could make himself pope; failing to do either of these things, he still believed that he could bend to the advantage of England the old-established authority of Rome.

The French treaty of August 1514,[2] brought about by the surprising volte-face, was of no long duration. Worn out, as is alleged, by his efforts to play the young gallant to his lovely English wife, Louis XII died on the last night of 1514. All was to do again. At first things promised well, although Mary's precipitate marriage with Suffolk removed a useful bargaining counter, for Francis renewed his English treaty on 5 April 1515. Yet soon the sky began to cloud. Albany, the French-speaking cousin and heir of the infant James V, appeared in Scotland, and before long Henry's flighty sister Margaret and her new husband Archibald Douglas, sixth earl of Angus, were driven over the border.[3] Francis, moreover, had supplemented his alliance with Henry by coming to terms with Charles of Burgundy in a treaty whereby the young duke, then much under French influence, promised to marry Renée, daughter of Louis XII, and to further the return of Navarre to France by Ferdinand.[4]

---

[1] *Letters and Papers*, III. i. 39.   [2] See above, p. 284.

[3] At Harbottle in October Margaret gave birth to a daughter, Margaret, who married Matthew, fourth earl of Lennox, in July 1544, and whose son Henry, Lord Darnley, had a claim to the English throne unvitiated by alien birth.

[4] 24 March 1515. Dumont, *Corps Diplomatique*, Amsterdam (1726), IV. i. 199.

Having thus immobilized potential enemies, the adventurous king of France hurried off south, crossed the Alps by chamois-tracks, descended suddenly upon the Milanese, and on 13 and 14 September defeated the renowned Swiss at Marignano. The pope hastened to welcome the conqueror at Bologna and his Medici kinsman in Florence reached out for Urbino with French aid. Venice welcomed the return of her old ally. The disasters of 1513 were avenged, and Italy lay at the foot of the conqueror. Henry was consumed with jealousy and rage; he felt that he was losing his place as *jeune premier*. As early as May 1515 he had set himself to show the Venetian Pasqualigo how much he excelled as an archer and horseman the spare-shanked Francis, who passed for a valiant man-at-arms; he took the ejection of Margaret from Scotland in September as an insult to himself, alleging that the French king had sworn to keep Albany in France, and when the news of Marignano came he was chagrined to discover that compared with this 'battle of the giants' his victory at Guinegate seemed to be a very small thing. Fierce as was his resentment he realized that for the moment he could do nothing. A new holy league was projected, but this vanished in smoke when Leo came to terms with Francis at Bologna; Henry made a fresh treaty with Ferdinand in October 1515, but he was no longer the youth of 1512 to be gulled into an invasion of France for the benefit of his father-in-law. Instead of waging war, he adopted, possibly at Wolsey's suggestion, an expedient which was as undignified as it was unsuccessful. In October 1515 the able Richard Pace was sent to hire 20,000 Swiss to serve Maximilian in an attempt to recover Milan, but though in March the uncertain emperor advanced to within nine miles of his objective he suddenly fled back into the Tyrol without fighting, and England had her expenses for her pains. The waste of money caused complaints, and it is possible that the retirement of Warham and Fox signifies an opposition to the policy of Wolsey; but the new cardinal went on his way unmoved.

The events of 1516 were unfavourable to his designs. Ferdinand died in January, and his grandson Charles, realizing that if he wished to secure Castile, Aragon, and Naples he must visit Spain and the Mediterranean in person, renewed his peace with Francis at Noyon in August[1] on terms which bound him even

[1] *Letters and Papers*, II. i. 699.

closer to the French king. He promised this time that he or, in the event of his death, his brother Ferdinand would marry Louise, the infant daughter of Francis or, failing her, another daughter yet unborn,[1] and again undertook to return Navarre. An English attempt to defeat this alliance by a new league was a complete failure.[2] Maximilian cheerfully consented to a treaty between himself, his grandson, England, and the pope and crossed Germany in state at the expense of his English ally. Arrived in the Netherlands he remarked to Charles, 'Mon fils, vous allez tromper les Français et moi je vais tromper les Anglais', took 75,000 crowns from the French king, and instead of breaking the treaty of Noyon adhered to it at Brussels in January 1517. In the following March Maximilian, Francis, and Charles made a fresh league of Cambrai for the partition of Italy;[3] from the councils of the great princes England was excluded.

King and cardinal were doubtless displeased by these developments, but the isolation of England and her impotence have sometimes been over-emphasized. It was understood not only by the English diplomatists—Tunstall, West, Knight, and More—but also at the Burgundian court, that the arrangement with France was a mere device to let Charles get safely to Spain. The treaty of Noyon rested upon impossibilities; Charles could not postpone marriage until the French babies grew up, and Spain would never countenance the return of Navarre. The cardinal was content to bide his time. The Burgundian ambassadors who came to England in 1517 were dazzled by a display of wealth, and were graciously accorded the loan they had come to seek.[4] No attempt was made to impede Charles's voyage down the Channel in September, and when the young monarch landed at Villa Viciosa near Santander he was lent a horse by the English ambassador, Spinelly. The English calculations were sound. As soon as he realized the state of affairs in his ancestral realm, Charles wrote to Francis explaining that it was beyond his power to restore Navarre, and a quarrel between Habsburg and Valois was imminent. Meanwhile, Francis, aware that Maximilian must soon die and uneasy about Italy,

---

[1] Treaty of Noyon (Dumont, IV. i. 224). If no daughter were born he was to marry Renée; in fact, Charlotte was born soon after the treaty was made.

[2] *Letters and Papers*, II. i. 757; *Spanish Calendar*, ii. 285.

[3] Ibid. ii. 1019.

[4] Ibid. 1087, 1095.

had been disconcerted by the threat of an Anglo-Burgundian *rapprochement* and set himself to gain the goodwill of England. He withdrew his support from Scotland; Albany, who went to France in June 1517 to renew the 'auld alliance', obtained only the ungenerous treaty of Rouen and was forbidden to recross the sea. In 1518, therefore, Wolsey found himself courted by both parties. The arrival of Campeggio to plead the necessity of a crusade and to collect money was made the opportunity for a display both of Wolsey's intransigence and of the respect due to a legate *a latere*. The cardinal was kept waiting for months at Calais, and when at last he entered London at the end of July with the instruments which made Wolsey, too, legate *a latere*, the splendour of his train owed much to the assistance of his English colleague. When, in September, Bonnivet and the bishop of Paris arrived to treat for the return of Tournai their embassy soon moved to a proposal of marriage between the dauphin and the king's daughter, Mary, which on 4 October became the subject of a formal treaty. By that time, however, the negotiations had blossomed into a magnificent treaty of universal peace. On 2 October the pope, the emperor-elect, the king of France, the king of Spain, and the king of England undertook to make common cause against the Turk and to include in their league all the other powers of Europe—Denmark, Scotland, Portugal, Hungary, all the Italian states, the Swiss confederation, and the towns of the Hansa.

The French treaty was obviously open to criticism—Mary was but two years old and the dauphin was an infant in arms; French pensions and dowries were not always paid; many Englishmen regretted the surrender of Tournai and the disbanding of its stout garrison. It was, however, of value, for Tournai was an expensive luxury of no use to England. As for the treaty of universal peace, it was a diplomatic triumph of the first order. At last the warring princes of Europe were rallied to a common cause and it was in London, and by the direction of the cardinal of York, that this great end had been accomplished. The fame of Wolsey resounded throughout Europe.

Yet the achievement was one of prestige rather than of reality. The dream of a united Europe vanished in a few months when, in January 1519, Maximilian's death brought to a head the rivalry between Habsburg and Valois. Francis spent money in vain among the electors—three million crowns according to

himself—and in May 1519, when it was apparent that some of
the German princes were seeking a third candidate, Pace was
sent to Germany to see how the land lay; but the result of the
election was a foregone conclusion and on 28 June Charles V
became king of the Romans. This boy of nineteen now represented in his own person not only the imperial dignity but the
Habsburg inheritance which had descended to him from Maximilian, the Burgundian territories which he had from his father
and grandmother, and the titles in Spain, Italy, and the rich
Indies which had come to him through his mother Juana. It
seemed that he bestrode the world like a Colossus and that
France must inevitably be crushed by his encircling might.
The forces of the two great adversaries were not, however,
unevenly matched. The dominions of Charles, each with its
crop of domestic difficulties, were far apart and not always well
disposed one to another; in Germany the ninety-five academic
theses of Martin Luther of 1517 had blossomed into a practical
dispute which already tore the land asunder; in Spain, the
advent of a foreign king provoked a discontent which issued in
1520 in the revolt of the Commuñeros: in Naples, where too the
northerners were disliked, there was still a French party which
hoped that the victor of Marignano would march his forces
south. Charles might lay claim to the Milanese and the duchy
of Burgundy, but already he seemed to have more than he could
hold. Francis, on the other hand, ruled a well-developed unitary
state with strong finances and a regular army, powerful in guns
and in heavy cavalry: especially when he held the Milanese
and controlled Genoa he could interpose an effective barrier
between Germany and Spain. From this equipoise two results
inevitably ensued—the hard struggle between Francis and
Charles for the command of Italy and the anxious endeavour
made by each to win the support of England, which could make
or break the contact between the Netherlands and the Iberian
peninsula.

The 'universal peace' had vanished as a dream, but England,
without great effort of her own, had come to occupy a most
enviable position in the world of politics. Wolsey revelled in the
situation. For some time each of the two monarchs had professed a desire to see the other's face and early in 1520 a proposal
was made from France that Henry and Francis should meet
personally near Calais, and the cardinal, enlisting the royal

enthusiasm by the announcement of a tourney in which the
two monarchs should challenge all comers, himself made the
arrangements on a lavish scale. On 31 May the king of England
sailed from Dover with his queen, his nobles, and a train of
beautiful ladies, and on 7 June, not without some suspicion on
both sides, the two monarchs met at a 'camp', or tilting-ground,
in the open field between Guisnes and Ardres. Until 24 June
the monarchs and their two courts competed in manhood and
splendour, and the story of the jousts, the dancing, the brave
clothes, the courtesies, and the pageantry is told in the enthusiastic pages of Hall; but the amity was as temporary as the
wondrous buildings and all the magnificence which, for a short
time, transformed a dull Picard plain into the 'Field of Cloth of
Gold'. The emperor Charles had come to England towards the
end of May. On the 26th he was met by Wolsey at Dover, and
next day Henry himself came to greet his visitor. The two
monarchs kept the Whitsun feast together at Canterbury and
Charles sailed from Sandwich on the same day that Henry set
off from Dover to Calais. He had given to Wolsey a pension of
7,000 ducats and a promise of aid when next the papal throne
should be vacant. After the brief interlude of the Cloth of Gold
Henry and his queen met Charles and his aunt Margaret at
Gravelines on 10 July, and next day the imperial visitors were
brought into Calais. The French, whose optimism had visualized
even the surrender of Calais, were greatly dashed by this
evidence of Anglo-Burgundian cordiality. Their suspicions were
well founded, for though there was public talk of the grand
tripartite confederation, Henry and Charles concluded on 14
July a treaty whereby they bound themselves not to make fresh
alliances with the French king for two years and to act together
in a congress to be held at Calais.[1]

England had ceased to be an arbiter and had become a
partisan. Her lapse was perhaps inevitable; she was tied to the
Netherlands by commercial necessity, and to most Englishmen,
though there was a French party at court, France was the
natural enemy. Yet it was hastened by one particular cause.
Wolsey wished to stand well with Rome, to commend himself
to Leo whilst he lived, and to succeed him when he died. Early
in 1520[2] France had offered her support at the next election,
but he knew that her help would be less efficacious than that of

[1] Hall, i. 220; *Spanish Calendar*, ii. 312.   [2] See p. 290 *supra*.

the emperor and he turned to a course which would gratify both Charles and Leo. This was the easier for him because at this time the papacy, where Giulio dei Medici, later Clement VII, had great influence, was striving to create a great confederation against France. In May Leo made a treaty with Charles; in April Wolsey gained an extension of his legatine power; in May Henry completed the *Assertio Septem Sacramentorum* which he presented to the pope in a golden cover, and in October the king's piety was rewarded by the proud title of *Fidei Defensor*. In these circumstances the cardinal, when he came to Calais in July 1521 to preside over a conference, did not come with an open mind. He may have hoped at first to effect a reconciliation, and Hall supposed that he continued his endeavours even after he had been supplied, during a fortnight's visit to Bruges, with proofs of French bellicosity. He may, perhaps, have hesitated a little, for the peace with France was of his own making and he valued the pensions; but in fact during his visit to Bruges he began negotiations whereby Charles was to marry Mary as soon as she became twelve, receiving a dowry of 400,000 gold crowns and promising a jointure of 50,000 gold crowns a year. By a treaty of 25 August Henry and Charles were to become allies against France; England was to sweep the Channel and to see that the emperor, who was to land at Dover *en route*, had free access to Spain; Charles, for his part, was to send a fleet to assist the English crossing to Calais; in the spring of 1523 the two princes were to undertake a formal invasion, Charles from Spain with 10,000 horse, 30,000 foot, and competent artillery, Henry, with a like force and with imperial assistance, from the north-east. The treaty was not formally ratified until 24 November,[1] but the conference at Calais, where Wolsey continued to sit until 27 November, was obviously a farce. France and the emperor were virtually at war. In September Bonnivet captured the key fortress of Fuenterrabia and in the north-east Bayard defended Mézières; but, on the other hand, the imperialists invested Tournai which fell on 30 November, and in November they overran the Milanese. Either side might have been deemed the aggressor, and England, where the nobles resented the execution of Buckingham and the people disliked taxation, was not enthusiastic for war; but

---

[1] *Spanish Calendar*, ii. 365; *Letters and Papers*, iii. ii. 760. The ratification did not include the terms of the marriage treaty.

Wolsey's mind was fixed. Even though, on the death of Leo, Charles made only a sham effort to secure his elevation to the papal chair, he persisted in pursuing the imperial alliance. The Scots were peremptorily ordered to expel Albany, who had returned in the preceding November; musters were taken; the fleet was mobilized in the Downs with Surrey as admiral, and provision was made for the reinforcement of Calais if this should be necessary. In March Francis distrained the goods of English merchants at Bordeaux. On 27 April Charles's general, Pescara, using a new discipline to maintain a continuous fire from his arquebusiers, broke the Swiss at Bicocca and before long drove the French from all but three cities in Lombardy; two days later Clarencieux delivered Henry's defiance of Francis at Lyons. By this time Charles was already in England. He landed at Dover on 26 May and was brought in a stately progress by way of Greenwich and Southwark to London, where he was given a great reception. Thence he passed by Southwark, Richmond, and Hampton Court to Windsor where, on 19 June 1522, treaties for the invasion of France, for the marriage of Mary, and for the extirpation of heresy were arranged. Three days later the monarchs were at Winchester hunting the hart and on 6 July Charles sailed from Southampton where thirty royal ships had been collected on the pretext of escorting the imperial visitor, though they had secret instructions to ravage the coast of France. Later in the month Surrey burnt Morlaix and ravaged the coasts of Brittany about Brest; in August he betook himself to Calais where lack of provision had hitherto prevented serious operations, and on the 30th marched out with an army of some 15,000 men, including 300 Spaniards sent from Charles and 400 'adventurers' who served of their own volition and without pay. He burnt and raided for some weeks, but he failed to take Hesdin for want of heavy guns, and his army was decimated by sickness. By 14 October he was back in Calais with a great booty of livestock, but that was all his gain.

Meanwhile there had been trouble on the Scots frontier, but there had been no serious fighting since the English were short of troops and Albany could not induce the Scots to risk another Flodden; on 17 September Dacre granted a month's truce, and although he was officially blamed for doing this without authority, Wolsey regarded his action as a *felix culpa* since the West

March was, in fact, indefensible and the warden's action could be disowned whenever the king pleased. Albany, whose army at once disbanded, sailed off to France on 25 October; the English government tried to take advantage of his absence and of the presence of Margaret to gain over Scotland without fighting. In November Henry prolonged the truce for three months and in January 1523 he proposed that if Albany were forbidden to return, Scotland should have a truce for sixteen years with a prospect of marriage between her young king and the English princess Mary, who was of course betrothed to Charles. These insincere proposals being rejected, Henry loosed Surrey upon the Scots with Dorset to command the East Marches and the experienced Dacre still in the west. During the course of the year the border was ravaged in a series of devastating raids, Kelso was burnt in June, and Jedburgh in September. Yet nothing was achieved; the Scots still resisted; the English lost 800 horses as they retired from Jedburgh and when Albany returned with men and guns in September he was soon able to attempt an offensive. In October he laid siege to Wark, but though his French gunners behaved well, the Scots declined to cross the Tweed and the expedition is mainly famous because George Buchanan, returned impecunious from his studies in Paris, marched in the Scottish host. On 20 May 1524 Albany left Scotland never to return.

In France, meanwhile, the fighting had been undistinguished. Hoping to enjoy the support of the constable Bourbon, driven to disaffection by the wanton conduct of Francis, the allies had planned a triple attack. Charles was to invade from the south, Bourbon from the east, and Suffolk from Calais. Charles contented himself with the recapture of Fuenterrabia; Bourbon brought nothing but his sword, and though after long delays Suffolk set forth from Calais in September with an army of nearly 20,000 men, of whom two-thirds were English, he achieved nothing. He and Henry had wanted to besiege Boulogne, but Wolsey, still anxious to gratify Charles, insisted that he should march south. Advancing resolutely the duke reached the Oise, only to find that he could expect no support from his allies and was compelled to retire precipitately. His retreat was well conducted and is often reported to have been 'without loss', but the *Chronicle of Calais*[1] reports that he had to

[1] Camden Society (1846), p. 34.

abandon his guns at Valenciennes and that it was 'but an ill jurney for the Englyshemen'.

Coupled with this lack of success abroad were the financial stringency and the resultant discontent at home, which have already been noted. Wolsey had paid dear for the imperial alliance and he had paid in vain, for when the pious and narrow-minded Adrian VI died in September 1523, to the relief of the artistic, joy-loving Rome, his successor was Giulio dei Medici, a cousin of Leo X. Wolsey had pressed his own claim hard—in 1521[1] he had suggested the use of imperial troops to make his election sure—but the emperor, if ever he had thought of aiding the Englishman, made no effort at all upon his behalf. Certainly he wished to gain the Medici; probably he knew that the cardinal of York would never become a facile tool. Wolsey understood very well what had happened. He still pinned his faith on Rome; perhaps he still hoped for the papacy, but no longer did he place reliance upon the promises of Charles, who had been the sole gainer from the alliance. Nothing had been done against the Turks; Belgrade had fallen in 1521 and Rhodes on the last day of 1522. Italy was now in danger—but Charles had increased his hold over Italy. So far from realizing his claim to the French throne, Henry had not won a foot of French soil; but under cover of his unsupported invasion, Charles had had successes south of the Alps and south of the Pyrenees. Wolsey began to think of changing sides. The fact that the new pope came to terms with Francis no doubt weighed with him, for he still trusted in the papacy, but other considerations entered too. With the disappearance of Albany, French influence had waned in Scotland; there was no need for further war. English Margaret, with the aid of her brother, 'erected' her son as king in his own right on 26 July 1524, and early next month she was given very definite hope that James should marry Mary and wear a double crown.[2] Whether Henry was sincere in this proposal may be doubted. Possibly he was, since he now knew that Catherine would give him no heir and a marriage between his daughter and his nephew would avail to keep the crown in the Tudor family. At all events, the whole tenor of England's policy towards Scotland makes it clear that Wolsey had no longer any fear from France

---

[1] William Bradford, *Correspondence of the Emperor Charles V* (1850), p. 26.
[2] Wolsey to Margaret, 2 August, *Letters and Papers*, IV. i. 240.

and no longer any scruple about breaking the treaty of Windsor.

Meanwhile England took no part in the continental war; she lacked money and probably she lacked inclination. When, in June 1524, Bourbon with imperial troops invaded Provence and besieged Marseilles, there was this time no invasion from Calais. A French envoy, Giovanni Giovacchino di Passano, known to the English as John Jochim or Joachim, was in London as early as June and Henry, though still speaking peace to Charles, refused to dismiss him at the emperor's request. On 12 December 1524 the pope, Francis, and Venice formed the new holy league, which was publicly proclaimed on 19 January 1525.[1] By that time it seemed clear that a change of sides would pay. Bourbon had recoiled from Marseilles in December and Francis, following him up across the Alps, swept all before him in Italy. His advent was welcomed not only by the pope, but by many of the Italian princes and it seemed that France would regain her ascendancy when suddenly, on 25 February 1525, she was utterly ruined by the disastrous battle of Pavia. Caught between the main imperial army and a valiant sortie by the beleaguered Antonio de Leyva, the French army fought with a desperate courage, but, perhaps owing to the king's intrepidity, the advance of the heavy cavalry masked the fire of the terrible French guns. La Palice, Trémouille, the hot-headed Bonnivet, and many other leaders were killed, and Francis was taken prisoner. Not without truth, he wrote to his mother: 'Madame . . . de toutes choses ne m'est demeuré que l'honneur et la vie qui est sauve.' His conqueror celebrated his twenty-fifth birthday by becoming in very deed emperor of the west.

Wolsey's first thought was not to restore the balance of power but to exploit the situation. In April 1525 he sent the privy seal Tunstall and Wingfield, chancellor of the duchy of Lancaster, to join Sampson at Madrid and urge not only the immediate invasion but the dismemberment of France. Henry and Charles should be satisfied in their just claims; Henry might become king of France; in any case Francis and his children were to be excluded from the throne. His proposals were refused. Charles had known all along of 'Joachim's' negotiations; he had heard of Wolsey's disparaging references to his family; he felt that he could now do without his English

<hr />

[1] *Spanish Calendar*, ii. 684, 692.

ally and was preparing to abandon the treaty of Windsor. On 26 March he wrote bluntly that if the English wanted to invade France they must do it themselves, and during the course of the year he gave his attention to Italy where his troops established a complete mastery. On 14 January 1526 he was able to exact onerous terms for the release of the French king.[1] By the treaty of Madrid Francis was to 'restore' to him Burgundy and its dependent territories, to surrender his claims in Italy, to abandon the suzerainty of Flanders and Artois, to give up Tournai, to reinstate Bourbon, to desert his allies, and to marry Charles's sister, Eleanor of Portugal. The freed captive, who had already denounced this treaty in a formal but secret instrument signed on the 13th, had no intention of carrying out its terms; but before he was liberated in March he had to surrender his two sons as hostages for his good faith and his defeat was patent to the world. His conqueror, now able to disregard Henry altogether, ignored his promise made at Windsor and in March gave his hand to his cousin Isabella of Portugal, who brought a useful dowry of a million crowns.

Wolsey, for his part, had disregarded Charles. Early in the summer of 1525 he had opened negotiations with Francis's mother, Louise of Savoy, from whom he accepted a secret present of 100,000 crowns, and on 30 August had concluded at 'The More', his house at Moor Park, a treaty which gave France peace at the cost of greatly increased pensions.[2] No doubt he was emboldened in his policy by the knowledge that Clement, who dreaded the ascendancy of any one power in Italy, was already making approaches to France. With the pope on his side Wolsey felt that the French alliance was entirely reputable, and when he heard of the treaty of Madrid he wrote at once to the queen-mother of France expressing the hope that her son would disregard promises to which he was bound neither in honour nor in conscience. The same opinion was expressed by some of the princes of Germany as well as by some of the Italian powers. Clement, therefore, did not hesitate to absolve Francis from his oath and to pursue his endeavour to re-create a holy league; on 22 May 1526 there was formed the league of Cognac[3] by which he, France, Florence, Venice, and Sforza of

[1] *Letters and Papers*, IV. i. 838; *Spanish Calendar*, III. i. 552.
[2] *Letters and Papers*, IV. i. 717; *Spanish Calendar*, III. i. 307.
[3] *Letters and Papers*, IV. i. 961.

Milan bound themselves to resist the designs of Charles. The
king of England, who was understood to have already given
his countenance, was named protector of the new confederation.
Despite its high patronage, the league of Cognac accomplished
nothing. Francis, anxious to make up for the boredom of his
captivity, gave himself up to pleasure and spent his money upon
buildings instead of upon armaments; Henry was short of funds,
and Wolsey, who in any case may have hoped to be *tertius
gaudens*, could not have precipitated England into a war
against the Netherlands even had he wished to do so. The pope's
Italian allies, after some initial successes, lost heart. Sforza
surrendered Milan in July and in September imperialist troops
under Moncada supported by cardinal Colonna occupied Rome
and hunted the pope into the castle of San Angelo. Henry, in
reply to papal appeals, sent the comforting message to the
successor of St. Peter[1] *oravi ne deficiat fides tua* along with a
promise of 30,000 ducats which, to the gratified astonishment of
the cardinals, actually arrived in Rome early in 1527.[2] Wolsey
was complimented by his fellow countrymen on having kept
England clear of the Italian disasters; but he was in fact
meditating a still closer alliance with France, and in March
1527 there arrived in London the bishop of Tarbes who, on
30 April,[3] signed at Westminster treaties uniting Henry and
Francis in a perpetual peace and binding them to make war
upon Charles as soon as the emperor refused to free the French
king's sons and to pay the debts of the king of England. By the
terms of this treaty Mary was to marry either Francis himself or
his second son, and in order to bring all to perfection Wolsey
proposed to go to France in person. Before he set out, however,
occurred an event which gave a new justification and a new
direction to his policy. On 6 May the imperialist troops, unfed,
unpaid, and many of them Lutheran, broke into the Holy City
itself. The hard-bitten Georg von Frundsberg had already
succumbed to apoplexy as he endeavoured to control his
mutinous landsknechts; Bourbon, who was in command, was
killed during the assault, and Spaniards and Germans, casting
discipline to the winds, gave Rome to a sack whose brutality
horrified the civilized world. The pope was again compelled to
take refuge in the castle of San Angelo and there he remained

---

[1] Ibid. iv. ii. 1137.   [2] Ibid. iv. ii. 1278.
[3] Ibid. iv. ii. 1382.

in effect a prisoner; for though Charles denied responsibility
for the outrage, he did not release the victim.

Wolsey saw his opportunity. Not only could he stir up
Christendom against the sacrilege but, on the pretext that
Clement was a prisoner, he might summon the cardinals to
France and have himself made vicar-general during the pope's
captivity. His negotiations had been impeded because the
marriage-alliances on which they hinged involved the validity of
Henry VIII's marriage to Catherine, and realizing full well
that Clement would now do nothing to the prejudice of his
captor's aunt, he thought he might gratify his master's wishes
and exalt his own dignity by taking the papal authority into
his own hands for the time being. The journey to France which
he undertook in July resulted in the ratification of the treaties
at Amiens on 18 August,[1] and later in the year a solemn
French embassy came to invest Henry with the insignia of the
order of St. Michael; but the design of a conclave came to
nothing, for only four cardinals joined Wolsey at Compiègne,
and Charles parried the attempt by allowing Clement to escape
to ruinous, insanitary Orvieto in December. Wolsey realized
that if he was to have his way in Europe he must appeal to
arms. Theoretically the treaty with France was meant to be the
foundation of a universal peace, and in fact France continued
to bargain with the emperor; as late as January 1528 the cardinal, modifying the procedure of 1521, tried to give the impression that he was the arbiter whose good offices had been
frustrated by the enormities of one party—this time the emperor
—but in reality he was determined upon war. He sent Clareci-
eux abroad with instructions which had not been seen by king
or council, and on 22 January 1528 the English herald, along
with Guienne, herald for France, gave a formal defiance.[2]

His threat was idle; England refused to fight. In Kent men
threatened to put the cardinal to sea in a boat with holes in it,
and all over the south of England there was sedition so serious
that the government was compelled to yield. In March it was
arranged that, war or no war, English trade with the Low
Countries should not be interrupted and in June a truce was
made between England and the Netherlands. In this bargain
Italy was not included and though Francis himself was ill and
indifferent, the French cause south of the Alps won a resounding

---

[1] *Letters and Papers*, IV. ii. 1519.   [2] *Letters and Papers*, IV. ii. 1703.

success in the beginning of 1528. The competent Lautrec overran the Milanese, established the authority of Francis at Genoa,
and forced the imperialists to quit Rome by a direct attack upon
Naples. Success attended his arms; Melfi was taken, Moncada
was killed, and by April Naples, invested both by land and sea,
was in extremity. French overweeningness and the Italian
climate robbed the conqueror of his victory. Francis alienated
the Genoese by his dealings with Savona, though as Wolsey
said, it would have been better to have lost six Savonas than to
have lost the goodwill of Andrea Doria, a great captain of a
great fleet. In July Doria came to terms with the emperor;
in August Lautrec died of fever; and the proud French army,
neglected by the government at home, was utterly destroyed at
Aversa. Early in 1529 a new French army was sent to the Milanese, but St. Pol who commanded it was beaten and taken at
Landriano and Clement VII, accepting the situation, reconciled himself to Charles at Barcelona on 29 June[1] in a treaty
which guarded the interests of his own Medici. Negotiations for
a general peace were already taking place in the Low Countries
between the mother of Francis and the sister of Charles, and
on 5 August was signed the 'ladies' peace' of Cambrai which,
for the time at least, brought to a close the long quarrel between Habsburg and Valois.[2] Taught by experience Charles
showed himself less grasping than in 1526. He made no claim
to Burgundy and, though he improved his position on the
north-east frontier of France, he was less exigent there than
he had been. He insisted that France should abandon her
claims in Italy, should pay him two million crowns, and should
assume liability for the debt of 290,000 crowns which he owed
to Henry.

Wolsey was thunderstruck by the news. He could not believe
that Francis and Charles would make peace and still less that
they should do so without consulting him. He had, however,
no real grievance. He had done nothing to help his ally. Charles
had been driven to accept peace not by diplomatic or military
pressure from England, but by the course of events in Germany
where the rise of Lutheranism and the advance of the Turks had
produced a situation demanding his immediate attention. At
the diet of Speyer in 1526 he had been compelled virtually to

---

[1] Ibid. IV, ii. 2541; *Spanish Calendar*, IV. i. 115.
[2] *Letters and Papers*, IV. iii. 2610; *Spanish Calendar*, IV. i. 162.

allow a system of *cuius regio eius religio* and though he had attempted to retract his concession by the second diet of Speyer (February to April 1529) his action had produced the 'protest' of five princes and fourteen cities which gave birth to the word 'protestant'. The religious troubles of Germany could not be ended by an imperial decree, and in any case the hands of the emperor were full. In August 1526 Lewis II, king of Hungary, had been beaten and killed by the Turks at Mohacz, and although the Habsburgs, with characteristic opportunism, had seized the chance to give the inheritance of the dead Jagiello to Charles's brother Ferdinand, the position of the new monarch was far from secure. Suleiman was as strong as ever and in September and October 1529 he was battering at the gates of Vienna. It was obvious that Charles should come to Germany with the prestige of the papacy behind him and on 24 February 1530 he was crowned by the pope at Bologna.

That Charles should have come to terms with his western opponents was natural enough, but when he did so he blew into the air the cobwebs of the English diplomacy. Wolsey had always tried to keep step with the pope and had always tried to exploit to the advantage of England the rivalry of Habsburg and Valois. The net result of all his efforts had been to destroy the balance of power, to give the Habsburg a complete ascendancy over the pope at the very time when Henry was counting upon papal aid to separate him from his queen, Catherine, who was the aunt of the triumphant Charles.

Wolsey was doomed. His downfall was attributed by some to his own ambition. According to Sander[1] he was devoured by a personal hatred of Charles who, when he needed English aid, had subscribed himself *filius tuus et cognatus* in autograph letters, and later, having disappointed him of the papacy, sent only letters in a clerk's hand signed merely *Carolus*. It is not to be denied that Wolsey hoped to wear the tiara and it is also true that as cardinal and legate he was always anxious to stand well with Rome. None the less he should not be blamed too much for the failure of his diplomacy. Peace with France was worth having. England was set free from the endless wars which cost so much and yielded so little and she was set free too from the constant trouble which had vexed her northern border. The Anglophile party in Scotland, led still by Angus, although

[1] Author of *De Origine ac Progressu Schismatis Anglicani Liber* (1585).

Margaret obtained a scandalous divorce from him in 1527, took complete control in Scotland; this was the period when 'none durst strive with the Douglas nor yet with a Douglas's man'. To peace at home, prestige abroad might well have been added. No one could have foretold the complete collapse of France at Pavia, and the complete ascendancy gained by Charles in Italy was to some extent the result of chance. That ascendancy established, the Anglo-French *entente* became more than ever justifiable, and the disappointing result of the new alliance was again a thing that could not easily have been foreseen. The union of two secular enemies might well have been expected to produce a political force of the first magnitude, and in making his 'diplomatic revolution' in 1527 Wolsey was actuated by the motives which later weighed with Kaunitz and Choiseul. The calculations of 1527, like those of 1756, were belied by the event; but they were not foolish calculations.

Yet the cardinal cannot escape censure altogether. The understanding with France was of his own making and he persisted in it in the teeth of a national opposition. His experience of 1523 should have shown him that England would not pay for a war in which she was not interested, and the average Englishman had very little interest in the mains of the Italian cockpit. Wolsey could not, it is true, proclaim abroad all the reasons which led him to seek the liberation of the pope, and it was not fair to him that his interventions in Italy should be ascribed to his own ambition. Yet his career in diplomacy and his attitude to his fellow subjects in England had been such that England was prone to suspect the worst in him. As legate he had interfered with the privileges and patronage of the bishops. As chancellor he had encroached upon the jurisdiction of the common lawyers. As prime minister he had suppressed the nobles and taxed the people at home, whilst abroad he had plunged into an unpopular foreign policy. He had aroused the animosity of England and gone on his way unheeding, secure in his position as representative of the all-powerful papacy and as minister of a resolute and popular king. Now, when the papacy was fallen from its high estate, he relied almost entirely upon his royal master, and his foreign policy had resulted in a situation wherein Henry was cheated of the hope which lay nearest to his heart. After Wolsey had fallen he was accused of many things, but it was 'the king's great matter' which brought him to his ruin, and after he was down it was the king who inherited his vast store of accumulated power.

'Of thys trouble I onely may thanke you my lorde cardinal of Yorke',[1] said Queen Catherine, according to Hall, when she appeared before the legates at Blackfriars; and according to Wolsey himself the outraged queen asserted as early as 1527 that the proposal of divorce was due to his 'procurement and setting forth'.[2] The same opinion was held by the imperial and the French ambassadors and by Charles V; it was held too by Poly-dore Vergil[3] and by Tyndale.[4] Both these men disliked Wolsey, but as Tyndale wrote in 1530 and professed to report common talk, contemporary opinion must have believed that Wolsey was the author of the divorce. In a formal treatise on *The Pretended Divorce between King Henry the Eighth and Queen Katherine*, written by Nicholas Harpsfield, it is bluntly stated that 'the beginning then of all this broil . . . proceeded from Cardinal Wollseye who first by himself or by John Longland, Bishop of Lincolne and the King's confessor, put this scruple and doubt in his head'.[5] Harpsfield wrote in the reign of the devout Mary, who was a daughter of Henry, but when the catholic account of the Refor-mation was put into final form in the *De Origine ac Progressu Schismatis Anglicani* of Nicholas Sander, the story took a rather different form. Sander wrote under that Elizabeth who surpassing in wickedness Jezebel, Athalia, and all examples of feminine depravity, was also a daughter of Henry, and he did not hesitate to enlarge upon the concupiscence of the wicked king whom he even represents as the father of Anne Boleyn. Yet for him too Wolsey was the arch-villain. Aware that he was disliked by Catherine, anxious to revenge upon her her nephew's deception about the papal chair, hoping to unite France with England against the emperor, the cardinal stirred up Longland to excite doubts in the king's mind about the validity of his marriage and suggested that Henry might marry Margaret, sister of the most Christian king. Henry, however, had fallen in love with

[1] Hall, ii. 148.　[2] Wolsey to Henry, 5 July 1527, *Letters and Papers*, IV. ii. 1471.
[3] Polydore Vergil, *Anglica Historia*, ed. 1570, p. 685.
[4] *The Practice of Prelates, Works of the English and Scottish Reformers*, ed. Russell, ii, 463.
[5] Harpsfield, Camden Society (1878), p. 175; for the phrase 'the king's great matter', p. 186. Richard Hall, in his *Life of Fisher* (E.E.T.S., extra series cxvii), adopts the same view: 'These and such other things lying hott boylinge in the Cardinalls stomacke against the Emperour.'

Anne Boleyn, whose mother and sister Mary he had already possessed, although Anne had a most unattractive *habitus corporis* and a *habitus mentis* more unattractive still, and finding that he could not obtain his desire without marriage, showed himself eager for divorce. Wolsey, though he did not like the prospect of Anne's elevation, cheered himself with the thought that Henry, if he grew tired of her, might seek a French bride after all, and, determined at all costs to retain the royal favour, forced himself to satisfy the desires of the king.

Apart from its essential error in representing the king's passion as the *fons et origo* of the English Reformation, Sander's story is both inaccurate and improbable; it was not Francis's sister Margaret but his sister-in-law Renée who was proposed for Henry VIII; Anne may well have been Henry's mistress before she became his wife; no one can believe that the heart of the English king, already well experimented according to the theory, was carried away by a fairy with an oblong face of jaundiced hue, a projecting tooth, an extra finger on her right hand, and a large wen under her chin. Anne was not a physical and a moral monstrosity. There is, on the other hand, no need to accept the portrait produced in the admiring days of Elizabeth; it is not certain that her sympathy with reform was anything more than the anti-clerical cast of mind proper to her family—though she and her father, unlike her uncle Norfolk, opposed the burning of heretics[1]—and the light which dawned in her big dark eyes was probably not 'gospel light'. Anne was a dashing young brunette who had served an apprenticeship in the gay court of France as maid of honour to Queen Claude. She was well connected; her father was of wealthy City stock, her mother was a daughter of the second duke of Norfolk, the victor of Flodden, and her grandmother a co-heir of the seventh earl of Ormond. When, on the outbreak of the French war in 1522, she returned to England, she entered the court circle as of right and, after having been admired by the poet Wyatt, who was already married, soon attracted the attention of the king. For long she held him at bay; his love increased with her resistance and the anti-clerical party saw in her an agent who might bring about the fall of the Colossus.

Wolsey may not have realized at first the extent of his master's

---

[1] *Letters and Papers*, v. 466–7; Pollard, *Henry VIII*, p. 192 n.; *Letters and Papers*, iv. i. 639 for Rochford's elevation.

passion, but he must have had an inkling of it and he cannot have regarded without misgiving the prospect of Anne's elevation. Sander's account of the part he played in the divorce is not entirely wrong, and it must be added that in a Latin tract written by a catholic, perhaps by Harpsfield himself before he produced his English work, there is little animus against the cardinal who is represented as the agent of the king's will.[1] Certain it is that Wolsey did not promote the divorce for the benefit of Anne, and it is by no means clear that he was anxious for a divorce at all. Upon his death-bed he asserted that he had often tried, upon his knees, to persuade the king from his will and appetite,[2] and Catherine herself had come to believe by September 1529 that the cardinal was coerced by the king.[3] Admittedly he wished to gratify his master; he was not very friendly with Catherine and he was very willing to put a slight upon Charles; he loved to pose as the man who could get things done at Rome and he knew that the granting of a divorce to a favoured prince was not without precedent. On the other hand, he must have been aware that, since Henry's marriage rested upon a papal dispensation, a divorce would be hard to obtain, and the events of 1526 must have shown him how easily the pope could be subjected to imperial compulsion. Himself he alleged that the question of the king's marriage had been raised by the bishop of Tarbes in the spring of 1527 when he was negotiating for a match between Francis and Mary, and as the English diplomatists had expressed doubts as to Francis's freedom in view of his recent betrothal to Eleanor, it is quite conceivable that the Frenchman riposted by querying the legitimacy of the English princess. At all events, soon after Tarbes's visit Wolsey began to make inquiries into the validity of the king's marriage[4] and in the subsequent negotiations with France the possibility of Henry's taking a French bride was considered. Anne's father, created in 1525 Viscount Rochford,[5] however, had already been involved in some discussion of the king's affairs, and it is at least possible that Wolsey, realizing that the king was

---

[1] *Le premier divorce de Henry VIII et le schisme d'Angleterre*, by Charles Bémont (1917).

[2] Cavendish, *Wolsey*, ed. Singer (1925), i. 321.   [3] *Spanish Calendar*, iv. i. 236.

[4] Early in April Dr. Richard Wolman was sent to consult Fox upon the matter (*Letters and Papers*, iv. ii. 1428–9) and on 17 May Wolsey, with Warham as his assessor, instituted a formal inquiry. *Letters and Papers*, iv. ii. 1426. *The Letters of Richard Fox* (1929), pp. 156–7.

[5] *Letters and Papers*, iv. ii. 1441 (iv. i. 639 for Rochford's elevation).

going to seek a new wife in any case, tried to make the best of the situation by directing Henry's eyes to France.

What is quite certain is that the idea of a divorce had been in the mind of the king long before the advent of Tarbes and long before he fell in love with Anne. From the very first there had been doubts about his marriage, not only in England and in Spain but even in Rome itself,[1] and in 1514, when Henry made his first 'diplomatic revolution', he had threatened to cast off the daughter of the perfidious Ferdinand and marry a French bride. Catherine was six years older than Henry and if her piety really showed itself in all the forms so much admired by Sander, she must have been a tiresome wife to a spirited young man, but the evidence is that she did her best to attune her life to her husband's way. Henry for his part behaved well to her and though he admired other ladies the story of his promiscuous amours is a myth.[2] To Elizabeth Blount he gave a son,[3] whose existence was long kept secret, but of other lovers nothing is said except in vague terms until reference is made to Mary Boleyn. Catherine bore him seven children, including four sons, but none of these survived for more than a few weeks save Mary, who was born in 1516. For years he gave his wife sympathy and hoped for better things, but he was seriously perturbed by the lack of a male heir; he must have known that the attempt of Henry I to set up Matilda had ended in disaster and that, if a woman could reign herself, his father had usurped the throne in 1485. By 1524 he realized that Catherine would not bring forth a prince and in that year he began to take steps to secure the succession. The suggestion that Mary should marry her cousin James of Scotland may have been seriously meant, but it came to nothing, and in 1525 the son of Elizabeth Blount was brought out of retirement, made duke of Richmond and Somerset and given offices two of which Henry himself had held as a young man—lord high admiral, lord warden of the Marches, and lord lieutenant of Ireland. Obviously regarded as the heir, Richmond was at one time much spoilt by Wolsey's agents, and in 1528 Campeggio was prepared to promote a match between him and his half-sister Mary, but by that time his chance was gone.

---

[1] Pollard, *Henry VIII*, p. 174, and p. 176 *supra*.
[2] But see Buckingham's accusation (*Letters and Papers*, III, i, cxxx).
[3] Henry Fitzroy (1519–36).

Henry had made other plans for the succession to the throne; he had fallen in love with Anne Boleyn and he wanted a legitimate son. It was he and not Wolsey who was the author of the divorce. For some time he had been convincing himself that the deaths of Catherine's children were God's vengeance upon an improper marriage, and though doubtless the wish was father to the thought, the king was none the less sincere. He had a conscience which would not permit him to do wrong; it always told him that he was right. He had been the paladin of the church; through all the kaleidoscopic changes in diplomacy he had remained constant to the papal alliance; with his own pen he had gained the style of Defender of the Faith. He was entitled to expect divine blessing. If a curse fell upon his nursery it was through no fault of his own and through no fault of his pious wife; the evil must lie in the nature of their marriage. The incestuous union must be dissolved, indeed must be pronounced to have been null from its beginning. It did not occur to him at this time that the pope would prove obdurate. The precedents were on his side,[1] and in the spring of 1527 both he and Wolsey felt that they could count on the papal favour. The only question was as to how that favour could be shown. Even as legate *a latere* Wolsey could not overthrow the papal dispensation on which the king's marriage was founded; but he could do much, and in one way or another, he thought, the thing could be accomplished. On 17 May he began a collusive suit by summoning the king before him to explain his marriage with his brother's widow and, though the matter was remitted to the consideration of the learned, on 31 May Henry told Catherine that they had been living in mortal sin and must separate. At the same time the king sought to convince the divines. At first he argued that Julius had exceeded his authority by dispensing with divine law[2] on insufficient grounds, but when Fisher stood

---

[1] Louis XII had been allowed to marry his brother's widow, Anne of Brittany; his own brother-in-law, Suffolk, had obtained a dispensation of more than one marriage before he was wedded to Mary, and in his case on the ground that a previous dispensation was invalid. His sister, Margaret, in March 1527, had been given a divorce from Angus in scandalous circumstances, largely through the instrumentality of Albany, who was a kinsman by marriage of the Medici. An even better precedent, probably unknown to Henry, was known to the Spaniards (*Spanish Calendar*, ii. 396); in 1437 a childless king of Castile, Henry IV, had been allowed to marry a second wife with the provision that if she too gave him no children he might return to his first wife.

[2] Leviticus xviii. 16.

resolute upon the plenitude of the power of the Keys, he altered his ground somewhat and alleged that the pope had been misinformed as to the facts. Even Fisher wavered; most of the divines professed to be convinced, and all was going well when, like a thunder-clap, came the news that Clement was a prisoner in the hands of Catherine's nephew. Wolsey had a momentary gain. To the shocked world it seemed that his anti-imperial policy was vindicated, and when he went to France in July he hoped to exploit the situation by obtaining for himself a delegated authority on the ground that the Holy Father was in captivity. The cardinals, however, would not dance to his piping; and the release of the pope in December blasted the whole design.

From that moment cardinal and pope were involved in a dilemma from which there was no escape. If Wolsey did not secure the divorce he would lose the favour of the king, by which alone he now stood; if he did secure it he would enthrone an enemy. The pope, too, was in a quandary; if he refused to grant the necessary Bull he would alienate yet another secular prince—there were schismatics enough in Germany—he would ruin Wolsey, and would pull down the pillar which upheld the English church; if he did accede to the English demand he would offend Charles, would destroy the position of the papacy in Italy, and would shake the prestige of Rome throughout the world. It was a complete impasse; but for the next two years pope and cardinal danced a solemn diplomatic quadrille, advancing and retiring according to the progress of the French arms in Italy.

To recount the long story of the bargainings is unnecessary. Its essence is that the pope was willing to let Henry have his way provided he himself were not made responsible, and that king and cardinal were determined that Henry's new marriage should be correct beyond suspicion. In the early summer of 1528, when Lautrec was sweeping all before him, the English cause prevailed. On 8 June Clement issued at last a 'decretal commission' authorizing Cardinals Campeggio and Wolsey to try the cause and to pronounce sentence. As Campeggio was protector of England at the curia his appointment was proper; as he was bishop of Salisbury he might be expected to favour the king. Clement gave a private undertaking that he would not revoke the cause to Rome, but at the same time he instructed the legate to induce Catherine to enter a convent and to use all possible means to avert a violent settlement. Gout and prudence prolonged the legate's journey, and before he reached England in October the French army had been destroyed in Italy. Caution was obviously necessary, the more since there was much sympathy for Catherine in England. She herself hotly denied that her marriage to Arthur had been consummated and indignantly rejected the suggestion that a union which had lasted for nearly twenty years was not good in the sight of God. Henry thought well to summon the notables to Bridewell on 8 November and justify his action in a speech which contained a great panegyric of his queen. The hesitant legate found cause for delay in the appearance of new evidence. This purported to be a copy of a brief sent by Julius to reassure the ailing Isabella at the time when the original Bull was granted on 26 December 1503, but free from the defects of the Bull itself. It came so appositely that the English insisted that it was forged; Burnet, writing long afterwards, considered it as spurious, and its authenticity has never been proved. Before the puzzled Campeggio had made up his mind the pope had convinced himself that the imperialists must win and had hastily sent to England to order the destruction of the decretal commission. In vain England and France tried to stiffen the pope with promises; in vain they hoped that his sudden illness would terminate fatally. Only on 31 May 1529 did the legates open their court at Blackfriars and proceedings had not gone far before Catherine, whose dignity moved everyone, including Henry, protested against the competence of the court and appealed to Rome. The proceedings continued, and though Fisher pleaded the queen's cause with courage, Wolsey hurried his reluctant colleague to a decision. No decision was obtained. The French had been routed at Landriano (21 June); pope and emperor had made peace at Barcelona (29 June); Habsburg and Valois were about to end their differences, for the time at least, at Cambrai (5 August 1529). On 23 July, when the king's proctor demanded sentence, Campeggio adjourned the court till October, on the ground that it must follow the rules of the Roman consistory of which it was a part. It never met again. Even before its adjournment Clement had already recalled the case to Rome. Catherine, as the event was to show, was not saved, but Wolsey was lost.

For some time it had been obvious that the power of the

cardinal was waning. Henry, who knew his antipathy to the Howards, had suspected that he was not really trying to gain the divorce; in 1527, whilst the cardinal was still in France, he had sent his secretary, Knight, to deal directly with Clement;[1] he had been told by Francis that the cardinal corresponded privately with Campeggio and the pope, and documentary evidence of this may have been provided by Gardiner when he came back from Rome in June 1529.[2] Wolsey had not been allowed to visit France in 1529 and had taken no part in the congress of Cambrai. When the king left London in the height of the summer,[3] he refused politely Wolsey's offer of entertainment at 'the More' and when, after a leisurely progress, the court settled at Woodstock, the cardinal was not invited to attend. Already the kites were circling round their prey. His clerical patronage was being disputed; his victims were stealing home from overseas; men whom he had made, as Gardiner and Brian, turned against him; libellous writings were abroad. He was present in council at the beginning of October, but although as chancellor he made out writs for a general election, these writs were promptly taken out of his hands and some of them were given to Norfolk—he was to have no chance of packing parliament. Making the best show that he could, he came to Westminster Hall in full state on 9 October, the first day of term, only to find that as the other councillors had gone to the king at Windsor, he must sit not in star chamber but in chancery; and whilst he presided there the attorney-general indicted him for praemunire in the king's bench. Wolsey was accused not of having accepted the office of legate, but of using his legatine powers in defiance of the statutes both of provisors and praemunire.

That he was guilty was obvious; it was obvious too that Henry had condoned his guilt, and therein lay the deadliness of the accusation. Plainly the king's face was turned away from his servant who, legate and chancellor though he was, must now underlie the common law. The apostle of prerogative was smitten to the heart, *cueur et parolle luy failloient entierement*.[4] Given the choice of appearing either before the king's bench or before parliament, he dared not face parliament. No doubt he remembered his last encounter with the commons; he knew that

---

[1] *Letters and Papers*, iv. ii. 1552.   [2] Pollard, *Wolsey*, pp. 239–40.
[3] Ibid., p. 237.   [4] Ibid., p. 244.

the bishops as well as the nobles were hostile in the lords and he dreaded an act of attainder which might take away his life. He preferred the jurisdiction which, under the statutes, could sentence him only to prison at the king's pleasure, with forfeiture of lands, of offices, and of goods, and he seems to have been allowed to suppose that he would recover Winchester and St. Albans without much delay.[1] On the 22nd he signed an admission of his guilt and on the 30th his attorneys, on his behalf, submitted to the king's mercy. He was sentenced according to the statute.

On the 18th he had been deprived of the great seal which, significantly, was given not to Warham, who indeed was debarred by his great age,[2] or to Suffolk, but to Sir Thomas More, and when parliament met on 3 November the new chancellor took part in a grand attack upon the fallen cardinal in which temporal peers and commons combined to produce a list of forty-four articles accusing him of various enormities, excesses, and transgressions of the law. The articles were signed on 1 December and presented to the king who took no action upon them; on 18 November he had already received the cardinal into his protection. Perhaps he had some regard for an old friend; perhaps he was not sure that he could dispense with an able minister;[3] certainly he had not let Wolsey go that Norfolk might reign in his stead. The cardinal, much to his relief, was allowed to retire to Esher and though he was deprived of Winchester, St. Albans, and York House, which became the palace of Whitehall, he was allowed to keep the archbishopric of York, a pension of 1,000 marks from the see of Winchester, and goods to the value of £6,374. 3s. 7½d., an enormous sum for the time. Other evidences of the royal favour were shown to him. He received rings and promises,[4] and when, in January 1530, he fell ill, the king sent Sir William Butts and three other royal physicians with orders to accept no fee from their patient.

Heartened by these good omens, Wolsey recovered his courage as quickly as he had lost it. Before long he was arguing that he must have at least £4,000 a year, and endeavouring to recover his pensions from abroad. He hoped to be recalled to the council, and when he was ordered to return to his benefice he

[1] Muller, *Stephen Gardiner and the Tudor Reaction*, p. 37.
[2] Warham died in 1532; he was born about 1450.
[3] *Spanish Calendar*, IV. i. 819.   [4] Pollard, *Wolsey*, p. 265.

proposed to go to Winchester instead of to distant York. Compelled to seek the northern see which he had never yet visited,
though he had been consecrated in March 1514, he halted at
Southwell, as near to London as he might be, on 28 April.
There, pleading poverty to the king but keeping a great household, he dazzled his flock by his magnificence and won their
hearts by his urbanity. Had he done no more, he might have
ended his days as a good bishop, but to him power was the
breath of life and even in his isolation he contemplated a return
to office. He magnified his position in the north and prepared
for a solemn enthronement in the minster on 7 November, to
which end he summoned convocation to York without awaiting
a royal mandate; he kept in touch with London and eagerly
welcomed news that his successors were not managing the king's
affairs over-well; he made contact with foreign powers, with
France, with the empire, and with the papacy. From France he
got nothing, though it was he who had made the French alliance
the keystone of English policy; Francis, still hostile to Charles,
had concluded that a king was a better ally than a fallen cardinal
and that Henry would be worth more to him if he were separated
from Catherine and married to Anne.

Before ever he went north, Wolsey's relations with de Vaux
—the 'Joachim' of the previous negotiations—were bad, and
the approach which he made to the imperial ambassador was
not, as he pretended, solely concerned with the payment of his
pension. Chapuys, advised that Henry and Francis intended to
support the Lutheran princes, did not spurn the overture,
especially since he knew that Catherine no longer regarded the
cardinal as her enemy but had already admitted his chaplain,
Bonner, to her counsels. How far the new intrigue went is
uncertain. It was stated soon afterwards that Wolsey had asked
the pope to excommunicate his master, but it does not appear
that he did more than persuade Clement to issue a solemn admonition to Henry to separate himself from Anne, and even this
the pope declined to do. At Bologna, however, he issued a brief
merely forbidding Henry to marry again *pendente lite*, and for the
suspicious king, egged on by Wolsey's enemies, this was enough.
Henry heard of the brief about 23 October. Information
obtained from letters taken with the Venetian, Agostini,
Wolsey's physician, showed that the cardinal was involved
with the papacy, and possibly the council feared that the

installation at York might have been made the occasion of some solemn demonstration, perhaps the reading of the papal brief.

On 4 November Wolsey was arrested by Northumberland and Walter Walsh, a groom of the king's chamber, at Cawood as he moved leisurely to his cathedral city. He was taken by easy stages to Sheffield Park where the earl of Shrewsbury treated him well for sixteen days, but on 22 November there arrived Sir William Kingston, constable of the Tower, with twenty-four of the king's guard, of which he was captain. Wolsey understood the omen and he was not deceived; Agostini's letters had told enough and the Tower, indeed, was his destination. All along he had asserted that he could justify his actions—'If I may come to mine answer I fear no man alive'—and now he said to his captor 'I see the matter against me how it is framed, but if I had served God as diligently as I have done the king, He would not have given me over in my grey hairs.'

Neither to his answer nor to the Tower did he come. He had been seriously ill for some time with an affection of the bowel and when he struggled into the abbey of St. Mary at Leicester, whose great stone walls spread for three-quarters of a mile in a timber-built town, he knew that his end was near. 'Father Abbot', he said, 'I am come hither to leave my bones among you.' 'This was upon Saturday at night and on Monday at eight o'clock in the evening he died.' During his last hours he was pestered for £1,500 in cash which the king could not find among his goods. Cavendish put into his mouth a long oration wherein he gave admonition about the dangers of Lutherans, Wycliffites, Hussites, and popular assemblies and warned Kingston of Henry's obstinacy and self-will. No dying man made such a speech, but with his last breath Wolsey said something which Cavendish and Kingston, belying a frightened yeoman of the guard, suppressed before the council; presumably he said something against the king. The great cardinal was buried in the Lady Chapel of the abbey; Chapuys, remarking that Richard III and he *gissent tous deux en une mesme Eglise*, stated that this church was already called the sepulchre of tyrants.[1] Richard, however, had been buried in the Greyfriars. In life and in death Wolsey resembled Richard less than his critics like to pretend: with all his faults he was not a man of blood.

[1] Bradford, *Correspondence of the Emperor Charles V*, p. 336.

His faults were many; he fell a victim to his ambition, his upstart arrogance, and, above all, his egoism. Yet he had great qualities, not least that mastery which, as the portraits of the age suggest, was a quality much admired by his contemporaries. It has been pointed out that his record is one of failure. His ascendancy was marked by the passage of no statute of importance. His schemes of legal reform were unaccomplished. Though he founded colleges at Oxford and Ipswich he was no enthusiast for the new learning. He was not, in fact, the defender of the church. On the contrary he made the church unpopular in England by his subservience to Rome and by subjecting laymen to his clerical government. He made no 'concordat' to fix the relations between England and the papacy. Moreover, he divided the ranks of the English church; he robbed convocation of its power and drove the bishops, oppressed by his legatine authority, into the arms of the king. His foreign policy, far from maintaining the balance of power, helped to destroy it, and turned the pope into the chaplain of the emperor.

The truth is that he stood midway between the old and the new. He had the ruthlessness, the administrative skill, the reliance on new men, the belief in positive law, and, above all, the absolute spirit of the renaissance prince; but his ideals were the old universalist ideals of the middle ages. Of the spiritual longings, even of the spiritual *malaise*, which produced the Reformation he had no real understanding, and he had no understanding either of the rising force of English nationality. He had no sympathy with the busy merchants. Although he attempted to reduce inclosures, it was partly because the new system of agriculture was a departure from the old and the known, and partly, perhaps, because he disliked other *arrivistes*. Probably he had a genuine feeling for the depressed country folk, but he had no idea of that alliance between prince and third estate that was one of the hall-marks of the new monarchy. Yet, though he was unseeing in an age of vision, an administrator rather than a creator, he was none the less a great man. He made his country famous abroad, and although his manipulations of the 'balance' were unsuccessful, he was not entirely wrong. As he predicted, the emperor made no effort to fight for his aunt's cause: France and Spain did neutralize one the other and, as the event showed, Henry was able to affect his 'Reformation' without any interference whatever from without. At home Wolsey welded into one the powers of the *regnum*, restrained in England by traditional checks, and of the *sacerdotium*, limited by no earthly bar, and the result was the creation of tremendous authority which he bequeathed to his master.

Profiting not only by Wolsey's example but even from his mistakes, turning to his own end the reaction in church and state against the overweening parvenu, harnessing to his chariot the lively spiritual and national forces which his proud servant had ignored, Henry VIII made himself an absolute king in his 'empire' of England.

X

ROYAL SUPREMACY

BETWEEN 1530 and 1534 Henry broke with the papacy. In the next six years he made the breach final by crushing active opposition and by seizing the lands of the monasteries, which he gave, usually at a fair price, to the English gentry. A great revolution was accomplished in a very short time and with remarkably little dislocation of the national affairs. It was achieved, it has been said, by king and parliament working together, and though this is true, there lurk behind the statement two compelling questions. Why did the king and the parliament wish to break away from Rome? How were they able to do it? It may be said at once that the motive force was that of Henry himself, never yet cheated of his desires and blessed with that convenient conscience; that the English king, whose bluff heartiness concealed a commanding intellect as well as egoism alternating unconcernedly between meanness and magnificence, revealed an uncanny ability to use parliament for his own ends; that the king's personal desires harmonized with the aspirations of the most active part of his people. Yet when all this is said, the ease with which the great change was effected demands explanation.

The first explanation lies in the atmosphere of the period. Revolution was in the air. Everywhere the facts were in rebellion against the theories and the waves of criticism were beating on the rock which, for so many centuries, had been a sure foundation of European society. Examined from a realist point of view the church no longer appeared as a divinely appointed reproduction of the ordered beauty of heaven. It seemed to be a human institution whose servants did not practice the righteousness which they preached and whose services and sacraments did not bring spiritual security even to those who obeyed its teaching to the best of their power.

The church was not in a healthy way; its standard of morality was not sufficiently above that of the world of laymen to justify the great privileges which it held. Often benefices had come to be regarded as the apanages of noble and gentle families; unspiritual men were appointed, and irregular marriages were not uncommon. Some of the parishes were attached to bishoprics or to religious houses which tended to appropriate the revenues and leave the parochial work to be done by vicars. The vicars, who were ill-paid, sometimes showed themselves exigent in the matter of tithes and dues, especially the hated mortuary dues, and excommunication was sometimes used to coerce those who resisted excessive demands. To say that the church sold the mercy of God for money would be too sweeping an assertion and it would be unjust to accept at their face-value all the criticisms launched by the reformers. Yet it would be unjust also to deprecate all mention of clerical shortcomings by an off-hand reference to the muck-rake, or to argue that the reformers had what is called a complex on sexual affairs. There is no need to reject as fabrication the evidence produced by Froude and Coulton about clerical depravity and clerical condonation of depravity in others.[1] The best men in the church were conscious of its weaknesses. The church was in need of reform, the question was whether that reform should come from within.

Rome gave no clear lead. The see of St. Peter had survived the shocks of the Babylonish Captivity and the Great Schism, but after the Council of Constance it had held its position with a difference; it had, in effect, come to terms with the national states which were emerging. Martin V had striven to regulate the relations between the papacy and the national monarchs; annates and 'reservations' had been abolished by the Council of Basel in 1435–6; the Pragmatic Sanctions of Bourges (1438) and Mainz (1439) were attempts to assert the privileges of national churches; England, which made no formal statement of her claims, was already entrenched behind the 'great *praemunire*'. The papacy, relying upon a competent and ever-increasing bureaucracy, had weathered the attack by making working arrangements[2] none of which became official until the famous concordat of Bologna made between Francis I and Leo X in 1516. The general effect of these was to place ecclesiastical patronage in the hands of the princes, while securing for Rome her formal power of confirmation as well as the fees

---

[1] Cf. 'The State of Affairs reported from Staffordshire' in *The Suppression of the Monasteries*, Camden Society (1843), p. 243, and the diocese of Chester (*Letters and Papers*, viii. 190).

[2] The indult granted by Innocent VIII to Scotland in 1487 is a case in point.

payable on promotion. These are usually called 'annates', although strictly the 'annates' were the perquisites of the papal curia whereas the 'common services', paid by prelates, were shared equally between the apostolic camera and the college of cardinals. The relations between the Holy See and the national states had become more and more secular; most of the 'supplications' which appear on the papal registers were concerned with livings, expectancies, and pensions. In Rome there developed a regular machinery whereby a cardinal-protector attended to the necessary business of each nation, and from Rome there still went forth a stream of papal officials and collectors, some of whom obtained high promotion in the lands which they visited. The princes, on the other hand, were able to secure the advancement of their own nominees; to reward their statesmen with rich benefices and pay their civil service with smaller livings. Under alien or absentee shepherds 'the hungry sheep looked up and were not fed', and though it would be false to suppose that the average man, using the religious observances to which his fathers had been accustomed, was actively discontented, it is certain that the lower orders were becoming apathetic, and that there was growing dissatisfaction among the wealthier and better-educated classes. A few foundations were still being made and doctrinal heresy was not very common, but there was a marked tendency to seek spiritual comfort in extra-ecclesiastical organizations, as in St. Ursula's *Schifflein* in Germany; there was a general dislike for the steady flow of bullion to Rome; there was revolt against the jurisdiction of the church courts, whose many refinements brought in revenue and yet, by their very multiplicity, bred contempt of canon law;[1] there was resentment against the financial exactions which were the inevitable result of the system of promotions and of the emphasis laid upon 'good works'—'no penny, no *Paternoster*'. Altogether the church was tending to become too secular. Its hold upon the peoples rested less upon true reverence than upon old tradition.

With the hour came the man. When Luther, 'justified by faith' but condemned by authority, went out alone, as he thought, into the dark, he found a great part of Germany waiting for him; when, in December 1520, he burnt the Bull of excom-

---

[1] 'Now-a-days there are so many laws, that whether a man do ill or well, he shall be taken in the law.' *Examination of Thomas Arthur*, 1527, *apud* Foxe, bk. viii.

munication, he kindled a fire which ignited the whole of Germany. National as well as religious zeal, already excited by the three great pamphlets,[1] were stimulated in the following years by tracts and hymns and by a translation of the New Testament which was finished in 1522. Not all of Luther's supporters were protestant; not all protestants were humanists —the German humanists at first regarded Luther's protest as part of a monkish quarrel and it was not till after the 'Leipzig disputations' of 1519[2] that they gave him their support; some of the enemies of the church were inspired neither by religious convictions nor by rationalism, but simply by restlessness and greed. Yet in all the imperial dominions revolution was in the air. In Spain itself and in Italy there were reformed congregations before 1530, and though the new ideas made little headway in the Mediterranean countries, they spread rapidly north of the Alps. Protestant churches were appearing not only in Germany, Switzerland, and Scandinavia, but even in France where, under the leadership of Lefèvre, *ceux de l'Évangile* were urging a return to the primitive church of the apostles.

The spiritual heads of Christendom, engrossed in their own affairs, had been unaware or contemptuous of the smouldering discontent, and the conflagration took them by surprise. The temporal princes were surprised too, and some of them were inclined to think that rebellion against pope and church might be turned to political ends. Charles himself, though he put Luther under the ban of the empire at Worms in 1521, sometimes hesitated in the course of his struggle with the pope, and in 1526 he was driven perforce to accept the compromise of Speyer.

In France the king, influenced by the Renaissance, and often hostile to the papacy, gave an uncertain protection to the reformers. Himself he had no real interest in religion, but his sister Margaret encouraged the translation of the New Testament, which was published in 1523; even after the attack on the images in Paris in 1528 had discredited the Reformers, Francis still thought they might be useful, and as late as 1536 Calvin addressed to him the great Preface to the *Institution of the Christian Religion*. His attitude, it is true, was dictated by con-

---

[1] *The Liberty of a Christian Man, To the Nobility of the German Nation, On the Babylonish Captivity of the Church of Christ*; all written in 1520.

[2] Debate with John Eck, professor at Ingolstadt, *De Primatu Papae*.

sideration of policy. He objected to the English demand for a general council in November 1533, only because a general council had already been promised by his rival, Charles. He was in the end surprised and possibly shocked when Henry actually severed his relations with Rome; but the threat of severance he had long regarded as a useful weapon in the political battle, and while his brother of England was content only to brandish the menacing sword, he looked on without disapprobation. From time to time he instructed the French cardinals to support the English cause at Rome, and in October 1532 he visited at Calais not only the recalcitrant king of England but the Lady Anne Boleyn, newly created marquess of Pembroke in her own right, who came to the meeting decked in the jewels which had belonged to Queen Catherine.

Catholic France no less than protestant Germany failed to see in Henry's quarrel with the papacy anything in the nature of a *casus belli*. In the atmosphere of the day there was no excessive reverence for the see of St. Peter, and for some years English ebullience excited no particular alarm. The relations between secular prince and papacy had for long been of a nature which has been defined as 'a sound business connection', and as criticism of Rome became more vocal the monarch who wished to alter in his own favour the terms of this business connexion could give to his action the appearance at least of popular support.

Henry's breach with the papacy was facilitated not only by the general atmosphere of the time but by the particular situation in the world of politics; he was very fortunate in the diplomatic background against which his great drama was played. The peace of Cambrai of August 1529 was a hollow affair. France, recovering from her wounds, meditated revenge. It was not until 1536 that she opened hostilities upon the emperor, but in the meantime she constantly thwarted his plans and willingly allied with his enemies. Charles, though he was crowned by the pope in Bologna in February 1530, was far from being lord of the world. Troubled on every hand by the enemies of Christendom and of Rome, he was not even secure of papal support. No idealist and apt to meet each question as it arose with the answer nearest at hand, he was neither inclined nor able to play the champion to his aunt in England. His hands were full.

In Italy, despite his astounding triumph at Pavia, or perhaps because of it, he was not secure. France had not abandoned her ambitions and some of her old allies, Venice, Florence, and even the discontented Milanese, looked to her for aid. The pope remembered his humiliations; he was a Medici and felt that French aid was necessary to secure Florence against imperial encroachment. As early as 1530 he was considering the possibility of marrying his kinswoman and ward, Catherine, to a French prince, and in October 1533 he himself took her to Marseilles where she became the bride of Francis's second son Henry, later Henry II. About this time France, strengthened by papal support, demanded that the Milanese should be returned to her, and though the death of Clement and the accession of Paul III, a Farnese, improved the position of the emperor, Charles felt himself compelled, on the death of Sforza in 1535, to offer Milan to France, though under conditions which France refused. He had secured Montferrat in 1533 from the duke of Mantua and he maintained an alliance with his kinsman, Charles III of Savoy; but the weak border duchy, unable to reduce protestant Geneva, was no secure defence for the imperial position in north Italy and was easily overrun when France, in 1536, opened hostilities once more.

Meanwhile Germany was torn by religious disputes. The Lutherans had united under the Augsburg Confession in March 1530, and in December of that year had created an instrument of political action in the league of Schmalkalde which, though never very effective, played a great part in politics for nearly two decades. At the diet of Nürnberg in 1532, Charles was compelled to promise toleration until a general council was held, and although the extravagances of the anabaptists of Münster discredited the protestants, by no means united among themselves, the Lutheran church steadily established itself in northern Germany. The religious difficulties were fomented by extraneous influences. It is not without significance that in 1534 the German protestants made a secret treaty with France; it is even more significant that, though Luther himself was resolute for the repulse of the Turks in 1532, some of his adherents held out a hand to the invader and all of them made their support of the imperial army dependent upon the grant of concessions to their religion.

The menace of the Turks, which threatened all Europe, was

for Charles a danger both on the Danube and in the Mediterranean. Though Suleiman had been repulsed from Vienna in
1529, he found an ally in Zapolya, who became his vassal king
of Hungary, and though he made a peace upon the *status quo* in
1533, he remained a constant and an active menace to the
empire. On the Barbary coast the corsairs, amongst whom
Khair-ed-Din Barbarossa was conspicuous, were at once independent pirates and servants of the Padishah; their fleets swept
the Italian coast, and in 1534 Barbarossa nearly succeeded in
carrying off from her castle of Fondi to the harem of Suleiman,
Julia Gonzaga, duchess of Traietto, esteemed the most beautiful
woman in Italy. Charles had his revenge when he captured
Tunis in July 1535, but Barbarossa remained at large and
continued his depredations from his stronghold at Algiers. To
the Turkish assault France was a party. Soon after Pavia, the
French government had made an approach to Suleiman and it
was said that the attack on Vienna in 1529 was due in part to
the machinations of Louise of Savoy and of Clement VII himself; in 1533 Francis received the emissaries of Barbarossa, and
in February 1536 he made a formal treaty with the sultan—the
first formal treaty of its kind.

   In these circumstances neither Habsburg nor Valois was
inclined to quarrel with England. Francis did not want an
attack from the north when his eyes were turned to the south;
Charles, facing perpetual discontent and a perpetual shortage
of money, had no mind to create fresh troubles for himself by
injuring the trade of the Netherlands. When, in 1535, Paul III
wrote to Francis urging him to stand ready to execute the Bull
which he had prepared against Henry, the French king merely
sent the bailli of Troyes to demand explanations from England,
and there was talk of a marriage between Mary and the duc
d'Angoulême. As for the emperor, he found the death of
Catherine on 8 January 1536 something of a relief, and on the
fall of Anne Boleyn in the following June was disposed to enter
into cordial relations with Henry. In any case, the invasion of
Savoy by France in March 1536 provoked a war between
France and the empire which lasted until the truce of Nice in
June 1538, and it was not until the two monarchs were reconciled at their famous interview at Aigues Mortes (July 1538)
that there was any possibility of a concerted attack upon the
grand schismatic.

In England itself, thus immune from foreign interference, both the atmosphere and the political situation were favourable to Henry's designs. For long the national spirit had resented interference from without. England had accepted the authority of Rome as the head of the universal church and there is no hint that an *ecclesia anglicana* consciously demanded privileges like those claimed for the Gallican church.[1] Yet even in the Constitutions of Clarendon it had been made plain that Englishmen would not permit an alien authority to deal with real property (advowsons) and although, as in other countries, church and state made a working arrangement as regards administration, *Praemunire* was a clear assertion that English law was sufficient for English folk. Not in the Statute of Provisors alone was there a protest against the loss of English money; the suppression of the alien priories in 1414 was a presage of more general schemes of confiscation; to Englishmen, obviously, the property of the church was no longer sacrosanct. The anti-clerical spirit which thus revealed itself in matters of jurisdiction and of finance had been greatly increased by the absolutism of Wolsey, and when the great cardinal was fallen there was evident, even amongst those who were unaffected by the new heresies, a desire to transfer into secular hands the wealth and the authority long enjoyed by the church. To these political tendencies was added the force of a genuine religious conviction felt, it is true, less by the magnates than by the middle class and by the poorer folk. The lollardy of Wycliff had been discredited by the extravagances of the Peasants' Revolt, but his influence still remained in the English Bible, in the demand for preaching, in the belief that priesthood depended more upon righteousness than upon formal orders, and in doubts about the doctine of transubstantiation.

To a rationalistic age it may seem remarkable that men and women would wager their lives upon a mystery of faith. Yet it was the miracle of the Eucharist which gave its tremendous sanction to the authority of the priest, and John Knox was not wrong when, in after years, he said that one Mass was more terrible to him than 10,000 armed enemies.[2] It had been asserted that the Lollards exercised but little influence upon public affairs,[3]

---

[1] Froude, *The Reign of Henry the Eighth*, ch. iii.
[2] *History of the Reformation in Scotland*, Wodrow Society, ii. 276.
[3] Froude, *The Reign of Henry the Eighth*, ch. vi.

and this is true; yet the accounts of the heresy trials in the reign of Henry VII show that men still professed the Wycliffite doctrine and, as appears from the cases of Butler[1] and Hunne, unofficial meetings for devotion were evidently held in the houses of substantial burgesses and country gentlemen. When, under the inspiration of the Renaissance, there arose a purpose to 'preach Christ from the sources', a tendency towards heresy revealed itself also in the academic world, especially at Cambridge where Erasmus had taught Greek. As early as 1521 a group of young dons were discussing the views of Luther, and the place where they met, the White Horse Tavern, gained for itself the name of 'Germany'. Most of the earlier reformers were Cambridge men, though William Tyndale began his education at Magdalen Hall in Oxford, and it is one of the ironies of history that it was Wolsey himself who stimulated the growth of the new ideas at Oxford when he sent young scholars from Cambridge to improve the teaching at his foundation of Cardinal College. The ties between England and Germany grew closer, and between 1520 and 1530 there developed a spiritual revolt of whose strength Wolsey, who showed himself contemptuous, was utterly unaware. Earnest young men who had been 'infect' at the university carried their gospel with them when they went forth to preach in London and in the provinces. At first they seemed to have thought that they would find support from the higher clergy, some of whom had been commended by Erasmus, and Tyndale, when he came to London in July 1523, was surprised to find that the humanist Tunstall frowned upon his proposal to translate the New Testament into English. With the help of Humphrey Monmouth, a wealthy cloth merchant, Tyndale escaped to Hamburg under the name of William Hutchins and came to Wittenberg where he completed his translation in 1524. Next year, he went on to Cologne accompanied by William Roy, a Cambridge man and a friar observant from Greenwich, where he began the business of printing. Driven from Cologne when the imprudence of Roy attracted the attention of the watchful Cochlaeus,[2] he sought refuge at Worms and there, in 1526, his New Testament appeared as 'a mean great book'. Soon afterwards a better (pirated) edition appeared in the Netherlands,

[1] Foxe, bk. vii.
[2] John Dobneck, a German theologian, a constant opponent of Luther.

and from a press at Antwerp Tyndale issued commentaries and controversial works; *The Parable of the Wicked Mammon* and *The Obedience of a Christian Man* appeared in 1528, *The Practice of Prelates* followed in 1530.[1]

Other refugees wrote dangerous books. The injudicious Roy, from whom Tyndale parted as soon as he could, was joined at Strassburg by another Observantine of Greenwich, Jerome Barlow, and the pair produced, probably early in 1528, *Rede me and be not Wroth* which, although it followed a Swiss dialogue in describing the death of the Mass, gained notoriety as an attack upon the cardinal so virulent as to call for a rebuke from Tyndale. Even more violent was the *Supplication for the Beggars*, written in 1527 by Simon Fish of Gray's Inn, which, after explaining to the king the enormities of the clergy, urged him to 'tie these holy thieves to the carts to be whipped ... till they fall to labour', arguing that until that were done the realm of England would always be pillaged by idle drones. *The Dialogue between a Gentleman and a Husbandman*, printed at Marburg in 1530, was another 'bitter cry' against the oppressions of the spiritual estate.

Even if it had not been accompanied by polemical works, the issue of the New Testament in English would have created alarm in the hearts of the orthodox. Tyndale, though he certainly used both the German translation of Luther and the Latin version which had appeared along with Erasmus's text, undoubtedly made use of the original Greek and gave his own colour to his translation. Unlike the authors of the modern revised version he did not always give each Greek word the same English, but he did endeavour to render the meaning of the Greek and in so doing he departed from the Vulgate and from the long-established connotations of the words and phrases used therein. Thus the words rendered in the Vulgate as *caritas, poenitentia, confiteri, presbyterus, gratia,* and *ecclesia* were rendered, or at least sometimes rendered, 'love', 'repentance', 'knowledge' or 'acknowledge', 'senior', favour', and 'congregation'. The net effect was to reduce the importance of works, to reject the sancrosanctity of the old tradition, and even to cast doubt upon the sacraments.

---

[1] The two controversial works were probably produced by John Hoochstraten at Antwerp, although no doubt to protect the printer they purported to come from Marburg. By a superb piece of effrontery an edition of John Knox's *Godly Letter* of 1554 purported to be 'Imprinted at Rome'.

Not unnaturally the arrival of this revolutionary book in England produced a sharp reaction. As early as 1521 Wolsey, with Warham and Tunstall in attendance, had presided at a holocaust of Lutheran books at St. Paul's, and in January 1525 proceedings had been taken against certain merchants of the Steelyard who possessed forbidden literature. When, early in 1526, the English New Testaments began to appear they found a ready market, for a well-organized 'Association of Christian Brethren' was at hand to aid in the distribution. Wolsey, whose leniency had been remarkable, was stirred into action, partly by the expostulations of the bishops, amongst whom Tunstall and Longland realized most clearly the issues which were at stake. On Shrove Sunday (Quinquagesima) 1526 Wolsey presided at an immense bonfire at St. Paul's, and in the autumn, possibly on 28 October, there was another burning of New Testaments which caused the gratified Campeggio to write from Rome that 'no holocaust could be more pleasing to God'.[1] At the same time John Hacket, an emissary sent to the Low Countries, was endeavouring to see to the annihilation— 'anychyllment'—of the new books there. He found the imperial authorities so slow to act that in the following January he was thinking of buying up the dangerous literature with a view to its destruction. Meanwhile in England Warham had been inciting his bishops to activity and on 26 May wrote to his suffragans asking them to contribute to a sum of £62. 9s. 4d. which he had spent on the purchase of prohibited books. The device was completely ineffective, though its failure was perhaps less dramatic than appears from the story of Foxe who made Tunstall the author of the plan and said that the money paid enabled the publishers to produce more New Testaments.

Not against books only was action taken. As part of the ceremony of Shrove Sunday, Robert Barnes, the Cambridge theologian, later a martyr, abjured his heresy; in the winter of 1527 Thomas Bilney, also a martyr in after-years, was sent to the Tower, from which he delivered himself by making his submission. Next year a grand attack was made upon the young Oxford scholars who had disseminated and studied heretical books; some were cast into prison where John Clark of Christ

---

[1] *Letters and Papers*, IV. ii. 1172, for the letters of Campeggio and Hackett (21 Nov.); ibid. 1158 for Warham's mandate to Voysey of Exeter (3 Nov.)

Church died; Robert Ferrar, the future bishop of St. David's, was driven to abjure. Yet Wolsey declined to take severe measures against Thomas Garrard of Magdalen, who had been the mainspring of the distribution of literature at Oxford, and John Frith of Christ Church, who had actually worked along with Tyndale, was dismissed from prison.

After the fall of the cardinal the government showed a more resolute attitude which has been associated with the rise of Thomas More, who had already entered the lists as the author of the *Dialogue concerning Heresies*, aimed against Luther and Tyndale (1528), and the *Supplication of Souls* (1529), which was a reply to Fish. To Foxe More appeared as the arch-persecutor. Recent historians have pointed out that More did not get the Great Seal until October 1529, and that though he held it until May 1532 he lost influence in February 1531, when convocation acknowledged Henry as 'Supreme Head'. They have pointed out too that a chancellor was not concerned with criminal law and that a layman could not deal with cases of heresy. They have represented that More was concerned to suppress not heresy but sedition and that among his alleged victims were ill-balanced persons guilty of unpleasant aberrations.

On the other hand, it is certain that More hated heresy as a thing which, if it were not checked, would imperil the souls of millions. Some of his criticism of Tyndale is surprising in a man who knew Greek, and proceeded on the assumption that since Tyndale was a heretic his work must be bad. His works abound in animadversion, sometimes of extreme violence[1]; heresy is the worst of all crimes; the man charged with it is not entitled to know the name of his accuser and must be punished in the public interest; heretics 'be kept but for the fire, first here and after in hell'; their burning 'is lawful, necessary and well done'. He was very confident in his own superior reason and he was not loath to employ authority. Although as chancellor he had no power to try heretics, yet 'he had both the power and the duty to take note of cases which were brought to his attention',[2] and by identifying heresy with sedition he made it a crime of

---

[1] It may be suspected that part of More's detestation of heresy arose from his fear that he himself might have promoted unbelief by his Utopian speculations. Gardiner later alleged that Erasmus, whom he had admired in his youth, laid the eggs which Luther hatched.

[2] F. M. Powicke, *The Reformation in England*, p. 57. Cf. Pollard, *Wolsey*, pp. 209–14.

which the council must take cognizance. It is not only in the pages of Foxe and Hall that he appears as the bitter foe of reformers. Tyndale noted him as 'the most cruel enemy of truth' and Stephen Vaughan, writing on 9 December 1531,[1] refers to his examination in cases of heresy as if such examination were a matter of course. From More's own works it appears that he took a personal part against heretics, and the fact that the clergy proposed to give him a 'reward' of £4,000 or £5,000 on his retirement—which he refused—is evidence that the bishops regarded him as their champion. If it is true that he lost sympathy with the government when convocation gave Henry the title of Supreme Head of the Church, it may be argued that he clung to office for the desire of power; and it is easy to see how Tyndale, thinking back upon the *Utopia*, came to regard his enemy as one who had sold his principles for money and authority.

Tyndale's suspicion was unjust. More had principles for which he was prepared to die, but amongst these principles was the belief that a timely severity towards heresy would prevent a greater severity at a later day. It is easy to see how a man resolute to preserve the faith would not hesitate to commend the taking of life. What is hard to understand is that a man who consciously mortified himself should not have realized that divine truth might be given to unlearned folk. Though he was all humility towards his Maker, More sometimes showed an intellectual arrogance towards his fellow men.[2] Yet the total number of his victims on any count is not large, and it may be that his rise to power coincided with persecution mainly because Stokesley and other bishops were anxious to employ the powers which they had recovered on the fall of the cardinal. Here it is important to notice that the ferment of heresy enhanced an anti-clerical spirit which was already stirring in England from political and economic causes, and that by their attacks upon men, many of whom were simple and pious, the bishops incurred an odium which was to weaken their cause in the day of trial which lay before them. The king was determined upon the divorce; the

---

[1] *Letters and Papers*, v. 265.

[2] In his life of Pico he recorded a story told by credible witnesses that the Queen of Heaven had appeared to the dying aristocrat and promised that he should not utterly die (*The English Works of Sir Thomas More*, i. 360); it does not seem to have occurred to him that the God on the Cross might come to the comfort of ignorant artisans who faced the flames.

pope opposed him; the English clergy, without the cardinal to act as buffer, found themselves beaten between the royal hammer and the papal anvil.

The divorce was not the cause of the Reformation, but it was the occasion, and it was the king who made the divorce. He was resolute to have his way, and if he had deluded himself into the belief that Wolsey had been the great obstacle he was speedily undeceived. The missions which he sent to Rome between 1529 and 1533, even when, as was usual, they were buttressed by French support, were uniformly unsuccessful and he found himself compelled to seek a δεύτερος πλοῦς, another way of doing things.

The first expedient was an appeal to the learned of Europe in whose hands there might be, pending a general council, an authority which could vie with that of the pope in the interpretation of scripture. In August 1529 there was brought to the king at Greenwich a young Cambridge scholar named Thomas Cranmer, who had forfeited his fellowship at Jesus College by marriage and who in a chance conversation with Edward Fox, the king's almoner, and Stephen Gardiner, the secretary, had suggested an appeal to the universities. The king and his advisers saw the possibilities which presented themselves. They did not act with undue haste, but before the year was done Cranmer was set upon the writing of a book. In November Richard Croke, a good scholar, was sent to Italy to seek new evidence (in Henry's favour) from the libraries, and to secure the support of the scholars. It does not appear that the book was ever printed, but it was circulated and the arguments it contained were presented to the learned at home and abroad. To theologians the point at issue was difficult, and it would be unfair to assert that the decisions given depended altogether upon political influence. Oxford and Cambridge, only after great debate, decided for the king; so did Paris, where Reginald Pole was active on the king's behalf, Orleans, Angers, Bourges, and Toulouse, not unmoved perhaps by the authority of Francis; so too did the north Italian universities of Ferrara, Bologna, and Padua, which had a great reputation for law. The Spanish and Neapolitan universities of course decided for Catherine; the Germans, who held lax views upon marriage—in 1540 Philip of Hesse was allowed to marry twice—were a disappointment to Henry, and Tyndale, himself a great king's man, pronounced against the divorce. None the less, the English king was able to show parliament a great consensus of opinion in favour of his cause.[1]

It was with the help of parliament that his victory was gained. The word of the universities might avail to counteract in the minds of the devout and the learned the hesitations of the pope, but it had no intrinsic power to settle the vexed question. Unable to persuade the head of Christendom to satisfy his desire, the English king took his own remedy with the aid of his own people. The 'Reformation parliament', which sat in seven sessions from 3 November 1529 to 4 April 1536, is a landmark in English history. It has been alleged that the parliament was 'packed', but there is little evidence that this was so. Certainly, the Crown used its influence in some boroughs, but so did nobles like Norfolk, whose policy was not always in tune with that of their royal master. In what would now be called the by-election of 1534 and in the 'general election' of 1536 Cromwell certainly showed himself very active, and though the total amount of packing which can be proved is not overwhelming, there is some justice in the complaint made by the 'Pilgrims' of 1536 that whereas 'the old custom was that none of the king's servants should be of the common house; yet most of that house were the king's servants'. Only a few, no doubt, were privy councillors, but others were councillors in the general sense, many must have held local offices of various kinds,[2] and many were justices of the peace. The speakers in parliament were invariably men who had commended themselves by loyal service to the king.

Yet the royal control over the commons was far from complete. In 1529 Cromwell certainly applied to Norfolk and to the king for a seat for Oxford or for one of the Winchester boroughs, but although he entered parliament it was for Taunton that he sat. John Petit, one of the members from London, in 1529 opposed the cancellation of the king's obligation to repay the forced loan; only in 1534 was a grant made, and then it was for a single subsidy and two fifteenths and tenths. The commons might appoint a committee of such of their number as were learned in the law to prepare their own bills.[3] They were not

---

[1] *Letters and Papers*, v. 83.   [2] See p. 199 *supra*.
[3] Hall, p. 766. Cf. W. S. Holdsworth, *Influence of the Legal Profession on the Growth of the English Constitution* (1924).

always amenable to royal promptings; in 1532 two members urged Henry to take back his wife, and a bill for wills and uses, dear to the royal heart, was rejected several times before it was passed at length in 1536. It may well be argued that the member of parliament in Henry VII's reign was more independent than his successor of the present day; he came to the house not as a delegate but as a representative, and he came unpledged to any programme. The true explanation of the harmony between king and commons was that both parties, in the main, wanted the same thing. The country gentry, as landowners, liked a settled authority and made common cause with the greatest landowner of all. The merchants were interested in trade; they were, on the whole, secular-minded; they had no particular reverence for the clergy; they had no desire to be drawn into European adventures on behalf of the papacy; if they supported the king it was because they approved the royal action, or at least did not disapprove violently enough to think resistance imperative.

With regard to the spiritual peers, Henry was extremely lucky. Between 1529 and April 1536 no fewer than thirteen sees were vacated by death and deprivation, and the men who filled the places were not unduly disposed to dispute the royal will. Of the other bishops two were nonagenarians, two were already committed to the divorce, and one (Llandaff) was the queen's confessor, George de Athequa, who could speak no English. Many abbeys, too, fell vacant about 1533, and the Crown hastened to put in its own nominees. Some of the churchmen, like Gardiner, were at first inclined to support the king in his quarrel with Rome, and when in 1534 they realized that matters were going too far they were powerless to resist. Altogether the king met with no great opposition from the spiritual estate. Nor did he meet with serious opposition from the nobles. A few, notably those of the blood royal, would not have been sorry to see their master take a fall, but many were inclined to welcome a diminution in the authority of the church, so proudly flaunted by Wolsey, especially as they hoped that some financial benefit might accrue to themselves.

Truly this parliament of England, understood by tradition to represent the public voice, inclined to share the royal opinion, and amenable to royal influence, was an instrument ready to the hand of the king. It was, however, an instrument which re-

quired skill in the playing. Some obvious advantages the sovereign had as a matter of course: he could summon parliament at
his pleasure; he could prepare a programme beforehand; he
could initiate legislation. It was the established custom that the
king with his councillors and judges should receive projected
legislation before a session opened, but though the 'Instructions'
sent by the king to Cromwell in 1531 were to be 'declared to the
council' they were to be 'undelayedly put into action', and
these instructions included the preparation of several bills.[1] The
king, it seemed, sometimes acted on his own initiative. Yet he
could not be sure that legislation would pass; as the events of
1523 had shown, he might encounter stiff opposition, and it
was part of Henry's greatness that he succeeded in getting
parliament to accept, without undue resistance, the great
measures which he presented to it. Wolsey had been a sort of
grand vizier; Henry has some claim to be regarded as the
first real prime minister of England. In the place of cabinet
he had, as his predecessors had had, members of his permanent council who could be used to do his business in parliament, and among them there appeared ere long one
who had a genius in that respect. Not without justice has
Thomas Cromwell been described as the 'first old parliamentary
hand'.

Cromwell was a man of the new age. Born about 1485, the
son of a Jack-of-all-trades at Putney who was distinguished both
by drunkenness and dishonesty, he left home to try his fortune
when he was about eighteen years of age, and for ten years lived
a wandering life. He learnt soldiering, probably in the French
army in Italy, banking in Florence, and business methods in the
Netherlands; when he returned to England in 1512 he brought
a practical knowledge of men and affairs and a useful connexion
with the English clothiers. He brought too the hard rationalistic
spirit which prevailed in the politics of Italy; he had little
belief in the omnipotence of the papacy and pinned his faith to
Realpolitik. Having made a good marriage he devoted himself
to law, business, and money-lending, and about 1520 entered
the service of Wolsey to whom he commended himself by his
knowledge of affairs and his ability to get things done. Was

---

[1] *Letters and Papers*, v. 196. Cf. T. F. T. Plucknett, 'Some Proposed Legislation
of Henry VIII', *Transactions of the Royal Historical Society*, 1936. See also *English
Historical Review*, lxiv. 174.

money to be found, was a lease to be arranged, was a monastery to be suppressed or a college founded, Cromwell was the man. When Wolsey fell, he adjusted himself with skill to the new situation. Whilst he gained the gratitude of the cardinal and possibly the respect of the king by seeming to maintain the cause of his old master, he was helping Norfolk and his friends to get a share of the spoils; when he spoke up for Wolsey in parliament he may have already known that Henry meant to show mercy to his humbled servant. It is not easy to determine the date at which he gained the complete confidence of the king. If it were admitted that his rise was a sudden thing consequent upon an interview at Westminster, where he advocated a royal supremacy adorned with the spoils of the church, it would have to be supposed that his position was established only in the year 1531, when he became a privy councillor.[1] Cavendish, however, and indeed Pole too, writing long after the event, represent that he commended himself to his royal master as soon as Wolsey was down, and there are traces of his fine Italian hand and of his actual penmanship in the attack on clerical privilege which marked the first session of the Reformation parliament. It is probable that his ascent was gradual; on the cardinal's fall Henry entrusted power to men whom he already knew. The elevation of Anne Boleyn's father, now earl of Wiltshire, to the office of privy seal bespeaks the rise of the anti-clerical party, but More as chancellor and Gardiner as secretary, though both king's men, were essentially conservative. Henry, in fact, had not yet made up his mind to a final severance from Rome. There was, as Foxe's pages attest,[2] a general belief that the king read the new books, and a very circumstantial story alleges that after a perusal of Tyndale's *Obedience of a Christian Man*, he said 'thys booke ys for me and all kynges to

[1] Foxe, Pole in his *Apology*, and Chapuys in a letter written in 1535 represent that Cromwell established himself with the king in the course of an interview held in the garden of Westminster Palace. They are, however, vague as to the date of this interview. Pole places it soon after Wolsey's fall, Chapuys after Wolsey's death, and Foxe, though he gives the date as 'about 1530', makes it synchronize with the 'supplication against the ordinaries' which was prepared at the end of 1531. Both Chapuys and Pole represent that Cromwell was made privy councillor as the result of this interview, which must have been in 1531. Cavendish, however, shows Cromwell as having several interviews with the king immediately after Wolsey's fall, and the similarity between the anti-clerical agitation of 1529 and that of 1532 is remarkable. Cromwell must have had influence in the commons and may have had influence before he obtained office.

[2] Foxe, bk. viii. (See the account of Simon Fish.)

reade'.[1] In 1531 he saved Edward Crome from the stake because his alleged heresy included a denial of papal supremacy; but with doctrinal heresy the champion of the seven sacraments had no sympathy, and when he summoned parliament in 1529, it was not with the intention of taking England outside the pale of the church. Yet it is not true that his sole object was to stabilize his finances—parliament removed his liability to repay the loan, but it made no other grant till 1534 when it gave one subsidy and two fifteenths and tenths, approximately £110,000 of which only £44,000 was paid before the end of 1536. What Henry wanted to do was to fortify himself with national support in his struggle against Clement, and the measures which he was prepared to take depended upon the reluctance of the pope to fall in with his wishes. The acts of the first session, which began on 3 November, were all popular. The usual economic legislation was passed and though the commons were allowed to launch an attack upon the spiritualty, their action touched the clergy in England rather than the pope in Rome and was based on empiricism rather than exact principle. Successive statutes remedied the abuses concerned with sanctuary, probates, mortuaries, the leasing of lands by spiritual men, pluralities, and non-residence. In each case, however, the evil was not excised altogether, but reduced within tolerable limits; the procedure against murderers and felons who sought sanctuary was made more severe; the fees for probates and mortuaries were fixed at moderate figures; and the number of pluralities to be held was reduced to four. These measures, which won the heart of the commons, served also to warn the pope as to what might happen if he failed to give satisfaction to the devout king of England, but for the moment there was no repudiation of the power of the keys. Parliament was dismissed on 17 December 1529, and for the next twelve months, whilst Henry importuned the pope and consulted the universities, no drastic action was taken. Before the year ended, however, the impatient monarch struck another blow. In December 1530 his attorney suddenly impleaded the whole of the English clergy for breach of praemunire in that they had recognized Wolsey as legate, and the convocation of Canterbury was allowed its pardon only on very hard terms. Besides paying a sum of £100,000 the clergy were

---

[1] *Narratives of the Days of the Reformation*, ed. J. G. Nichols (Camden Society, 1859), p. 56.

compelled to recognize the king as 'their singular protector, only and Supreme Lord, and, as far as the Law of Christ allows, even Supreme Head'. This was in February 1531, and later in the year the convocation of York made its peace with a grant of £18,840. Parliament, which had reassembled on 16 January 1531, duly ratified the pardon to Canterbury; but it was at pains to obtain for the laity of England, and that gratuitously, a complete indemnity against any offence committed in the recognition of the cardinal's legatine power. After having passed an act which declared murder by poison to be treason, and a number of economic acts including the famous act against beggars and vagabonds, it was prorogued on 30 March.

When it met again in 1532 the atmosphere was more tense. Clement, who in January 1531 had forbidden Henry to remarry, had shown no sign of yielding; Catherine had refused to withdraw her appeal; Thomas Cromwell had become a privy councillor. The fruits of a new policy soon appeared. On 18 March the commons presented to the king a supplication against the ordinaries which, if not designed by the new councillor, must certainly have been supported by him, since several drafts of the document bearing his handwriting are still to be seen.[1] Though orthodox in doctrine the petition was a violent attack upon clerical jurisdiction on the ground both of its cost and its partiality; not only were specific abuses denounced, but complaint was made on the general ground that the clergy in convocation legislated without the consent of king and laity. Convocation, in reply, began to prepare a programme of useful reforms, but before long it was asked to consider recommendations that its power of making ecclesiastical laws should cease.

In their replies, which were drawn up by Gardiner, the clergy referred to the help that the church had given to kings in time past, and, while they insisted that they should make ecclesiastical laws, proposed that these laws should not become operative without the royal licence. The king rejected the proffered compromise out of hand, and withdrew his favour from

---

[1] Mr. Elton adduces evidence to show that the supplication began as a genuine petition from the commons in 1529, and was handled by Cromwell when he was a member—but allowed to drop; he argues that the first weeks of the session of 1532 were devoted to discussion, and rejection, of Henry's bill for uses and primer seisins, and that when the commons were in an ill temper a petition, ready made by Cromwell on the basis of the 1529 grievances, was timeously produced. G. R. Elton, 'The Evolution of a Reformation Statute', *English Historical Review*, July 1951.

Gardiner, who lost what chance he had had of succeeding to the see of Canterbury when Warham died in August 1532. On 10 May convocation was bluntly told that it must agree to three propositions to which the reluctant clergy were driven to consent by a fresh threat from their imperious master. On 11 May Henry sent for the Speaker and a dozen members of parliament to hear of a grave constitutional issue of which he had just taken cognizance. He had discovered, he said, that the clergy of our realm 'be but half our subjects, yea and scarce our subjects'. In proof of his assertion he produced copies of the oath of the prelates made to the pope on their consecration and commanded that they should be read in parliament. The threat was enough. On 15 May convocation made its surrender in a document known as the 'submission of the clergy'. By the terms of this the clergy must make no new constitutions, canons, or ordinances without the royal licence; the existing body of ecclesiastical law must be reviewed by a committee of thirty-two, half clerical, half lay, all chosen by the king; the laws approved by the majority of the committee must receive the royal assent before they became valid. Plainly convocation had yielded up its independence. Next day More abandoned his office, and the seal was at once entrusted to another layman, Sir Thomas Audley, who was given the dignity of chancellor in the following January. Thenceforward, the fate of the spirituality of England was in the hands of the king. For the clergy it might be argued that they had of their own accord begun to remedy the abuses of the church courts; against them it must be argued that their last recorded act was a petty illiberal prosecution of a dead body for alleged heresy and that, though their ranks included men of ability and force, there was little promise of any well-directed or generally supported scheme of reformation. What stands out clearly is that the grievances of the commons against the ordinaries, which were in fact rejected by the lords, had been made a pretext upon which control over clerical legislation was transferred not to parliament but to the Crown.

Whilst his struggle with the English clergy was taking place Henry launched his first assault upon the privileges of Rome. He promoted in parliament a bill restraining the payment of annates, which he forced through in the face of discontent amongst the commons and of the opposition of all the bishops. The payments to Rome on presentation were limited to 5 per cent.

of the net revenue of any benefice; to forestall retaliation it was
enacted that if the necessary Bulls could not be obtained from
Rome, a bishop should be consecrated by the archbishop of his
province and an archbishop by a commission of two bishops
appointed by the king. This was almost a declaration of independence, but its stringency was alleviated by the provision
that the king should have power until the Easter of the following
year to declare by letters patent whether the act should be
operative or no; obviously there was a hint that if Clement
could come to terms with Henry his revenues from England
might yet remain secure.

Parliament was prorogued on 14 May, and before it reappeared in February 1533, events occurred which brought the
crisis nearer. Warham died on 22 August 1532, and the king
determined to have as his successor not Stephen Gardiner,
who had retired to his see under the royal displeasure, but
Thomas Cranmer, who had suggested the appeal to the
universities. The choice was significant. Gardiner, as his life
and works attest, was a stout Englishman and a resolute king's
man, but he had no sympathy with doctrinal reform. Cranmer,
who early in 1532 had been sent to Germany to gain support
for the 'divorce', had married Margaret, the niece of the theologian Osiander of Nuremberg; he had hearkened to views upon
the Eucharist and upon justification which, though less advanced than those of Luther, were not in accord with the
doctrine of Rome. Henry had no intention of departing far, if
at all, from the Roman doctrine; but in his determination to
secure a separation from Catherine, he was content to woo the
support of the reformers both in England and in Germany.
Perhaps because Cranmer passed for a moderate man and
certainly because Clement was still anxious to ally with France
and England, no opposition to the promotion was offered at
Rome; the necessary Bulls, no fewer than eleven in number,
were issued between 21 February and 2 March; the customary
payment of 10,000 marks was remitted and the new archbishop
was formally consecrated on 30 March, taking the usual oath of
obedience to the pope, but qualifying it by a formal assertion
that he was not bound to anything contrary to the law of
God or to the king, realm, laws, and prerogatives of England.

From the royal point of view Cranmer's advancement was
timely, for Henry was following a course which demanded the

help of a canonically chosen yet obedient archbishop. Knowing that Anne was pregnant, he had married her secretly about 25 January 1533, and as soon as parliament met he hurried through an act in restraint of appeals[1] which made it possible for the English primate to decide 'the great matter' without reference to Rome. The act asserted that the realm of England was an 'empire' governed by one supreme head and king, unto whom the whole body politic, spiritual and temporal, owed obedience; that the king, under God, had plenary power to 'yield justice' to his subjects without restraint of any foreign prince or potentate; that, when any cause of divine law or spiritual learning came into question, it should be declared and interpreted by that 'part of the said body politic called the spiritualty, now being usually called the English church'. The assumptions underlying this act are even more important than its provisions. The state is regarded as an organic whole; the English church, though it is given a corporate being, is identified with the clergy and invested only with the function of declaring divine law.[2] Like the temporalty, the spiritualty, as part of the body politic, owes obedience to the head. The king, presumably, is bound by his office to do righteousness, but his absolute power is not officially limited by the authority of the scriptures and still less by the authority of convocation. The position was made quite clear when the new archbishop, immediately after his consecration, presided over a meeting of convocation which decided, John Fisher, the bishop of Rochester, almost alone protesting, that Henry's marriage with Catherine was against divine law. Armed with this decision, Cranmer obtained licence from Henry to try the cause, and on 10 May opened his court in the quiet priory of Dunstable; Catherine refused to appear, and sentence was given in the king's favour on the 23rd. Five days later Cranmer pronounced that Henry's secret marriage with Anne was valid, and on 1 June 'the lady' became queen. In a belated effort to protect his own interest and the dignity of the church, the pope prepared on 11 July the sentence of the greater excommunication, though he postponed its issue until September. Henry, who on 9 July had confirmed the act

[1] 24 Henry VIII, c. 12.

[2] The attitude of the English government resembled that of Luther in 1520. The *regnum* was magnified because the *sacerdotium* was stripped of its power, but to the *sacerdotium* was left the power to distinguish truth from error. J. W. Allen in *Tudor Studies*, p. 91.

against annates by letters patent, was confident that the pope would lack a champion; Francis was his ally and Charles would rather see Calais English than French. He was unshaken by the papal thunder and in November 1533 the pugnacious Edmund Bonner, late chaplain to Wolsey and future bishop of London, bustled into the presence of his holiness at Marseilles, to give the English reply in an appeal to a general council. Francis, who had regarded his cousin's operations as a means of coercing the pope, was aghast; Henry, still protesting that he would not be less but more Christian in separating himself from Rome, was determined that the breach should be complete. It was, in fact, already complete, but a great series of acts in 1534 defined the position of the English king with clarity.

During the spring session of 1534 there was passed an act for the submission of the clergy,[1] which gave statutory validity to the submission made by convocation in 1532 and at the same time ratified and slightly altered the act of appeals of 1533; henceforth appeals from the archbishop's court were to go not to the upper house of convocation, but to chancery, and the king was to name a commission under the Great Seal 'like as in case of appeal from the Admyrall Courte'. An act in restraint of annates[2] made absolute the conditional act of 1532 and defined the procedure to be used in the election of bishops. Applying to himself a privilege accorded to the founders of abbeys and their heirs, the king, as representative of the founders of English bishoprics, demanded that on a vacancy he should send to the chapter a licence to elect; this licence would contain the name of a person who must be elected with 'all speed and celerity' under the threat of the penalties of praemunire. Elections to abbeys were to be made in the same way. Another act[3] placed in the hands of the archbishop of Canterbury the power of dispensation, that 'ever springing fount of oil playing upon the ecclesiastical machinery'. Although it asserted that there was no intention to use the power for causes contrary or repugnant to the holy scriptures and the laws of God, it put that power under the authority of king and parliament. Two-thirds of the profits were assigned to the king and his

[1] 25 Henry VIII, c. 19. The act provides for a commission of thirty-two persons to examine existing canons, but this commission was never appointed, hence the old canons save where contrary to the law or the royal prerogative remained in force.

[2] 25 Henry VIII, c. 20.   [3] 25 Henry VIII, c. 21.

officers; the visitation of monasteries 'exempt' from episcopal inspection was given, not to the archbishop, but to the Crown. Finally, an act of succession[1] confirmed Henry's marriage with Anne and established the right of the issue of this marriage; it set forth severe penalties against all who opposed the marriage either overtly or secretly, and empowered the king to exact from any of his people at any time an oath to maintain the whole provisions of the act under pain of misprision of treason. The words of the oath were not included in the act, but they were formulated on the last day of the session and the oath was taken by members of parliament, ministers, and others before it was given statutory sanction in a second act of succession[2] which was passed when parliament reassembled in November.

During the winter session, which lasted until 18 December, were passed three other acts of the first significance. First came the great act of supremacy[3] which summed up and emphasized all the claims made by the English king. He is and ought to be supreme head of all the church of England, called *Anglicana Ecclesia*, and shall have united to the imperial Crown of this realm 'as well as the style and title thereof' all the prerogatives 'to the said dignity of supreme head of the same church belonging and appertaining'. He shall have full power to visit, redress, reform, correct, and amend all errors, heresies, abuses, and enormities which by any manner of spiritual authority may lawfully be reformed and redressed. The king in fact claimed a spiritual jurisdiction. His vicar-general could preside in convocation; his visitors could supersede the authority of bishops; his injunctions would be binding upon the clergy; he, in parliament, could define the faith, and henceforth heresy might be prosecuted by special commission or even at common law as well as in the courts of the church. The king was not giving his protection to an independent national church; he was asserting that England was a sovereign state in which king in parliament was supreme over all things ecclesiastical as well as temporal. Another act[4] not only annexed to the Crown the first-fruits of benefices (condemned in the annates act of 1532 as 'grounded upon no just or good title' and 'importable') but gave the Crown a new revenue in the shape of an annual tenth of all clerical income. Royal commissioners were empowered to compound

---

[1] 25 Henry VIII, c. 22.   [2] 26 Henry VIII, c. 2.
[3] 26 Henry VIII, c. 1.   [4] 26 Henry VIII, c. 3.

for the value of first-fruits of benefices, and to make a new
valuation for the tenth, noting the gross income and allowing
deductions on certain specified grounds—fixed charges to
which lay and spiritual persons were liable, fees for administration, and alms. All that the clergy got in return was the remis-
sion of the sums still due in respect of the 'pardon'. The third
great measure[1] had for its object the securing of the royal title,
but it took the form of an extension of the old statute of treasons
of 1352. This statute, which demanded an overt act of a
specified kind and the testimony of two witnesses, contained a
phrase providing for a possible enlargement by parliament of
the crime of treason, and at various times fresh definitions had
been made. In the crisis which followed the year 1529, treason
had already been extended empirically to cover poisoning and
offences under the act of succession, and ever since 1531 Cromwell, later assisted by Audeley, had been trying to prepare a bill
for the 'augmentation of treason'. No fewer than five drafts,
altered and corrected, survive to show the difficulties which
the secretary encountered in preparing a bill which would be
accepted by a suspicious parliament, but at last he managed to
get through an act 'whereby diverse offences[2] be made high
treason', including amongst these the malicious wish, will, or
desire to deprive the king and queen of the title or name of their
royal estates and the slanderous publication of writing or words
describing the king as heretic, schismatic, tyrant, infidel, or
usurper. To make word or mere intention as culpable as an
open deed was a monstrous thing; but the first effect of the act
was to silence all criticism of Henry's arrogation to himself of
spiritual power. It had, however, another effect and possibly
another purpose; it enabled the government to enforce the
oath of succession under penalty of death. In the act of succession failure to take the oath involved only the pains of misprision of treason—forfeiture and imprisonment; now it could
be argued that the subject who refused to take the oath was
endeavouring to deprive the king of his title and even to brand
him as a schismatic.

The reason for this severe measure was that the king, though
obviously he must have had much support, had encountered

---

[1] 26 Henry VIII, c. 13.
[2] Isobel D. Thornley, 'The Treason Legislation of Henry VIII (1531–1534)' in
Transactions of the Royal Historical Society, 3rd series, vol. xi (1917).

considerable resistance in carrying out his policy. The legislation of 1534 has, for convenience, been set forth in an unbroken sequence, but it did in fact excite opposition in the country at large, for many Englishmen felt that Catherine was Henry's wife and their queen. The restlessness which prevailed in the south of England came to a head in the cult of Elizabeth Barton, the so-called Holy Maid of Kent. This poor woman, as the victim of epilepsy, had fallen into trances in which she had uttered prophecies, and some of these prophecies had been fulfilled. Before long pilgrims began to resort to the country church which was the scene of her revelations; the parish priest, Richard Masters, reported to Canterbury, and Warham mentioned the matter to the king who referred it to More. More found nothing but what 'a right simple woman might speak of her own wit', but Dr. Edward Bocking, one of the commissioners sent by the archbishop to investigate, appears to have exploited the political opportunity. Elizabeth was received into the convent of St. Sepulchre at Canterbury, where her fame and her claims grew apace. Sometimes her influence was used for moral ends but, no doubt because she perceived amongst her consultants a real anxiety about Henry's marriage, she acted as did Joan of Arc, gained the royal presence and delivered an admonition. What may have begun in vanity was developed by skill; Elizabeth was *endoctrinée*. She made touch with the papal ambassadors and professed to know by revelation that Henry would cease to be King within a month and would die within six months if he put away Catherine and took another wife.[1] Henry, who had been unmoved by papal displeasure, was sensitive to English criticism, especially when he discovered that the nun had been in communication with the ladies of Salisbury and Exeter, whose houses represented the old Yorkist stock. In July the government determined to proceed against the nun; two months later, some of her confederates were arrested; on 23 November she and half a dozen accomplices confessed their impostures at St. Paul's Cross, and in the same month Chapuys wrote to his master that the alleged crime of Elizabeth Barton would be used to ruin the party of the queen. His prognostication was correct. When, on 21 February 1534, a bill of attainder

---

[1] *Letters and Papers*, vi. 587. *Letters relating to the Suppression of the Monasteries* (Camden Society, 1843), pp. 14–34. *Transactions of the Royal Historical Society*, New Series, xviii (1904).

was introduced into the house of lords against the offenders, the names of Fisher and More were included, though for misprision and not for treason. More, who had refused to hear politics from the nun and who, after her confession, referred to her as 'the wicked woman of Canterbury', was denied an opportunity of making a public defence; but his name was struck out of the bill because the king's advisers assured their incensed master that unless this were done the bill might not pass. Fisher was found guilty of misprision of treason and fined £300. Elizabeth and four of her clerical supporters were executed at Tyburn on 20 April.

The illustrious victims were not to escape. The situation was too tense for compromise. On 23 March 1534 the pope gave his decision in favour of Catherine.[1] Almost at the same time parliament passed the act of succession and on 13 April More was summoned to Lambeth. There he was invited to take the oath, and plied by arguments from Cranmer, who wished to save him; but though he was prepared to swear to the succession itself and to abstain from advising others against the oath, he refused to take it in the form in which it was presented. He was committed to the Tower where, during a long imprisonment, he occupied his time in writing the *Dialogue of Comfort* and the *Treatise on the Passion*, which show his spirit at its best—gentle, resolute, and claiming for itself a liberty of conscience which in other days it had denied to 'heretics'. Though resigned, he was lawyer enough to defend himself with skill, and even after the new statute made it possible for his offence to be reckoned as treason he, and Fisher who was also a prisoner in the Tower, both made the point that since there was no malice in their hearts their refusal of the oath did not bring them within the terms of the act. The king, however, would have no mercy. In the spring of 1535 he was aware of a growing unrest in the north, and he was stung to fury when, in the following May, Pope Paul III created Fisher cardinal. The victims were tried by special commission and the judges held that the word 'maliciously' in the act was of no effect. Fisher was beheaded on 22 June, More on 6 July; five valiant heads of houses, three of them Carthusians, two months earlier had suffered the full penalties of treason, and now three other Carthusians were condemned to brutal execution. All died with constancy, and More

[1] *Letters and Papers*, vii. 150.

with a gay courage which commands our whole-hearted admiration; but to many Englishmen of the day loyalty to king and country were of the first importance and it is significant that in the pages of Hall[1] Fisher appears as 'a man of very good life, but therein wonderfully deceived', who had maliciously refused the king's title of Supreme Head, while More is shown as one whose learning and great wit were mingled with taunting and mocking, a man of levity who could not forbear his jesting even on the very scaffold.

The attitude of mind expressed in these comments goes far to explain the royal supremacy. England would obey a master and especially a master who expressed in himself the full-blooded life which pulsed in the veins of a prosperous middle class. Long before, More had said to 'Master Cromwell', 'if a lion knew his own strength hard were it for any man to rule him'. The royal lion had discovered his strength, and though perhaps he knew it not, that strength was founded upon his alliance with the most effective part of his people. In overthrowing the pope the king had come to rely instinctively upon the support of his subjects, and the Tudor monarchy, though it appeared to be, and in some sense was, absolute, yet presaged the rule of the king in parliament.

It was not to the people of England alone that the lion showed his strength. Henry was already thinking of an 'empire' which should include England and Wales and perhaps Scotland too. Over Scotland he did not for a moment advance a claim of suzerainty, partly perhaps because James V was his nephew, partly because he did not wish to antagonize France. In May 1533 he concluded a truce at Newcastle which became a permanent peace exactly a year later, and thereafter he set himself to lure James into his own quarrel with Rome. The insignia of the garter carried north by Lord William Howard, a selection of effective texts presented by Dr. Barlow, hints as to the wealth to be obtained from church lands, were all used in vain, and even though James married Madeleine of France on 1 January 1537, Henry did not withdraw his friendship. He was incensed because the Scots harboured some refugees from the Pilgrimage of Grace, but even when, after Madeleine's premature death, James in June 1538 married Mary of Guise, whose 'opulent

[1] Hall, ii. 264.

beauty' had been brought to his own notice, he still preserved at least the semblance of amity. For Scotland and for Scotsmen he had little regard as the event was to show, but he waited upon opportunity to display his mind and his power.

From the affairs of Ireland, however, he was not able to withhold his hand. Surrey, who had been recalled in 1521, had believed that order could be established only by force, and had calculated that for a rapid conquest a regular army of 6,000 men would be needed. Henry and Wolsey, engrossed in their European adventures, declined to provide the necessary money and fell back on the time-worn policy of playing one great family off against another. At first a Butler was trusted, but before long reliance was placed upon the far more powerful Kildare, under whose rule Irish influences spread all over the Pale. In 1526 Kildare was summoned to London, where he was kept for two years as a hostage; during his absence Ireland, under vice-deputies who were his kinsmen, became more unruly than ever, and in 1529 the emperor had sent a mission urging the other great Fitzgerald, Desmond, to rebel. Aware of the danger, the government in 1529 tried to sweeten Irish opinion by allowing Kildare to return, and by giving the office of lord lieutenant to Richmond who was given as deputy Sir William Skeffington, master of the ordnance. The competent 'gunner' did his best, but Irish obstruction was as effective as usual, and by 1532 Kildare became lord lieutenant again. He used his power to prosecute his feud with the Butlers, and in 1533 John Allen, master of the rolls, came to England with complaints so serious that the great earl was summoned again to London and imprisoned in the Tower.

He had appointed as deputy his eldest son Lord Thomas, and when a rumour arose that Kildare had been executed this 'silken Thomas', who was gay and popular, cast off his allegiance to England on 11 June 1534. He declared himself to be the pope's man and sent for aid to Paul III and Charles V; John Allen, archbishop of Dublin and chancellor, was murdered on the beach at Clontarf; Dublin was besieged; and the rebel chief, who became tenth earl of Kildare when his father died, from natural causes, in the Tower, seemed to have Ireland at his feet. The storm, however, blew itself out with surprising rapidity. The Butlers were proof against the offers of Kildare;

Dublin held out; in October Skeffington and Sir William Brereton landed with fresh forces. When the castle of Maynooth fell in the spring of 1535, the army of 'silken Thomas' melted away, and the young earl surrendered to Lord Leonard Grey, who had arrived as marshal to the English army. Grey, whose sister was the second wife of the ninth earl of Kildare, almost certainly promised the captive his life; the Irish council and Henry's servants in Ireland were urgent that the promise should be kept if only for politic reasons. Kildare, however, had already been attainted under an act of 1534[1] and as soon as Lord Leonard Grey, who became deputy in January 1536, had suppressed the rising—mainly by craft and diplomacy—the government took action. Five of the young earl's uncles were treacherously arrested; the attainder was renewed on 17 July 1536 and on 3 February 1537 'silken Thomas' and his uncles suffered the traitor's death at Tyburn.

It is an ugly story, and it has an ugly sequel. Grey, who became deputy when Skeffington retired, to die almost immediately at the end of 1535, continued at first a vigorous policy and conducted a successful campaign in the south and west, though much hampered by lack of pay. On 1 May 1536 he assembled a parliament which sat in various places and with many adjournments until December 1537, in which he succeeded in carrying, not without difficulty, acts which imposed upon Ireland Henry's 'reformation'. The pope's authority was repudiated; the king was made supreme head of the Irish church, appeals to Rome were quashed, first-fruits were made payable to the Crown, and, in the face of strong resistance, the abbeys were suppressed. This would seem good service, but in order to gain his ends the deputy showed himself conciliatory to the Geraldines, and for that reason quarrelled not only with the Butlers, but with his own council. As a commission sent to review matters in 1537 proposed to reduce his force to 340 men, it is hard to see how he could have adopted any policy other than conciliation, but his complacence went too far and when he was in England on a visit in 1540 he was sent to the Tower; an attempt to attaint him failed, but when fresh evidence was brought he pleaded guilty to a charge of high treason and in June 1541 he was executed. Sir Anthony St. Leger, who succeeded him as deputy, managed to draw the malcontents to the king's side by offering

[1] 26 Henry VIII, c. 25, and 28 Henry VIII, c. 18.

them a share in the lands of the suppressed monasteries and investing with peerages some of the king's persistent enemies, most of whom renounced the pope. On 30 December 1540 the deputy and council advised Henry to put an end to papal pretensions by himself assuming the title of king of Ireland, and statutes to this effect were passed by the Irish parliament in the summer of 1541. That parliament had already given him the title of head of the Church in Ireland, which was used in the preamble to the Irish statute of 1540, and in the English Act of Succession of 1544 Henry appeared as king and head of the Church in both England and Ireland. It is significant that the first Jesuit mission appeared in Ireland the following July. Henry had 'scotch'd the snake', not killed it. His monarchy, like other English attempts to coerce Ireland, rested in the end on no surer foundations than the rewarding of the king's enemies and the mutual animosities of the Irish leaders. They were enriched with the lands of the abbeys, but the 'island of the saints' still turned her eyes towards Rome.

In his policy towards Wales Henry was more successful. There, too, Celtic animosities flourished, but there the Tudor started with real advantages. As king, he controlled the 'principality' established by Edward I, which had been 'shired' into Anglesey, Caernarvon, Merioneth, Flintshire, Cardigan, and Carmarthen. Through his father he inherited the prestige of an ancestor who had died for Owain Glyndwr and the attachment of the old Lancastrian domains, Pembroke and Glamorgan, Kidwelly and Monmouth; from his mother he inherited something of the power and influence of the Yorkist Marcher lords and this had been consolidated when the great lordships of Sir William Stanley were forfeited to the Crown. In deference to Welsh sentiment Henry VII had called his eldest son Arthur, and sent him to hold court at Ludlow; but in pursuance of English policy he trusted the representative of a great Celtic family. When Henry VIII ascended the throne, Sir Rhys ap Thomas wielded the power in Wales. He seems to have ruled fairly well, but no doubt he considered the interest of his friends and kinsmen, and when, in 1525, he was succeeded by an impetuous grandson, Sir Rhys ap Gruffydd, complaints of partiality and injustice soon came to Wolsey's ear. The inevitable end was that, after the usual hesitations, the young chief was executed on Tower Hill on 4 December 1531, and with his fall the English government found itself compelled to renew the machinery of state which had fallen into serious disrepair.

Disorder was rife; the Council in the Marches was ineffective and its president John Voysey, bishop of Exeter, was either negligent or incapable. A great change came when, in 1534, a new president appeared in the person of Rowland Lee, recently become bishop of Lichfield and Coventry, who was a man after Cromwell's heart. The parliament of that year passed several laws[1] for the establishment of good order in Wales and the Marches, and the bishop showed himself a more than resolute administrator. Whether he really hanged 5,000 offenders may be doubted, but he certainly established the authority of the law, checking violence, fortifying castles and harbours, preventing large assemblies, and punishing jurors who refused to do their duty. Not all the energy of the council, however, could prevent malefactors from escaping from the shires into the lordships, and a remedy was found in an act of 1536[2] which divided all Wales into shires. The six existing counties were little altered; the earldom of Pembroke and the lordship of Glamorgan, which had been in the king's hands since the death of Jasper Tudor, were now formally recognized as shires; four new counties were added, Denbigh and Montgomery in the north, Brecknock and Radnor in the south. Each of the shires was to send one knight to parliament and, with the exception of Merioneth, one burgess from the shire town. Monmouthshire, which was placed under the jurisdiction of the English courts, was treated as an English shire and given two knights and two burgesses. The lord chancellor was empowered to appoint two commissions, one to delimit the new shires and divide them into hundreds, the other to scrutinize the laws and customs of Wales.

The adjustments made during the next half-dozen years were summed up in a comprehensive act of 1543.[3] Sheriffs, constables, and coroners were given the powers enjoyed by English officials of the same name; justices of the peace were to be appointed on the English model, and the authority of the feudal courts, preserved in 1536, was confined within the limits imposed by

---

[1] 26 Henry VIII, cc. 4, 5, 6, 11.
[2] 27 Henry VIII, c. 26.
[3] 34 & 35 Henry VIII, c. 26.

English practice. English common law was to govern questions of land-tenure, and primogeniture now took the place of partible succession. Four circuits were established, each to be presided over by a justice who twice a year was to hold 'the king's great sessions' in each of the shires attributed to him, in accordance with English law. Appeals were to go to Westminster, except those arising from purely personal actions, which were to be decided by the Council in the Marches, now given statutory recognition for the first time. That council, in spite of all the new arrangements, retained its authority both judicial and administrative. Besides its powers on appeal, it had an original jurisdiction equal to that of the 'great sessions' and it could, moreover, determine any causes which his majesty pleased to assign to it. Administratively, it was articulated into the Tudor system. It was directly responsible to the Crown, and in close touch with the privy council. Its president, who was normally lord lieutenant of all the Welsh and of some of the border counties, was responsible for musters and defence, for supervising royal officials and juries, and for dealing promptly with every action likely to lead to a breach of the peace.

The obvious purpose of the royal policy was to assimilate Wales to England. It took little account of considerations of geography, history, race, and language and it ignored the traditional divisions of the Welsh church. Whether such a policy could have been carried out by any king save one whose descent commended him to Welsh sentiment may well be doubted, and it was not, in fact, completely carried out. The old social organization remained and family alliances and family feuds continued to play a great part in politics. Yet the new system established itself. Its great merits were that it gave Wales a better order than she had enjoyed for years and at the same time provided opportunities to the Welsh gentry both in England and in Wales itself. Modern criticism, not uncontradicted,[1] has regretted the attack on Welsh nationality, as prejudicial to the best interests of Wales; and the aggrandizement of the squires, who became justices of the peace, at the expense of the peasantry; but at least it must be said that the life of the

---

[1] W. Garmon Jones, *Welsh Nationalism and Henry Tudor*, Cymmrodorion Society Publication, 1918; Sir J. F. Rees, *Tudor Policy in Wales* (1931); W. Rees, *The Union of England & Wales* (Transactions of the Cymmrodorion Society, 1938) contains a useful map and a transcript of the 1536 act.

ordinary Welshman was made more secure, that England was set free from an embarrassment which might have been dangerous in troublous times, and that Wales in her future development did preserve her individuality, her language, and her proud consciousness of being herself.

XI

THE FALL OF THE MONASTERIES

THE meaning of the royal supremacy speedily showed itself in action. Although the pope, shocked by the deaths of the martyrs, dated in August 1535 the Bull of excommunication already prepared, and sanctioned another Bull depriving Henry of his realm—never actually launched— Henry, secured by the mutual animosities of Charles and Francis, proceeded steadily upon his way. When, in January 1536, Queen Catherine died, he felt that his path was made clear, and Cromwell bluntly told a servant of Chapuys that relations between England and the empire would now be easier. The epitaph of the dead queen was written in the words of him who had long been her husband: 'God be praised, we are free from all suspicion of war.'

While these arrogant acts expressed his master's determination, Cromwell was taking measures to exploit the new financial assets of the Crown and at the same time to rivet upon England the bonds of the new authority. On 15 January 1535[1] the king formally assumed by ordinance the title of Supreme Head on earth of the Church, and Cromwell, as vicar-general, was free to exercise in his own person powers even greater than those which Wolsey had enjoyed. These powers he used first of all in the realm of finance, for the question of finance was pressing. The Reformation parliament, which did so much for Henry's power, made him no grant of consequence and he was reluctant to apply coercion; yet money was urgently needed. Although the government kept out of war, it could not avoid some outlay upon the defences. In 1533 alone the northern border cost nearly £25,000; the Geraldine rebellion in Ireland[2] cost, in 1534, £38,000 more than the Irish government could supply; fortifications at Dover and Calais were expensive;[3] the suppression of the Pilgrimage of Grace cost £50,000; the king was building palaces, and the upkeep of the court became steadily dearer as the value of money fell. After 1534 France paid no more pensions, and the customs revenue, owing mainly

---

[1] *Letters and Papers*, viii. 18.   [2] Ibid. 295 and ix. 72. Dietz, 104, 140.
[3] e.g. ibid. viii. 423.

to a fall in the wool-subsidy, persistently declined. The situation was not as bad as it may appear, for some surpluses were quietly tucked away in the royal coffers;[1] yet Cromwell, who was more apprehensive than his master of trouble from abroad, was anxious to prepare for the evil day, and to him the spoliation of the church appeared as a solution, perhaps the only solution, of his financial difficulties. It is not easy to discover at what exact date he resolved to obtain the possession of the monasteries, but he had from the act of first-fruits and tenths a right to assess clerical revenues, and on this pretext, early in 1535, he set about making a valuation of incomes which would serve in good time to facilitate a seizure of capital.

In accordance with the provisions of the statute, the chancellor prepared commissions for each shire, dated 30 January,[2] empowering the persons appointed to acquire all the information needed to ascertain the true value of every benefice, and to specify the amounts of all deductions which might properly be allowed.[3] Exact arrangements were made for the methodical conduct of the inquest, and the commissioners were ordered to make a complete return to the exchequer under their seals along with the local returns of the sub-commissions no later than the octave of Trinity (30 May). As a rule, the bishop was the only cleric on the commission for his diocese, but he was to be chairman.[4] It proved impossible to complete the task within the time, but the work was done with ruthless efficiency. The commissioners pared to the lowest limit the deductions permitted by the act—Gardiner even proposed to deprive the College of Winchester of the revenues which it enjoyed,[5] and the government tried to assuage the loud discontent by asserting that once the gross value was established, the deductions could be fixed by arrangement. The returns, some of which were not made until September, were drawn together in a general report, the *Valor Ecclesiasticus*.[6] The evidence is not now extant in its

---

[1] Dietz, pp. 142, 219: e.g. from the court of augmentations.
[2] *Letters and Papers*, viii. 40.
[3] Cf. *Valor Ecclesiasticus*, ii. 289, for the text of the commission for Devonshire.
[4] For details as to the operations of the commissioners see Savine's book mentioned in note 6.   [5] *Letters and Papers*, viii. 244 and ix. 369; Savine, p. 11, n. 1.
[6] This was published in six volumes by the Record Commissioners, 1825–34. The best exposition is that made by A. Savine, *English Monasteries on the Eve of Dissolution*, vol. i of the 'Oxford Studies in Social and Legal History', ed. Paul Vinogradoff. He is mainly concerned with the value of the monasteries (cf. Dietz, ch. xi and app.; Tables III, IV, V, and VI).

entirety, but the gaps can be made good from other sources, particularly from the *Liber Valorum*, a working hand-book in which the net income and the tenth of every benefice were recorded for official use. According to the most reliable computation, the total annual income of the church after deduction must have been about £300,000 a year.[1] Under the act, therefore, the king might expect to obtain about £30,000 per annum from the tenths alone as well as a sum, not easily fixed, from first-fruits. The records of the court of first-fruits and tenths show that in fact his receipts from tenths between 1535 and 1538 averaged about £29,500 a year, while the first-fruits produced a mean of £16,500. An increment of nearly £50,000 a year was obviously a rich prize for a government whose ordinary revenue had dwindled to little more than £100,000 a year.[2] Yet even this welcome increase of revenue did not satisfy the anxieties of Cromwell, nervous of foreign invasion and increasingly made aware that Henry would never join whole-heartedly with the German princes. What Cromwell wanted was less a permanent revenue than a large sum of ready money, and he or his master wanted also to unite the English laity beneath the royal supremacy. Both ends could be attained by confiscating some of the lands of the church and giving a share of the spoils to the nobles, the gentry, and the wealthier members of the mercantile class. To a scheme of spoliation the lands of the monasteries were the most accessible. They were alleged to possess one-third of all the land of England, though this estimate is probably an exaggeration,[3] and as nine-tenths of their acres were already leased or let, expropriation might be effected without undue disturbance if the old tenants were allowed to keep their holdings under the new lords.

There were in England, excluding the commanderies and the

---

[1] The work of Savine shows that the gross income from the monasteries was nearly £163,000, of which over £121,000 was temporal, i.e. derived from land. The net yields work out at about £136,000 and £110,000 respectively. As the receipts from the first-fruits and tenths fell to something over £18,000 after the dissolution of the monasteries, the monastic tenths must have yielded rather less than £12,000 a year, and as this figure tallies fairly well with the gross total of the incomes already set forth, it appears that the tenth was approximately a real tenth. On this assumption the taxable income of the secular clergy must have been over £180,000.

[2] Dietz, p. 138.

[3] Simon Fish said more than one-third; Gairdner seems to accept one-third; Gasquet accepted 2,000,000 acres; Dr. Liljegren estimated them as one-quarter of the whole and Mr. Pickthorn one-sixth as a minimum (see Savine, p. 80, and Pickthorn, *Henry VIII*, p. 377).

preceptories of the Order of St. John of Jerusalem, 563 religious houses, containing perhaps 7,000 religious men, 2,000 nuns, and 35,000 laymen of various kinds. The visitations, which should have been triennial, had not been carried out with regularity, and conflicts of authority had availed to hinder investigation. The evidence is far from complete, but in the reports which survive there are many examples[1] of grave financial irregularities, neglect of rules, indiscipline, and immorality. Even making allowance for the mutual antipathies of the various deponents, it is hard to avoid the conclusion that often standards were low and that abuses of various kinds were accepted as inevitable.

The monastic ideal had lost much of its old appeal and much of its old fruitfulness. Since 1400 only eight new houses had been founded, and some of the old were declining in numbers. No chronicles were being written. The development of printing, it is true, had removed the need for manuscript chronicles, and it is possible that the monks bought some of the devotional books which were printed, but there was no monastic author of great distinction. Little was being done for learning, for the monastic schools were used mainly to train novices and choristers, though a few boys of good birth, sent to the abbots as pages, received instruction; from most houses, too, few scholars were sent to the universities. The primary duties of a religious house—the offering of prayers, the giving of alms, and the exercise of hospitality—were not performed with the old zeal. In order to maintain the numbers required for the singing of the offices, persons who had no real vocation were sometimes allowed, and occasionally indeed coerced, into taking the vows, and the nunneries were sometimes used as havens for superfluous daughters. The almsgiving, for the most part, was the bestowing of doles on holidays, mainly in food and drink, and the grant of daily rations, sometimes with board, to a few persons.[2] Not more than 5 per cent. of monastic income went upon charity, and the benefit was often given to

[1] See 'Morton's Register' by Canon Claude Jenkins in *Tudor Studies*, p. 64; 'The Monastic Legend' in *Medieval Studies* (London, 1915) by G. G. Coulton, and ch. iii of Mr. Baskerville's *English Monks and the Suppression of the Monasteries* (London, 1937), which uses the Norwich visitations edited by Jessopp, the *Collectanea Anglo-Premonstratensia* (Camden Society, 3rd series, vols. x and xii); and the unprinted Lincoln Visitations.

[2] From the Lincoln Visitations it would seem that some of the beneficiaries had bribed minor officials to secure their places.

The Sees of England, showing the modifications made by Henry VIII.

friends or relatives of the religious and to 'abbey lubbers' rather than to the deserving poor. Even where it worked well, the system, like others of its kind, perpetuated pauperism instead of curing it.

Certainly the houses entertained guests who were accommodated in halls above the western side of the cloister or in separate buildings and, especially in the north, where the inns were few and bad, the traveller might find welcome shelter. But, as was inevitable, the good fare attracted visitors whose journeys were, in the modern jargon, 'not really necessary', and some nunneries tended to become 'homes of rest' for ladies of a certain age who made no pretence at profession. To assert that the religious houses were all corrupt would be false; in some, especially the larger, religion was maintained in purity and in dignity. It would be wrong to deny that they played an important part in the community; hospitality was exercised in places where otherwise there would have been none, and the monks were slower than others to turn their land from tillage to pasture. Yet their life was becoming increasingly secular. Sometimes they served as banks and repositories of records. Their abbots were men of business and in the country the houses maintained close relations with the neighbouring gentry, whose ancestors perhaps had been among their founders and who expected to act, not gratuitously, as bailiffs, agents, and protectors. The country lawyer no less than the country squire was already dipping into the rich pie of the monastic estates. The houses were merging into the county and complete secularization would be less of an upheaval than might at first appear.

Already schemes of confiscation were in the air. The ecclesiastical authorities themselves, conscious that in some houses things were not as they should be, had not hesitated to apply old foundations to new uses. In 1497 Bishop Alcock of Ely had expelled the nuns of St. Radegund, Cambridge, on the ground of their dissolute conduct, and used the money to found Jesus College in their buildings. Soon afterwards Fisher of Rochester had suppressed two nunneries on the ground of misconduct to help in the foundation of St. John's College, and in 1518 Wolsey, having secured a Bull authorizing him to reform the monasteries, had suppressed twenty-one houses, some of them not so small, for the benefit of his colleges at Ipswich and Oxford. As in other

matters, Wolsey's example served as a precedent for the growing powers of the Crown. In 1531 there was talk of a general
scheme of confiscation. In 1532 the suppression of Christchurch,
Aldgate, an embarrassed house of Austin canons, showed how
the royal machinery could best be used; in 1534 the small
order of Friars Observant was dispersed and their seven houses
were acquired by the Crown; in the same year there was a
suggestion that the Crown should seize all the wealth of the
church and make the clergy paid officials.[1] When Cromwell clad
himself in the panoply of the vicar-general he had a victim
almost helpless before him and the weapons of execution ready
to his hand. On 21 January he obtained a commission[2] empowering him to hold a general visitation, and while the superseded bishops stood impotent, he prepared for an investigation
whose real aim was not reform but suppression. He may have
hoped, under pretext of reform, to make monastic life so uncomfortable that houses would surrender of their own accord,
but he was resolved in any case to force through a large measure
of confiscation.

It was not until July 1535 that the general visitation began,
and though its methods have been condemned it did not
differ greatly, at least in form, from earlier visitations. The
bishops themselves had sometimes professed to examine two
houses in a single day; they had insisted upon a deferential
reception, had proceeded upon information obtained beforehand, and had examined individuals privately. The eighty-six
articles of inquiry and the twenty-five injunctions[3] used by
Cromwell's agents were in no way different from those used by
episcopal visitors. It was not in the form, but in the intention
of the investigation that the novelty lay. The aim was to obtain
a conviction, and the men employed—lawyers, some clerical
and some lay—knew what was expected of them. Dr. Richard
Layton, later dean of York, of prurient mind but of marked
efficiency, boasted that 'there is neither monastery, cell, priory
nor any other religious house in the north, but either Doctor
Lee or I have familiar acquaintance within ten or twelve miles
of it, so that no knavery can be hid from us in that country'.
Plainly he was determined to find evil, but though his censures

[1] E. Jeffries Davis, 'The Beginning of the Dissolution: Christchurch, Aldgate,
1532', in *Transactions of the Royal Historical Society* (1925). Dietz, p. 114.
[2] *Letters and Papers*, viii. 24.   [3] Burnet, part i, bk. iii.

upon Oxford may express the gratification of a Cambridge man, it cannot be denied that he effected there a real reform.[1] If he 'set Dunce in Bocardo' and abolished him with all his 'blind glosses', he approved the good lectures which he found in some of the colleges and established new lectures in Greek and Latin with suitable stipends; if he suppressed the lecture in canon law he provided for the teaching of civil law instead. Dr. Thomas Leigh, a layman, was 'a conceited young don' who later became a master in chancery and was eventually knighted. The arrogance expressed in his 'satrapike countenance' was offensive; he may occasionally have taken bribes; he condemned Layton for being too lenient, but he seems in the main to have behaved fairly according to his narrow legal mind. Dr. John London, warden of New College, who distinguished himself by exposing impostors, was a conservative at heart who in after years became a denouncer of heretics and suffered humiliation for the perjury into which his headlong zeal had led him. Dr. Bedyll was a fellow of New College and a clergyman. Dr. (later Sir) John Tregonwell was a layman; so was Dr. (later Sir) John Ap Rice, who was a henchman of Leigh. The commissioners did their work with immense speed, and though they occasionally commended the houses which they visited, the general effect of their *comperta et detecta* was to drive half a dozen small houses to voluntary surrender, and to produce a damning report upon many of the rest. It is not quite certain, in spite of the statement of Latimer, that the evidence so carefully collected had a decisive effect in parliament, but the king had no difficulty in having passed a bill which he presented to the commons on 11 March 1536, whereby all religious houses of less than £200 a year were 'converted to better uses' while the 'great and solemn monasteries wherein (thanks be to God) religion is right well kept and observed' were saved for the time being. There is obvious difficulty in believing that righteousness tallied so exactly with annual income. The differentiation was made for practical reasons. The king gained the support of the upper house by granting a few licences of absence; the 'great and fat abbots' allowed themselves to be persuaded; the laymen of both houses,

---

[1] *Suppression of the Monasteries* (Camden Society, 1843), p. 70. 'Dunce' is a pun on the name of Duns Scotus, a famous Franciscan schoolman (d. 1308). 'Bocardo' was one of permitted forms of syllogism as taught to medieval students in the mnemonic poem known as 'Barbara'; it was also the name of a prison in Oxford.

anxious to avoid taxation and hoping to share in the spoils, readily accepted an act which would replenish the royal finances without inconvenience to themselves.

According to the returns of the *Valor Ecclesiasticus* there were 372 houses in England with an annual income of less than £200, and from other evidence it seems there were 27 more in Wales, but a surprising number of houses gained respites,[1] and those which were condemned were not treated over-harshly. The abbots and priors were given good pensions; those religious whose vocation was still strong, were transferred to surviving houses, and the remainder were allowed to take 'capacities' and become secular clergy. The lands and properties of the suppressed houses were transferred to the Crown; a 'court of the augmentations of the revenues of the king's Crown' was erected to deal with the new assets which were annexed, according to a fresh valuation which sometimes exceeded that of 1535. The jewels, ornaments, and plate were sent to the Jewel-House in London; the lead was stripped from the roofs and cast into pigs; the bells were taken from the towers to be sold later; the movable crops and stock were sold to pay the debts of the houses and the houses and grounds were given to farmers or to royal nominees who had as a rule to pay for their acquisitions upon the new and higher valuation.[2] Between April 1536 and Michaelmas 1538, the Crown sold lands to the value of nearly £30,000 and goods to the value of nearly £7,000. It acquired £1,000 in fines for leases and kept in its hands rents, which amounted to £27,732 during the whole period. The net gain, though considerable, was not enough to close the gap between revenue and expenditure, and there can be little doubt that the larger monasteries were already destined to be destroyed. Before the process was completed, however, the Crown had to face difficulties of a serious kind.

The suppression of the monasteries was only the supreme example of the royal power which was revealing itself in other ways. The last session of the Reformation parliament passed at length, though in a modified form, the statute of uses, so long desired by the king, and the general effect was to increase the

---

[1] Only about 220 houses were immediately attacked (Baskerville, p. 144).

[2] At first land was granted on very easy terms to the king's servants, but before long it was being sold at a good price. Dietz, p. 148; Baskerville, p. 287; Fisher, app. ii. Of the 1,593 grants of monastic land made under Henry VIII only one in forty was a gift.

king's revenue by laying upon *cestui que use* the burden of feudal incidents.

Meanwhile, outside parliament Henry was asserting his claims with a persistent energy. All the bishops were made to sign a formal renunciation of the power of the Holy See declaring that the papacy was not ordained by God, but set up by men. Gardiner was employed to justify the execution of Fisher by his pen, and his *De Vera Obedientia*, which was completed in the autumn of 1535, was only the most distinguished of several pamphlets issued to support a king who more than any of his predecessors understood the value of propaganda.[1] In the year 1536 the fall of Anne Boleyn provided a grim commentary on the meaning of the absolute power commended by the pamphleteers.

Anne had a miscarriage on 29 January, and Henry, feeling no doubt that God had declared against the marriage, was evidently more sorry for himself than for his wife. On 24 April he signed a commission to the chancellor, the two dukes, and a few noblemen to inquire into every kind of treason, and on 2 May Anne was sent to the Tower. On 10 May in Middlesex and on 11 May in Kent she was indicted before the grand jury for treason on the ground of her adultery with half a dozen men including her own brother Lord Rochford, and on 15 May she appeared before a tribunal presided over by her uncle Norfolk as lord high steward. Though she asserted her innocence and though no witness was heard she was condemned without a dissentient voice; and on 19 May she was beheaded by the executioner of Calais, specially summoned, since he was very skilful with the sword. That Anne had been reckless in her conduct seems certain, but it is very doubtful if she was guilty of the enormities attributed to her, although the juries were supplied with circumstantial evidence which probably convinced them.[2] The weakness of the case against her appears from the triviality of some of the charges made—for example, she had laughed at the king and at his dress; of her alleged paramours only one, Mark Smeton, the musician, confessed to any guilt and he had been cruelly tortured and given hope of pardon;

---

[1] See F. le van Baumer, *The Early Tudor Theory of Kingship*, Yale Historical Publications, 1940.
[2] For the process against Anne see *Wriothesley's Chronicle*, i, app.

the queen herself, as she took the last sacrament, affirmed upon her soul that she had never been unfaithful to her husband. Moreover, even if she were guilty of gross misconduct it was only with difficulty that her crime could be construed into treason. One moral of her fall might be that Nemesis follows hard upon Hybris, for Anne had behaved very arrogantly to Catherine and Mary; another might be that Cromwell was suspicious of the Howards on one side and of the extreme protestants on the other; but the most obvious moral of all is that no one dared dispute the might of the king. It is significant that to all the parties concerned, including the victim herself, the proceedings were justified as being according to law and that the law was in fact the will of Henry.

Two days after the execution of the queen Cranmer declared that her union with Henry had been invalid from the start—perhaps on the ground of her pre-contract with Northumberland, perhaps because of Henry's relations with her sister Mary—though if Anne were not Henry's wife she should not have been charged with adultery—and on 30 May the king was privately married to Jane, daughter of a simple knight, Sir John Seymour of Wolf Hall in Wiltshire, but descended, like all Henry's wives, from Edward I. The new queen, to whom Henry had already paid addresses, modestly repelled, was not particularly beautiful, but she was not unintelligent and, though rather reserved, of a kindly disposition. Although she was not opposed to the protestants, she was at pains to reconcile Henry with his daughter Mary, and she won the goodwill of the catholics. She was perhaps the happiest of Henry's queens and she it was who gave him the anxiously awaited son.

The new marriage at once brought to the fore the question of the succession, now more urgent than ever since the duke of Richmond was obviously dying—he died on 23 July 1536[1]—and parliament was summoned again for 8 June with the promise that it should not sit for long. Cromwell, who realized more and more that his imperious master must stand upon his own resources, was at some pains to pack the house of commons and either because his fears had been vain or his manœuvres had been successful, the assembly when it met showed itself most accommodating to the royal policy. Its acts were few and some of them had no lasting effect, but taken together they represent

[1] *Letters and Papers*, xi. 65, 70.

an immense assertion of the royal power. The act of succession[1] which cancelled the two acts issued after the Boleyn marriage, not only declared the rights of the issue of the new union, but also empowered the king to 'give, dispose, appoint . . . the imperial Crown of this realm . . . for lack of lawful heirs of your body . . . to such person or persons in possession and remainder as shall please your Highness', and himself to nominate councillors and nobles to have the governance of his successor should that successor be a minor when he attained the throne. The act was never to be altered or repealed by 'any act heretofore made or hereafter to be hadde done'. It might seem that subservience could hardly go farther; yet an act for minorities,[2] passed about a fortnight later, authorized the heir or nominee who succeeded as a minor to revoke by letters patent when he attained the age of twenty-four any act made before he reached the age, not of eighteen, as might have been expected, but of twenty-four. Its intention plainly was to maintain the authority of the dead king until his successor was not only of legal age but enriched with six years' experience, perhaps, of sovereignty. Another act extinguished the authority of Rome more emphatically than ever and yet another preserved the rights created under existing dispensations 'in all such cases only as may be dispensed with by the archbishop of Canterbury by authority of the laws and statutes of this realm'. The attainder of Lord Thomas Howard, Norfolk's brother, on the ground that he had a purpose of marriage with Margaret of Scotland, Henry's niece, was another emphatic declaration that the king was supreme while he lived and competent to provide for the succession after his death.

Yet in all this magnification of the power of the Crown one feature of the first importance presents itself. All was done by parliament; and though Henry, in the act for minorities particularly, expressed great suspicion of parliament which might 'at suche tyme as the Kinges of the same shall happen to be within age, havyng small knowlege and experience of their affaires', make laws 'to the great hindraunce and derogacion of the imperiall Crowne of this Realme, and to the universall damage of the commonwelth of the Subjects of the same', he was in fact exalting the prestige not of the Crown alone but of the Crown in parliament. The act of succession took no effect; the minorities act was repealed in 1547 quite easily in the parliament of a

[1] 28 Henry VIII, c. 7.   [2] Ibid., c. 17.

ten-year-old boy and re-enacted in a different form equally unavailing in practice.

For the moment, however, the king was using parliament as
his instrument in constitutional affairs and at the same time he
used convocation to enforce his authority over things spiritual.
The doctrine of the church was in sore need of definition. In
spite of the efforts of Barnes and Edward Fox in Germany in
1535 and the spring of 1536, Henry would not accept the
Augsburg Confession or come to terms with the protestants of
Germany. He was determined to assert the spiritual independence of his own church; he held suspect the Lutheran views
about marriage; he would not support subjects against their
monarch; and he had no wish to quarrel with the emperor who
was inclined to be more cordial after the deaths of Catherine and
of Anne. Cranmer, who had before him Melancthon's summary
of the recent conversations at Wittenberg, inclined towards
reform, and he had upon his side vigorous prelates like Latimer
(Worcester), Shaxton (Salisbury), Goodrich (Ely), Edward Fox
(Hereford), Hilsey (Rochester), and Barlow (St. Asaph). In
spite, however, of the absence of Gardiner, who was sent to
France as ambassador in October 1535, he was stoutly opposed;
the discussions amongst the prelates held in his house in the
spring of 1536 failed to produce agreement and the violence
of some reforming preachers created alarm in the minds of
moderate men.

To the uncertainties the Crown was prepared to offer a
solution. When convocation met on 9 June it soon found that
Dr. William Petre, as the proctor of the vicar-general, took
the chair, and Cromwell himself subsequently presided in person.
Cranmer's annulment of Anne's marriage was obediently ratified,
but thereafter debate ran high. The lower house presented to
the upper a complaint against sixty-seven errors and abuses
mostly of a protestant tendency, although the complainants
avoided any suggestion that they were on the side of Rome, and
the argument was only quelled when, on 11 July, Edward Fox
produced a book of ten articles approved by the king himself.

Reginald Pole, who was made a cardinal in the following
December, found in the articles little to condemn except their
royal authorship. They asserted the traditional doctrine of
transubstantiation; they lauded good works, the use of images,
prayers to saints and prayers for the souls departed. On the

other hand, although no pronouncement was made as to the number of sacraments, only three were mentioned, baptism, penance, and the eucharist; the whole Bible and the three creeds were to be taught to the people; the use of works, images, and prayers for the dead was limited by caveats, and the abuses connected with the pope's alleged power to deliver souls from purgatory were expressly condemned. Although the formal preface stated that the king who 'had in his own person taken great pains and study about these things' had given the examination of them to the 'bishops and other learned men of the clergy', and although the articles were accepted by convocation,[1] they were essentially the work of Henry himself and to no one else did they give satisfaction. The protestants were disappointed, the catholics saw the bold beginnings of schism, but the king remained unmoved. Convocation, ere it was dissolved, was made to counter the papal summons of Henry to appear before a general council in 1537 by the declaration that no general council could be convened without the consent of all Christian princes 'especially such as have within their own realm . . . *imperium merum*'. The supreme head of the English church was claiming an absolute veto in the affairs of Christendom and withdrawing his country from full fellowship with the church which claimed to be universal.

It was not long before the nature of the royal authority in spiritual affairs was made abundantly plain. In August Cromwell issued a series of 'injunctions' designed to improve both the conduct of the clergy and the worship of the people. The ten articles were to be declared by every parish priest; sermons were to be preached at stated periods against the usurpations of Rome; relics were not to be exhibited for gain, and a good life at home was to be preferred to pilgrimage; children were to be taught the Lord's prayer, the creed, and the ten commandments in English, and a Bible in Latin and English was to be set up in the chancel of every parish church before 1 August 1537. The material as well as the spiritual life of the people came under consideration. Provision was made for the diligence of the clergy, for the education of hopeful scholars at the expense of incumbents, for the repair of buildings, for the setting of children

[1] Burnet (bk. iii) says they were signed by Cromwell, Cranmer, 17 bishops, 40 abbots and friars, and 50 archdeacons and proctors of the lower house including Polydore Vergil and Peter Vannes.

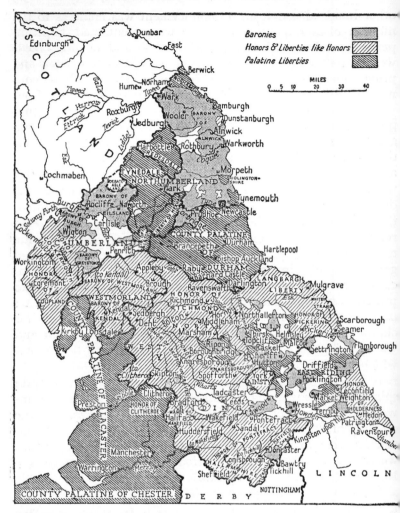

The North of England showing the Palatinates, Honours, Liberties and Baronies about the time of the Pilgrimage of Grace.

to apprenticeship, and for the diminution of the number of holidays. As thus expounded, the official religion of England did not condemn the Mass and it did not condemn good works; but emphasis was laid upon the words of the Scriptures and upon the merits of the simple Christian life.

The changes, explicit and implicit, were by no means popular throughout England, and in the autumn of 1536 Henry was confronted by the great crisis of his reign. This was the rising in the north known as the Pilgrimage of Grace. It was due to causes religious, social, economic, and political, and in essence it was a demonstration in favour of the old days and the old ways.[1]

The authority of the Crown in the north had long been some-what uncertain. During the fifteenth century the kings had established some order in Yorkshire, but in the Marches they had left the wardens in control and when, in the days of the Tudors, the Percies and the Nevilles were distrusted, the warden-ships fell into the hands of minor gentlemen whose authority was not equal to the task, especially as they were not very well supported from the south. The machinery fell to pieces; the quarter sessions ceased to be regular; in 1521 there were no sheriffs in Cumberland and Northumberland, and when Surrey went up to confront the Scots in 1522 he found all in disorder. The remedy he suggested[2] was the establishment of a council like that in the Marches of Wales, and his proposal commended itself to Wolsey who liked to control all from the centre. When the resistance to taxation in 1523 and 1525 emphasized the need of restoring order, the government, determined not to have recourse to the old nobility, followed a policy already adopted in Ireland and in Wales.[3]

In 1525 the nominal authority, with the title of lieutenant general north of the Trent, was given to the duke of Richmond, and the real power to a body of hard-working lawyers, finan-ciers, and administrators, some lay, some clerical, who, though officially the duke's household, constituted in fact a council.

---

[1] The best account is found in M. N. and Ruth Dodds, *The Pilgrimage of Grace*, 2 vols., 1915.

[2] Rachel R. Reid, *The King's Council in the North*, p. 95.

[3] Henry VII had made his son Henry lord lieutenant of Ireland on 12 September, 1494, and his son Arthur justice of the Welsh Marches in 1493. In September 1525 Mary was sent to establish her household at Ludlow.

This council, which resembled that of a great nobleman, exercised both administrative and judicial authority in all shires north of the Trent save Durham, but its attempt to control the Marches did not succeed. In December 1527 Northumberland was made warden of the East and Middle Marches and next year Lord Dacre was made warden of the West March. In each case some of Richmond's council were attached to the warden, but in effect the council's administrative authority was limited to Yorkshire. In matters of justice it was more effective. As a court of equity it was well liked because it diminished, though it did not altogether extinguish, the constant interference of Westminster; but as a criminal court the gentlemen of the north thought scorn that a parcel of nobodies dependent upon the cardinal should interfere with their feuds and riots.

In 1529 there was a formal protest led by Darcy against the rule of spiritual men, but though on Wolsey's fall Henry recalled his son, the council remained, now under a president as there was no longer a lieutenant, and the first president was Cuthbert Tunstall, bishop of Durham. Tunstall, who lost credit for his support of Catherine of Aragon, could not control the north, and in 1533 the office of lieutenant was revived and given to the sixth earl of Northumberland, who surrounded himself with a council containing his own friends and household servants as well as officers appointed by the king. Henry's acquiescence in the aggrandisement of a magnate was due perhaps to his difficulties, but it may have been an attempt to lure Northumberland to his own destruction. 'Henry the Unthrifty', steeped in debt and at odds with his own family, was persuaded in 1535[1] to disinherit his brother, Thomas Percy, and surrender his lands to Henry in return for £1,000 a year. Meanwhile, though his council still existed, it had no real control and its members, hostile to Cromwell and afraid of losing their old liberties, were inclined to resist the Crown rather than to execute its orders. In the troubles of 1536 many of the councillors threw in their lot with the 'Pilgrims'.

The north was seething with discontent. The treatment of Northumberland aroused not only resentment but apprehension in the minds of men whose 'liberties' were threatened. The clothiers of the West Riding particularly were vexed by insistence upon the execution of an act for the true making of woollen

[1] Reid, p. 119.

cloths[1] which they had hoped to evade; the commons were galled by inclosures and an increase in rents and fines; the clergy, set in their old ways, resented the innovations. To the burden of grievances the suppression of the monasteries was the last straw. It stood for all the things which were hated in the north—the increase of the central power, the rise of upstarts, the new economy, the destruction of hallowed tradition. The monasteries were popular in the north. Now Lincolnshire and Yorkshire were to lose more than forty houses apiece; in fact from Lincolnshire alone the commissioners collected more than £8,756 during the first six months of their operations and rumour said that the larger monasteries would soon be dissolved too.

It was in Lincolnshire that the trouble began. No fewer than three commissions were busy there in the autumn of 1536; one for the collection of the subsidy voted in 1534, one for the suppression of the monasteries, and one to communicate Cromwell's injunctions to the clergy. On 1 October the suppressed irritation burst into action in a riot at Louth. This was in origin popular and largely inspired by religious motives, as appears from the fact that before long neighbouring Horncastle took the field under a banner bearing the five wounds of Christ and the Host as well as the plough and the horn. But though the first leader was a shoemaker, Nicholas Melton, whose followers swore men to be true 'to God, the king and the commons for the wealth of Holy Church', it is significant that the gentry allowed themselves to be coerced very easily, that the first persons attacked were the representatives of the three commissions, and that the demands sent to the king by one of the captured commissioners for the subsidy expressed the view of taxpayers rather than of an irresponsible mob. These demands, which were repeated with variations throughout the insurrection, were that the king should suppress no more abbeys, should impose no more taxation, should surrender Cromwell to the people, and get rid of the heretical bishops.

It seems probable that the insurgents really supposed that Henry, whom they acknowledged as Supreme Head, would accede to their requests; but the king, not unnaturally, replied by denouncing his 'sworn servants' for their breach of trust and prepared to suppress a rising whose weight he did not at first realize. Whilst he was penning his answer, however, the rebels

[1] 27 Henry VIII, c. 12.

advanced in great force and occupied Lincoln. Rumour said that they mustered 40,000 men, including 16,000 harnessed men, and 700 or 800 monks and priests; certainly, the gentry, willingly or unwillingly, had come in. At Horncastle on 5 October and again on 8 October at Ancaster definite 'articles' were drawn up increasing the former demands; the statute of uses was to be repealed; firstfruits and tenths were no longer to be paid to the Crown, and a free pardon was to be given to all. At this juncture messengers from Beverley and Halifax arrived announcing that their countries were up; and a general advance was prevented only by the efforts of the gentry, who dreaded an outbreak of mob violence, and insisted that they must await the answer of the king. In a rebellion to hesitate is often to fail. Distrust awoke between gentry and commons, and this was increased when Henry's uncompromising message to the commissioners was brought to Lincoln on the 10th. Next day Lancaster Herald appeared from Shrewsbury, who was at Nottingham, bidding the concourse disband, and when his proclamation was read aloud on the 12th the vast host began to melt away. Some of the gentry helped to put down the disorders, others came into the king's forces suing for pardon.

These forces, despite the high tone taken by their commanders, were inadequate. At the first rumour of disorder Shrewsbury had collected the Derbyshire levies at Nottingham, but though he spoke of 100,000 men, he had but very few and no money to pay even the few he had. Henry had summoned his main army to gather at Ampthill on the 16th, had called Norfolk out of retirement and had meanwhile sent Suffolk as his lieutenant to face the rising in Lincolnshire. By 10 October, however, Suffolk had only 900 men and if at this time the insurgents had pressed on, he might easily have been overwhelmed. Even when he was reinforced by the arrival of men and guns he had to dismiss 2,000 for lack of arms, and to the king and his commanders it was welcome news that the Lincolnshire rising had collapsed. Norfolk was sent back to his own county. Suffolk was ordered to Lincoln, which he reached on the 17th; Shrewsbury was told to be ready to advance into Yorkshire; on the 19th the king sent a haughty answer to the Lincolnshire articles, denouncing the commons of 'one of the most brute and beastly'[1] shires in England for daring to give

---

[1] 'beastly' means 'stupid'.

their prince advice about government, asserting that all his
actions were approved by parliament and alleging that they
themselves had been loud in denunciation of clerical wealth.
He bade the rebels disband at once, handing over their leaders
for punishment. His answer was skilful, as calculated to set
against one another gentry and commons, laymen and clergy;
but before it was delivered he had discovered that it was too
soon to talk about punishment.

Yorkshire had risen; and before long Yorkshire was joined by
almost the whole of the north. While the Lincolnshire army was
melting away, the malcontents in the East Riding had found a
leader in Robert Aske, at once a country gentleman and a
London lawyer, whose eloquence, resolution, and skill gave to
the movement purpose and unity. By 13 October a large host
gathered at Wighton Hill where the new-made captain gave his
followers the name of 'Pilgrims', and though an attempt on Hull
failed, York was occupied on the 16th. From York, Aske issued
new articles similar to those of Lincoln, though religion was
given a greater prominence; he forbade pillage, promulgated
a form of oath, and before long rallied to his side the whole of
the county. Northumberland was ill at Wressell; Tunstall fled to
Norham; Lord Darcy of Templehurst, old and ill provided, who
should have maintained the king's cause, surrendered Pontefract
on 21 October. Archbishop Lee came in with him, and when on
the same day Lancaster Herald appeared at the castle, he
found Aske keeping state like a great prince with many of the
gentry about him. Many of the members of the council placed
themselves at the head of the revolt and they it was who
organized its machinery, including a great council which was
to contain representatives from every wapentake.[1] By the 24th,
there mustered at Doncaster a force of some 30,000 men, most of
them well armed on horseback, in whose ranks were to be seen
not only the banner of St. Cuthbert, but representatives of many
of the families who had fought at Flodden. The king's case
seemed desperate. The earl of Suffolk was held in uneasy
Lincolnshire; Lancashire had risen in a revolt which, more than
elsewhere, had a religious spirit; Shrewsbury, who had advanced
too far, was isolated south of Doncaster, and though Norfolk,
who deplored Shrewsbury's rashness, readily found men to serve

[1] Reid, pp. 137–8, 140. The wapentake was the northern equivalent of the
hundred.

under the popular white lion, he lacked equipment and his troops were far in the rear. Reluctantly and, as he told Henry, without intention to keep the promises he might be compelled to make, he suggested a conference. The pilgrims sent to him by word of mouth five articles in the same strain as the Lincolnshire demands and a discussion was held at Doncaster Bridge on 27 October. As a result the duke promised to take the articles to the king along with two representatives of the pilgrims and agreed to a truce during which both armies should disband. Two days later, he wrote to the council lamenting his necessity but telling them bluntly that his weather-stricken host was in no case to fight, and that, except for his own retinue and that of Exeter, his men could not be relied upon, since many of them believed that the insurgents were in the right. He was not exaggerating; at the very start of the Lincolnshire rising Lord Hussey had been unable to bring in his tenants and the movement was much more aristocratic now. Indeed, it is possible that Norfolk himself, seeing a chance to ruin Cromwell and the new men, may have shown himself more sympathetic to the rebels than appears from the written evidence; certainly it was upon his promise of the king's 'most gracious free pardon' that the northern host withdrew.

Henry was furious, but he made the best of the situation. On 2 November he received the duke and the northern representatives at Windsor. Three days later he himself prepared a skilful answer, traversing the argument of the five articles and demanding the delivery of ten ringleaders. Second thoughts convinced him, however, that he was not able yet to take the high hand, and only on 17 November did the envoys arrive back in Yorkshire with a message which was not a reply to the articles, but a verbal invitation for 300 pilgrims to meet Norfolk again at Doncaster. Delay, the king knew, would be upon his side.

The northerners for their part, saw the danger. Aske strove hard to keep his host together, and though the 'great council' of the pilgrims which reassembled at York on 21 November, agreed, through its executive, to meet Norfolk on 5 December, it did so only under conditions. To the wrath of the king it demanded guarantees of safe-conduct, and decided before the meeting was held to gather at Pontefract in order to put its demands into final form. As the clergy in the north were to hold something like a convocation at the same time an obvious attempt was being made to reproduce in Yorkshire the arrangements usual at Westminster. The articles evolved from this great council, which met on 2 December, were in spiritual affairs mainly an assertion of what had been asked before, though there was now a demand for the restoration of benefit of clergy, of sanctuary, and of the old liberties, and, while the royal supremacy was admitted, it was claimed that matters touching the *cura animarum* were reserved for Rome. Some of the consitutional demands, however, were new, and the emphasis laid upon parliament is remarkable. Parliament was to be reformed, the influence of the king's servants in the commons was to be reduced; there was to be greater freedom of speech; royal interference in elections was to cease and old boroughs were to be re-enfranchised. Spiritual matters should be dealt with by convocation and not by parliament, and the lords should be supplied with copies of bills laid before the commons. Mary was to be made legitimate by statute and the stature empowering the king to declare his successor was to be repealed. A full pardon was to be given by act of parliament and parliament was to meet soon, either at Nottingham or at York.

The pilgrims were, in fact, demanding a free parliament, and evidently it was hoped that Lee in his sermon would justify the moving of war against a prince for good cause. But either the archbishop's courage failed him or the arrival of the safe-conducts for the Doncaster meeting enabled him to avoid the issue, for all he said was that persons calling themselves pilgrims should not behave as rebels. On 6 December ten knights, ten esquires, and ten commoners met Norfolk and his council at Doncaster. Aske showed himself very reverential towards the king and it was agreed that all the articles should be remitted to parliament, the condemned abbeys being allowed to stand meanwhile. No promise was made as to the place and nature of the new parliament; the pilgrims saw nothing in writing except a full and free pardon; there was no guarantee that this pardon applied to Lincolnshire—the envoys seem to have been satisfied by Norfolk's word. Not without difficulty Aske persuaded the insurgents to disband. On the reading of the pardon by Lancaster Herald he renounced the name of captain and tore off his badge of the five wounds; all then said 'we will wear no badge or sign but the badge of our sovereign lord', and on 8 December the pilgrims dispersed to their homes.

The crisis was past and the decision was in the king's favour. Yet there had been a moment of great danger, for the rebels had been in touch with the Netherlands and Paul III had resolved to send the newly created Cardinal Pole as legate *a latere* to the Low Countries for the purpose of aiding the revolt. Before Pole arrived, however, the insurrection was at an end; Francis and Charles vied with one another in refusing countenance to the pope's emissary, and Henry was left to exploit the opportunity presented to him by the last heavings of the sea of discontent. These were not in fact serious, for they were in the main supported only by the commons who felt themselves betrayed by their leaders, but their occurrence allowed the king to assert that his pardon had been rejected.

As early as January 1537 there was an ill-judged attempt upon Scarborough and Hull by Sir John Bigod and John Hallam which was suppressed by Aske and his friends; there was trouble in Richmondshire, for which the monks of Jervaulx were blamed, and in February there was a rising of the common folk of Cumberland and Westmorland under 'Captain Poverty'. Norfolk, who saw the difficulty of proceeding by indictment against persons who would testify one in favour of the other, displayed the king's banner and proceeded at Carlisle by martial law. Summary executions were carried out all over the north and more care was taken to see that every affected area witnessed some horrible example than to be sure that the most guilty suffered. When, later, juries were found necessary at York, care was taken to constrain them, and with their aid convictions were readily attained. In March there was held a great assize against the Lincolnshire prisoners who had been excluded from a general pardon announced in the preceding November, and thirty-six poor wretches, including only one gentleman, were condemned. The more important victims were dealt with specially. Hussey was beheaded in Lincoln and Darcy on Tower Hill; Lady Bulmer was burnt, and in July Aske, who asserted that he had been promised pardon by both Henry and Cromwell, was hanged in chains at York. In all, according to the records, 216 persons suffered the death-penalty, and though others were summarily executed or died in prison, Henry's vengeance was mild compared with the holocausts which followed the suppression of rebellion in other lands. It was, however, sufficient for its purpose; not only did it terrorize the north, but it served to justify the creation of new machinery to coerce the over-mighty subjects.

In January 1537 the ill-defined authority of the 'king's lieutenant and council in the north parts' was reduced to precise form by the erection of the 'council in the north'. Norfolk was lieutenant, the bishop of Durham was continued as president, and John Uvedale as secretary, the earls of Westmorland and Cumberland and five knights were included, but there were added three tried common lawyers and three civilians. So, in the accustomed Tudor way, administrative power was handed over to a hard-working, middle-class committee, paid a regular salary and in constant touch with the council at Westminster. When Norfolk returned to London in September, the royal authority was firmly established in the north.[1]

There was now no lieutenant, and the nobles disappeared. The royal authority was exercised by a council which, on 15 October 1537, held its first meeting at York. Its officers were the lord president, the vice-president, four or more councillors learned in the law, the secretary, the king's attorney, a registrar, and some three dozen minor officials. The council had its own signet, and in all probability its members took the councillor's oath. It had jurisdiction over the five northern counties including Durham, had the highest administrative power, could inflict any penalty short of death and although itself it might appeal to the privy council for advice, its suitors had no appeal from its decision. Its members were required to attend each of the four general sessions, though this obligation was soon relaxed, and since in the sessions they could inflict the death-penalty they had, in some sense, a power greater than the members of the star chamber. Here was an instrument of power in the hands of the Crown, and this power was more effective because provision was made for a nucleus of permanent members in close touch with the council attendant upon the king's person. The new council did not misuse its authority. Its officers were instructed to grant their protection to 'the poorest man against the richest lord' and it did in effect act as a court of requests.[2] It redressed the grievances connected with 'intakes' or inclosures

---

[1] Pickthorn, p. 502; Reid, p. 151.
[2] Pickthorn, p. 502; Reid, p. 160.

and excessive gressums[1] and, by the year 1540, it had established itself in the goodwill of the ordinary folk in the north.

The king was stronger than ever. He owed his victory, it is
true, to the dissensions of his enemies, to the ill weather which
had made the Don unfordable, and to something very like
treachery, since he had made promises which he hoped to
break. He owed it, too, to the resolution of his commanders, and
it is one of the ironies of history that the pilgrims were defeated
by the nobles whom they proposed to elevate in place of
'villein councillors'—Norfolk is the type of the old aristocracy,
contemptuously pitiful of the 'poor caitiffs' who have lost all
yet personally witnessing every execution done under martial
law. Something too the king owed to his administrators, who
produced in a sudden emergency troops which could take, and
above all keep, the field. Most of all he owed to himself, to his
unshakeable courage and his instinctive wisdom. With real
statesmanship he calculated that the prestige of the Crown,
which stood for good order, would defeat the assaults of rebellious subjects, and that, with the example of Germany in their
minds, catholic nobles could be trusted to resist a popular rising
even though that rising were made under the banner of the old
church. The event proved him right. His success was gained in
part by bluff; but had he been compelled to deploy his full
power, it is probable that, though he must have suffered serious
setbacks at the start, the might of the well-organized south, and
the all-important guns would, in the end, have prevailed even
against 'all the flower of the north'—the phrase is Norfolk's.

From a period of great difficulty Henry emerged a clear
winner. His prestige was enhanced; his machinery of government was increased by the creation of three new bodies
intimately dependent upon the privy council—the council in
the north, the court of augmentations, and the court of first-
fruits and tenths. The failure of Pole's mission had, for the time
at least, improved his position abroad, and when, on 12 October
1537, a son was born to him, it seemed that the fortunes of his
throne were fully established. His abounding joy was saddened
by the death of his queen, to whom he paid the compliment of
remaining single for more than two years, recognizing that he
was no longer the disappointed Christian bachelor but the
afflicted Christian widower.

---

[1] That is, fines on the renewal of leases.

The political wisdom which had served him in the night of trial did not desert him in the day of success. Until he was threatened from without he made no extravagant exhibition of the great power he now possessed, and to his people he offered a religious settlement which conceded something to both sides. The events in the north had convinced him that the old religion was still strong, and the *Institution of a Christian Man* issued in September 1537, which, though inspired by the conservative spirit of convention and generally called 'The Bishops' Book', was circulated by royal authority, marks a return to the traditional views. The four lost sacraments now 'found' again were formally expounded, and emphasis was laid upon the fact that justification through the merits of Christ did not dispense with the necessity of good works.

The project of making common cause with the German reformers was allowed to die. In vain Cranmer studied Bugenhagen's *Pia et vera Catholica et consentiens veteri Ecclesiae Ordinatio*; in vain four doctors and four prelates, under a royal commission, conferred with the ambassadors of the German princes in London (1538). Henry had gone as far towards the Lutherans as he meant to go. Yet, in one thing he joined them, namely in an attack upon the anabaptists whose excesses had rendered them suspect to protestants and catholics alike. On 1 October 1538 Cranmer and eight others were commissioned to receive back repentant anabaptists and to punish the unrepentant; on 22 November all anabaptists were banished by proclamation, and before long some unfortunates were burned. Most spectacular was the fate of William Nicholson or Lambert, who was accused not only of anabaptism but of errors concerning the Sacrament, the Incarnation, and the interpretation of the Scriptures. This resolute heretic was tried in the palace of Westminster in the presence of the king, who himself reasoned with him; having withstood even royal argument he was condemned and, very soon after his trial, dragged through the streets of London to be burned at Smithfield.

Yet while the king was at pains to emphasize his orthodoxy and while many of the bishops were conservative, the hand of Cromwell was still dominant and the assault upon the Roman church did not slacken. In 1538 a grand attack was made upon images used to encourage superstition. The rood of Boxley in Kent and the blood of Hailes in Gloucestershire were exposed as frauds; a wooden image named Darvell Gadarn, who could pull souls out of hell, was dragged from north Wales to burn beneath the Observant Friar Forrest who had once been confessor to Catherine of Aragon and who could not, in spite of temptation, quit his old allegiance to Rome; many images were removed, and in some cases the suppression of the pilgrimage-saints was very profitable to the Crown. Pieces of gold and silver work were found in most shrines. Winchester supplied a great cross of emeralds, Chichester three caskets of jewels; and the greatest prize of all was the spoil of St. Thomas of Canterbury, which yielded to the Crown two great chests of jewels 'such as six or eight strong men could do no more than convey one of them' and, according to Sander, some twenty-four wagon-loads of varied treasures besides.

The destruction of Becket's shrine had more than a financial significance; it was a definite defiance of the pope. The *rapprochement* between Charles and Francis had reawakened Roman hopes, and the ill-judged enthusiasms of Cardinal Pole had plunged his family into destruction. In August his brother, Geoffrey Pole, had been arrested and though in the end he was contemptuously pardoned, there was obtained from him evidence against his mother, the countess of Salisbury, his brother Lord Montague, and his kinsmen, the marquess of Exeter and Sir Edward Neville. The crime of these offenders was rooted in their connexion with the White Rose, but their arrest was a hint to foreign princes that intervention in English politics was a dangerous thing. When the pope, shocked by the violation of Becket's tomb, threatened drastic action Henry was unmoved. He issued a proclamation declaring that Becket was a 'rebel who fled to the realm of France and to the bishop of Rome to procure the abrogation of wholesome laws'.[1] In December he executed Montague, Exeter, and Neville, and incarcerated the countess of Salisbury, and the issue on 17 December 1538 of an order that the Bull of 1535 should now be executed had no effect.

It is notable that amongst the grounds alleged for the execution of the Bull was the necessity of waging war against a corrupt English version of the Bible. In September 1538 Cromwell issued a fresh set of injunctions ordering, amongst other things, that there should be set up in every parish church a large

---

[1] *Letters and Papers*, XIII. ii. 354.

English Bible which the clergy were to admonish the people to read. As early as 1535 even the conservative bishops had proposed to make an English version, and Gardiner had translated the gospels of Luke and John from the Greek, but what Cromwell may have had in mind was the version published by Miles Coverdale in October 1535, which he had himself virtually commissioned and which rested upon Luther as well as upon the Vulgate. By August 1537, however, Cranmer had found a new Bible dedicated to the king by 'Thomas Matthew', who was really John Rogers, the proto-martyr of Mary's reign. It was in part a reissue of Coverdale's Bible, but the beginning of the Old Testament and the whole of the New Testament were copied from the work of William Tyndale who, being dead, yet spoke, and spoke with royal authority; for this was the Bible which, shorn of the polemical notes supplied by 'Matthew', was introduced into the churches in 1539.

In this atmosphere the suppression of the larger monasteries went on apace. The ground was already prepared. The disturbances of the north had given a pretext to accuse the heads of various houses and though some were spared on the ground that they had been coerced, or because their offences had been committed before the general pardon, or for causes not ascertained, several abbots and priors were executed. The abbots of Kirkstead (Cistercian) and of Barlings (Premonstratensian) were hanged for their share in the Lincolnshire rising. The abbot of Fountains, who had been deposed in 1536, paid for his intrigues to recover power with his life, and dragged with him to destruction his Cistercian brother of Jervaulx. The abbot of Whalley, another Cistercian, was hanged for harbouring the king's enemies in Lancashire and the prior of Bridlington, an Augustinian, was brought to Tyburn by the delation of some of his canons. The abbot of the great Cistercian house of Salley, on the other hand, seems to have been pardoned in spite of his share in the Lancashire rising; the prior of the Carmelites at Doncaster was certainly pardoned; and though a few monks and canons were condemned, many seem to have gained immunity in one way or another. Yet the executions, which took place between March and May 1537, had their effect. The abbot of the great house of Furness in Lancashire led the way in surrendering his house to the Crown early in 1537. The priory of Lewes gave itself up in November and next month the abbey of Warden in Bedfordshire followed suit. In January 1538 Leigh and Layton began a new visitation, this time producing to the doomed houses a ready-made document of surrender to the king, which in many cases was signed without undue resistance. The abbots and priors knew that the more readily they surrendered, the better might be the pensions granted to themselves and the less subject to criticism might be the arrangement made with their friends about lands, livings, and offices.

When parliament met in April 1539, there were present only seventeen abbots as compared with the twenty-eight of 1536, and though three houses sent proxies, Glastonbury's proctor was Cromwell himself. The clerical estate, its position thus weakened among the lords, was unable to oppose the king's will. One of the first acts of the session was to assure to Henry and his heirs in perpetuity all the fruits of surrenders made or to be made—the temporary holder of office was allowed without cavil to alienate for ever the property of a corporation. It is, of course, idle to suppose that the surrenders, in spite of the wording of the act, were entirely voluntary, and proscription was used to supplement the statute. Already in 1538 the abbot of Woburn and the prior of Lenton near Nottingham had been executed for criticism of the royal supremacy, and in the autumn of 1539 the abbots of Glastonbury, Colchester, and Reading were hanged. 'the abbot of Redyng to be sent down to be tried and executed at Redyng'[1] ran one of Cromwell's memoranda. It is possible that all three abbots had supported the 'pilgrimage' with money and influence; Reading and Glastonbury at least had been involved in the Pole conspiracy, and obviously it would be easy to secure delations on the subject of the royal supremacy from discontented monks. Yet it is difficult to dismiss the memorandum as the private note of an official who knew beforehand what the evidence must be,[2] and the fate of Abbot Cook must be regarded as an ugly example of the *droit administratif* employed by the ruthless vicar-general. Aware at once of the possible benefits of complacency and the dangers of recalcitrance, the heads of other houses made what terms they could. In March 1540 Christchurch, Canterbury, and Rochester, which still resisted, were dissolved by commission, and with the

---

[1] *Letters and Papers*, xiv. ii. 139.
[2] Pickthorn, *Henry VIII*, p. 374; Baskerville, p. 177. See Froude, *Henry VIII*, ch. 16 for the evidence.

surrender in the same month of Waltham Abbey the last of the monasteries was gone.

The forty-three commanderies and preceptories of the Order of St. John were confiscated by statute in the same year,[1] and with their fall the process of confiscation was virtually complete. When Henry died, the chantries alone remained, and on them too the monarchy had already shed the light of its 'damned disinheriting countenance'. They were dissolved by an act of 1545, but though commissions to survey were appointed in February 1546 few had actually been put down before the king died, and most of the chantries survived until they were abolished by a new statute passed in the first parliament of Edward VI.

The suppression of the monasteries left the king with two great problems, the problem of the religious and the problem of the land; both he settled without great difficulty entirely in his own favour. The heads of houses were generously treated. In some cases where a cathedral church had been served by religious, as at Carlisle, Durham, Ely, Norwich, Rochester, Winchester, and Worcester, the monks or canons simply became a chapter of secular clergy with their old head as dean. Some abbots and priors became bishops; others, unprovided with benefices, received large pensions. Although there was now no provision for drafting monks and canons to other houses, the evicted brethren were by no means left to starve; some were given pensions, not unhandsome; some were given benefices; some were given both. In many cases they were absorbed without difficulty. A few sought dispensation from their vows and founded county families. The friars, whose foundations had mostly been poor, got no pensions, but for them too livings were usually found.[2] The process of resettlement was made easier because a few monastic churches were taken over for parochial use,[3] and because the establishment of the secular church was slightly increased. According to a document written by Henry VIII with his own hand it was thought 'most expedient and necessary that more bishoprics, collegiate and cathedral churches' should be established[4] from the resources of the suppressed houses, and

---

[1] 32 Henry VIII, c. 24.

[2] For the fate of the religious see G. Baskerville, *English Monks and the Suppression of the Monasteries.*

[3] Rose Graham, *An Essay on the English Monasteries* (1939), p. 37.

[4] *Letters and Papers*, XIV. i. 404, 530–1, and XIV. ii. 151–3.

at one time the king considered the erection of thirteen new
sees upon a county basis. In fact, however, he created only six,
Westminster, Oxford, Chester, Gloucester, Bristol, and Peterborough;—and Westminster survived only till 1550. All things
considered, however, the great transition was made with surprising ease, and the rank and file of the spiritualty, soothed by
Henry's doctrinal soundness, acted in the main as loyal servants
of their supreme head.

It was not only service but wealth which the king obtained,
and this wealth was not devoted entirely to the promotion of
religion. The new foundations were established, the pensions
were duly paid, but the great bulk of the lands and revenues
acquired were appropriated to the royal use. Most of them did
not remain in the king's hands for long, for the depreciation of
money, the growth in the royal expenses, and the conduct of a
war with France played havoc with the finances. Henry managed to retain an income which averaged about £40,000 a year
during his last eight years,[1] but between 1539 and the end of the
reign lands were sold to the value of nearly three-quarters of a
million pounds. A survey of all his transactions[2] shows that over
the whole period relatively little went in free gifts to peers,
courtiers, and officials. Some was sold to peers, and in one way
and another Essex, Norfolk, Rutland, Audley, Wriothesley,
Hertford, Suffolk, Shrewsbury, St. John, Russell, Lisle,
Wharton, Clinton, Sussex, and Howard all got land of value
more than £200 a year; of the commoners, the officers of
the court of augmentations and the gentlemen of the privy
chamber fared particularly well; Sir Ralph Sadler, the king's
secretary, is said to have become the richest commoner in
England. Other courtiers and officers benefited, but the great
bulk of the land went to grantees, generally described as knights,
esquires, and gentlemen, who represented very often families
already interested in some suppressed house. It is noteworthy,
however, that among the purchasers of land were lawyers and
physicians as well as citizens of London and other persons
engaged in trade. Some of the purchasers, including probably
the syndicates of London merchants, may have been speculators,
and other grantees soon disposed of their gains or of some of
them. Altogether a fair proportion of the population was in-

---

[1] Dietz, app., Table III.
[2] Fisher, app. ii, based on the work of Savine.

volved in the new economy. The king, in effect, was binding to himself and to the royal supremacy not only the gentry of England but the rising middle class. The bond was strong; the practical hands which seized the church acres would not let them go. The land settlement was the sheet anchor of Henry's reformation, and even Mary, who gave back to the old church so much of what it had lost, was unable to restore the lands and revenues which had passed beyond the control of the Crown.

From the point of view of political, social, and constitutional history the destruction of the monasteries was no stupendous crime. It produced an upheaval less than might have been suspected because it was only the final act of a destruction which was wrought from within. From the point of view of spiritual experience, however, it represents a great breach with the past. Despite the secular tone of some of the houses, despite the willingness of some monks to accept their release, despite the good provision made for the victims and the field of useful activity offered, there must have been many honest souls to whom eviction was a tragedy. Those who were valiant for truth suffered death which was often very cruel; even the secular-minded must have thought long upon the good order and the solemn offices which had constituted their daily life; and to those who had a true vocation the end of the monastic system must have seemed the prevailing of the gates of hell. In setting forth the ease with which the new dispensation was generally accepted, it would be unjust to forget the sufferings of honourable men and women who were true to their vows.

XII

IMPERIUM MERUM

I F the destruction of the monasteries made plain the meaning of the royal supremacy, no less plain became the meaning of *imperium merum* in the last decade of Henry's reign. Though condemned by the pope and designated as the object of a catholic crusade, Henry not only defied the threat of invasion, but carried the war to France and to France's ally, Scotland. At home, he married and discarded wives at will, crushed the remains of aristocratic opposition, prescribed the religion of his people, and all the while raised from his subjects vast sums in subsidies and in forced loans. Although most of the forms of the constitution were observed and its machinery even developed, his ministers were his obedient servants and the whole conduct of the state was the expression of his imperious will.

To catholic Europe Henry's position must have seemed precarious when, on 17 December 1538, the pope ordered the execution of the Bull of excommunication drawn up in 1535. The ambitious Alessandro Farnese who, between 1534 and 1549, wore the tiara as Paul III, cherished the constant hope of reuniting Christendom. In 1536 he had issued two Bulls, one for reform, the other for a general council to meet in Mantua in May 1537, and though this council was postponed, first to Vicenza for 1538 and later to Trent for 1542, the possibility of a common catholic effort seemed to be drawing near. In 1538 a league was made against the Turks, and when, in that year, Charles and Francis drew together, it seemed to the pope and to Cardinal Pole that something could now be attempted against the presumptuous king of England.

Henry had endeavoured to keep the breach between France and the empire open by offering simultaneously to each of the contending powers a marriage with his own royal person. Whilst he discussed with Charles a match between himself and the widowed duchess of Milan, the emperor's niece, he was also suggesting that, since he was disappointed of Mary of Guise, given to his nephew James, he might wed some French lady to be selected by himself from a bevy of beauties paraded at Calais. There is no authority for the story that Christina spurned the overtures on the ground that she had only one head, but the French ambassador certainly pointed out that the ladies of his country were not to be inspected like ponies; Henry's proposals were, in effect, rejected by both parties who, by the treaty of Toledo of 12 January 1539, agreed to make no new alliance without mutual consent. It may well be that Henry's superb egoism was unshaken by this rebuff, but Cromwell was apprehensive, and king and minister alike thought a display of English power would not be untimely.

England stood to her defence; musters were held, the coast was fortified, a fleet of 150 sail was concentrated at Portsmouth, and on 8 May the armed might of London was exhibited in a grand review of 15,000 men magnificently accoutred who, after traversing the City, marched past the king at Westminster. Foreign observers were suitably impressed; not every king could trust his citizen-soldiers with fire-arms provided at their own expense, and not every capital could exhibit at the same time both hearty enthusiasm and good discipline. Meanwhile, Henry buttressed his military preparations with diplomacy; he wooed the peace-loving Netherlanders by proclaiming on 26 February that strangers should pay, except for wool, no greater customs than those paid by the king's subjects; he played on the fears of the duke of Urbino, threatened by Farnese ambition; he opened negotiations with William of Cleves, brother-in-law of the Lutheran elector of Saxony who, in June 1539, increased his power by the inheritance of Gelders. He had, however, no mind for an alliance with the German protestants, whose theologians had spent to no purpose five months in London during the summer of 1538, and whilst he prepared for war he set himself to show the catholics that war was unnecessary; he exhibited his complete orthodoxy, and the statute of six articles served to remind the world that England's quarrel was with the decision of a pope and not with the holy catholic church.

The crisis passed, but Cromwell still thought that England needed an ally abroad, and on 4 October was signed at Hampton Court a treaty for the marriage between Henry and Anne, sister of the duke of Cleves, who, despite his family connexions, was no Lutheran but an 'Erastian'[1] like Henry himself. It is probable that this alliance was due to an increasing cordiality between

---

[1] The use of this word is convenient, but, in fact, the opinions of 'Erastus' (Thomas Lieber) were not published till after his death in 1583.

Charles and Francis which revealed itself conspicuously when the emperor was allowed to cross France, and even to visit Paris, on his way to quell rebellious Ghent in December 1539; but it is not clear from the evidence that England was aware of the Franco-Imperial *rapprochement* when the Cleves treaty was made, and it is possible that even in his defence Henry meditated attack. It was believed in the Low Countries that the Cleves match was approved by the Most Christian king, and that the Tudor, relying on the French policy of encouraging Netherlandish discontent, hoped to unite the Ghentish insurgents, Cleves, Francis, and himself in a coalition against the emperor. Whether conceived as a measure of defence or as a measure of counter-attack, the marriage turned out to be unnecessary, since the two great catholic monarchs were soon at variance again, but it served to ruin its maker. When, on New Year's Day 1540, Henry met Anne at Rochester, he was greatly disappointed, and though he married her on 6 January he was ill content. For a few months the balance swayed uncertainly, but on 10 June Cromwell was arrested; in July, first convocation and then parliament annulled the royal marriage; on the 28th Cromwell was beheaded and when, about a fortnight later, Henry produced as queen Norfolk's niece, Catherine Howard (whom he had married privately on the very day of Cromwell's execution), it seemed to the world that the cause of conservatism had triumphed.[1] The protestants were shaken. To them the execution of Cromwell seemed comparable to the execution of Anne Boleyn and presaged a return to Rome. Henry, however, was not going to return to Rome unless on his own terms. Luther read the omens correctly: 'Junker Henry means to be God and do as he pleases.'

He did as he pleased. Francis and Charles were falling apart; by July 1541 they were on the brink of war and again Henry exploited the situation. Both sides sought his aid, and because the English council inclined to the imperial alliance for reasons of trade, France must make the better offer; she proposed a match between the duke of Orleans and the Lady Mary (August 1541), and suggested in February 1542 a meeting between Henry, Francis, and James. The astute Tudor realized that France would no longer give effective support to Scotland and seized

[1] For the fall of Cromwell and its domestic implications see p. 414.

his opportunity. Early in 1540 he had sent the competent Sadler north to counteract the influence of Beaton[1] and to point out the practical benefits which James might reap from the confiscation of church lands. For the moment his lure was unsuccessful, but when Sadler went to Scotland again in 1541 he was able, Beaton being in France, to extract from James a promise to meet his uncle at York in the following September. Henry kept tryst, but James did not.

That the English king should have undertaken, for the only time in his life, the long journey to the northern capital may be regarded as a proof of his anxiety to consolidate the British Isles in the face of a possible attack; but it must be regarded too as a proof that he felt himself quite secure in the south, and it is significant that his resentment at being brought on a fool's errand did not hesitate to express itself in arms. All was ready. Norfolk had mobilized the north in the summer of 1541, and in August 1542 Sir Robert Bowes crossed the border to raid Teviotdale. He was beaten at Haddon Rig, and though the dead bore silent witness as to the place of the encounter, Henry insisted that the Scots had been the aggressors. He made no declaration of war, but prefacing his attack with a haughty declaration which asserted the old claim of suzerainty, he launched the redoubtable duke of Norfolk against Scotland with instructions to march upon Edinburgh. Roxburgh, Kelso, and a score of smaller places were burnt, but the invasion was an utter failure. Weather-beaten, short of food and beer, and harassed by a watchful foe, the host returned with little honour but with many mutual recriminations.

The Scottish counter-attack was a disaster. Dividing into two portions a large army which he had collected south of Edinburgh, James sent one detachment to make a feint on the eastern border, where it accomplished nothing, whilst another force of 10,000 men was to attack along the western route. Stopped by greatly inferior English forces in a marshy ground between the Esk and the Sark, the Scots gave a very poor account of themselves. The nobles were infuriated because the king had entrusted the command to his favourite, Oliver Sinclair; many of them

---

[1] David Beaton, nephew of James Beaton, archbishop of St. Andrews (1522–39), became bishop of Mirepoix 1537, coadjutor to his uncle 1537, cardinal December 1538, archbishop of St. Andrews 1539, chancellor and legate *a latere* 1543; he was the champion of Roman catholicism and the French interest in Scotland; murdered 1546.

believed that they had been already delated by the churchmen
as heretics, and the protestants suspected that they had been
put in the van with sinister intention. Solway Moss was the most
disgraceful field that ever Scotland fought. Two earls, 5 barons,
500 lairds, and 20 guns were among the trophies taken by the
English. James, of brittle spirit and shaken frame, was lying ill
at Falkland when he heard the news, and on 14 December he
died of the shame of it. His heir was the little girl born in
Linlithgow on 8 December, concerning whom he made the
famous prophecy[1] which differs from others because, having
proved false, it has survived to become a commonplace. From
the unlucky lass the throne of the Stewarts did not pass; it
was the fate of her son the king of Scots to wear the crown of
England too.

In the spring of 1543 this outcome was still upon the knees of
the gods. To Henry it seemed that England and Scotland might
now be united on his own terms. For him the year 1286 was
come again. As Edward I had been the great-uncle and the
nearest male kinsman of the maid of Norway, so he was great-
uncle and nearest male kinsman to Mary Stewart, and he, like
his progenitor, had a son named Edward of suitable age for
marriage with the Scottish queen. Not only opportunity but
the means of exploiting the opportunity lay ready to his hand;
Angus, once the husband of Margaret, held the warlike
Douglases at his disposal; the Scottish protestants, who had
suffered under the resolute Beaton, looked to him for deliverance; now in the prisoners of Solway Moss he had potential
agents whom he at once sent back to their country as 'assured
Scots' pledged to promote his marriage project. Some of them
were pledged in great secrecy to secure for him immediately the
suzerainty of Scotland, and there was even a design to seize the
person of the baby queen. Events in Scotland inclined him to
use persuasion rather than force; an attempted *coup de main* by
Beaton failed; Arran, the heir-apparent, a weak man and inclined toward England, was proclaimed regent; and when the
estates met on 12 March they not only secured the right to have
the scriptures in the vulgar tongue, but appointed ambassadors
to treat for the English marriage. Henry, though he regarded the

---

[1] 'It came with a lass, it will pass with a lass'—the Stewarts came to the throne
through the marriage of Marjorie, daughter of Robert the Bruce, to Walter
(Fitzalan), 6th hereditary steward of Scotland.

elevation of Arran as a breach of his rights, let the question of
suzerainty rest for the time being, and on 1 July concluded at
Greenwich two treaties, one for peace and the other for the
marriage between Queen Mary and the prince of Wales. On
25 August the treaties were ratified at Holyroodhouse; protestant sentiment ran high and as Grimani reported in October
the cause of Rome seemed to be lost in Scotland.

The legate was mistaken.[1] Henry overplayed the magnificent
hand which fate had dealt him. His arrogant pretensions and
his seizure of Scottish ships in the Thames alarmed even his own
allies; Arran yielded to French pressure; Beaton recovered
power. A parliament which met in December revoked the
English treaties on the ground that Henry had broken the peace
which was their ostensible end, renewed the 'auld alliance'
with France, and passed stringent laws against heresy. The
English monarch was furious. He had still a party in Scotland,
but he disdained to use diplomacy. The preamble to the
subsidy act passed in the spring session of 1544 renewed the
claim to suzerainty in uncompromising terms, and in the spring
Hertford was launched upon Scotland with instructions to
burn and destroy, putting man, woman, and child to the
sword wherever resistance was offered. On 4 May he landed at
Newhaven; he burnt Edinburgh, Holyroodhouse, and Leith,
devastated the country round about, and marched back to
Berwick leaving a trail of ruin behind him. The English king
found a new ally in Lennox, whom he betrothed to his niece,
the Lady Margaret, but the effect of his severity was in the end
to unite all Scotland against him. Angus himself changed sides
and on 27 February 1545 inflicted a sharp defeat upon the
English at Ancrum, near Jedburgh. A few months later a
French army arrived under Lorges de Montgomerie and although in the following September Hertford destroyed the
great abbeys of the Tweed valley, Kelso, Melrose, Dryburgh,
Roxburgh, and Coldingham, the havoc wrought by his host,
which included foreign mercenaries, did not affect the main
issue of the war. In his wrath the English king condescended to
support plots for the murder or kidnapping of Beaton, and it is
just possible that George Wishart, the valiant martyr whom
Beaton burnt on 1 March 1546, was an emissary between the
English council and the Scottish enemies of the lustful but

[1] Cf. Sadler, *State Papers*, ii. 560, for Scottish opposition.

competent cardinal. The murder of Beaton in his own castle of
St. Andrews on 29 May was an act of protestant vengeance, but
its perpetrators, perforce, looked to England for aid, and as long
as the English fleet held the sea they held the castle against all-
comers. When, on Henry's death, the English ships were withdrawn, the 'Castilians' soon succumbed to French gunnery.
The French, it is true, did not make themselves popular in
Scotland, but the grand result of Henry's interference had been
to reduce the English party there to nothing at all.

Henry's failure in Scotland was due in part to the fact that
both his interests and his efforts were preoccupied by a war
with France. This was, diplomatically speaking, quite unnecessary since the renewal of hostilities between Charles and
Francis put Henry once more into the position *tertius gaudens*.
Both sides sought his aid. Francis once more offered the hand
of his younger son, Orleans, to the Princess Mary, and his
proposal was supported by Norfolk. Gardiner, however, and
the majority of the council favoured an alliance with the imperialists which commended itself to the interest of the mercantile
classes and to the sentiment of the country at large. Henry had
no need to go to war, for with France and the empire at one
another's throats, he was free to settle his affairs with Scotland
which then promised well; but age had not dimmed his natural
bellicosity. Though he was grown gross and suffered from an
open ulcer in his leg, he still saw himself as the *jeune premier* of
1512; he still believed in his suzerainty over France; he still
dreamt of conquests overseas; he still concerned himself with
the things of war.

In 1542 he was sending to Vienna to get kettle-drums for
mounted men along with other drums and fifes, and it was not
to pageantry alone that he gave his attention; since 1535 he
had been casting even the largest guns in England; in 1543 he
patronized Peter Bawd and Peter van Colin, two foreigners,
who devised mortars and shells, and when his army did cross the
sea in 1544, it was equipped with ovens which cooked and mills
which ground as the wagons which bore them kept their place
in the line of march. On the navy too he cast his eye and in-
vented, as a reply to the French galleys, a long row-boat which
had not only a bow-chaser but a short gun on each side under
the half-deck. His own martial spirit no less than the diplomatic ability of Chapuys persuaded him to war, and on 11

February 1543 he concluded a treaty whereby he and Charles pledged themselves not only to give mutual aid and defence but to attack France jointly within the next two years. In the following July some 6,000 men under Sir John Wallop served for 112 days in the army of Charles, who beheld with admiration the prowess of the northern horsemen, and on the last day of December Gonzaga concluded at Hampton Court arrangements for a grand attack in the following year wherein each of the monarchs should invade France with 35,000 foot and 7,000 horse. In the preparation for the campaign Henry showed his old enthusiasm, but though he crossed to Calais on 14 July, he did not assume command of his forces, which had arrived before him. One army under Norfolk and Russell masked Montreuil; the other under Suffolk beseiged Boulogne which capitulated with the honours of war on 14 September. On 18 September Henry made a state entry, but on that very day Francis and Charles signed the peace of Crêpy. For Charles it must be said that, having beaten Cleves, he had advanced as far as St. Dizier only to find that England had no intention of joining in an attack on Paris and that he endeavoured to comprehend Henry in the peace he made. Henry, however, would not be comprehended. Despite French offers of pensions and arrears he refused to give up Boulogne, and he succeeded, not without difficulty, in keeping his conquest even though he was left to fight unaided. He himself returned home on the last day of September and after his departure the French made strenuous efforts to recapture a port which was so valuable to them. In the summer of 1545, 20,000 Frenchmen appeared before the town, and when an English fleet of 160 sail under Lisle approached the mouth of the Seine in June it found ready to meet it a force of more than 200 French ships, including 26 galleys. After an indecisive action the English were driven by the weather to Portsmouth; the French under Annebault followed and in some skirmishing in the Solent Lisle, who failed to lure his enemy within the range of the shore guns, suffered some losses. The great *Mary Rose*, with 400 men, was lost because the gunports were left open when she attempted too sharp a turn in narrow waters, and even the *Great Harry* itself was at one time in difficulty. The French, however, completely failed in attempts upon the Isle of Wight, which was well defended, and though they appeared off the Sussex coast, where they burnt Brighton,

they soon drew off in their turn with nothing done of consequence. In the autumn, Lisle was burning villages on the
Norman coast, while du Biez from Montreuil ravaged the pale
of Calais. Meanwhile about Boulogne there was hard fighting
in which the English had not always the advantage, and this
continued until the very eve of a peace which was made at
Camp, near Ardres, on 7 June 1546.[1] By the terms of the treaty
England was to recover her pensions, 94,736 crowns a year
during Henry's lifetime, 50,000 crowns in perpetuity, and
10,000 crowns for the commuted tribute of black salt if found
to be perpetual. She was to keep Boulogne until Michaelmas,
1554, after which Francis was to recover the town with its
fortifications and to be released from the arrears of pension due
to 1 May 1546 in consideration of a payment of two million
crowns. Scotland was to be comprehended, on the same general
terms as those set forth in the treaty of 1515. Experience had
shown that pensions were more readily promised than paid,
and even if they were paid regularly, an annual income of
about £21,000 a year was a poor return for the temporary
possession of a town which had cost the country £13,000 a
month in time of war and was very expensive to maintain even
in peace.

First and last the war with France and Scotland cost nearly
£2,200,000,[2] and its effect was to ruin the royal finances. The
preparations made by the cautious Cromwell were rendered
quite insufficient owing to the increased costs of war and the
general rise in prices. The 'ordinary' revenue could not meet
ordinary expenditure, and direct taxation did not avail to meet
the new requirements, though it was imposed upon a scale
hitherto unknown. Between 1540 and 1547 there were granted
six fifteenths and tenths and three subsidies; the fifteenths and
tenths remained constant at something over £29,000, but the
yield of the subsidies reflected both the prosperity of the
country and the results of inflation; that of 1540 produced over
£94,000, that of 1543 about £183,000, that of 1545 nearly
£200,000. The total amount brought in was about £650,000
and as the payment of each grant was spread over several
years, the lay subjects of the Crown paid direct taxation annually. The richer of them contributed also to forced loans,

---

[1] *Letters and Papers*, xxi. i. 507.   [2] Dietz, p. 147.

which were not meant to be repaid, and to 'benevolences', which were theoretically free gifts. The 'privy seals', which the government issued in 1542, were really royal bonds in blank which commissioners in the shires were authorized to fill in and present to persons who lent to the king; along with the blanks and with instructions for the use of discretion, the commissioners were given lists of persons to be approached and information as to their assessment in the subsidy books. The loan of 1542 produced more than £112,000 and the parliament of 1544 released the king from his obligation to repay it. After various devices had proved ineffective, Chancellor Wriothesley issued, in January 1545, instructions for the raising of 'benevolence money', although 'benevolences' had been explicitly condemned by an act of Richard III. Amongst those who protested were two bold aldermen. One of them, Sir William Roach, was confined in the Fleet until he bought himself out. The other, Richard Reid, was sent to fight the Scots with a following provided at his own charges. It is significant that when Reid fell into the hands of the enemy and had to redeem himself by ransom, some, at least, of his contemporaries seem to have regarded his fate as amusing rather than tragic. The 'benevolence', imposed more widely at a lower rate than the loan of 1542, brought in nearly £120,000 and in 1546 the Crown levied a contribution on the same lines, whose exact yield does not appear.

While laymen were thus taxed, the clergy did not escape. First-fruits and tenths continued to bring in a slowly declining income which never fell below £27,000 a year, and this was supplemented in 1540, 1542, 1543, and 1544 by clerical subsidies which averaged something under £20,000. Never had England seen such taxation; but the sums raised, even when increased by the useful profits of forfeiture, were quite insufficient for the needs of the Crown. In an attempt to close the gap, the government, as we have seen, began to alienate land, and before the reign closed it had sold about two-thirds of the monastic property for a sum which fell little short of £800,000. That it should prejudice the future welfare of England to pay for present necessities was a serious thing, but far more serious was its debasement of the coinage.

Some debasement was justifiable on the ground that most of the continental currencies had already been debased; Richard

Gresham who, with his brother John, aided the finances of the king, had not yet formulated his law, but men of wisdom saw that if the English coinage was better than that of its neighbours it would take wings and fly. In 1489 Henry VII had raised the value of the old noble from 6s. 8d. to 8s. 4d. and had decreased the amount of bullion in some of his other coins. In 1526 Wolsey had enhanced the value of the English currency by increasing the price of gold from 40s. to 45s. an ounce and by issuing new coins of lighter weight.[1] In 1542 the government, not without justification in view of the state of foreign currency, made yet another debasement, using this time a new method. It left unchanged both the price of bullion and the weight of the piece, but it introduced into each coin an increased amount of alloy. The success of the expedient led to its use upon a disastrous scale and in 1544 began a general debasement whose object was simply the obtaining of money. The value of gold and silver was raised, the standard of fineness was lowered, and the king, who coined at great profit his plate and bullion, reduced at the same time the burden of his debt; even when he bought bullion at the market price, he made a vast profit on his currency and between May 1544 and the end of the reign he drew from his Mint no less a sum than £363,000.

To one more shift the desperate government was driven. In 1544 it began to borrow abroad in the Low Countries upon a large scale. The regent, Mary of Hungary, was astonished to learn that the rich king of England was hard pressed, but she did not persist in her objection, and Henry's agent, Stephen Vaughan, was left to do the best he could. His task was not easy; there was money enough in the Netherlands but it was not easy to borrow. Economic development and the growth of ocean navigation had made the Netherlands the financial centre of Europe. Firms like the Fuggers and the Welsers of Augsburg, who had prospered from the Venetian trade and had mined silver and copper in Saxony and Austria, were now installed at Antwerp and casting hopeful eyes upon the bullion of America. Such firms had become money-brokers and international bankers; but though they were prepared to lend they demanded heavy interest, sometimes as high as 14 per cent. The lenders drove hard bargains in the exchanges and, not unnaturally, refused repayment in the debased English currency; as

[1] Dietz, p. 175.

sureties they would not be content with governmental promises or always with the bonds of staplers and merchant-venturers, but demanded the guarantee of Italian houses in London. They might advance sums on the lead which the king had stripped from the monasteries or offer a loan in fustians to be sold by the king at the best figure obtainable; but if they bought English goods at all it was at a low figure, and an attempt to exchange lead for alum to be resold in England at a profit resulted only in loss to the Crown. The English government was hard put to it to repay, and it did not always repay exactly to time, but repay it did. When Henry died, none of his debts abroad was actually due, but he left to his successor an obligation to repay advances amounting to about £100,000 Flemish—about £75,000 sterling. In his borrowings, as in his sale of lands and his debasement of the currency, Henry left an evil legacy behind him; for him it may be said that if the government was well-nigh bankrupt, the country prospered fairly well.

The king, who led his people, for the most part unprotesting, into an unnecessary and expensive war, at the same time established himself as absolute master at home. The country was not without internal dissensions, in many of which the element of religious dispute was inextricably blended with aristocratic discontent, with the choice of queens, and with the quarrels of rival ministers; but over all the controversies the authority of the king asserted itself with unvarying success. The religious issue demands separate consideration; here it suffices to say that the catholics made common cause with political conservatism at home and papal pretensions overseas, while the protestants strove to obtain royal support for further innovations in England and for an alliance with the German Lutherans. To neither party did Henry bow. With utter impartiality he burnt heretics and executed noble malcontents; although he professed to act in conformity with the law or in deference to his people's wishes he cast off his wives not for political reasons but for his own pleasure. The completeness of his authority appears in the terms of the act of succession of 1544 which followed upon his sixth and last marriage.[1] This act gave the Crown, failing Edward, or heirs of the recent marriage, first to Mary and then to Elizabeth, but under definite conditions which were to be

[1] 35 Henry VIII, c. 1. Henry married Catherine Parr on 12 June 1543.

set forth by the king either in letters patent or in his will. In the case of failure of his own issue either by death or by breach of conditions the succession was to pass to Henry's own nominee.

The monarch who thus treated the Crown as a real estate vested in himself had obviously no doubt whatever of his right to choose his own advisers, and power was steadily concentrated in the hands of servants competent to execute his will. Behind the façade of the Tudor council there was a bitter dispute between the new men and the old aristocracy and Henry, in this as in other matters, held the balance between the contending forces. At the beginning of 1538 the number of effective councillors seems to have been about a dozen[1] and, by a process familiar to students of the history of the privy council and cabinet, real authority lay in the hands of an inner ring of the 'privy council'.[2]

In this Cromwell was then pre-eminent. Cromwell could manage parliaments, and the act for precedence[3] which was passed in the spring of 1539 emphasized the superiority of the king's officers over those who held their rank by mere hereditary right. A grim commentary upon the significance of the measure was the act of attainder[4] which was passed in the same parliament against Poles, Nevilles, Askes, and their accomplices, though many of the condemned were already dead, though the countess of Salisbury and the marchioness of Exeter were already in prison, and though Reginald Pole was out of range. High rank might give no privileges, but it certainly brought danger; good service, on the other hand, could give high dignity. On 18 April Cromwell was made earl of Essex,[5] and on the following day he received the office of lord chamberlain. A few weeks previously Sadler and Wriothesley, who had both been his followers, were made joint secretaries; in May Sampson of Chichester was imprisoned as a papalist; other bishops were threatened; Tunstall himself wavered before the attack and denied that he had counselled Sampson to hold by the old usages of the church; a fresh treason bill was introduced; on

---

[1] Pickthorn, p. 411.
[2] Norfolk was said to have complained that he was 'not of the most secret, or as it is there termed, the privy privy council', but the repetition of the word 'privy' may be a mere slip.
[3] 31 Henry VIII, c. 10.   [4] Ibid., c. 15.
[5] The title was available since Henry Bourchier, the second earl, was thrown from his horse and killed on 12 March 1540.

1 June three persons were attainted for 'a treason' which was in essence denial of the royal supremacy. It seemed that the new men were triumphing everywhere, but with a kaleidoscopic rapidity the scene changed. On 10 June Cromwell was arrested in the council chamber at Westminster by the captain of the guard; Norfolk tore the George from his neck; Wriothesley seized the Garter. On the same day his goods were confiscated; on the 29th a bill of attainder was passed against him, and though, probably without much hope, he testified to the unreality of the Cleves marriage, he was beheaded—at Tyburn, not in the Tower—on 28 July.

The story of his fall abounds in morals for the philosopher, the lawyer, and the historian. The nemesis which overtakes the parvenu demands no commentary, but there is matter for consideration in the nature of the legal procedure adopted against the doomed man. What exactly was meant by statements that he perished by his own bloodthirsty laws and that he was the first man attainted and never called to answer? It is not the case that hitherto persons attainted had always been allowed to speak in their own defence, and Cromwell certainly did not invent the process of attainder. Coke asserts that[1] Cromwell himself helped to drag from the reluctant judges an opinion (which Coke thought wrong), that attaint by parliament overrode the right of the defendant to come to his answer, and the story is perhaps true. At all events it contains the gist of the matter. Cromwell had made parliament omnicompetent and parliament slew him. It was parliament which had passed the act of 1534, drafted by himself, which increased the connotation of treason, and it was parliament which decided that the things laid to his charge were treasonable. There was no remedy, and he knew it. At the moment of his arrest he cast his bonnet to the ground and asked indignantly if this was the reward of good service, and if any man there could call him a traitor; but later, though he denied that he had ever been a sacramentary or an embezzler and above all that he had ever been unfaithful to his king, he submitted to the sentence of parliament: 'I am a subject and born to obey laws.' Doubtless, he had carried his correspondence with the Lutherans too far to please the king; doubtless, like other public servants, he had done things in his office for which a private man might be

[1] Pickthorn, p. 433.

condemned, but we may let the matter of his treason go. His offence was that he had lost the confidence of the king; he had confiscated the wealth of the monasteries; he had magnified the power of the Crown, and now he was superfluous to a master who more and more liked to balance the various parties of his realm and to ride supreme above them all.

Cromwell's fall was less sudden than the public supposed. During the spring of 1540 his power had been sapped from within. Henry's distaste for Anne had been increased by his growing affection for Catherine Howard, niece of Norfolk and cousin of Anne Boleyn, which was sedulously fanned by 'wily Winchester'.[1] The appointment of Wriothesley to serve as secretary along with Sadler may represent a success for Cromwell's enemies, since Wriothesley was a conservative at heart; certainly the consecrations of Bonner to London and Heath to Rochester, no less than the arrest of Barnes and other protestants 'by the king's own commandment', show that Gardiner and his friends were gaining the ear of the king. When the king's favour was gone Cromwell was lost. The grimy-fingered mechanic who had oiled the wheels of the chariot of state was crushed beneath it, while his ungrateful master rode triumphant on. Cromwell had Tudor thanks for his pains, and history, no less than Henry, has dealt hardly with him. He has been shown as one who was brutal in success and abject in defeat, and one who, though he played the protestant during life, professed to be a catholic at the moment of death. Yet, it is possible that the cry 'for mercye mercye mercye' in his last letter may have been an appeal to escape not death but the grisly mutilation which preceded death for a traitor who was not of noble birth, or the fire, which as Burnet hints, might be the doom of a heretic. That he took bribes and used the machinery of government to assist his own friends is certain; but it must not be forgotten that the treasure he left behind him was far less than had been expected and that he used some of his wealth, as Stow attests, to maintain the poor in London. He denounced the lewdness of monks, but he himself was not lewd. Certainly he was a realist of the new age; he knew the value of trade; he understood money and he was in touch with men in every walk of life. Justly has it been said that though he followed

[1] *Letters and Papers*, xvi. 270.

the maxims of *Il Principe* too much,[1] he gave insufficient attention to the precepts set forth in *Il Cortigiano*; he made the prince too great, he studied the man too little. It was the monarch, not Henry, whom he magnified; his god was the state, and though, after the manner of the time, he identified king and state, it was not the king alone but the king in parliament whom he exalted. In after ages the place which he gave to parliament had a great and growing significance, but for the moment *le nouveau messie c'était le roi* and the immediate effect of his work was to make Henry absolute.

The French ambassador, Marillac, thought early in 1541 that Henry regretted Cromwell;[2] certainly he had no other grand vizier; if his opponents supposed that Norfolk or Gardiner would seize the reins of power, they were speedily disillusioned. Henry rid himself of Anne of Cleves swiftly and easily. On 6 July the lords invited the commons to join in an address to the king on the subject of his marriage; the commons agreed and the disquieted monarch issued a commission to convocation to try the matter. On 9 July the united convocations of Canterbury and York, having been skilfully led by Gardiner, pronounced the royal marriage invalid on the ground that there was a possible pre-contract on the part of the bride, that there was a certain 'lack of inward consent' on the part of the bridegroom, and that the union had never been consummated. 'This was the greatest piece of compliance that ever the king had from the clergy', wrote Burnet. Consoled by the promise of £4,000 a year for life and of a precedence above all the ladies of England except the wife and daughters of the king, Anne went cheerfully into a retirement where she could wear 'new dresses every day'. On 12 July the marriage was annulled by act of parliament and the king was free to seek another bride.

A fortnight later, on July 28, Henry wedded Catherine Howard, but although the royal alliance with the great catholic house awakened high hopes in some quarters there was, in fact, no great conservative reaction. Cranmer, who almost alone had dared to say a word for his old ally, still enjoyed the royal confidence. During the remainder of his reign the king pursued a steady course, keeping the balance between parties and

---

[1] Baskerville, p. 130. He had a copy of *Il Cortigiano* (1528) and promised to lend it to Bonner.
[2] *Letters and Papers*, xvi. 285.

striking down with an iron hand all who were even suspected of disputing his authority in church and state. Early in 1541 Sadler and Wyatt were in the Tower for a short time, but so too was Wallop who was a conservative, and Wriothesley and Pate, no friends to the fallen vicar-general, were in disfavour too; before long the two secretaries were in office once more, and all save Pate again enjoyed the king's grace.

While the king preserved his faithful servants, he had little mercy upon aristocrats who incurred his displeasure. An abortive conspiracy in the north led to the execution of old Lady Salisbury on 27 May 1541.[1] In June Leonard Grey was beheaded for his alleged misconduct in Ireland, and immediately afterwards Lord Dacre of the south was condemned and hanged for the murder of 'a simple man', though many of the nobles tried to save him. Before the year closed, England was given a more startling exhibition of the royal justice. On his return from the north, whither he had gone to quell the last stirrings of revolt and to meet his nephew James, who did not appear, Henry was given, through Cranmer, evidence that his queen had been guilty of impropriety before her marriage, and before long there was evidence pointing to her adultery with her cousin, Thomas Culpepper, after she had become queen. At the end of November two of her alleged paramours were indicted for high treason in the counties where their offences were said to have occurred. On 1 December they were tried at the Guildhall of London and condemned to be hanged; on the 10th they were executed. For the queen a more formal condemnation was reserved. On the fourth day of a parliament which met on 16 January 1542, a bill was read attainting of treason Catherine Howard and Lady Rochford, who had abetted the affair with Culpepper, and of misprision the duchess of Norfolk, who had been a careless guardian of the queen's youth, along with Lord William Howard and others. Led or directed by the chancellor, the lords behaved with extreme care, assuring themselves in advance of Henry's approval of their proposed procedure, and suggesting that the king might give his consent in absence by letters patent. Henry, who in the lords and before a committee of the commons had spoken eloquently upon other matters, gave his assent to the lords' proposals, and, that obtained,

---

[1] Wriothesley, i. 124. Marillac in *Letters and Papers*, xvi. 411, would make the date 28 May.

parliament proceeded without delay. On 11 February the bill of attainder had been hurried through the lords and commons accompanied by another bill, made necessary by Lady Rochford's 'frenzy', for proceeding against traitors who had lost their reason. Two days later both women were beheaded in the Tower; Lady Rochford was not mistress of herself, but Catherine died like a daughter of the white lion. The duchess of Norfolk and several other persons who were condemned for misprision were very soon released.

Catherine had been foolish and probably worse than foolish, but the statute under which she suffered was remarkable both in its form and in its matter. It was in effect a request to the king to grant the attainder approved by parliament, and that by royal letters patent; it laid down the doctrine that any unchaste woman marrying a king of England without declaration of her unchastity should be guilty of high treason. Somewhat shaken by his experience Henry remained unmarried for nearly eighteen months, but on 12 July 1543 he wedded again, and again his bride was a woman much younger than himself. The chosen lady was Catherine Parr, who was sought by Sir Thomas Seymour and who had already been widowed twice. Her second husband had been a catholic leader in the Pilgrimage of Grace, but she herself was suspected, not without cause, of a leaning towards the new religion; her advent did not affect the royal policy though she seems to have exercised a moderating influence all round. She showed herself kind to the king's three children, and being endowed, not only with wifely experience, but with a kind heart and common sense, managed to survive her imperious lord. This was an achievement; for Henry grew ever more arbitrary with the passing years, and his court, as his health deteriorated, became the scene of quarrels which were personal as well as political.

Just as Cromwell's party had survived his death, so Norfolk and Gardiner, though somewhat under a cloud after the fall of Catherine, still kept their places; the last years of the reign were marked by a struggle of the factions which became more open as the king's illness grew upon him. If the position of the 'protestants' was strengthened by Henry's last marriage it was weakened by the absence on military duty of Hertford and Lisle and by the deaths in 1545 of Suffolk, Poynings, and the king's physician Dr. Butts. Gardiner and his friends, seeing the king very devout in his religious observances, made several attempts to pull down Cranmer, but failed in their purpose, partly because the king always upheld his archbishop in the crisis and partly because the 'catholics' were divided amongst themselves. Norfolk favoured an alliance with France, Gardiner an alliance with the emperor, and when, on the making of peace in June 1546, Hertford and Lisle returned to the council, they were able to make their presence felt. In October Lisle prejudiced his position by striking Gardiner in the council, from which he was in consequence excluded for a time. On the other hand, Gardiner alienated the king by a refusal to exchange certain lands with him, and for some two months between November 1546 and January 1547 he did not appear in the council.

Henry, now seeing death not far away, seems to have concluded that his son's future would be safe only in the hands of the soldiers, one of whom was the boy's own uncle Hertford, and the ruin of the conservative party was completed by the indiscretions of the young earl of Surrey. Surrey, who shares with Wyatt the credit of having introduced into English poetry the graces of Italy, was a headstrong youth, proud of his illustrious descent and inclined to regard himself as above the law. He was married to Frances Vere, daughter of the fifteenth earl of Oxford, and his sister was the widow of the duke of Richmond. He had served as cupbearer to the king at the age of fourteen and subsequently he had received many marks of the royal favour. He was at one time the companion of the Duke of Richmond, and there had been some thought that he might marry the Princess Mary. In the ceremonies of the court he had taken a conspicuous part and on St. George's Day, 1541, he had been elected a knight of the Garter, but his quarrelsome temper and overbearing manner had led him into difficulties. In 1537 he rendered himself liable to lose his right hand by striking Edward Seymour 'within the verge' of Hampton Court, and owed his pardon to the good offices of Cromwell; in July 1542 he was imprisoned for a short while in the Fleet for challenging a rival to single combat, and on 1 April 1543 he was summoned before the privy council for having eaten flesh in Lent and gone about breaking the windows of London with a catapult; the first offence he denied, for the second he again suffered a short imprisonment. Having served an apprenticeship against the Scots and with Wallop in 1543, he acted as marshal to his father in the vanguard of Henry's army in 1544 and was wounded in a desperate attempt to storm Montreuil. Next year he took command at Boulogne, where his conduct displayed all the valour of his race but not all of its military ability, and in 1546 he was superseded by Hertford. After his recall he was accused of having used his office for the benefit of his own friends. Proudly he denied the charge;—'For there be in Boulogne too many witnesses that Henry of Surrey was never for singular profit corrupted, nor never yet bribe closed his hand which lesson I learned from my father'; but he was piqued and disappointed, and as Hertford's ascendancy declared itself, he showed his ill will very openly. Unluckily for him the affairs of his family were at the moment much involved. His father had abandoned his mother in favour of Elizabeth Holland, and an attempt by the Howards to recover prestige by a marriage between the duchess of Richmond and Hertford's brother, Sir Thomas Seymour (later the admiral of the Princess Eliza-beth's acquaintance), had come to nothing. With the duchess, who inclined towards the protestant party, he was himself at odds, and when the loyalty of the house was impugned, its defence was prejudiced by the mutual incrimination of its members.

The attack was begun when Sir Richard Southwell, once Cromwell's man but latterly an intimate of the Howards, alleged that he knew things touching the fidelity of the earl, who did not improve his position by offering to fight the accuser in his shirt. Ever since 1543 the council had been aware that Surrey vaunted his royal descent through Thomas of Brother-ton from Edward I, and had professed a right, through the Mowbrays, to the arms of Edward the Confessor. Although the Heralds' College had declared against him, he had not hesitated to emblazon on a panel at Kenninghall an escutcheon display-ing the leopards and the cross along with his own arms. He had talked loosely of his father's right to a regency when the ailing king should die, and had by his own rashness put weapons into the hands of his enemies. On 12 December Norfolk and Surrey were arrested; on 7 January 1547 both were indicted for high treason at Norwich, the son for displaying the royal arms, the father for failing to reveal the offence. On 13 January Surrey was tried at the Guildhall and though his fierce defence evoked the sympathy of those who disliked the new men, a message delivered by Paget from the king settled the mind of the hesitant jury. The earl was condemned and on the 19th he was beheaded on Tower Hill. His father was attainted in parliament, and on the 27th condemned to suffer on the morrow; but ere the morrow came Henry himself was dead and in the resultant uncertainties the duke's life was spared. He languished in prison, however, throughout the reign of Edward VI; for the time at least, the power of his great house was gone.

The fall of the Howards was rather the symptom than the cause of the 'protestant' ascendancy in the council at the time of Henry's death. To some extent it is true that, as the king's strong hands lost their grip, power passed to those who could get it, and Hertford and Lisle were the fighting men. Yet the king's word was of weight until the very end and the situation in the council reflected his imperious will. He was determined to secure the succession of his son, and he dared not rely upon a balance of parties when his own command was removed. He had to choose between them. He had no love for the new dogma, but he was influenced by Catherine Parr and still more by Cranmer, whom he had long trusted. Probably he felt that the young prince was safer in the hands of laymen than in the hands of the bishops, for he may have had some suspicion that churchmen, who were so zealous for transubstantiation, might in the end be zealous for the pope as well. Always there was the fear that, for devout catholics, Mary was the legitimate successor of her father. To the very English king of England the rights of the son came first, and even upon his death-bed he exercised his *imperium merum*.

In the affairs of the church as in the affairs of state Henry had his own way. 'It appears plainly that the king acted as if he had a mind to be thought infallible', wrote the seventeenth-century historian of the English reformation.[1] The act of supremacy had given the king the right to amend all enormities, including heresies, which might be amended by spiritual authority, and Henry never doubted that he could deal with doctrinal questions as with any other ecclesiastical matter. In his view 'that part of the body politic now usually called the English church' was in subjection to him as much as any other

[1] Burnet, bk. iii.

part, and his views were accepted by most of his people. These views were not entirely new. The idea of a sovereign state had been set forth by Marsiglio of Padua and it is significant that in 1534 Cromwell helped to pay for the publication of an English version of the *Defensor Pacis*.[1] To the men of the Renaissance the nation-state seemed more important than the universal church. Reginald Pole, it is true, discharging his conscience in the *De Unitate Ecclesiae* of 1536, preached the traditional authority of Rome, and Tyndale in the *Obedience of a Christian Man* (1528), which so gratified Henry, did not make the royal authority absolute—the duty of obedience to the king, though everywhere emphasized, was implicitly limited by the presumption that the king's law will be God's law.[2] One recognized Rome, the other perhaps the human conscience, but both stood for continental opinions. In England itself the doctrine developed by Saint-German[3] in 1523 turned to an almost complete 'Erastianism' which was accepted both by Gardiner and Cranmer, and even the uncompromising Henry Brinklow preached 'non-resistance'. In the *De Vera Obedientia Oratio* which Gardiner produced in the autumn of 1535, it is categorically asserted that 'princes ought to be obeyed by the commandment of God; yea, and to be obeyed without exception'; and that the 'church of England is nothing else but the congregation of men and women, of the clergy and of the laity united in Christ's profession'.[4] Gardiner, it must be noted, on the one hand defended the execution of Fisher and, on the other, repudiated the arguments of Bucer; for him the contempt of human law, made by rightful authority, was to be punished more heavily and more severely than any transgression of the divine law.[5] Hardly less definite was Cranmer. When in 1542 he was required to answer seventeen questions, some of which concerned episcopal authority, his reply to one was couched in guarded terms 'this is mine opinion and sentence *at this present*',[6]

---

[1] *Letters and Papers*, vii. 178.   [2] Tyndale, *Works*, ii. 273 (edition of 1828).

[3] Christopher Saint-German 1460(?)–1540, an Oxonian and a barrister, published in 1523 *Doctor and Student*, a law-book written in Latin and translated into English in 1530 and 1531.

[4] Muller, *Stephen Gardiner and the Tudor Reaction*, p. 62. See edition of *Obedience in Church and State* by Pierre Janelle (1930). For Brinklow's view see the conclusion of his *Lamentacyon*.

[5] Janelle, p. 175. It has been questioned if Gardiner was sincere.

[6] Italics mine. The seventeen questions were posed by an ecclesiastical commission appointed by the king who was considering a revision of *The Institution of a Christian*

but it was on another point quite clear. 'All Christian princes have committed unto them immediately of God the whole cure of all their subjects' in things spiritual as well as in things temporal and all ministers ecclesiastical as well as civil 'be appointed assigned and elected and in every place by the laws and orders of kings and princes'. The form of commission granted to bishops made the king's view absolutely plain.

Even the royal commission, however, recognized that some spiritual authority was conveyed by ordination, and it has been argued that the control of the Crown was therefore not complete; it has been suggested that the royal power was limited, in theory by the overriding law of God, and in practice by the presence of convocation. There is, however, no evidence that at this period 'the English church was regarded as an independent shrine of the divine law'.[1] It was accepted no doubt that the law of God should inform the commands of the temporal authority, but as Saint-German taught, it was for 'the king's grace in his parliament to expound scripture, and so decide what the irrefragable law of God is; for the king with his people have the authority of the Church'. As for convocation, it certainly did much, but it was, in the end, the mouthpiece of the king. Parliament was called upon to justify every decision of consequence which it made, every grant of money which it made, and that the members of convocation realized their impotence appears from the fact that in 1547 the lower clergy appealed to be joined in association with the commons in parliament. Although the act of 1534 for the submission of the clergy had authorized the preparation of a code of canon law, and although Cranmer from time to time worked upon the task, no code was produced in the day of Henry VIII; and when at last, after various endeavours, it was produced by John Foxe in 1571 for information of the commons it was refused royal authority. There was, in short, no room for two codes in England. The lawyers knew that the fundamental principles of canon law were implicitly accepted by statute and by the decisions of the law courts, and they saw no need for a conflict of authorities. Heresy, for instance, which had always been tried by the church courts, had become punishable by statute under Henry IV and

---

Man; Cranmer presided. (Strype, *Memorials of Cranmer*, i. 112, giving the date as 1540; Burnet corrects to 1542.)

[1] Powicke, *The Reformation in England*, p. 50.

cognizable by the royal officers under Henry V. When, in 1534, Henry was in process of breaking with the papacy, an act was passed[1] defining more exactly the duties of the sheriffs and others in regard to heresy, and it was no great change in practice when, by the 'Act of six articles' of 1539, heresy became a felony at common law. In theory, however, the altered status of heresy bespeaks a new situation; things spiritual, like all other things, were now under the control of the Crown.

Although by the year 1539 Henry had no doubts as to the entirety of his 'empire', he indulged in no extravagance. He was resolute not to acknowledge the pope, and he was content to see some reforms in worship—the use of an English Bible, for example—but he adhered firmly to the traditional doctrines of the church. The middle way which was enjoined by his theological predilections was also commended to him by his innate political sense. Both at home and abroad circumstances made moderation desirable. If he behaved with discretion he might, in the days when the general council was pending, induce European princes, even the emperor, to share his views, and with their aid either coerce the pope to modify his demands or else drag half Christendom into rebellion against the Holy See.[2] At home, while he was ready to use the new men against the aristocracy, he was not prepared to accept the new theology and become either a Lutheran or a 'sacramentary'. Very much aware of his own, somewhat convenient, conscience, he showed little regard for the consciences of other people. It seemed to him that if he produced a moderate and realistic solution of the ecclesiastical question his subjects would readily accept it, especially if he dealt lightly with differences of opinion which did not issue into action. The basis of his calculation was wrong, for both catholics and protestants were prepared to die in the cause of faith, and his attempt to administer with the advice and by the agency of representatives of both parties was foredoomed to failure. It was impossible to get a compromise by arrangement, and what happened was that the king imposed his will by authority, veering from one side to another according to the exigencies of policy or diplomacy, but always returning to his

---

[1] 25 Henry VIII, c. 14.

[2] In 1533 Henry offered to lay his case before the general council, and his subsequent denunciations of the council were protestations that the assembly called by the pope was not a true council.

own path. When he could he buttressed his decisions with the authority of the clergy in convocation; he had no thought of giving them a power equal to his own, but he could conveniently represent them as the delegates of an overworked sovereign in spiritual affairs. His common sense told him that a consensus of opinion amongst the clergy was a political force of some magnitude, and his imperious dexterity enabled him to turn that force to his own end. The assent. of convocation gave an increased respectability to measures which he approved or which he altered to meet his own desire.

His desire was above all things for unity. He did not doubt his right to force his subjects to believe as he believed; he tried to do so; and he had some measure of success because his doctrine, though it varied in presentation according to political requirements, remained fairly constant throughout.

Very instructive is the story of the act abolishing diversity of opinions,[1] usually called the 'statute of six articles'. In the spring of 1539 the political situation demanded an authoritative pronouncement which would allay unrest at home and disarm hostility abroad. Accordingly, when parliament assembled in April the chancellor at once drew its attention to the need for religious unity. At the instance of the king an episcopal committee, containing members of both parties, was appointed to deal with the matter; this failing, Henry was at once ready with six articles of his own which were very conservative in tone. These had already been approved by Norfolk, and when they were opposed by Cranmer and his friends in the lords, Henry came down in person and 'confounded them all with God's learning'.

Convocation agreed with the opinion of the supreme head, and the completed act was so worded as to give the impression that the king in parliament was merely providing for the punishment of opinions declared to be erroneous by the clergy. In a sense this was true; no new penalties were prescribed by the act, which was merely declaratory; nothing was made heresy which would not have been thought heresy in the year 1401; but heresy now became a felony. To deny transubstantiation was to incur the death penalty. Communion in one kind for the laity, the celibacy of the clergy, the permanence of religious vows, the benefit of private Masses, and the use of auricular

[1] 31 Henry VIII, c. 14.

confession were all commended; to repudiate any one of them was to become liable to loss of property and liberty for the first offence, and to death for the second. To deal with offenders, there were to be appointed for every shire commissioners who might proceed upon the sworn information of only two lawful persons.

The act was followed up by a proclamation enjoining uniformity and moderation, but its passage was a severe blow to the protestants. Latimer and Shaxton resigned their sees, and each spent a year in custody; Cranmer was compelled to put away his wife; in London 500 persons known to be careless of ceremonies, to read the Bible in church, or to revile priests, were arrested on suspicion of being the kind of persons condemned by the new law. The catholics were jubilant, but before long they found that their rejoicings were premature. Cromwell remained in power, and the 500 suspects were pardoned by the king; certainly Gardiner gained more influence in the royal counsels—but the papalists were not freed from their cruel dilemma.

The 'whip with six strings' was kept for show rather than for use, and an act of 1540[1] mitigated its severity by removing the death-penalty for all offences save the denial of transubstantiation. The chronicles[2] show some burnings of protestants in this year, but they also show some hangings of papalists. Even the fall of Cromwell made little change in the general situation. On 30 July, the day after his execution, three of his protégés, Barnes, Jerome, and Garrard, were burnt at Smithfield; but along with them suffered Featherstone, Powell, and Abell, who were hanged for denying the royal supremacy and the validity of the king's separation from Queen Catherine.

These executions, however, were exceptional. Protestants as well as catholics were able to benefit from the general pardon which covered all offences committed before July 1540; but now and again, when conservative influence prevailed, the whip was brought into action. In 1543, largely through the activity of Dr. London, three heretics were burnt at Windsor; a fourth victim, John Marbeck, the famous musician who did so much for English church-music, was saved only by a

---

[1] 32 Henry VIII, c. 10. An act of 1544 (35 Henry VIII, c. 5) limited its arbitrariness, and made its procedure conform to the usual course of law.

[2] e.g. Wriothesley, i. 118, 119; cf. Hall, ii. 308.

royal pardon. Again, in 1546, there was a series of prosecutions, in the course of which the valiant Anne Askew was condemned, and after having been racked by the hands of Wriothesley and Rich, burnt at Smithfield (16 July), along with three persons of lower degree. Anne's supreme confidence in her faith and her courage under torture have deservedly given her a high place in the martyrology; yet it must be remembered that her troubles were in part matrimonial, that she was truculent to her accusers, and that the accusers themselves made great efforts to induce her to recant. Efforts of the same kind had been successful in the cases of Crome and Shaxton; Latimer by boldness and good fortune escaped; Sir George Blage was pardoned by the king. On the evidence it seems clear that the government, which was perhaps milder than some of the ecclesiastics, did not go about to seek victims, and that the king himself, who was said to have pitied the Windsor men as 'poor innocents', was reluctant to have persons of standing persecuted for their opinions, however fiercely he punished their alleged 'treasons'.

Yet although Henry tried to hold the scales of justice even, he knew that a mere balance of opposing forces was not enough, and strove to find a formula that would satisfy all. One of the acts of Cromwell's last parliament[1] provided for the preparation of authoritative articles on faith and ceremonies by the archbishops, bishops, and the best theologians, who were to define and decree 'according to Goddis wourde and Christes gospell by his Majesties advice and confirmation by his lettres patentis'. These articles (not yet stated) were to be believed and accepted by all the clergy of England, under penalties (not yet appointed), 'provided always that nothing ordained or provided by this act should be repugnant to the laws and statutes of the realm'.

It was three years before the promised articles appeared, and during the interval the position of the conservative party improved. The renewal of war between Francis and Charles removed the danger from catholicism abroad while the fall of Cromwell weakened protestantism at home. Certainly the new ideas were spreading and the bishops noted the growth of heresy with alarm, but the discretion of the protestant champions was not always equal to their courage, and sober opinion was sometimes shocked by their extravagances. The king became convinced that the Bishops' Book had gone too far; he felt that

[1] 1540, 32 Henry VIII, c. 26.

free disputation by unlearned men was improper, and began to think that unrestricted reading of the Bible was a dangerous thing. Gardiner afterwards alleged that in 1541 he sought a reconciliation with Rome through the good offices of Cardinal Granvelle, the minister of Charles V, and perhaps he did; but, though the catholics were for a time misled, he had no intention of surrendering his supremacy or of abandoning altogether his stoutest allies in the conflict with Rome.

He still gave his confidence to Cranmer, and it was Cranmer who, officially at least, took the lead in establishing the standards of creed and practice so much desired by the king. Gardiner, it is true, had a much larger following than his archbishop in convocation, and Henry, who knew well the importance of carrying the clergy with him, allowed the rival parties to debate freely upon the great issues. Yet behind the contending clergy loomed the figure of the king, alive to all that went on, ready to interfere decisively when the occasion arose. It was at his instance that the convocation of 1542 discussed the question whether the Great Bible should be retained, although that Bible had been formally authorized, and it is noteworthy that the question was raised by Cranmer who had himself written a very 'protestant' preface to the version. Most of the bishops believed that the text should be corrected according to the Vulgate, especially Gardiner, who produced a list of about a hundred words requiring retranslation, but when it seemed that the 'catholics' must prevail, the archbishop suddenly announced the king's pleasure that the task of revision should be entrusted to the universities and nothing more was heard of the matter.

Henry, however, was still full of suspicion as to the effect of promiscuous Bible-reading, and in the spring session of 1543 there was passed an act[1] which made his wishes plain. By this were condemned not only heretical books but also all 'craftye false and untrue' translations of the Bible including that of Tyndale. The reading of even the approved version was severely restricted. In church none was to read save persons appointed by king or ordinary; at home noblemen and gentlemen might read to their families, substantial merchants and gentlewomen might read to themselves; but other women, prentices, serving-men, and persons of base degree were not to read at all. For

---

[1] 34 and 35 Henry VIII, c. 1; described as an Act for the Advancement of True Religion.

the breach of these regulations penalties were imposed, but far more severe were the penalties prescribed for preaching against the king's doctrine—which, it was promised, should be set forth in 'a certaine forme of pure and sincere teaching';—for the third offence a clergyman was to be burnt alive, a layman was to be imprisoned for life. Explicit in this act is the promise that the Crown would provide an orthodox creed, and the positive side of the royal programme had already revealed itself in convocation. For some time Cranmer had been collecting from bishops and doctors opinions which now he submitted to the king. His majesty gave his comments with confidence. There had been much discussion as to the nature and number of the sacraments; Henry declared for the traditional seven. The clergy had concluded that the bishop had his appointment from the Crown, but his 'ordering' *per manuum impositionem* in the apostolic manner; Henry could see no reason for any distinction and thought that all power came from himself. Convocation accepted the royal amendments, many of them added to the text by the king's own hand, and though the 'Booke off Relligion' was read to the peers in the council chamber on 5 May[1] and approved, in advance, by parliament[2] before it was published on 29 May, it was essentially the King's Book. So it was generally called, though its official title was *The Necessary Doctrine and Erudition of any Christian Man*. According to the preface to the Latin edition published for the benefit of foreigners in 1545, it was a true statement of the old catholic doctrine free from both the leaven of papacy and the poison of heresy; and it was in fact a well-arranged presentation of the orthodox views except on the point of the papacy and the papal power of indulgence.

This 'Third Confession' of the English church represented a triumph for the conservatives, but Henry was not minded to run the catholic course altogether. It was in July 1543 that he married

---

[1] Dasent, i. 127. Cf. *Letters and Papers*, xviii. i. 312, where council, in mentioning the 'true and perfect doctrine' set forth by the king for all his people, says that it has been confirmed by a law made in parliament.

[2] 34 & 35 Henry VIII, c. 1. In the famous royal preface Henry explicitly claimed that the book had been seen and liked by 'the lords both spiritual and temporal, with the nether house of our parliament'. Burnet, who goes astray in his dates, says that this preface was added by Henry two years later and though it appears in all the copies of the 1543 edition in the British Museum it does not appear in the Latin version of 1545. Burnet also says that the completed book was never before convocation, but its 'articles' were discussed there. Wilkins, iii. 868.

Catherine Parr who favoured the protestants and, if Foxe is correct, on one occasion came near to paying dearly for opposing her husband in ecclesiastical debate. The story, though it rests on the authority of Cranmer, may not be true in detail, but it carries a moral; the king was not going to be hurried into violent action to gratify the catholics. He still supported Cranmer, who had long ago aided him against the pope and who, in the debates of 1542 in convocation, had, along with Barlow, declined to set the *potestas ordinis* of the bishop against the overriding power of the Supreme Head. In his theology Cranmer did not entirely agree with his master. He believed that the clergy might marry; he had argued that extreme unction and 'orders' were not true sacraments; his view of the Real Presence probably fell short of an acceptance of transubstantiation. Yet in many ways the attitude of the two men towards traditional belief was much the same, and three several times, in 1543, in 1544, and in 1545, the king delivered his servant from accusations of heresy by personal intervention of a most vigorous kind. The story of the ring told in the play of *King Henry the Eighth* is, except for the date, very largely true.

In spite of occasional waves of catholic reaction,[1] highest when Henry was abroad, or when Hertford and Lisle were absent from council on military duties, Cranmer had his own way in the matter of liturgical reform. In the convocation of 1542, which established the Sarum Use, he had advocated a simplification of ceremonies; and in 1543 he managed to suppress a comprehensive *Rationale* prepared by convocation, which adhered to the old way. Thereafter, such changes as were made were, in the main, of his own devising. Not all he planned could he accomplish. A book of homilies designed to counteract extravagance in preaching was begun in 1539, but though it was presented to convocation in 1543, it was not published till the day of Edward VI; an attempt, made in 1545, to carry out the long-promised codification of church law ended in the usual impasse. Yet one thing he achieved which set its mark upon the church of England for all time; he gave the English church its

---

[1] For protestant disappointments and apprehensions see the *Supplications* dedicated to Henry in 1544 and 1546 (Early English Text Society, 1871) and Brinklow's *Complaynt of Roderyck Moss* (Early English Text Society, 1874). The ceremonies at the rededication of the Greyfriars Church on 3 January 1547 must have alarmed protestants. *Chronicle of the Grey Friars of London* (Camden Society, 1851), p. 53.

litany. The use of English in primers, or books of private devotion, was by no means unknown before the Reformation, and after Henry's quarrel with the pope the primers of Marshall and Hilsey[1] had appeared. In June 1544, on the eve of his invasion of France, Henry ordered the preparation of an English litany, and in the following October Cranmer set his hand to produce a more musical version of his own, in making which he took, he said, 'more than the liberty of a translator'. In June 1545 the new litany appeared, along with other devotional exercises in a primer which was set forth 'by the King's Majesty and his Clergy'. Its use was made obligatory in all acts of public worship by injunctions issued in August, and no doubt its grace and dignity commended it no less than the royal authority.

It seems that at one time Cranmer nearly persuaded the king to a wholesale condemnation of ceremonies involving the use of crucifixes, of images, and of bells, and though he did not succeed in his attempt, it is clear that as Henry drew near to his death the influence of the protestants gained the ascendancy. In 1545 a heresy bill, apparently more severe than the Six Articles, was thrown out by the commons, after a stormy passage through the lords; in 1546 the French ambassador, along with others, believed that Henry was about to change the Mass into Communion and to urge the Most Christian king to do likewise. Yet when on 8 July 1546 the king issued what was to be the last of his proclamations[2] on religion, its effect was to condemn protestant books and the versions of the New Testament by Tyndale and Coverdale.

Even under the gentle ministration of Cranmer, Henry would not abandon his middle way. His attitude is expressed in the speech which, according to Hall,[3] he delivered to parliament when he prorogued it on Christmas Eve 1545. This was an eloquent appeal for Christian unity, based on the thirteenth chapter of the First Epistle to the Corinthians, lamenting that, the 'moste precious juel the worde of God is disputed, rymed, song and jangeled in every alehouse and taverne'. The king lamented also the manifest lack of Christian charity among the churchmen and laity alike, 'some be too styff in their old

---

[1] Dixon, *History of the Church of England*, ii. 360–1.
[2] *Letters and Papers*, xxi. i. 611. There is no mention in the Acts of the Privy Council.
[3] Hall, ii. 356–7.

Mumpsimus, other be to busy and curious in their newe Sumpsimus'.[1] The protestant martyrs who died at the stake in such exaltation that they did not seem to feel the flames, the valiant priests who suffered the cruel penalties of treason, might have wondered to hear this defence of toleration from the lips of their persecutor. Were these brutal executions the outward signs of an inward Christian charity?

For Henry it must be said that to some, probably to most of his victims, the idea of toleration would have appeared strange; many of them must have thought that the only truth permissible for all men was the truth which they themselves accepted. Henry himself, secure in the majesty of his supreme headship, took the same view. He could not coerce all men into his way of thinking by argument or by brutality, yet this king, devoid of feeling, avid of glory and a greedy extortioner, accomplished a great work, and that without the wholesale slaughter and destruction which accompanied religious changes in other lands. He was no protestant, yet he broke the authority of Rome; before he died the Bible, the Creed, the Lord's Prayer, the Ten Commandments had been authoritatively published in English.

Religion is not fashioned from without, and the power even of a king is limited; yet Henry VIII has no small claim to be regarded as the founder of the church of England,[2] though that church may trace its history far back beyond his day, and early discarded his Caesar-papalism. Probably he would have regarded himself as its preserver, and not without some justice, for between the church which he found and the church he left there was real continuity. Under Henry's direction, the Reformation in England was a very English thing.

'*Ista enim Respublica opus regis est*';[3] so said Lord Chancellor Audley in the oration which opened the parliament of 1542. The last decade of Henry VIII witnessed no startling innovations in the system of government, but what development there

---

[1] Camden, in his *Remains concerning Britain*, says that Richard Pace first gave currency to the story of mumpsimus; he may well be correct, but the story was in general use. See the note by Nichols in *Narratives of the Days of the Reformation* (Camden Society, 1859), p. 141.

[2] Burnet thought Henry rather than Luther merited the title of the 'postilion of reformation' as he 'made way for it through a great deal of mire and filth'.

[3] *Lords' Journals*, i. 165.

was attested a steady growth of the royal power. The king himself was the mainspring of the state machine, and under his hand he kept all the apparatus of an administrative despotism. Dependent on the Crown were administrative institutions which, between them, touched a great part of the life of England. The council in the Marches of Wales, the council in the North,[1] the court of Augmentations, the court of First Fruits and Tenths, the court of Wards and Liveries (1540), which exploited the king's feudal rights, the court of Surveyors (1542), which supervised the crown lands, the Stannary courts of Cornwall—all these contained, along with officials to do the routine work, some members of council, and with the council all were in constant touch. It was the council which dealt with the lord-deputy in Ireland and with the captain of Calais, and which exercised an unceasing vigilance over the justices of the peace—on every commission of the peace there was at least one councillor.

Over jurisdiction no less than over administration the council held great sway. The courts of common law, it is true, maintained their independence, but the prerogative courts were in very close touch with the conciliar machinery. The chancellor and the lord privy seal, who supervised the court of requests, were members of the council; and so was the archbishop of Canterbury who had great power in the church courts; the members of the court of delegates, the ecclesiastical court of appeal created by the act for submission of the clergy (1534), were royal nominees and some of them at least might well be councillors; the court of star chamber was the council itself sitting publicly to dispense justice in certain cases, mainly connected with the preservation of order, and in its own hands the council retained an undefined power for use in emergency.

While its authority thus stretched out over all England, the council gradually acquired a more definite organization at headquarters. The Ordinances of Eltham of 1526[2] had done something to distinguish between the effective and the nominal members of council; of the twenty members specially mentioned fourteen were officials: a distinction was made, too, between the council in London and that about the king's person, but the

---

[1] A council in the West is mentioned in a subsidy act of 1540 (32 Henry VIII, c. 50) but it did not last long: see the article by Caroline A. J. Skeel in *Transactions of the Royal Historical Society* (1921).

[2] *Letters and Papers*, IV. i. 864. On the text of the Ordinances see A. F. Pollard in the *English Historical Review*, xxxvii (1922), 358, n. 7.

members 'in attendance' on the king did not include the chancellor, the treasurer, and the privy seal who, partly because they
were heads of writ-issuing departments, were almost essential
to the council. For this reason, although the importance of
having the king's ear is obvious, it is impossible to identify this
council 'in attendance' with the later 'privy council' which
gradually took shape out of the vague nebula of councillors
which surrounded the throne. At one time this inner ring
represented the personal following of a great minister, but after
the fall of Cromwell there was no grand vizier. In what was
now called 'his highnes Pryvey Counsaill' a great place was
held by Gardiner, who, when he went off to Germany as
ambassador, was able to leave his follower Paget as clerk of
council; Norfolk, however, had great influence too; so too,
at times, had Cranmer; later Hertford and Lisle gained an
ascendancy. There was, though possibly not by design, a balance
of parties which helped to place the direction of the council, in
fact as well as in theory, in the hands of the king. As most of
the members—twenty in 1540—were officials, Henry had about
him the germ of a 'cabinet'. From 1540 the privy council had
its own clerk,[1] usually resident with the members in attendance
on the king, who kept in definite form a register of its decisions.
The process of definition was, however, slow. The 'privy council'
might still be in two places at the same time—in London or with
the king; it had no control whatever over the private dealings
of the king with individual ministers; it saw no need to minute
all its transactions and left much of its important business
unrecorded; it had until 1556 no seal of its own and for long
issued its orders under the Great Seal, the privy seal and the
signet, whose custodians were among its members. It was still,
in form at least, a consultative rather than an executive body.
None the less, it was gaining executive power, and more and
more the privy councillors in attendance at court became the
effective core. They began to authenticate documents with the
king's stamp, or with their own signatures, if there were enough
of them; and before long they were able to issue their instruc-

---

1 10 August 1540. *Letters and Papers*, xv. 487. Cf. prefaces to Nicholas, vii. and
Dasent, i; also *Bulletin of Institute of Historical Research*, v. 23. There is extant no
true record for the period 1435–1540. Even after 1540 the register omits all reference to much important business (e.g. the proceedings against Catherine Howard);
it never records debates, discussions are never minuted, and it is probable that there
were meetings with the king which were never recorded at all.

tions over the name of the secretary. The secretary, who was in origin the king's own servant, had gradually gained an official status,[1] though the title 'secretary of estate' was not used till 1601, and the elevation of an officer, who was necessarily at the king's hand and who controlled correspondence, provided the privy council with a member whose name was sufficient in itself to authenticate documents. More and more the council took to sending forth its orders without the authority of any seal. It issued proclamations, which were given, in 1539, the force of laws; it authorized the issue of 'privy seals' for the raising of money; it advised the king upon high matters and took upon itself almost all the work of government. Yet all it did, in fact as well as in theory, was done in the name of the king. Henry attended its meetings, not always but so often that his absences were noted as a reason for not sitting; he rated his councillors like schoolboys if they displeased him, and if the evidence of the register may be believed, he often issued proclamations without consulting the council at all. Of the 200 proclamations issued by Henry VIII only thirty-six purport to be made by 'the advice and consent of the council', and in the last seven years of the reign only three proclamations appear in the records of the privy council. The arrangements made for the enforcement of proclamations are significant. By the act of 1539 execution was entrusted to a dozen lords of the council, two selected from a quorum of important officers; an act of 1544 reduced the minimum to nine.[2] The proclamation upon religion of 8 July 1546 not only lessened the number to four but prescribed punishment of the body of the offender 'at his majesty's will and pleasure'.[3]

While council was thus a potent instrument in the hand of the king, parliament became his effective ally. Its constitution altered little except for the franchise of the Welsh boroughs by an act of 1543, when at the same time the county and borough of Chester attained representation. The upper chamber became the 'house of lords', and there are references to the 'nether chamber'.

[1] The importance of the secretary was recognized in the Ordinances of Eltham, 1526, where he was one of the councillors 'in attendance'; and in the act of precedence 1539, as well as in division of the office between Wriothesley and Sadler in 1540. [2] 34 & 35 Henry VIII, c. 23.

[3] There is no record that this proclamation was before the council at all. Wilkins, *Concilia*, iv. 1.

Although names were not yet certain, the constitution of parliament was, however, fixed, and the skilful Cromwell had already worked out a system whereby the king might use the assembly for his own ends. He 'packed' parliament judiciously before it met; he was ready with a programme, and he 'managed' the debates. The parliament of 1539 witnesses the success of his method; as he wrote to the king he and the other councillors would bring 'all thinges so to passe that your Majestie had never more tractable parlement'.[1]

It is, however, uncertain how far the process of 'packing' went. Even in the parliament of 1539 Gardiner had his own nominees who would normally be opposed to Cromwell, and a representative from Calais, which might be supposed to have been a government stronghold, boldly opposed the Crown; the bill for proclamations was debated long by the commons before it was accepted in an amended form; in 1545 a heresy bill was thrown out altogether; and the statute of wills (1540) which permitted landowners to devise land held in socage, and two-thirds of land held by knight service, though it was passed with the royal approbation, may, in fact, have been more acceptable to the gentry than to the Crown. Parliament was not a 'lion under the throne' which roared when the king pressed the hidden spring. It was, however, susceptible to 'management' and Henry understood 'management' very well. Much could be done through the lords; for there were no longer abbots, the bishops were the king's nominees, while the majority of the lay peers were of fairly recent creation and held church lands. Important legislation was often introduced in the lords, and as appears from the complaint of Darcy in 1536[2] the lords claimed some right to scrutinize all bills introduced in the house of commons which touched the royal prerogative.[3]

Even without the aid of the lords the Crown could sway the debates in the commons by the agency of its servants who sat therein. It was Sir William Kingston, comptroller of the household, who silenced the critic from Calais in 1539; in 1540 Wriothesley and Sadler, the secretaries, were exempted from the obligations of the precedence act because they might do the king better service in the nether house, 'where they now have places instead of in the upper house of parliament'.[4] In the

---

[1] Merriman, ii. 197, 199.   [2] Dodds, i. 360.
[3] *Letters and Papers*, XII. i. 410.   [4] Ibid. xv. 180.

account of Ferrers'[1] case it is bluntly stated that the commons mustered in their ranks 'not a few as well of the kyng's privy council as also of his privy chamber', and, indeed, Ferrers was himself 'a servant of the king'.

'Packing' and still more 'management' aided the king to get his own way in parliament, but the real secret of Henry's success was that he understood well the sentiments of the commons, who helped him not upon constraint but on the whole with goodwill. The goodwill, indeed, sometimes seemed like subservience; apart from voting the king great subsidies, parliament showed itself extremely ready to gratify the royal pleasure; it was parliament which in the end ratified the king's separation from Anne of Cleves in 1540, and it was parliament which in 1542 attainted Catherine Howard, allowing the king to give his assent by letters patent. Other acts of attainder attest the readiness of parliament to please its master. More remarkable still are the proclamations act of 1539,[2] and the act releasing the king from his debts in 1544,[3] for in each of these cases parliament seemed to surrender rights which were the essence of its being—the right to legislate and the right to control the Crown with the weapon of finance. In neither case was the surrender so abject as it appears. The proclamations act[4] gave to proclamations issued by the king, with the advice of his council, whose names are set forth, the force of acts of parliament. The preamble, however, suggests that this power could be used only when there was no statute, and no time to make a statute; care was taken to see that there was no prejudice to rights under the common law except as clearly set forth in the proclamation, and except in the case of heresy; and provision was made that trials under the act should be fair. With some justice it can be argued that the act implicitly recognized the paramountcy of statute law; that it ensured that proclamations should be within the common law, and that its general effect was to bring offenders within the range of conciliar jurisdiction. It has been said that the statute 'was concerned not at all with the legality of proclamations but merely with the manner of trying offenders against them'; and it is true enough

---

[1] *Tudor Constitutional Documents*, ed. J. R. Tanner, p. 580. For Ferrers see p. 440 *infra*: for the presence of the king's servants in the house cf. the complaint of the Pilgrims in 1536. Dodds, i. 359, p. 391 *supra*.

[2] 31 Henry VIII, c. 8.   [3] 35 Henry VIII, c. 12.

[4] E. W. Adair in the *English Historical Review*, xxxii. 40.

that proclamations were issued freely both before the passing
of the act and after its repeal. The act might be justified on the
ground that it gave to proclamations, which the king was going
to issue anyhow, some semblance of parliamentary authority,
and possibly its very existence gave the opportunity for later
ages to dispute the legality of proclamations not issued within
its terms. Yet, when all has been said, the act was dangerous,
and it created uneasiness at the time. Gardiner seems to have
thought[1] that it gave proclamations power contrary to statute;
it was passed with difficulty, and it was repealed as soon as
Henry died.

The act which relieved the king from his liability to repay
certain loans might also be justified on the grounds that it gave
parliamentary sanction for a situation which was already
irremediable—Henry, in any case, was not going to repay his
loans. It must be added that the parliament of 1544 did not lack
a precedent, and that it behaved more cautiously than its
predecessor of 1529. The act of 1529 had released Henry from
his debts without qualification; that of 1544 absolved him from
repaying loans as from 1 January 1542, and endeavoured to
act equitably by providing that such lenders as had already
been repaid should return the money to the king.

It was not only by individual acts of complacence such as
these that parliament commended itself to Henry; for him
parliament was a means whereby he could make his country an
accomplice, often a willing accomplice, in action dictated by his
own fierce spirit, or by his own selfishness,—and upon an institution which he found so useful he cast a favourable eye. During
his reign the commons were established more firmly than ever
in their house, and in their traditional privileges. An act of
1515 authorized the speaker to license the absence of members,
and ordered the clerk to keep a register of names and atten-
dances; the commons gained control of their own assembly, and
though their register was at first of small account it was, in all
probability, the parent of the *Commons' Journals* which began
in 1547.

In 1512 the case of Richard Strode,[2] a burgess for Plympton,

---

[1] Pickthorn, ii. 417, n. 1.

[2] Strode was delivered by a writ of privilege out of the exchequer, as he was one
of the collectors of a fifteenth for Devon; but a special act (4 Henry VIII, c. 8) was
passed in his favour, and his case was made a precedent.

who had been imprisoned by a stannary court for speeches made in the commons, gave an opportunity to vindicate the right of freedom of speech, and although in 1515 the speaker merely asked for personal access and personal immunity, the idea grew that in the commons speech must be free. Early in 1532 the king instructed his representative with the pope to explain that in the English parliament debate was absolutely free, and that the Crown had no power to limit its discussion, or control its decisions.[1] As Henry was at that time using the annates act as a lever against Rome, his enthusiasm for the rights of the commons may be to some extent discounted, but in 1542 Sir Thomas Moyle, as speaker, requested not only access for the whole house, but liberty for every member of the house freely to speak his mind.[2] In reply the king granted *honestam dicendi libertatem*—and freedom of access—though he said that the commons, when they wished to approach him on matters of great difficulty, should not all appear but should send delegates.

In the following year the commons made a dramatic assertion of their immunity from arrest, when they delivered George Ferrers,[3] a burgess for Plymouth, from the Counter in Bread Street, to which he had been consigned at the suit of one White, for a debt, although he was not the actual borrower but only a surety. The case has peculiar importance because the commons proceeded, not by passing a special act, or by getting a writ of privilege, but by sending their serjeant with no more authority than that of his own mace, and generally by acting with a very high hand. Having secured the support of the lords and the judges, they committed to the Tower and to Newgate the civil officers who had detained their member and resisted their serjeant, refusing to hear them, or the recorder of London on their behalf. Henry went out of his way to uphold their action. 'We be informed', said he, 'by our judges, that we at no time stand so highly in our estate royal as in the time of parliament.' The king, however, who saw that White's rights were secured, and before long knighted Rowland Hill, one of the sheriffs of London, was not actuated entirely by a love of liberty, or even by a desire to win the hearts of the commons. The underlying

[1] *Letters and Papers*, v. 415, also *State Papers*, vii. 361, for the complete Latin text.
[2] *Lords' Journals*, i. 167.
[3] The story rests on the narratives of Hall and Holinshed.

principle of the case was that the highest court in the land should not be impeded by any action of an inferior court, and Henry was very conscious that parliament was his own high court wherein he himself might preside; Ferrers, moreover, was his servant—a page of his chamber—and the serjeant's mace, which was broken in the fracas, carried the royal crown. The high speech on Ferrers's case goes on to say of parliament that therein 'we as head and you as members are conjoined', and Henry's attitude is made very clear in the oration delivered by Chancellor Audley at the opening of parliament in January 1542. The king, said the speaker, has summoned the principal parts of his realm, the prelates, and lay magnates, and the commons, called definitely the three estates—*tanquam universum corpus Reipublice Anglicane*—in order to know the affections and the qualities of all, with a view to removing superfluities in the existing law, and making new laws, if they should be needed, 'by the common counsel of all, and the authority of his Majesty'.[1] For Henry parliament was an adjunct, perhaps a necessary adjunct, to his own supreme power. Yet, when another king should arise who lacked supreme power, the commons might put a different interpretation upon the nature of their authority and of their privileges. From the instruction of Holofernes the pupils were to draw a moral which the master did not intend to convey.

The 'mere empire' which he had erected upon the foundation laid by his father, Henry kept till the day of his death. By the end of 1546 it was plain to all that this day was drawing very near. The king had made his usual progress in August and September, but after staying in his various palaces in and near London, he returned to Whitehall on 3 January 1547 a dying man. No longer could his weak legs support the bulk of his body and he must be borne in a wheeled chair or in some such machine. He lost, perhaps, some of his grip upon affairs; but his mind was still strong, he was still wilful, still arrogant, still merciless to his enemies—and still full of courage. To foretell the king's death was treason, and for long none dared inform him of his condition, till at last Sir Anthony Denny, chief gentleman of the chamber, took his courage in both hands.

---

[1] 'hos tres ordines, seu status: . . . communi omnium Consilio et sue Majestatis Auctoritate'. Ibid. 165.

Henry received the news quietly, and after sleeping awhile sent for Cranmer; before the archbishop could arrive he was already speechless, but when Cranmer spoke of the sure mercies of Christ, his dying master grasped his hand in token that he heard and believed.

So, confident and unafraid to the last, died a great king of England. His death was kept secret for two days in order that the council might prepare to execute the will, whereby the dead monarch thought to impose his commands upon the realm of his successors. On 30 January Wriothesley, hardly able to speak for tears, gave the news to parliament, who received it with lamentations. The secretary, Sir William Paget, read part of the will, and the chancellor declared parliament dissolved.

My body, said the king in his will, being mere cadaver when the soul is gone, might rest anywhere; yet for the grace and dignity which God has called us to, it shall be buried in the choir of our college of Windsor. And there, with great pomp, King Henry VIII was laid in a tomb beside his queen, Jane Seymour.

Henry VIII was brutal, crafty, selfish, and ungenerous; even in the splendour of his youth he had let Empson and Dudley perish, and as the years passed, what there was in him of magnanimity was eaten up by his all-devouring egoism. His triumphant ride through life carried him unheeding over the bodies of his broken servants, and though he had an outward affability for use at will, he was *faux bonhomme*. He was an utter realist; his principle was expediency; his god was himself in the state which he personified. Yet his people accepted him for what he was; there is no evidence that his cruel hangings and burnings produced either popular resentment or even overmuch resentment in the hearts of his unhappy victims—the king was the king, and the law was the law.

The respect, nay even the popularity, which he had from his people was not unmerited. He gave to them, or let them have, the things which were most desired by most of those who were politically conscious; and he kept the development of England in line with some of the most vigorous, though not the noblest forces of the day. His high courage—highest when things went ill—his commanding intellect, his appreciation of fact, and his instinct for rule carried his country through a perilous time of change, and his very arrogance saved his people from the wars which afflicted other lands. Dimly remembering the wars of the Roses, vaguely informed as to the slaughters and sufferings in Europe, the people of England knew that in Henry they had a great king.

XIII

ECONOMIC DEVELOPMENT

IN the field of economics, as in all other fields, the early sixteenth century witnessed the transition, or part of the transition, from the medieval to the modern world. This transition was neither sudden nor violent. Phrases like 'the agricultural revolution' or 'the reorganization of industry' are convenient, but they suggest changes cataclysmic or deliberate; and the suggestion is not altogether fortunate, for the changes which occurred were due to the speeding up of processes which had begun long before the Tudor had mounted the throne of England. Yet there is significance in the fact that changes occurred, simultaneously and with increased velocity, in every sphere of human activity, and the changes in the world of economics have been ascribed, not without justification, to the triumph of a certain spirit in man's approach to his agricultural, industrial, and mercantile affairs. That spirit would represent the victory of individual enterprise over communal effort, of competition over custom, of capitalism—and not unnaturally its growth has been connected with the reformation in religion. In the ecclesiastical world, too, may be witnessed the rejection of old custom, the assertion of the individual conscience, and the development of creeds which seemed to set the virtues of thrift and industry above those of resignation and charity. Few authorities would now ascribe the rise of capitalism to a triumph of Calvinism, but many would assert that Calvinism and capitalism are alike children of the same active, achieving, self-satisfying spirit.

The theory has obvious limitations: capitalism was present in the middle ages, at first under various disguises, later quite openly, long before Calvin was born. Again, Calvinism can hardly be said to have produced either the economic man or the free individual. How much free enterprise in the spiritual world was permitted in the city of Geneva? How much liberty did the soul enjoy in the trammels of determinism? New presbyter was but old priest writ large. What is true in a world of religion is true in a world of economics. As the forces which produced the great change were there before the 'period of transition' began, so did they outlast the convulsions of the agricultural and economic 'revolutions'. In the new world which emerged much of the old remained.

It is with these considerations in mind that the economic history of sixteenth-century England can be understood. The change which occurred was not a sudden thing, and it did not make all things new. Nevertheless there was a change.

The England of the middle ages was an agricultural country; the towns were small and all of them, including London, had their 'fields'. The great part of the population lived in little agricultural communities, called manors, each set in the midst of unfenced arable fields, usually three, but sometimes two, to which were attached some meadow-land and some rights in the surrounding pasture and woodland. The fields were divided into strips, often a furlong in length, which were tilled by various cultivators, and the essential feature of the manor was that the lord kept in his own hand a 'demesne', sometimes in the form of strips, which he farmed with the aid of his dependants. These dependants were of various grades; some had a score or more of acres—thirty acres had at one time been a common villein holding—others held only a few acres, others a mere cottage with a surrounding close; but all owed services of some kind. There were various forms of tenure. Some men were freeholders, some were copyholders, some had leases, and some were mere tenants at will; but each owed to the lord the service which was proper to his tenure, and each had rights, more or less defined, in the common fields after the crop was reaped, in the pasture, and in the surrounding waste.

In the manor the economy of life was an endless round of ploughing, sowing, reaping, threshing, and grinding, of shearing, spinning, and weaving, of milking, churning, and cheese-making, supplemented by the daily activities proper to stock-breeding and poultry-keeping, and by interludes of fruit-picking, brewing, killing, and salting, which came round with the revolving seasons. A country life is far less dull than the towns-man supposes, and for medieval folk it was relieved by fairs, feasts, and holidays, mostly connected with the church, vexed by occasional plagues or civil wars, and excited from time to time by the mustering of stout archers to serve in the king's pay. Apart from these exceptional interruptions it was varied by the incidents of ordinary economy. The manor was never quite self-sufficing; almost everywhere iron and salt and millstones had to be imported from without, and as life became less simple the contacts with the great world became more frequent, and more intimate. Along with the coloured cloth and the spices new ideas came in.

At no time did all manors conform precisely to the same pattern, and as the years passed the shape of each was inevitably altered by the chances of time and circumstance. The changes which occurred varied with the operation of chance, but the manor in most places retained its essential form and underwent the same sort of development. Money payments tended to take the place of the predial services owed by the lord's dependants; the lord used the money to hire labourers to work on his demesne; hired labourers became available partly because, with the natural increase of families, the peasant's holding became too small to support all its occupants. If any advance were made into the surrounding waste it was made by the lord himself or by some substantial landowner; it was the lord or the substantial landholder who bought up the land of peasant families which had got into difficulties; the general tendency was towards the creation of the big estate, not infrequently held by leasehold—the demesne itself was sometimes leased to a practical farmer by the lord.

By the year 1500, therefore, life in the manor had lost much of its old simplicity and much of its old stability; if in many cases the peasants had shaken themselves free from burdensome personal services, they had lost their old security of tenure. The leaseholder was at his lord's mercy when his lease expired, and must pay a fine which might be very heavy on renewal, besides being compelled, perhaps, to promise an increased rent. The position of the copyholder depended much on the nature of his 'copy'. If it were a true extract of the manorial roll, hallowed by long tradition, then he was fairly secure, though something depended upon the original terms; but if, as was often the case, it was some covenant representing a tenure lost in the mists of antiquity, then he was much less safe. His rent, it is true, might very well be fixed, but, on the other hand, he might be liable to pay a fine every time the estate passed from one holder to another. As for the cottar or the landless labourer, his position obviously depended upon his chance of employment at a reasonable wage, and upon the continued possession of rights in the common which enabled him to feed his animals and his poultry and to gather wood for his fire.

Upon a society which thus contained in its midst the seeds of its own destruction there had already begun to operate a powerful solvent from without. This was 'inflation'—a steep fall in the purchasing power of money. The needs of an expanding exchange had created a demand for new currency which had been met by the exploitation of the silver-mines in Germany and in the Tyrol. The European bullion was probably abundant enough to meet the necessity required, indeed it may have been too abundant, and with the influx of precious metals from America, which after 1530 became a flood, prices all over Europe rose to unheard-of levels. All over Europe the effects of inflation presented themselves. The great banking houses made money;[1] the princes, unable to carry on a far more expensive government with the old resources, debased their currencies; the merchants and the manufacturers increased their prices, wages tended to rise, and persons with fixed incomes found themselves in difficulties. In England the rise in prices was enhanced, for a time at least, by the disbursement of Henry VII's treasure, and it was accentuated still more towards the end of Henry VIII's reign by the great depreciation of the coinage.[2] The Tudor was slower than some other monarchs to begin the process of devaluation, but having begun he carried it farther than most.

Upon the little world of the manor the effects of the inflation were quickly visible. The landlord, who was a purchaser of manufactured goods from without, was the first to feel the rise in prices, and the least able to discount it by ordinary economic process, since much of his land was let at fixed rentals, or held by copyholders whose obligation was established by old tradition. His only remedy was to seize the opportunity provided by the expiry of a tenancy when a copyholder died, to increase the fines payable on renewal, and in the case of leases to increase the rent. Only by accommodating himself to the new situation could he hope to survive; if he were too poor or too conservative to alter his ways, he must in the end be expropriated, and his successor would be less likely than himself to

[1] For the rise of the Fuggers see Richard Ehrenberg, *Capital and Finance in the Age of the Renaissance*, translated by H. M. Lucas.
[2] See p. 412 *supra*.

regard the 'old custom of the manor' where it was not supported by legal sanction. The successor might be a competent farmer who had profited from long tenure of his land at a far too easy rate, or he might be a merchant anxious to invest his accumulated wealth in founding a family, or he might be a more opulent lord of the neighbourhood. Whoever he was, he was in all probability one who was determined to make his land pay.

With regard to the meaning and the use of land new conceptions were arising. No longer was it regarded solely as the stable basis of an ordered society; it was becoming a commodity to be exchanged and used for gain like any other commodity. During the middle ages, and especially during periods of civil war, land had changed hands often enough, but the attainder of a great baron did not as a rule affect the economy of his estate or the lives of the agricultural community. Quite different was the effect of this new traffic in land. A land-market was coming into being, and people who made purchases, which might even be speculative, did so in the belief that this transaction would be profitable to themselves. It is easy to exaggerate the speed with which the new system established itself; old custom dies hard; country folk are conservative, and in spite of all changes much of England still tried to live on in the old way. Yet the advance of the spirit of competition and the growth of a land-market created conditions which presaged the end of the medieval world.

From the economic development thus outlined arose the great question which agitated rural England in the sixteenth century—the question of inclosures. It was perhaps in the nature of things that the holder of a growing estate should wish to have his land in one compact piece instead of in scattered strips; it was in accordance with the spirit of the times that he should wish to farm it efficiently; it was almost inevitable that in his desire for efficiency he might disregard the welfare of his poorer neighbours and even ride rough-shod over their rights when these, though very real, rested rather upon custom than upon law.

The word 'inclosure' did not always mean the same thing, and its connotation was not necessarily evil. It might mean an attack upon unbroken land. If the lord, relying upon the statute of Merton (1235), 'approved' some of the 'waste' he might well confer a benefit upon the community, since he improved the productivity of the area, and gave more opportunities of employment to his dependants; if, however, he 'imparked' his new-won land, he did less for the public good; and if he took in so much land that he deprived his tenants of their opportunities of feeding their animals—including their plough-oxen—he not only robbed them of their customary prerequisites, but also ruined their husbandry.[1] Again, if the lord collected the scattered strips of his demesne, or if the tenants, by some system of exchange[2] among themselves, consolidated their own holdings, the inclosure would be rather of benefit than of detriment to the manor,[3] since the community was enriched by an improved agriculture. Even if, in the process of consolidation, the strips of penurious peasants were absorbed by a prosperous neighbour, little harm was done, except to the spirit of the expropriated landholder; for it may be assumed that the tenant who lost his holding was already in difficulties, and his chances of gaining employment as a hired labourer were possibly improved. The abandonment of the wasteful strip-culture might mean an economy of man-power in some directions, but, on the other hand, the more intensive tillage and the need for hedging and ditching would provide new opportunities of employment. When, however, the land inclosed was used, not for tillage, but for pasture, the whole economy of the manor was threatened. Depopulation was almost bound to follow, for a shepherd and his dog would suffice where a dozen men had tilled the soil.

Inclosure for pasture was well known in England long before the beginning of the Tudor age. Quite early in the fifteenth century there are instances of the destruction of houses to make sheep-runs, of the casting-down of hedges by an irate peasantry, and of a shrinkage of cornland even upon demesnes which were increasing in size. Evidently inclosure was a profitable thing; inclosure for tillage paid well, but inclosure for pasture paid far better. The export of corn was forbidden when prices were high, but there was no restriction upon the export of wool, and the weavers of Flanders and Italy were very willing to take all the wool which England could supply. When, towards the end of the fifteenth century, England began herself to manufacture

[1] Latimer's sermons of 18 January and 8 March 1549.
[2] Lipson, i. 137; Tawney, *The Agrarian Problem in the Sixteenth Century*, p. 165.
[3] Lipson, i. 140, quoting Tusser, *Five Hundreth Pointes of Good Husbandry united to as many of Good Huswifery*, and Fitzherbert, *Boke of Surveyinge*.

good cloth for export, the demand for wool became greater tha
ever. It has been computed that when land was inclosed fo
agriculture its annual value increased by 31 per cent., and tha
the annual value of land inclosed for pasture exceeded that c
inclosed arable land by 27 per cent. So, despite the plaints c
the commons the practice of inclosing for pasture went steadil
on, and before Henry VII mounted the throne of England it ha
already attracted the unfavourable notice of the Crown. Prob
ably the government felt, with an instinctive wisdom, tha
it was a serious thing to meddle with the foundation of rura
society; certainly it believed that if the country were given ove
to pasture it would lack its natural defenders. Not only woul
the peasants be too few in number to protect the soil, but the
would be too poor in manhood. Shepherds were not highl
regarded: 'We do reken that shepeherdes be but yll artchers.'
It was felt that husbandmen, 'bred not in a servile or indigen
fashion but in some free and plentiful manner', made the bes
soldiers. 'What comen folk in all this world may compare witl
the comyns of Ingland, in ryches, in fredom, lyberty, welfar
and all prosperytie? . . . What comen folke in all this worlde i
soo mightty, and soo strong in the fylde, as the comyns c
England?'[2] So boasted an author of 1515 in comparing th
English and the Irish. Fortescue had already contrasted th
qualities of the free-living English with those of the servil
French, and certainly some of the achievements of Englisl
troops, judged merely by the standard of physical performance
compare favourably with those of the soldiers of other lands.

Two acts of 1489[3] make the view of Henry VII's governmen
very plain. The first expressed the fear that the Isle of Wight
depopulated owing to the concentration of holdings in a fe
hands, would be left defenceless before an invader. The othe
dealt generally with 'the pulling down of towns'. To the wilfu
waste of houses and 'towns' the king attributed great 'incon
veniences'—the growth of idleness, that breeder of all mischief
the decay of husbandry, one of the greatest commodities of th
realm, the decline in religion, and the enfeeblement of th
national defence. It was therefore enacted that every owner c

---

[1] *Certayne causes gathered together wherein is shewed the decaye of England only by th
great multitude of shepe* (Early English Text Society, 1871), p. 100.

[2] *State Papers*, ii. 10. Summarized in *Letters and Papers*, ii. i. 371.

[3] 4 Henry VII, cc. 16 and 19.

a house or houses which had carried twenty acres of arable land within the last three years, should maintain and restore the buildings proper to the work of agriculture, under penalty of forfeiting to the king, or to his lord, one-half of the issues of the land until the fault was amended.

Sheep, however, were, in the words of Fitzherbert, 'the most profitable cattle that any man can have',[1] and against the prevailing economic tendency the legislation of the Crown was of little avail. Inclosure went on, and though the ecclesiastical landlords were perhaps more considerate than laymen,[2] they too, according to More in the *Utopia*, joined in the remunerative occupation of sheep-breeder. 'Sheep', he said, 'have become so great devourers and so wild that they eat up and swallow down the very men themselves. They consume, destroy and devour whole fields, houses and cities.'

The inefficacy of the act of 1489 appears from the fact that the fourth parliament of Henry VIII virtually re-enacted it, and in so doing seemed to legislate *ut de novo*. It is true that the act of 1515[3] specifically ordered the reconversion of pasture to tillage as well as the rebuilding of decayed houses, but the evils inherent in an increase of pasturage had been well understood in the earlier act, and the real difference between the two acts lies not in their terms, but in their execution. Thomas Wolsey, who had no lack of courage and no particular love for the gentry, set himself to do what parliament had failed to do.

On 28 May 1517[4] he issued seventeen commissions, covering in all thirty-five counties, by the terms of which the persons appointed were to inquire in each shire, or group of shires, what towns and hamlets, houses and buildings had been destroyed since Michaelmas 1488; what and how much land had been converted to pasture; what new parks had been made and what additions had been made to existing parks. Persons found by the inquest to have proceeded contrary to the act of 1515 were impleaded in chancery, and if found guilty were forced to give recognizance of their liability to pay one-half of the issues of the misused land. Most of them managed to make some compromise, but the cardinal persisted in his activities. The records of

---

[1] *Boke of Husbandry* (1534, ed. W. Skeat), n. 42.
[2] *Domesday of Inclosures*, i. 48 (ed. I. S. Leadam, 1897).
[3] 6 Henry VIII, c. 5. For drafts of a bill against engrossing and a proclamation against the turning of arable into pasture see *Letters and Papers*, I. ii. 1493, 1494.
[4] Ibid. II. ii. 1054.

the court of exchequer show that fines were exacted; two proclamations of November 1526 show the government still determined to bring offenders to justice, and proceedings were going on in 1527.[1]

To the student of economic history the proceedings of Wolsey's commission are important because of the evidence which they supply as to the nature and the extent of the inclosures. By some scholars this evidence has been held to prove that the grievances arising from inclosure were greatly exaggerated by contemporary writers, and that modern historians have been misled. In certain counties, Kent for example, and the south-western shires where Celtic influence was strong, inclosure had long been common; in the north inclosure was very rare. It was the midlands which were most affected by the developments of the late fifteenth and early sixteenth centuries, especially Berkshire, Buckinghamshire, Northamptonshire, Oxfordshire, and Warwickshire, which did much to supply the London corn-market. Yet from the extant figures[2] it appears that in these five counties the total areas inclosed averaged only about $1\frac{1}{2}$ per cent. of the total areas of the hundreds upon which the returns were made, and that, of the land inclosed only a proportion—rather less than three-quarters—was specifically shown as inclosure for pasture. The total displacement of population was not large, and as inclosure for tillage demanded labour, by no means all who lost their holdings lost their employment too; in Berkshire some 560 persons were evicted, and of these only 80 were displaced from their labour.

These statistics seem impressive, but they may be discounted upon several grounds. The returns from each county are not complete; they were obtained from juries who may have been reluctant or may have been intimidated; it is not safe to assume that all inclosures not specifically stated to be for pasture were necessarily for tillage; finally, in estimating the amount of disturbance, the proportion of the inclosed land to the area of the whole hundred is not relevant—what should be known is the proportion of the inclosed land to the arable acreage of the whole hundred. It is true that the eviction of a few families

[1] *Letters and Papers*, II. ii. 1546, for a decree of 12 July 1518 ordering all offenders who could not prove their inclosures to be of public benefit to pull down all inclosures made since 1485 under a penalty of £100. Cf. Steele, *Tudor and Stuart Proclamations*, nos. 106 and 107; Leadam, op. cit. i. 14.

[2] *Ibid.*, p. 72.

well known to some sympathetic writer might lead him to exaggerate the miseries of the times, especially as there was no machinery for dealing generously or even justly with displaced persons, but it is patent that in the midlands inclosure went on upon a scale which made it a real grievance.

To the reality of the grievance the action of the government testifies no less eloquently than the lamentations of the men of letters. Proclamations of 1526, 1528, and 1529 ordered the casting down of hedges and the opening up of inclosed lands; an act of 1534 forbade anyone to keep more than 2,000 sheep or to hold more than two farms;[1] an act of 1536[2] emphasized the government's determination. Yet the mere repetition is proof that the efforts of the government were in vain. Everything tended towards the creation of large leasehold farms, and it may well be supposed that the ultimate execution of statutes and proclamations lay in the hands of the local justices who were themselves inclosers.

The gravity of the situation and the concern of the government are attested by the acts against vagabondage. In all the acts, the supervision of persons moving about the country without following regular employment is entrusted to the justices of the peace, but the measures to be adopted vary in a definite progression. The first, passed in 1495,[3] ordered local officials to set in the stocks all vagabonds, idle and suspect persons, and thereafter to eject them from the town; it also ordained that all beggars, not able to work, should return to their own hundreds and beg there. The act of 1531[4] made a distinction between the impotent poor who could not work, and the sturdy and valiant beggars who would not. It ordered all impotent beggars to obtain a licence from a justice of the peace to beg in their own area, under penalty of the stocks if they went outside their limits; and it made liable to the stocks or to the whip, persons who begged without a licence. With regard to able-bodied beggars, it was of great brutality; all persons, men and women alike, found vagrant outside their own areas, were to be whipped at the cart-tail and sent off home by the shortest route with a certificate of their punishment, and with the promise of another flogging if they wandered from their direct route. Arrived home, the delinquent was 'like a true man'

[1] 25 Henry VIII, c. 13.   [2] 27 Henry VIII, c. 22.
[3] 11 Henry VII, c. 2.   [4] 22 Henry VIII, c. 12.

to set himself to work at some honest occupation. The act of 1536[1] developed the machinery of its predecessors as regards both able and impotent beggars. Children from five to fourteen might be put to masters under penalty of whipping if they refused to work or deserted after reaching the age of twelve; lusty beggars upon a second offence might suffer the loss of part of an ear, as well as a whipping; returned vagrants who did not set to work in their own parishes, might be put to forced labour. On the other hand, local officers, mayors, and church-wardens were authorized to gather 'charitable and voluntary alms' on Sundays and holy days, for the support of the impotent poor, and preachers were ordered to exhort their hearers to exercise charity in their wills; vagrants pursuing their way home to their own parishes were to be given a free meal by the constable for every ten miles of their journey.

It is true that some elements of common sense appear in the distinction between the impotent poor who could not work, and the valiant ne'er-do-well who would not. It is true also that gangs of sturdy beggars who might terrorize lonely little communities were a real danger to society. None the less, the cruelty of the act is as appalling as its inefficacy. It rested upon the medieval notion that every man had a permanent position in society, which was determined by the place and circumstances of his birth, and that for the generality of mankind unremitting toil was the appointed portion. Statutes of the realm denounced idleness and prohibited the common man from playing games, or at least from playing most games;[2] the moralist preached that man is not born to live in idleness and pleasure, but labour and travail, 'non otherwyse than a bird do fle';[3] and more eloquent than the generalities are the actual regulations made by parliament and amplified by the local authorities. By an act of 1495[4] from the middle of March to mid-September an artificer or labourer was to begin his work before five in the morning and end it between seven and eight in the evening, having half an hour for breakfast and an hour for his dinner, except between the middle of May and the middle of August when he was to have an extra half-hour for sleep at dinner-time. During the winter months he was to begin in the spring of the

---

[1] 27 Henry VIII, c. 25.   [2] See, for instance, 11 Henry VII, c. 2.
[3] The opinion attributed to Pole by Thomas Starkey in the *Dialogue* with Lupset, Early English Text Society (1871–8), p. 78.   [4] 11 Henry VII, c. 22.

day and go on till it was dark.[1] In accordance with these
provisions trade gilds made their local rules appointing, like
the cappers of Coventry, a twelve hours' day which might rise
to fourteen hours in the summer. Wages, no less than hours,
were subject to strict supervision; an act of 1515[2] fixed the
labourer's wage at 3d. a day for half the year and 4d. a day for
the other half, with opportunities for extra earnings at harvest-
time; skilled artisans got 6d. a day for half the year and 5d. a day
for the other half, and the efforts of combinations of workmen
to increase the rates of pay were vigorously suppressed. A man
who had seen the world and perhaps served overseas might not
wish to return to his dull parish: Thomas More, like other non-
combatants, was very hard upon the soldier home from the
wars. Even if he did return to his own parish, there was no
guarantee that he would find congenial work to do, or any work
of his choice; he might be put to forced labour. Why should a
man of spirit return to uncertainty or a life of hopeless drudgery?
Needless to say, the act failed of its purpose; an act of 1542[3]
complains that the statutes against vagabonds have not been
carried out because certain officers 'hath bene verie remyse and
negligent', and orders the justices of the peace to divide themselves up among the districts, two at least to each district, and
hold sessions every quarter, six weeks before the general quarter
sessions. But this attempt at increased efficiency was unavailing,
and in the day of Elizabeth, the sturdy beggar was as great a
menace as ever.[4] Justices, it appears, were not always willing to
prosecute companies of vagabonds who made themselves at
home in their barns, knowing that even if they could muster
enough force to prevail for the moment, their stacks and barns
might soon be destroyed by some mysterious fire.

Like the progress of inclosure, the growth of vagabondage
must not be exaggerated, but both were symptoms of the age,
and both were stimulated by the fall of the religious houses. The
charity of the monks has been overstated, and what there was,
was not always wisely directed; the leniency of the monastic
landlords has been overstated too. None the less, the religious

---

[1] Lipson, i. 396.

[2] 6 Henry VIII, c. 3; an act of the following session (7 Henry VIII, c. 5) made
allowance for the higher cost of living in London.

[3] 33 Henry VIII, c. 10. The earlier acts preceded the fall of the monasteries.

[4] Awdeley's *Fraternitye of Vacabondes* and Thomas Harman's *Caveat for Commen
Cursetors vulgarly called Vagabones*.

houses represented a conservative element in society, and their disappearance coincides with the advance of new ideas concerning the land, and the people who worked upon the land. The medieval economy was passing away, and the new economy was struggling in the pangs of birth.

There was great opportunity for some as the bands of caste were broken—the story of the Pastons makes this plain—but for others there was great unhappiness. It is impossible to dismiss the evidence of so many writers of different shades of opinion who all tell the same tale, though their language and their presentation vary. The catholic More regrets the all-devouring sheep, so does the protestant Latimer.

Fitzherbert and Tusser, writing as farmers and economists, recognize the value of inclosure and try to discount the consequent hardships. The protestant critics, like Henry Brinklow, are concerned to prove that the ills under which England laboured were due to the fact that the reformation had stopped half-way, but though they tend to confuse ecclesiastical with economic issues, and denounce all sorts of abuses of church and state, they are quite clear about the economic evils of the time.

Henry Brinklow, in the *Complaynte of Roderyck Mors* (1542),[1] applies to economic questions some of the criticism which he directs against the church. Clerical wealth should be used for the relief of the poor, for the establishment of a physician and a surgeon in every town, and for the provision of free schools wherein Greek and Latin should be taught. Even the old monks used to help the poor a little, but the new lords, among whom are some parsons, have inherited all the avarice of the old church. Some of them are shepherds, some of them inclose parks, forests, and chases; some of them add farm to farm. If a priest should be content with one benefice, surely a farmer should be content with one farm. In *A Supplication of the poore commons* of 1546, which may be by Brinklow, the same arguments are used. The old clergy, 'the valiant and sturdy beggars', against whom Fish had fulminated, have been succeeded by 'a sturdy sorte of extorsioners'. No hospitals have been founded; no relief is given to the poor; no man can get a farm, tenement or cottage, except upon payment of a huge fine, which probably he is unable to produce, since for many years his rent has been

[1] Early English Text Society (1874). Cf. the same Society's edition of *The Supplication of the Poore Commons* (1871).

excessive. The new lords break the old leases though they have two or three lives to run; they call in the old covenants and profess to find flaws in them. Simony no less than usury is rampant; the principle of 'no penny no *paternoster*' is generally accepted, and tithes work out at 13.75 per cent.

In the works of Thomas Starkey, who conducted Henry's dealings with Reginald Pole, there is mention of economic grievances, and in the *Dialogue*[1] which the author makes Pole conduct with Lupset, there is much talk of rural depopulation; it is the orthodox Pole who is made to defend inclosures, while his opponent denounces the idleness of all classes. Becon's *Jewel of Joy* refers to 'gentlemen shepe-mongers', and the 'setting of beasts above men'. In *Certayne causes . . . wherein is shewed the decaye of England only by the great multitude of shepe* (1550-3),[2] the hated inclosures are made responsible for most of the ills which vex the land. The more the sheep the dearer the wool on account of the monopoly; the dearer the mutton because it is in increased demand, since beef, corn, poultry, and eggs are vanishing with the old husbandry. People are being driven to want because, for every plough put down, 6 persons lose their employment, and food for $7\frac{1}{2}$ persons is ungrown. Suppose that in Oxfordshire only 40 ploughs are lost, 240 persons lose their living—and the number of ploughs lost may well be double. Upon another computation if 1 plough were lost in each of 50,000 townships of England, at least 300,000 persons would lose their employment, and though the author does not say so, 375,000 more would lose their bread.

The figures are not to be trusted. It is not clear why the townships of England are reckoned as 50,000, though this wildly excessive figure was used by other writers; there is no proof that every township lost a plough; and there may be reduplication of the reckoning of the mouths unfed. Yet however faulty be the author's arithmetic in this instance, the weight of the indignant testimony makes it clear that the transition from arable farming to pasture was causing real hardship. Even in the day of Henry VIII men dared to voice their grievances, and when the fierce old king was gone the grievances became more urgent and more vocal. Robert Crowley, scholar and printer,[3]

---

[1] Early English Text Society (1869), p. 97.
[2] Early English Text Society (1871).
[3] For Crowley's works see Early English Text Society (1871).

pleaded the cause of the poor commons against greedy land-lords and cowardly clergy with clarity and vigour; Latimer was more outspoken than ever. The commons endeavoured to take the law into their own hands.

Whilst the country-side was vexed by the emergence of a new economy, urban society had also lost its old security. Manufacture and trade had burst out of the narrow local limits of their origin and were tending towards a wider economy. In medieval practice, the town, besides being an agricultural centre, had been a place of exchange, with the privilege of holding markets and fairs, and a place which turned the raw material of the country-side into commodities locally consumed, as well as producing, in many cases, some speciality which would appeal to the outside market;—Bristol, for example, specialized in points, and Coventry in blue thread. The trade of the town was at first in the hands of the wealthier burgesses, who formed the gild merchant, and whose function was to protect their monopoly. By 1500, however, the gild merchant, though it might retain a separate existence formally, was, in fact, usually identical with the town—London was not the only place where the mayor sat in the gildhall. The manufacture of the town was controlled by the various craft-gilds, each of which exercised a monopoly in the production of its peculiar ware.

England had long possessed a good name for her manufactures, and the number and variety of her craft-gilds is astonishing. As gilds sometimes broke up into several different units and on the other hand sometimes joined with kindred 'misteries' to form a single body, it is not easy to state definitely the number of craft-gilds which existed at any given moment. It appears, however, that London had sixty-six distinct gilds, and York forty-seven. These cities were exceptional, but the records of Coventry attest both the variety of occupation and the closeness of organization which obtained in a medium-sized town; we read of bakers, barbers, brakemen (crushers of hemp, &c.), butchers, cappers, card-makers, card-wire-drawers, carpenters, chandlers, drapers, dyers, fullers, girdlemakers, haberdashers, leather-tanners, masons, painters, pewterers, saddlers, shearmen, smiths, tailors, tilers, and weavers. Over the whole business of production the gild exercised supervision. It fixed hours of work, wages, and prices; it punished faulty workmanship; it settled disputes between members and acted as a friendly society; by its maintenance of altar-lights and the performance of a set piece in the annual pageant it proclaimed itself to be, under the protection of its proper saint, a separate corporation within the machinery of the parent town. The craft-gild became 'an organic but strictly subordinate department of civic administration, supported and controlled by the municipal government, which always retained a reserve of power while delegating to them the supervision of trade and industry'.[1] The gild was self-governing, at least in theory. None save the man who had served a long apprenticeship could become a journeyman, and every journeyman was supposed to have the right to qualify as a master by producing a 'masterpiece' to the satisfaction of the officers of the gild, and so to become eligible for office himself.

Long before the Tudor period began, however, the old simplicities had disappeared and the government of the gilds had become oligarchical. The appearance of new methods made capital necessary. As industry became more complicated it began to suffer from 'over-organization', and the craft tended to split into subdivisions which were too small. The maintenance of the altars and the contribution to the annual pageant became burdens too great to be borne; in 1490 it was reported that Canterbury had abandoned its pageant altogether, and in 1494 the rulers of Coventry were admonishing craftsmen to do their duty to the gild in this and other matters. The enforcement of discipline was unpopular; in many places there was great reluctance to accept office, and the whole machinery of the craft-gild became slack.[2]

Most of these difficulties were connected with a development which corresponds in a general way with the movement which was taking place in agriculture; the benefits of the gild were being concentrated in the hands of the rich and the enterprising, and the ordinary craftsman had no longer a full share in its administration.

In spite of the efforts of the government, made during the fifteenth century, to prevent the drift to the towns, and in spite of the efforts of the towns to raise the age of apprentices and

---

[1] Lipson, i. 384.
[2] Yet the Coventry drapers paid for mending the battlement of their pageant as late as 1540. *Archaeologia*, xciii (1949), p. 57.

limit their numbers, there was, by 1500, an excess of labour
which was convenient for the incipient capitalism of the day.
The ideal that every apprentice might eventually become a
master had long proved itself unrealizable, and there had
appeared a class of journeyman destined to remain wage-
earners. In the fourteenth century there were already disputes
between masters and workmen over wages and hours, and there
had arisen 'yeoman gilds' which, while they remained within
the ambit of the craft, represented the wage-earners against the
employers. The masters strove hard to suppress these gilds, and
where they did not succeed in doing so they managed, as a rule,
to bring the journeymen under their control. There came into
being the institution known as the 'livery company' wherein the
yeomanry, while still remaining within the gild, became definitely inferior to those who were 'of the clothing'.[1] The distinction
in clothing represented a real distinction in status; by raising
the fee for entrance of apprentices and limiting their number
the masters succeeded in excluding all save the sons of substantial burgesses; by the exaction of heavy fines and fees from
men who wanted to set up as masters they maintained power in
the hands of a narrow clique; and for this clique they endeavoured to secure all the advantages of their monopoly. It is
significant that about the year 1500 the process of subdividing
the gilds had given place to a process of amalgamation.[2] In
1479, for example, the white tawyers in London were absorbed
by the leathersellers, in 1498 the pursers and glovers united to
form a single body which joined the leathersellers in 1502, and
in 1517 the leathersellers absorbed the pouchmakers. The resultant combination was strong, yet it did not rank as one of the
twelve great livery companies of London. The hall-mark of
these companies was that the livery men were a 'select body of
industrial capitalists inside the craft gilds, who controlled the
craft from within',[3] and arrogated to themselves all the trading
functions of the society.

A good example of the development may be seen in the story
of the tailors. Established ever since 1326 in a monopoly of their
craft in London, they received from Henry VII in 1502 a
charter which, in recognition of the fact that they sold throughout all England, gave them the title of 'Merchant Taylors'
and allowed them to recruit into their ranks anyone whom they

---

[1] Lipson, i. 428–9.   [2] Ibid. 424.   [3] Ibid. 431.

pleased. While the livery thus became merchants the yeomanry were gradually deprived of their right to trade, and in spite of some support from the government[1] became a body of small masters who, unlike the old master-craftsmen, had no direct connexion with the consumer.

Along with the rise of the livery must be noticed the appearance of purely mercantile societies which had no real connexion with handicrafts at all. Conspicuous amongst these were the grocers, the mercers, from whose ranks had sprung the vintners, the haberdashers, and the all-important drapers. These trading societies, which in the provinces took the form of an amalgamation of various small companies (whereas in London the companies remained distinct), claimed the sole right of exercising merchandise. In 1480 the merchant adventurers of Newcastle excluded craftsmen from retail trade, and though the latter resisted stoutly they were defeated in 1516 when a decree of star chamber forbade a craftsman to engage in trade unless he first abandoned his craft. Between the ordinary craftsman and the wealthy manufacturer or trader there was a great gulf fixed.

The cause of this gulf was the advent of capitalism, and that in turn came into being as the result of the operation of economic forces which are readily discerned, particularly the increasing complication of manufacture and the steady widening of markets. The important tin-mining industry of Cornwall and the lead-mining industry of Derbyshire, the Mendips, and Cumberland had long been conducted on a capitalist basis. The miners had their privileges, especially the tin-miners whose corporations, 'the stannaries', had their own laws and courts, but even in the thirteenth century they were already wage-earners. In the fourteenth century 'Abraham the Tinner' employed 300 men, and the bishop of Bath and Wells developed his rich lead mine with the aid of foremen and hired workers. Capitalism therefore was not new, but one of the features of the economic development of the fifteenth and sixteenth centuries was its application to the woollen industry.

During the fourteenth century this had expanded very rapidly, and it was deliberately fostered by the protectionist

---

[1] The livery did not prevail without a struggle; in 1507 the livery of the Founders Company endeavoured to deprive the yeomanry of the right to trade, but the yeomanry appealed successfully to the exchequer court and the star chamber.

policy of the government. Edward IV, in 1464,[1] had forbidden the import of foreign cloth, and three years later had prohibited the export of unfulled cloth.[2] In his reign the export trade in wool, which during the wars of the Roses had fallen to 35,000 cloths, rose to 62,000. The Tudors, in this as in other matters, followed the Yorkist precedent and encouraged the cloth industry by all means in their power. During the reign of Henry VIII the export figures showed a steady rise from about 85,000 cloths to over 120,000, and the average annual figure was upwards of 98,000.[3] True, this export consisted almost entirely of plain undyed cloth, partly because the Flemings, who were great buyers, insisted upon dyeing themselves, but even so the expansion of the woollen trade spelt prosperity for many areas.[4] Before long the trade became wholesale. In London, Blackwell Hall, really Bakewell Hall, became the scene of a weekly market for cloth brought from the country, and in Norwich, Bristol, York, Ipswich, Southampton, Beverley, Coventry, Northampton, and Winchester separate markets were reserved for the sale of cloth alone.

A business so widespread and so prosperous soon burst forth from the simple organization of the gild and the narrow limit of the town. The 'clothier' became a producer on a large scale who employed men engaged upon all the processes of manufacture. Factories were not unknown. The most famous was that established in the reign of Henry VIII by John Winchcombe whose achievements, even if they did not reach the magnificence of the ballads, were real. 'Jack of Newbury' was not alone in his enterprise, William Stumpe was a considerable figure too, who not only bought the abbey of Malmesbury from the king, but rented Osney abbey near Oxford in 1546 and installed, in each of the conventual buildings, a large number of looms. Tucker of Burford, who employed 500 workers, sought, in 1538,[5] to establish himself in the abbey of Abingdon, and all over the country from Kendal in the north to the Cotswolds, Somerset and Wiltshire and East Anglia, wealthy clothiers began to flourish. From the evidence it is plain that their looms were often 'in their houses',[6] but often too, perhaps

---

[1] 4 Edward IV, c. 1.
[2] 7 Edward IV, c. 3, confirmed by 3 Henry VII, c. 2, and 3 Henry VIII, c. 7.
[3] Schanz, *Englische Handelspolitik gegen Ende des Mittelalters*, ii. 18.
[4] Lipson, i. 459.   [5] *Letters and Papers*, XIII. i. 113, 154.
[6] Lipson, i. 478, 479.

usually, workmen still lived in their own homes, each doing his proper work upon the material as it was delivered to him with apparatus supplied by the capitalist.

With wholesale production came the use of simple machinery —'instruments of iron' instead of shears, engines for straining or stretching cloth, and water-mills for fulling. These innovations of private enterprise were unpopular. No more popular was the advent of alien workmen, some of whom were imported by private enterprise, even although the new-comers helped to develop industry; it was, for example, the introduction from France of hat-making in 1543, and of russel weaving[1] from the Netherlands, which revived the fortunes of Norwich, ruined by the decline in the worsted trade.[2] Most unpopular of all among the gildsmen and the municipalities was the steady exodus from the towns. The new industrialist found it more convenient to operate in the country where he was free from the regulations of the gild, especially if he could get water-power; many new towns and villages arose, especially in Suffolk, Somerset, Gloucestershire, and Wiltshire and in the West Riding of Yorkshire, while old-established towns like Ipswich, Bridgnorth, Coventry, and Beverley fell upon evil days. The cloth trade, in a word, shook itself clear from the restrictions of local economy, and became a wholesale business operated by men who understood competition better than custom. The new clothier found a natural ally in the grazier who supplied wool in bulk, and in their alliance may be seen the close connexion between the 'revolution' in industry and that in agriculture.

The rise of the new capitalist was viewed with some doubt by the government whose attitude to the developments in industry was much the same as its attitude to the developments in agriculture. Conscious of its duty above all things to maintain good order, it was not very partial to change, and much of its action seems to be an ineffectual attempt to hold back the hands of the clock. Yet it was essentially realist and accommodated itself to the movements of the time. Just as in its handling of agricultural problems it could not afford to quarrel with the

[1] A wool fabric, white as well as red, whose name may be derived from Ryssel, the Flemish name of Lille. The russel weavers of Norwich were incorporated by an act of 1554 (1 and 2 Philip and Mary, cap. 14).

[2] Worsted was made of long combed wool instead of the short, carded wool used for cloth.

country gentry, so, in its treatment of industry, it was unwilling to offend substantial taxpayers.

The basis of its action lay in the medieval notion that it was the king's duty to see that every man was maintained in his appointed place in society;—a surprisingly large proportion of Tudor legislation deals with industry and trade. Yet while it did not hesitate to deal with the minutiae of economics, it was in no haste to disturb existing arrangements which guaranteed the ordered life of the community. It made sporadic attempts to maintain the industrial supremacy of the towns. In 1523[1] it gave to Norwich the supervision of the worsted trade of East Anglia even as exercised by Yarmouth and King's Lynn, and the right to dye, colour, and calender all the clothes made in the surrounding area; in 1534[2] it passed an act forbidding the making of cloth except in the five towns of Worcestershire; and in 1543[3] it gave to York a monopoly of the manufacture of coverlets, authorizing the York gild of coverlet-makers to search throughout the shire. Yet it did not resist effectively the rise of the new capitalism, and its attitude to the gilds is instructive. On one hand it occasionally upheld the authority of the towns against recalcitrant craftsmen, but on the other it steadily brought local organizations under the purview of the central authority, often through the agency of the justice of the peace.[4] An act of 1437,[5] bidding the gilds submit their ordinances to the justices in the counties, and to the 'chief governors' in the towns, had, it may be supposed, lost much of its meaning because it was only in the towns that the gilds flourished, and there their interests were often identical with those of the municipality.

Over civic and economic corporations both, the Tudor government asserted its control. In 1495[6] the shearmen of Norwich were compelled by statute to obtain the approval of the mayor and aldermen for any ordinances they made. An act of 1504,[7] premising that the gilds consulted their own profit to the damage of the people, supplemented the powers of the city magistrates with the far more weighty authority of the state; all gild ordinances must henceforth be approved by the chancellor,

[1] 14 & 15 Henry VIII, c. 3.   [2] 25 Henry VIII, c. 18.
[3] 34 & 35 Henry VIII, c. 10.
[4] e.g. 1 Henry VII, c. 5; 22 Henry VIII, c. 5; 24 Henry VIII, c. 4; 24 Henry VIII, c. 6. There are many other examples.
[5] 15 Henry VI, c. 6.   [6] 11 Henry VII, c. 11.
[7] 19 Henry VII, c. 7.

treasurer, and the chief justices of both benches, or by the justices of assize in their circuits. There is evidence that, formally, the act was carried out; in 1511 the ordinances of the worsted weavers of Norwich and the neighbouring counties were approved by the chancellor; in 1536 the butchers of Oxford presented their ordinances, already approved by the mayor, to the justices of assize. Yet the act did not avail to limit seriously the authority of the crafts over their members. In 1520 and in 1526 the cappers of Coventry revised the hours of labour, and in the former year they revived an order made in 1496[1] prohibiting journeymen from making caps for any save their masters; the Founders' Company in London in 1507[2] endeavoured to prevent the yeomanry from selling their own wares 'at their own liberty'; the municipality of Coventry, in 1518,[3] forbade the daubers and rough masons to make a craft and forbade the journeymen of all occupations to make societies of their own. The gilds fixed the fees of apprentices, and when in 1531[4] the Crown, in this case supporting the action of the municipalities, at least in Oxford and Coventry, reduced these fees to 2s. 6d., the masters soon found a counter by making the apprentice swear that he would not set up as a master without their licence, or by demanding a heavy fee from anyone who proposed to set up for himself.

All these circumventions of the spirit of the statute went unchecked, and though, as with the Founders' Company, the law-courts on some occasions opposed the oligarchy of the gild, on others they supported it, as in the case of 1516 from Newcastle already cited.[5] The act of Edward VI, passed in 1547,[6] confiscating the property of all religious gilds, has been taken as a proof that the government was opposed to gilds altogether. That act, however, was only the corollary to the act of 1545[7] dissolving chantries, which had not yet been carried out, and its effect was merely to rob the gilds of that part of their revenue devoted to religious or 'superstitious' purposes. The Merchant Taylors, for example, lost about a quarter of their total income of £440; but their corporation remained, and to this day remains. The decline of the gilds was due, as has been shown, more to economic causes than to the action of the government.

---

[1] Lipson, i. 398. [2] Ibid. 429. [3] Ibid. 394.
[4] 22 Henry VIII, c. 4. [5] p. 461 *supra*.
[6] 1 Edward VI, c. 14. [7] 37 Henry VIII, c. 4.

The truth is, that the government, in its determination to 'manage' the national economy, tended to use the old institutions and proceeded on the assumption that the gilds were as important as ever. Many of its acts exemplify the principle, emphasized by the gilds themselves, and inherent in their being, that one man should follow one trade. Some of these acts, those for example preventing worsted-makers and dyers from doing the work of calenders,[1] seem to be dictated, in the main, by a desire to maintain what is called nowadays 'full employment'; but others, as for instance the acts forbidding butchers to be tanners,[2] brewers to be coopers,[3] cordwainers to be tanners,[4] and vintners to assess the price of wine,[5] had an obvious basis in common sense and consulted the interest of the consumer.

The interest of the consumer, it is fair to say, was carefully regarded in much of the Tudor legislation; the government seems to have realized that with the development of wholesale trade slipshod methods might creep in, and took upon itself the supervision heretofore exercised by the gilds. Abuses, it may be supposed, were corrected according as they drew attention to themselves, and the various acts appear confusedly in the records interspersed with legislation on all kinds of other matters. Some of them were concerned with food and drink. Meat was to be sold by weight, the seller producing scales, at fixed prices— beef and pork at a halfpenny the pound, mutton and veal a halfpenny and half a farthing; heads, necks, &c., were to be cheaper, and the lord chancellor, justices of assize, justices of the peace, and mayors could raise and lower prices.[6] The wholesale prices of wines and the contents of vessels were fixed by one act,[7] and another,[8] in imposing penalties for non-compliance, empowered justices of the peace to enter the houses of wine merchants and sell to customers at the proper price. The bodily health of the subject was guarded by the law of 1512[9] which provided that none was to practise as physician or surgeon in London until he had been approved by the bishop; the status of the professional physician was made secure by an act of 1523[10] which made the physicians of London into a

---

[1] 25 Henry VIII, c. 5.   [2] 22 Henry VIII, c. 6.   [3] 23 Henry VIII, c. 4.
[4] 19 Henry VII, c. 19: cf. 2 Henry VI, c. 7, and 1 Henry VII, c. 5.
[5] 3 Henry VIII, c. 8.   [6] 24 Henry VIII, c. 3.   [7] 23 Henry VIII, c. 7.
[8] 24 Henry VIII, c. 6.   [9] 3 Henry VIII, c. 11.
[10] 14 & 15 Henry VIII, c. 5.

corporate body, and empowered this body with the right to approve all outside the city who wished to practise medicine, except graduates of Oxford and Cambridge. The fees of the London watermen were fixed by parliament,[1] and the prices to be charged for the repair of bridges were definitely established.[2]

Against over-charges of vendors, due precaution was taken. An act of 1512,[3] which prohibited the import of foreign caps, contained a full statement of the qualities of English caps available, and the prices which should be paid for them. A cap of the finest Cotswold was to cost 2s., and to be marked 'C'; Leominster caps, to be marked 'L', were to be sold at prices ranging from 3s. 4d. to 1s. Care was taken, too, to protect the customer from concealed dishonesties of workmen and producers. A bridle was placed upon the practices of itinerant tinkers, and the whole business of making pewter and brass was put under the control of the gilds, who were to appoint searchers.[4] The mayor of London and the master of the tallow-chandlers were authorized to search out and punish vendors of impure oils,[5] and in 1533[6] the leather trade was brought under the supervision of the curriers' company, with the power of the justice of the peace looming in the background. A whole series of acts is directed against the trickeries practised in the making of cloth. Conspicuous is an act of 1512[7] against the deceitful working of woollen cloth, which covered the whole process of manufacture, and tried to guard against short weight, short measure, undue shrinkage, and undue stretching. The fact that in 1515 another act[8] of the same general purport appears on the roll may be taken as proof that dishonesties persisted, and they were still going on in 1532 when penalties were imposed upon persons who wound wool unwashed or mixed it with sand and stones to increase the weight of the fleece.[9] In the following year provision was made, with a wealth of detail, against the untrue dyeing of woollen cloth.[10] Two acts[11] dealt with the proper making and content of the 'white straits' of Devon; another regulated the activities of Suffolk clothiers;[12] and an act of 1514,[13] to avoid deceits in worsteds, forbade English makers to adopt the method

---

[1] 6 Henry VIII, c. 7. [2] 22 Henry VIII, c. 5. [3] 3 Henry VIII, c. 15.
[4] 4 Henry VIII, c. 7. [5] 3 Henry VIII, c. 14. [6] 24 Henry VIII, c. 1.
[7] 3 Henry VIII, c. 6. [8] 6 Henry VIII, c. 9. [9] 23 Henry VIII, c. 17.
[10] 24 Henry VIII, c. 2. [11] 5 Henry VIII, c. 2, and 6 Henry VIII, c. 8.
[12] 14 & 15 Henry VIII, c. 11. [13] 5 Henry VIII, c. 14.

of dry-calendering used by 'strangers beyond the sea' with deceitful intent.

Upon the operations of 'strangers' the government kept a vigilant eye, though less forbidding perhaps than the native merchants would have wished. It was the prince's duty, in the renaissance state, to make his people self-sufficing, and the Tudors were true monarchs of their age; political convenience and the realistic appreciation of competence led them to make some concessions to aliens in the public interest, but in the main they were resolute to protect native industries. An act of 1439 which compelled aliens to live in the houses of recognized 'hosts', and to sell their wares within eight months, was still in force, but though some of the towns[1]—Yarmouth, for instance, in 1491, and Newcastle in 1548—tried to assert the control of the 'hostmen', the government seems to have been content to let the matter slide. Nothing seems to have been done in response to a pathetic petition from the crafts of London professing that Englishmen were ruined because 'strangers' used their 'mysterys' and especially exercised retail trade. The immediate effects of the Evil May Day riots of 1517 was to harden the heart of the government. Against aliens, however, who came not to trade but to work, action at last was taken; an act of 1523[2] which prohibited alien craftsmen from taking alien apprentices, limited the number of journeymen whom they might keep, placed them under the supervision of the native gilds and compelled them to put special marks upon their wares. This act was ill observed, and in February 1529 star chamber, on the complaint of the London merchants, made a decree which, in the following session, was embodied in a statute.[3] Aliens were now permitted to keep only two alien servants; those exercising handicrafts were to pay all charges that subjects had to pay, and were to assemble only in the halls of their several companies; all were to swear allegiance to the king; denizens alone were to set up shops, and all the provisions of the statute of 1523 were confirmed and made perpetual.

In the industrial economy of the day the woollen industry was pre-eminent, and many statutes witness an endeavour to keep

---

[1] Lipson, i. 529.   [2] 14 & 15 Henry VIII, c. 2.
[3] 21 Henry VIII, c. 16; cf. Schanz, ii. 598, for the complaint of the London cordwainers.

it in English hands. Very early in his reign Henry VII re-enacted Edward IV's law[1] that unfulled cloth was not to be exported; this act was confirmed by himself in 1487,[2] and was renewed by his son in 1512.[3] In 1523[4] Englishmen were forbidden to sell broad white woollen cloths to aliens, who would take them overseas to be dyed, unless such cloth had been exposed for sale for eight days in Blackwell Hall, and had failed to find a purchaser. The attempt to keep the dyeing in England was not successful; but at least it was made. In the linen trade, too, advantages were given to the native producers; by an act of 1529[5] the importation of dowlas and locheram[6] was restricted in the interest of the linen drapers of London, and in 1533 it was enacted[7] that every person occupying land for tillage should sow yearly a quarter of an acre of flax or hemp for every sixty acres he had under tillage. The purpose of this act was frankly stated; it was to check the unemployment resulting from the importation of linen cloth.

On the ground that alien merchants evaded the laws which limited tanners, curriers, and cordwainers each to their own business, an act of 1512 forbade all aliens to buy leather except in the open market, and put them under the control of the gild of curriers;[8] when, however, the curriers attempted to create a monopoly in their own favour and insisted that 'strangers' should buy curried leather only, and from them alone, a relaxation was made in favour of the subjects of the emperor and of the prince of Castile.[9] Even so, 'strangers' were still compelled to buy leather only in open market or in fairs. An incidental hit at foreigners was made in an act of 1515[10] which provided that every merchant carrying goods from Venice was to bring in ten bowstaves of good quality for every tun of wine that he imported.

It must be noticed, however, that in all its efforts to promote the welfare of the subject the government was severely realist. An act of 1531[11] asserted that brewers, bakers, surgeons, and

---

[1] 7 Edward IV, c. 3. [2] 3 Henry VII, c. 11. [3] 3 Henry VIII, c. 7.
[4] 14 & 15 Henry VIII, c. 1. [5] 21 Henry VIII, c. 14.
[6] Coarse linens named after Daoulas and Locrenan in Brittany.
[7] 24 Henry VIII, c. 4. [8] 3 Henry VIII, c. 10.
[9] 5 Henry VIII, c. 7. Charles, the grandson of Ferdinand and Isabella, was, under various titles, lord of the Netherlands from the death of his father, the archduke Philip, so that from 1506 until Charles's accession to the throne of Spain in 1516, the Netherlands were subject to the Prince of Castile.
[10] 6 Henry VIII, c. 11. [11] 22 Henry VIII, c. 13.

scriveners of alien birth did not fall under the restrictions imposed upon handicraftsmen; evidently in these occupations the aliens were of use to the community. The welfare of the state was paramount. The subject, though benefited by the laws, found himself liable to restrictions imposed in the national interest; he must not sell horses abroad, for example, especially to the Scots;[1] he must avoid the easy way of the crossbow and the hand-gun;[2] he must not wear apparel too fine for his estate.[3]

England, for all her energy, could not be self-supporting, and for centuries she had met her needs by an active foreign trade. She imported wines from Gascony and the Levant; silk, cotton, and spices from the Mediterranean; oil, leather, and iron from Spain; fur, iron, grain, and, above all, the materials for shipbuilding from the Baltic; from the Netherlands not only the linen and herrings of Holland, but all kinds of manufactured articles whose number and variety grew as life became more comfortable and more ornate. In return she exported wool, hides, corn, salt, coal, lead, tin, meat, and fish; as well as cloth, pewter, worked alabaster, and various artifacts mainly of metal and leather.

Even after England began to produce in bulk it was some time before the merchant as distinct from the manufacturer appeared, and for some time trade was largely in the hands of foreign merchants; but by the end of the fourteenth century the wool-trade, so closely allied to production, was in English hands to the extent of 75 per cent.,[4] and in the fifteenth century the 'pure merchant', 'adventuring beyond the seas', who had no direct contact with manufacture, became important. The foreign merchants were unpopular, partly because, even as late as 1450, the idea persisted that they brought frivolous trifles and luxuries into the country, in exchange for English goods of solid value. The foreigners, however, were not easily ousted, for they were powerful and well organized. The Flemings kept a house in London, but the main rivals to the English merchants were the Italians and the German merchants of the Hanse.

The Italians had behind them powerful finance, a virtual

---

[1] 23 Henry VIII, c. 16.
[2] 3 Henry VII, c. 3; 6 Henry VIII, c. 13; 14 & 15 Henry VIII, c. 7.
[3] 1 Henry VIII, c. 14.   [4] Lipson, i. 569.

monopoly in Levantine wines and currants, and a capacity to absorb English goods; besides these economic advantages the most powerful of them, the Venetians, had the benefit of state support. From 1317 on there sailed every year a great fleet, known as the 'Flanders galleys', part of which came to English ports, especially Sandwich, Southampton, and London. Yet even more dangerous competitors were the Easterlings of the Hanse, partly because they brought commodities which England could not well lack, partly because, as 'the men of the emperor', they might usually be regarded as allies by an England always at variance with France; they were established in London as early as the twelfth century, and in 1320 occupied the Steelyard near the Thames which remained their head-quarters for centuries. Despite their reception in England they gave no reciprocity but managed, in a long contest, to exclude English merchants from the Baltic and the Scandinavian countries; at the same time they acquired from the Crown privileges of great value, which were secured to them by the so-called 'Treaty of Utrecht' of 1474. This was really a payment made by Edward IV for the support of the league, and its effect was to cancel some gains made by the irate English merchants. The men of the Hanse, it is true, virtually accepted their exclusion from Boston, but, on the other hand, they established themselves more firmly than ever in their favoured position in London. They paid, for example, only six-sevenths of the customs paid by native Englishmen on exported cloth, and less than a third of the duty paid by other aliens;[1] they escaped too, from paying 'poundage' on other exports and imports, and though Henry VII, at one time, cancelled their privileges on the ground that they did not observe the bargain of Utrecht, he restored their position in 1504.[2] Meanwhile, however, the English merchants had been improving their position.

It has been calculated that by the close of Edward IV's reign native merchants controlled 88 per cent. of the wool-trade, and 59 per cent. of the export in cloth, about 65 per cent. of the merchandise paying tunnage and poundage, and more than 75 per cent. of the wine-import.[3] A table of shipments from

[1] Gras, *The Early English Customs System*, corrected by Lipson, i. 537 (note).
[2] 19 Henry VII, c. 23.
[3] Lipson, i. 569; Eileen Power and M. M. Postan, *English Trade in the Fifteenth Century*, p. 406; Gras, pp. 111–18.

various ports shows an immense preponderance of native enterprises, more marked in the out-ports than in London, though even in London the 'strangers' were losing ground. In spite of the gloomy picture given in the *Libelle of Englyshe Polycie* in the mid-fifteenth century, English shipping was good and strong. The Cannings of Bristol may be paralleled by the shipowners of other ports;[1] in the fiscal year 1503–4 John Tanne of Lynn made twenty ventures, and all were conducted in English ships. The advance of the English merchants was due, partly to the support of the Crown, and partly to their own improved organization.

Because, for political reasons, they wished to control trade, and because they wanted to keep watch upon their customs revenue, the English monarchs of the fourteenth century had decreed that all staple goods should be exported through definite channels. After 1370 there was only one staple town, that at Calais, and there appeared a company called Merchant Staplers which was, in effect, a public corporation containing at first all 'the merchants of the realm' engaged in foreign trade.[2] This company before long tightened its organization, limited its attentions to wool, made itself responsible for the custom and subsidy in wool and for the payment of the Calais garrison. Secure in its position, it was apt to behave in a high-handed manner which from time to time produced ill feeling in the Low Countries, but it commended itself by its uses, and its fall was due not to political but to economic reasons. England began to manufacture her wool at home, and power passed to a company whose fortunes were founded not upon wool but upon cloth.

In origin, the Merchant Adventurers were nothing more than 'all the merchants of the realm', and they never became more than a loosely bound 'regulated'[3] company whose members had

[1] The Cannings were a family of merchant princes at Bristol where there was a strong 'fellowship' of merchants; during the fifteenth century they had risen to eminence. In 1461 William Cannings the younger had ten ships afloat and is said to have had 800 men employed in them as well as 100 workmen in his yard. See *Town Life in the Fifteenth Century* by Mrs. A. S. Green, i. 84, 89, 107, and E. M. Carus-Wilson, 'The Overseas Trade of Bristol', in *Studies in English Trade in the Fifteenth Century* (ed. Power and Postan). Cf. Gras, op. cit., s.v. Tanne.

[2] Lipson, i. 565.

[3] Apparently the Merchant Adventurers and the Merchant Staplers were in origin one and the same (ibid. 571). A regulated company was one whose membership was open to all who would pay the fees. It had a monopoly of trade within its proper area; but the trade was not conducted by the company as a whole, though members might join together for one or more 'enterprises'.

depots at various ports in England, and spheres of influence in Germany and Scandinavia as well as the Low Countries (Holland, Zeeland, Brabant, and Flanders); and as the Englishmen were driven from Germany and Scandinavia by the competition of the Hanse, it was the trade in the Netherlands which was their main activity. The Netherlands, however, could absorb almost the whole of the English product and, especially after Antwerp had fallen heir to Venice, could supply most of England's wants. The economies of the two countries were largely complementary, and although there was rivalry, especially after England began to make and dye her own cloth, the occasional quarrels served to convince both sides that mutual understanding was profitable. It was no great disadvantage that the Adventurers were compelled to concentrate upon this narrow but fruitful field.

At the same time their power in England was concentrated too. The trading companies of London, in the forming of which the Mercers had a great part, steadily took control, and the merchants in other ports, though they kept their local organizations, were regarded as being affiliated to the company of Merchant Adventurers. Bristol, where the Merchant Adventurers were organized in 1467, kept her independence and obtained a charter from Edward VI in 1552 which excluded all artificers from foreign trade;[1] the merchants of Newcastle, in 1519, agreed to pay £8 a year in lieu of sums due by individual members. Theoretically, any member who paid the fees and obeyed the authority of the company could enjoy its privileges, but, in fact, the London company began to assume authority. In 1505 Henry VII granted the Merchant Adventurers a charter whereby they were to appoint a governor with twenty-four assistants, and though this governor sat overseas, usually at Antwerp, it is plain that the influence of London still predominated.[2] The first of the Tudors was too cautious to put all his eggs into one basket. When, in 1497, the Merchant Adventurers of London endeavoured to exclude the merchants of the outports from foreign trade by the imposition of a fine of £20, the

---

[1] Lipson, i. 573.

[2] W. E. Linglebach, 'Merchant Adventurers of England', in *Transactions of the Royal Historical Society*, N.S. xvi. 35. Confirmed by W. P. M. Kennedy in the *English Historical Review*, xxxvii. 105. The reference of a dispute to London, *Acts of the Privy Council*, N.S. i. 51–2, shows that though formal decisions were taken in Antwerp London had great power.

king by statute[1] reduced the admission fee by two-thirds, and when, in 1504, the Adventurers tried to absorb the Staplers by force, a star-chamber decree ordered each company to respect the rights and the authority of the other.[2] The day of the Staple, however, was passed; the Adventurers disregarded the pronouncement of the star chamber, and by 1527 the Staplers were complaining to Wolsey that they now numbered only 140 exporters of wool instead of 400;[3] it was the Merchant Adventurers who were to fight the battle against the Hansa, and establish the commercial prosperity of England.

On the whole circumstances were favourable. Henry VIII was less interested than had been his father in the development of customs, and his visions of glory were expensive. Yet he was fully alive to the importance of English shipping, and statutes of 1532 and 1540[4] renewed the 'navigation act' of his father in an emphatic manner. He did not make treaties purely for the benefit of English trade, but he did not neglect his merchants, and was by no means blind to the repercussions of diplomacy upon commerce. One instance will suffice. The heavy war-taxation of 1525 drove the clothiers of Suffolk to dismiss their workmen, and when, in 1528, the Netherlands trade was interrupted by war, the government made strenuous efforts to avert the danger of unemployment. Wolsey bluntly ordered the merchants of London to keep on buying even though the best market was gone, so that the clothiers would still produce and the artisans still make a living.[5] The incident may have convinced Henry, if indeed he needed convincing, that dislocation of trade was a bad thing, and that a good understanding with the Netherlands was essential. Happily for England his policy, despite his separation from Catherine, and his consequent *entente* with France, rested, in the long run, upon an alliance with the Habsburgs.

To the English merchants the Netherlands became more important than ever when Antwerp became the financial and commercial capital of north-west Europe, and their other competitors fell out of the race. The great maritime republics of Italy were sorely stricken by the advance of the Turks, by the opening of a sea route to India, by the discovery of America,

---

[1] 12 Henry VIII, c. 6.   [2] Lipson, i. 578.   [3] Lipson, i. 578.
[4] 23 Henry VIII, c. 7, and 32 Henry VIII, c. 14.
[5] *Letters and Papers*, iv. ii. 1868, 1881; Lipson, iii. 303.

and by the endless 'Italian Wars'. After 1533 the 'Flanders Galleys' are heard of no more; Italian merchants remained in England, and Venetian ambassadors continued to write their knowledgeable reports upon trade, but no longer could Venice sponsor state enterprises to the northern seas. The merchants of the Hansa were shaken by the upheavals in Germany and Scandinavia which followed upon the reformation, of which the stormy history of Lübeck provides so great an example, and though they were still active, their fortunes were upon the down-grade. In 1552 Edward VI revoked their privileges,[1] and the grounds on which he did so are illuminating. The Hansards, he alleged, gave their protection to persons and to goods not entitled to enjoy it; they denied to English merchants the reciprocity promised at Utrecht, and their commerce was now prejudicial to the welfare of the state. These grievances were not new; what is important is that the government now felt able to defy the powerful Hanse. English commerce could stand upon its own feet, and although the Steelyard recovered the favour of the Crown in the days of Mary, it lost its position finally in the reign of Elizabeth.

The victory of the English merchants over their rivals bespeaks a prosperity which shines in the confident pages of the chronicles and reveals itself in the increasing yield of the subsidies. Yet the evidence of the customs accounts seems to indicate that there was a decline rather than an increase in English trade during the reign of Henry VIII, and the matter requires explanation. In the last year of Henry VII the customs brought in more than £42,000; in the last year of Henry VIII they yielded less than £35,000, and that despite the great devaluation of the currency.[2] The decline might be attributed to the king's ambitious diplomacy, to the cardinal's disregard for finance, to the dislocation produced by the reformation, and to the unfortunate experiments with the coinage. Certainly all these factors operated in a manner prejudicial to the economic welfare of England, and yet the decline in English prosperity, as shown by the customs accounts, was far less than a casual inspection suggests. To understand the situation one must examine briefly the nature of the customs.

During the fourteenth century the customs of England had been consolidated into a system which worked well enough in

[1] *Acts of the Privy Council*, N.S. iii. 487.　　[2] Schanz, ii. 46 and 59.

practice though its nomenclature is confusing to modern students because the terms used did not always bear the same meaning, and because the accounts of an impost might show receipts as derived from different sources although they were in fact paid in a single sum. The phrase 'custom and subsidy' is a case in point; the old custom on wool was 6s. 8d. a sack, but it was in fact never paid without a supplementary subsidy which raised the payment to £2 for native merchants and twice as much for aliens.[1] Beside the custom and subsidy on wool, which, as has been shown, was long dealt with by the Staple, the government relied on the custom on cloth which rested on a grant of 1347, and the subsidy of tunnage and poundage which had been granted on all exports and imports since the days of Edward III, at a rate which in Tudor times was usually 3s. on the tun of wine, and 1s. in the pound value of all other goods;—aliens always paying more except in the case of the Hansa whose privileges have already been discussed. The wool duty was administered at Calais; all others were collected either at London or at one of the 'out-ports' of which there were only sixteen. In each of the ports (save that Exeter and Dartmouth, Plymouth and Fowey were administered together) there was a simple machinery operated by a collector, a controller, a searcher, and a surveyor, with a small corps of minor officials—clerks, weighers, crane-keepers, and others; this machinery dealt with the 'members' (smaller ports) dependent upon each main port so that the whole coast-line was covered. No doubt there was smuggling and evasion, but, on the whole, the system seems to have worked fairly well.

A scrutiny of its operations during the reign of Henry VIII reveals some salient facts. The export of wool steadily declined, from 8,469 sacks in 1509–10 to about 4,700 in 1546–7.[2] The export of cloth showed notable increase,[3] as did also the value of the goods which paid tunnage and poundage. In the light of the figures it is possible to reconcile the apparent discrepancy between the fall in the customs returns and the apparent prosperity of England. The great weight of the customs fell upon wool, the export of which was declining; the custom on cloth was relatively light and the export of cloth was increasing; the yield of tunnage and poundage depended, in part, on the Book of Rates issued by the government which, unlike a modern

---

[1] Gras, p. 80.   [2] Schanz, ii. 84.   [3] Ibid. 104.

tariff, set forth not the duties payable by various categories of goods, but the prices which were set upon various goods for the purpose of assessing customs.[1] It is evident that unless the Book of Rates kept pace with the devaluation of the currency, the income derived from tunnage and poundage would not be commensurate with the increase of trade. The fall in the customs returns indicates, not necessarily a diminution of trade, but a redistribution.

From a study of the accounts other truths emerge. London was acquiring an ever-growing predominance; English merchants were steadily extruding 'strangers'; most important of all the government did not 'meddle'. There is no evidence of an attempt to exploit the increased export of cloth or the increased turnover in general merchandise, as there was no real effort to support the declining fortunes of the wool-trade, remunerative as it was in customs dues. The king and his advisers certainly thought it their duty to oversee the lives of all their subjects; yet with an instinctive realism they allowed the commerce of England to prosper in its own way.

[1] Portions of such a Book of Rates are found in Arnold's *Chronicle*, and the Book of Rates for 1507 has been printed by Professor Gras, app. C, p. 694. The Books of Rates issued in 1536, 1545, and 1550 seem to have been substantially copies of the Book of 1507; no general revision was made till 1558.

XIV

THE YOUNG JOSIAH

To historians the reign of Edward VI has always seemed of great consequence because of its place in the development of the English church. The 'political' reformation of Henry VIII was succeeded by a 'doctrinal' reformation. The compromise made by the father was abandoned by the son, and England passed to a complete protestantism.

That so great a change should take place when the royal supremacy was vested in a little boy is a remarkable phenomenon which has justly been regarded as the central theme of the period. Yet, while all historians unite in recognizing its importance, they differ much in estimating its significance. It has been regarded as the victory of truth over error, and as the victory of error over truth; as the inevitable consequence of the march of events, and as the chance fruit of ambitious enterprise; as the outcome of a great spiritual impulse, and as a side-issue of a far-reaching economic development. Each of these explanations has something of truth; none contains all the truth. Before the significance of the doctrinal reformation can be apprehended the background of attendant circumstances must be examined.

Henry's design of prolonging his authority into the new reign by the instrument of a will came to naught; the royal power passed to his young son and in his name it was exercised. 'Ve tibi O terra ubi puer est Rex'; Hugh Latimer used this text in his second Lenten sermon of the year 1549. He discounted it by another 'place'. 'Beata terra ubi Rex nobilis'; but in the new monarchy the prince was the mainspring of the state, and it was a serious thing for England that the successor of a powerful and resolute man should be a boy of nine years. Certainly the boy was of more than usual ability; he had been trained to his office; he was already aware of his royalty; but he was only a child.

From the moment of his birth, at Hampton Court, on 12 October 1537, he had been surrounded with all the attention due to the long-expected heir to the English Crown. The utmost care was taken to keep his food free from pollution and his person free from infection.[1] At first he was brought up 'among the women', but at the age of six he was handed over to the men, and his household, which had been considerable when he was only eighteen months old, was given a definite organization when his father went to France in July 1544. By the head of his household, his 'governor', he was taught deportment and horsemanship; as early as 1546 he played a part in the reception of the French ambassador, and at his coronation he seems to have behaved with dignity. Attention was paid to his physique, but although, especially in the spring of 1551, he was riding, running, and shooting, he never had the strength of his father or a great interest in sport. Most of his time, and evidently much of his interest, was devoted to his studies; he was of quick intelligence, and he was well taught. His principal tutor was John Cheke of Cambridge, a protégé of Butts, who was able to challenge Gardiner on the pronunciation of Greek, and he had for other teachers Richard Cox, later bishop of Ely, and Anthony Cooke, the father of the four learned daughters.[2] From these masters he learned English, and Latin and Greek, according to the principles inculcated by Erasmus, and though some of his written work probably owed something to his masters, much of it seems to be his own, including most of the famous *Chronicle*— Burnet gave it the name of *Journal*—which he began in 1550. His penmanship he learned from Roger Ascham; John Belmayne taught him French; but though he is said to have had a German tutor, Randolph, he does not seem to have learned German, and the story that he was 'not unversed' in Italian and Spanish lacks support. It is improbable that the famous composer Dr. Christopher Tye was his musical preceptor, but he was taught the lute by Philip van Wilder.

Along with him were educated other youths of high degree, scions of the great old houses and of those recently ennobled, but his only real friend seems to have been Barnaby Fitzpatrick, a cousin of the tenth earl of Ormond. No doubt he had in infancy, from his nurse 'Mother Jak', something of the motherlove of which fate had deprived him, but the general impression given by his writings is that his youth was lacking in human

[1] For Edward's youth see the original papers included in the Biographical Memoir by J. G. Nichols in the *Literary Remains of King Edward VI* (Roxburghe Club, 1857).

[2] Cooke was an exile during Mary's reign; his daughters married William Cecil, Nicholas Bacon, Thomas Hoby and later John Lord Russell, and Henry Killigrew.

affection. He had perhaps a certain kindness for his sisters, possibly because they too were royal. His letters to them and to his stepmother express some regard, as when he asked Queen Catherine to beseech Mary no longer to attend 'foreign dances and merriments'. Poor Mary! it had been better for her if she had danced more.[1] He had the Tudor power; he understood the theological discussions of the day, and his observations on economic affairs show much good sense. Upon the ordinary events of political life he was well informed, and though he may not have comprehended all their meaning, he certainly grasped much of their immediate significance. Always he was conscious that he was a king; never did he betray signs of personal regard for men who served him and who used his royalty to compass their own ends. He may have suspected the motives of his kinsmen and ministers; he certainly noted their risings and fallings, their triumphs and their executions without sign of emotion. For the insurgent peasants he showed no sympathy at all. From the moment of his father's death he was king.

The work of ruling the land, which was in those days the personal concern of the king, could obviously not be done by a child of nine; it must be done in his name, and the question was as to who should wield the royal authority until the child should grow up. Henry had endeavoured to provide for the minority by his will, but the word of a dead king could not prevail against the authority exercised in the name of a living monarch. The question was settled by the mere fact of power; the council was the general repository of executive power and at the moment of Henry's death the council was dominated by a clique which, though not truly protestant, was determined at least to maintain the 'reformation' which Henry had made, if not to carry it farther. Cranmer, who had convinced himself that Henry had meditated a more complete severance from Rome, was of the party, but he was no politician, and at first direction lay with vigorous fighting men like Hertford and Lisle and with calculating politicians like Paget. Throughout the whole reign of the boy king there was a constant struggle for power between ambitious politicians whose attitude to

---

[1] The duchess of Feria, Jane Dormer, afterwards alleged that Edward was greatly attached to Mary; it is true that in the *Journal* the statement that Edward had suffered his sister's Mass 'against my will' is struck through with the king's own pen. *Literary Remains*, i. ccxxxv and ii. 308.

religious and economic questions was affected, not only by their own predilections, but by their search for popularity, and in many cases by their desire for personal gain. Hertford became duke of Somerset; Lisle, earl of Warwick.

The first device was the establishment of a protectorate under Somerset, who, as the king's uncle, had some title to the office. But he proved unequal to his hour, and in the autumn of 1549 he was overthrown by Warwick who enlisted Catholic support. In February 1550, however, Somerset was released from the Tower, and for some eighteen months the rivals maintained an uneasy alliance, during which Warwick steadily gained an ascendancy over the young king whom he persuaded that opposition to his own plans was treason against the royal person. In October 1551 Warwick was made duke of Northumberland and Somerset was sent to the Tower. The ex-protector was beheaded on 22 January 1552 and thereafter Northumberland did all for himself in the name of the king who now appeared personally on the political stage. When it became apparent that the king could not live, Northumberland sought to perpetuate his power by persuading the dying boy in the interests of protestantism to 'devise' the Crown to Lady Jane Grey.

It may be taken as proof of the solidity of the Tudor system that, in spite of all, the monarchy survived in the Tudor line. Survive it did, though it lost prestige both abroad and at home as the successive 'governments' faced with inadequate strength difficulties political, constitutional, economic, and religious.

The political dangers which threatened from without were not those which would suggest themselves to a student of political theory. United catholicism made no effort to strike a blow for Mary, the true legitimate issue of the late king in the eyes of catholics. Cardinal Pole, it is true, was urging Pope Paul III to regain lost England; the long-expected council, summoned *causa unionis* as well as *causa reformationis*, had met at Trent;[1] Charles V established himself against the German princes

---

[1] The council having been summoned by Paul III to Mantua for 1537, Vicenza for 1538, and Trent for 1542, met at last at Trent from December 1545 to March 1547; it continued its session at Bologna from April 1547 to September 1549. It held a second session at Trent from May 1551 to April 1552, at which protestantism made a faint appearance. Its final session, from January 1562 to December 1563, reformed the church which remained to the Holy See.

whom he beat at Mühlberg in April 1547; but if there were any
dream of a catholic crusade it faded at once before the hard
realism of the times. The pope, who as a Farnese had ambitions
in Italy, recalled the council to Bologna in March 1547. The
infuriated emperor endeavoured to settle Germany himself,
and in May 1548 issued the Augsburg *Interim* which conceded
marriage to the priests, and the Cup to the laity; so doing he
alienated some of his friends without satisfying his opponents,
and whilst the council of Trent continued its ineffectual sessions,
Germany hovered on the brink of a religious war. This broke
out early in 1552 when France came to terms with the discontented princes.

From the heads of Christendom, then, schismatic England
had little to fear, but the very realism which rendered their
threats innocuous brought danger from another quarter—from
France. Henry VIII had made peace with France in June 1546,
and when he died the *convenances* were politely observed in Paris
with a dignity equalled in London when he was followed to the
grave by Francis I, who died on 31 March. The accession,
however, of a young and bellicose king augured ill for the
maintenance of peace. Henry II was a prince of the 'new
monarchy'. He, who allied with the protestants abroad whilst
he persecuted them at home, sought to draw his profit from the
differences between the pope and the emperor, and meditated
reviving his father's projects in Italy and Germany. At the same
time he turned his eager eyes towards the north; he had no
mind to overthrow the government of Edward VI for the
benefit of Spanish Mary, but he saw in the weakness of England
a chance for the aggrandizement of France. From the first he
meant to regain Boulogne without delay, and if possible to
expel the English from Calais. Later, when the presence of
Mary Stuart in France and the decline in Edward's health
brought new opportunities, he cherished far greater ambitions.
He plotted in Ireland; he did his best to make Scotland a
French province; he supported, underhand, attempts to debar
or expel Mary Tudor from the English throne. In his dreams
all the British Isles would fall beneath the sway of France.

At the moment when Somerset took power these vast designs
lay in the unknown future, but the protector had immediately
upon his hands a question with which France was almost sure
to be involved. This was the question of Scotland.

Henry VIII had refused to include Scotland in his treaty with France, and it was believed that on his death-bed he had charged Hertford to settle once for all with the perfidious neighbour, still pledged, as he believed, to give her daughter in marriage to Edward of England. To resolute action Somerset was urged, not only by the instructions of his dead master, but by political necessity, for the English cause was all but lost in Scotland. The desperate 'Castilians', who had held St. Andrews whilst the English ships kept the seas, succumbed, on 21 July 1547, to the artillery brought by a French fleet under Leo Strozzi; the prisoners were taken to France, and some of them, including John Knox, who had joined the garrison about Eastertide, were sent to the galleys. Somerset was not slow in his reply, he endeavoured to prevent, or at least to postpone, French intervention by offering the immediate cession of Boulogne, and while France, perhaps as a consequence of Charles's victory at Mühlberg, still held her hand, he launched a powerful attack upon Scotland. Having obtained, on 11 August, a patent constituting him 'lieutenant and captain-general of all the warres, both by sea and by land', he collected a formidable army of about 18,000 men including trained gunners, mercenary hackbutters, and an unusually high proportion of heavy cavalry.[1] Supported by sixty ships under Lord Clinton, of which thirty-five were warships, and confident that the English match had much support among the Scots themselves, he crossed the Border on 4 September. He made no declaration of war, presumably on the ground that hostilities had been going on ever since the Scots had repudiated the treaties of Greenwich, but prefaced his assault with a proclamation in which, saying nothing of the claim of suzerainty, he announced that he came merely to obtain the performance of a marriage covenant, and would use no force save against the 'hinderers of so Godly and honourable a purpose'.[2]

To oppose him the Scots had gathered, with the aid of the fiery cross, a large army, which was, however, conspicuous only from its numbers; its equipment excited the derision of the

---

[1] An account of the *Expedicion into Scotlande* written by the military judge William Patten (1547) is reprinted in *Tudor Tracts* in Arber's 'English Garner', 2nd ed. 1903).

[2] *Tudor Tracts*, p. 76, not verbally identical with the *Epistle Exhortatorie* in the *Complaynt of Scotland* (E.E.T.S., 1872), in which Somerset enumerates the extraordinary benefits offered by England.

professional soldiers in the English ranks, and its leaders were divided among themselves. The ineffective regent, Arran, distrusted the queen mother, Mary of Guise; and he and she alike were suspected by the protestants. Some leadership there was, however, for the Scottish host was posted very strongly near Musselburgh, with a marsh on its right, the sea on its left, and the river Esk along its front. Somerset prepared to attack, but he was spared the necessity by the action of his enemy, who, on the morning of the 10th, suddenly crossed the stream, and advanced in echelon from the right with the object of seizing Falside Brae which commanded the English position. The spearmen of the van, commanded by the hard-fighting Angus, roughly beat off a charge of the English cavalry under Grey, who was wounded, and for a moment it seemed that the Scots had gained the day. The English however, 'rally'd as they are wont to do; (and this is a distinguishing Character and Faculty they possess above most Nations)'.[1] Artillery was successfully established on Falside Brae, and beneath its blast the Scottish phalanx was shaken; the centre failed to support the van; the Highlanders on the left were enfiladed by an English galley which came inshore; before long the whole army dissolved in a dismal rout, in which thousands were slain, and about 1,500 prisoners taken. The victor, who conducted his campaign with what seemed like clemency to these rough times, spared the town of Edinburgh, and retired almost at once; but before he departed he made provision for an effective bridling of Scotland by the occupation of strong points. Sir John Luttrell was established on the island of Inchcolm in the Firth of Forth, and Sir Andrew Dudley in Broughty castle which commanded the Tay. The occupation of Home castle, and the fortification of Roxburgh, controlled the East March; and when, in the spring of 1548, a strong earth-work of modern design was erected at Haddington, the English gained a *point d'appui* only eighteen miles from Edinburgh.

As he had marched north Somerset had dreamed that though he had succeeded well he had yet done nothing, and the event proved that his dream had issued from the gate of horn. Some Scotsmen, especially those of the East March, became 'sworn Englishmen', but the main effect of his victory was to drive

---

[1] Patrick Abercromby in the preface of his translation of Jean de Beaugué's *Histoire de la guerre d'Ecosse: pendant les campagnes 1548 et 1549*. For original French text see Maitland Club (1830).

Scotland into the arms of France. According to Patten, Huntly, taken prisoner at Pinkie, said 'I . . . haud well wyth the mariage, but I lyke not thys wooyng', and the *de haut en bas* spirit evident in the writings of the English explains their ill success. At the end of May 1548 the sieur d'Essé arrived at Leith with 6,000 men and guns; Arran joined him with 5,000 more; the siege of Haddington was formed, and on 7 July a somewhat thinly attended parliament, convened in the abbey of Haddington, resolved to send the little queen of Scots to France to be married in due time to the dauphin. Towards the end of July Mary boarded a French galley at Dumbarton and, sailing round the 'back of Ireland', landed at Roscoff in Brittany; almost immediately she was received at Saint Germain-en-Laye with the dignity due to a future dauphiness. After her departure the war in Scotland raged with increasing bitterness; Somerset, departing from the mild principles expressed in his *Epistle Exhortatorie*, renewed the claim of suzerainty; Essé was superseded by Termes who, in June 1549, arrived at Dumbarton with 1,300 men. The English held on with tenacity, and though they slowly lost their strong-points, they kept Haddington until they deliberately evacuated it in September 1549; even after its fall they maintained a desultory warfare.

The English withdrawal was due to the outbreak of war with France. Somerset had tried hard to keep the peace.[1] In vain he had proposed that in return for the immediate surrender of Boulogne, and a cash payment, the French king should promote the Anglo-Scottish marriage, and should let England keep Ambleteuse, along with some places near the pale of Calais. Unofficial hostilities began in which the French repaid the depredations of English pirates by an aggression on land which steadily became more serious. In August 1548 the English council meekly swallowed a defiance that was almost a declaration of war, perhaps in the vain hope that Charles would be induced to guarantee Boulogne as well as Calais, but their complacence merely postponed the issue. A year later war was declared. In August the French captured Ambleteuse; swept

[1] A. F. Pollard, *England under the Protector Somerset*, pp. 138–42. Châtillon raided the Boulonnais in December 1547 and there were no reprisals; in September 1548 Somerset told the French Ambassador that his master's letter was practically a declaration of war.

into the Boulonnais, and seized some of the defences of Boulogne itself. The English, though ill prepared, resisted manfully, and for months beat off all attacks against a fortress which Henry VIII had taken in six weeks. Neither side, however, was anxious to continue the struggle. Poverty and domestic uncertainty weakened the hand of England; France, with her eye on an opportunity to attack the emperor, was willing to agree with one enemy before she assailed another. Peace negotiations were begun in January 1550, and brought to a successful conclusion on 29 March by the treaty of Boulogne, the terms of which reflect the superiority of France over distracted England. Boulogne was surrendered at once instead of in 1554 as stipulated by the treaty of 1546, and that for only 400,000 crowns; what was far more serious was the fact that England withdrew from Scotland, which during the next decade was in great danger of becoming a French province.

During the remainder of the reign England cut a sorry figure in European politics. Warwick, driven by his own ambition, stood aloof from the emperor, who would naturally support the claims of Mary. Charles did, in fact, intervene to secure her in her Mass, whilst at the same time he banned the use of the English Prayer Book in his dominions. His interference was resented, as was his alleged indifference while Boulogne was being lost; and when, hard pressed in 1552, he demanded help from England under the treaty of 1542, no help was given him. Among the various reasons alleged, some of them frivolous, was the unworthy argument that the young king was not bound by his father's treaty.[1] More relevant was the plea that he had sworn an amity with the French king which he could not break; for England, digesting as best she might her losses at Boulogne and in Scotland, was now deeply committed to France. On 19 July 1551 there was signed at Angers a treaty whereby Edward resigning his claim to the hand of Mary of Scotland, betrothed himself to Elizabeth of France, and the year witnessed a great exchange of civilities. In June Henry was made a knight of the Garter, in July Edward was invested with the insignia of St. Michael, and in October Mary of Guise, returning from a visit to France, was given a magnificent reception.[2]

---

[1] See 'King Edward's Journal' in *Literary Remains*, ii. 432.
[2] Wriothesley, ii. 60; *Literary Remains*, ii. 362.

For cultivating France Warwick could quote precedents from the reign of Henry VIII, but these he followed only with a difference. Henry, confident in his strength, had stood four-square to the world in defiance of pope and emperor, whilst France was relatively weak; now France was strong, and the power of England despicable. Perennially short of money, the government neglected to maintain its armed forces. The little king took an obvious pride in his ships, and the navy he inherited from his father was considerable,[1] fifty-three vessels including thirteen ships and fourteen galleys. Efforts were made to maintain its strength, but on the close of the French war sailors were discharged. Piracy seems to have thriven, and in 1552 Sir Henry Dudley was unable to insist that de la Garde, who appeared with a French fleet off the Isle of Wight, should give the accustomed salute. The troops from Boulogne were soon dismissed,[2] and it is significant that a good part in quelling the internal disturbances was played by mercenaries. What money could be devoted to the maintenance of men-at-arms was spent upon the 'bands' of certain nobles; in October 1552 bulwarks on the sea-side were being abandoned as superfluous; garrisons in Ireland, Berwick, and Guisnes were being reduced;[3] sensible plans for new defences remained unaccomplished, and the situation was such that France assumed a complete superiority.

Henry was supreme in Scotland; he was conducting, through George Paris,[4] negotiations with the Irish which he denied in a very off-hand manner, and his people were saying that once they had settled with Charles the capture of Calais would not be a seven nights' work. It may be true enough that Calais was saved only by Henry's war with Charles, which for a while succeeded brilliantly; certain it is that the English government became alarmed, and in September resumed negotiations with the emperor.[5] Cecil prepared a characteristic 'reasoning' on the question which concluded 'the treaty to be made with th' emperour, and by th' emperour's means with other princes', with the cautious proviso that the emperor's acceptance must be secured before anything was attempted against France. This document the king put into his desk, and although Morison's

---

[1] Ibid., pp. 323-4, 581, 582.   [2] Ibid., p. 267.   [3] Ibid., p. 544.

[4] P. F. Tytler, *The Reigns of Edward VI and Mary*, i. 292, 302, 351.

[5] *Literary Remains*, pp. 457, 541; Hardwick, i. 48, Oct. 1552. Sir Richard Morison: once of Wolsey's household, later servant of Cromwell, and propagandist for Henry, was ambassador to Charles, 1550-3, with Ascham as his secretary.

unofficial *démarches* at Speyer were well received, no concrete result emerged. Northumberland, who already may have realized how fragile was Edward's health, would not ally with the cousin of Mary. England continued to court the dangerous friendship of France.

While the affairs of England thus went ill abroad, her domestic politics were marked by uncertainty and deterioration. Somerset's power had been founded upon the support of a group of nobles and officials who had hoped to share in the spoils. He lacked the dread presence of Henry VIII, and after his erection had established the new order his confederates became steadily more restive. He was at no pains to cultivate a party; the friends gained by his protestantism were, many of them, feeble of power and anti-authoritarian in sentiment; the popularity which his economic policy gave him with poor folk alienated the rich and the mighty, and, like other liberal policies, produced not gratitude but discontent when its hopes were frustrated.

The first important revolt against his authority took shape in a conspiracy led by his own brother Thomas, an ambitious man who married Catherine Parr early in 1547 and after her death, on 5 September 1548, made advances to the Princess Elizabeth with whom, when she was in his wife's care, he had enjoyed a somewhat gross familiarity. He won over Dorset by assenting to a plan whereby Edward VI was to marry Lady Jane Grey; he took from Sir William Sharington part of the illicit profits of the Bristol mint; he connived at the piracy, which as lord admiral he was charged to suppress; worst of all he made the little king his friend, with flattery, pocket-money, and promises of greater liberty.[1] His obvious ambition was ignored by the protector until the exposure of Sharington in January 1549 brought to light his dishonesty and his recalcitrance. He refused to come privately to his brother; and even after he was lodged in the Tower declined, on 23 February, to answer before the council, on the ground that he should have an open trial. No trial was given him; a few days later he was attainted in parliament and on 20 March he was beheaded. Following the precedent of the action against Catherine Howard, the council

---

[1] *Acts of the Privy Council*, ii. 248–56; Pollard, *England under Protector Somerset*, pp. 190–9.

had obtained the king's leave to proceed without troubling the protector. But though Somerset's face was thus saved, it was patent to all that he had let the admiral go to death without a fair trial. Certain councillors drew their own conclusions; a man who would not defend his own brother might not be very capable of defending himself; a king who made no effort on behalf of one Seymour, might not put himself out to help another. When, later in the year, the discontents aroused by the protector's religious and social policy came to a head, his former allies had no scruple in overthrowing his authority, and in seizing power for themselves.

In June 1549 there was a serious rising in the West Country against the use of the new Book of Common Prayer and the religious innovations generally. Exeter stood a siege of six weeks before Russell, reinforced by German and Italian mercenaries, was able to effect its relief on 6 August, and it was not until 17 August that the rebel host was completely scattered. Meanwhile there had been a rising in Oxfordshire, quelled by Grey, who hanged the offending priests from their own church steeples, and a far more dangerous rebellion, not religious but economic in origin, had broken out in Norfolk. Although the disturbance began in a county whose catholic duke was a prisoner, in a region close to Kenninghall where Mary resided, and on the festival of the translation of St. Thomas, there seems to be no clear evidence that the rising was due to catholic discontent. Famous both for its sheep and for its agriculture, Norfolk was pre-eminently a land of small arable farms, and though it may not have suffered as much as some other areas from ordinary 'inclosures', it felt more than most the assertion of doubtful manorial rights, the truncation of the commons, and the rise in rents and prices. To these evils a sturdy and independent peasantry was not disposed to submit, especially when rumour told them that the protector was on their side; on 1 June 1548 he had issued a proclamation against inclosures and on 14 June 1549 he had offered a general pardon to those who, having pulled down inclosures, would now submit. There had been several disturbances in the county before, in the second week of July, a tumultuous assembly at Wymondham found a leader in Robert Kett.[1]

---

[1] S. T. Bindoff, *Ket's Rebellion 1549* (Historical Association Pamphlet, 1949), reviews the evidence clearly; on the whole sympathetic with the insurgents.

Kett was a well-to-do tanner and landowner whose action in having his own fences pulled down along with those of his hated neighbour Flowerdew commanded popular support. Men rallied to him from twenty-four of Norfolk's thirty-two hundreds, from some of the towns such as Norwich and Yarmouth, and even from Suffolk. With a force which eventually mustered 12,000 men, he dominated Norwich, from 12 July to 26 August, from a great camp on Mousehold Heath where, with a council of 'governors', he maintained good order. Among the demands made by his followers was 'that all bondmen may be made free, for God made all men free with his precious blood shedding'; and although this was a somewhat vain echo from the German Peasants' war of 1524, since personal serfdom was a small thing in England, it expressed a prevalent feeling that men were entitled to a liberty which the 'good duke' would willingly grant if he was not coerced by evil counsellors.

The steady refusal to accept the royal pardon, which was offered on three several occasions, no less than the absence of murder and manslaughter, may establish a presumption that the insurgents considered themselves no rebels. Rebels, however, they were. As early as 22 May[1] a proclamation which denounced inclosures had promised punishment for outrageous 'attemptates' by those who took the law into their own hands, and ordered sheriffs to suppress unlawful assemblies. On 8 July a fresh proclamation contained instructions for the quelling of disorder. By this time there had appeared traces of organization among the malcontents which alarmed the government,[2] and as the halting measures of the protector failed to produce peace the decision necessarily passed to the sword. On 30 July Northampton occupied Norwich without difficulty, but he was soon driven out, and it was not until 23 August that the more resolute Warwick appeared outside the city. On the 27th Kett, either in obedience to an ambiguous oracle or because his supplies were being cut off, moved from his camp to neighbouring Dussindale where his ill-armed host was speedily cut to pieces. It is said that the numbers killed on the field amounted to over 3,000, and vengeance followed at the hands of the government and the returning lords. In December, after they

---

[1] Steel, p. 36, suggests 17 May.
[2] Sir Thomas Smith to Sir William Cecil, Tytler, i. 187.

had been condemned for treason in London Robert Kett was hanged at Norwich castle, and his brother William from Wymondham steeple.

Like other unsuccessful rebellions the East Anglian rising produced a sharp reaction; its main political result was to discredit Somerset and to exalt Warwick. In October the protector was deposed by a cabal in which Wriothesley, now Earl of Southampton, and Arundel had a part, and which was at first believed to presage a catholic reaction. The main sentiment of the council, however, was dislike of Somerset's 'liberal' views, and its natural leader was Warwick whose object was to create an authoritarian government exercised by himself in the name of the little king. Either because he realized that Edward, though only an infant, was already of importance, or because he felt that an alliance with the catholics might produce intrigues on behalf of Mary, the unscrupulous adventurer determined to run the protestant course. Somerset was released from the Tower in February 1550, peace was made with France in the following March, in June Anne Seymour was married to John Dudley, Viscount Lisle, and the government set out on a policy of advanced 'reform' in religion.

For the moment it seemed that the two rivals would work hand in hand, but Warwick's ideas were very different from those of the late protector. Both men were ambitious, both enriched themselves at the public expense; but Somerset favoured the empire and some accommodation with Mary, while Warwick looked to France. Somerset had courted popularity while Warwick sought for power. Somerset had encouraged a 'liberty' which ended in disorder; Warwick, as the champion of order, aimed at an autocracy. He deliberately built up a party, and used the pretext of local disturbances in 1550 and 1551 to create the germ of a standing army. He allowed his tried friends to recruit 'bands' of men-at-arms—paid for at the royal expense—and by giving permanence to the temporary commissions of lieutenancy freely granted during the disorders, he originated a system whereby the lord lieutenant supplanted the sheriff as controller of the armed forces of the shire. Meanwhile he set himself to cultivate the young king who, in August 1551, was brought personally to sit in council. In October the ruling clique garbed itself in the decorations of success. Warwick became duke of Northumberland, and Dorset duke of Suffolk; Henry Grey had entered the royal circle by his marriage with Frances Brandon, and had enjoyed a marquisate since 1530, but John Dudley had no such claim to a title which at that time was enjoyed by two other men only, Norfolk and Somerset. Pliant Paulet was promoted from earl of Wiltshire to be marquis of Winchester, Herbert was made earl of Pembroke; some of Warwick's kinsfolk, and some of his soldiers were honoured by knighthood. At the same time opposition was swept ruthlessly away, and on 16 October Somerset was arrested on a very doubtful charge of conspiracy.

His doom was certain. In November it was decreed that documents bearing the royal signature needed no countersignature by councillors before they passed the Great Seal, and the king was now in the hands of the duke's enemies. The chancellor, Rich, foreseeing trouble feigned illness and before long resigned, to be succeeded by Thomas Goodrich, bishop of Ely, but Northumberland did not hesitate. On 1 December Somerset was condemned for felony, and executed early in the morning of 22 January 1552 in order to make sure that parliament, summoned for the following day, should not interfere on his behalf. Next month four knights (Vane, Arundell, Partridge, Stanhope) followed him to the scaffold. Paget, who had been confined to his house for alleged misconduct in his diplomacy, was, in April 1552, stripped of his garter on the ground of his insufficient birth, and in the following June was fined £8,000[1] for defalcations committed as chancellor of the duchy of Lancaster; he was succeeded in the Order of the Garter by Sir Andrew Dudley, and as chancellor by Sir John Gates, a 'trusty' of Northumberland's, but not reliable in any other sense. The whole story of Northumberland's machinations[2] is told in the *Journal* kept by the unloved and unloving boy who was persuaded that he was king of England in fact as well as in law, and who, for all his ability, was naïve enough to take at face value the reasons given him for the abasement of some and the exaltation of others. So firmly convinced was he of the existence of a conspiracy that he recorded his uncle's death in these terms: 'The duke of Somerset had his head cut of apon Towre hill betwene eight and nine a cloke in the morning'; and he explained that Sir Andrew Dudley, who was promoted, had

---

[1] Afterwards reduced to £6,000: Dietz, p. 181.

[2] *Literary Remains*, p. 415 (for appointments), cf. p. 409 for rewards to his friends.

'indetted himself very moch by his services at Guisnes'.[1] So, guilelessly, he reveals the process by which military, executive, financial, and even judicial authority was concentrated in the hands of Northumberland's friends, and reveals, incidentally, how very much he himself was in the hands of the flatterer who treated him as a man of the world.[2]

It is time to regard the 'constitutional' background of these factional contendings. The accession of a king who was a minor confronted England with a difficulty which had been known before; and England attempted to meet it by the use of precedent—by erecting a protector who was a close kinsman of the infant monarch. The clique which was in power when Henry died kept his death a secret until all their arrangements were made. Hertford, after arranging things with Paget, escorted Edward from Hertford castle, first to Enfield where Elizabeth was residing, and then to London where he was lodged in the Tower. On 31 January thirteen of Henry's executors assembled there, and after taking a solemn oath to carry out the will at once broke it, or at least modified it, by naming as protector the earl of Hertford who had been given no particular place in Henry's testament. According to Burnet Wriothesley protested on the ground 'that the late king intended they should be all alike on the administration',[3] but the councillors resolved to give their oath to the new king precedence over their oaths to execute the will, and on 13 March, perhaps because France had expressed some doubts about the government, decided to ask from the king a commission signed by the royal hand which, their own signatures having been appended, might pass under the Great Seal and be a complete warrant for their use of power. This commission, which was issued on 21 March, ratified all things done by Hertford as protector, and granted him full power and authority 'until such time as we shall have accomplished the said age of eighteen years'.

[1] Ibid., p. 461.
[2] Cf. the king's schemes for modifying the Order of the Garter which omitted the name of St. George, and pledged the knights to the defence of 'the truthe holly conteined in the Scripture'; the king wrote as if the new order was thoroughly established in April 1552, but in fact it was not. *Literary Remains*, p. 511. He referred to the French order which was conferred on him as that of 'Monseigneur Michael'.
[3] Froude, without citing the evidence, makes him allege the failure of other protectors in support of his opposition; for the recorded transactions as to the protectorate see *Acts of the Privy Council*, ii. 7, 63, 67.

This public instrument, which emphasized the blood relationship between king and protector, endeavoured to represent the
elevation of Hertford as the natural thing, while the minutes of
the privy council (in which the 'executors' and the 'assistants'
under the will now appeared as nominees of Edward) were so
framed as to produce the impression that everything was being
done in the normal regular way. None the less, the affair was a
coup d'état carried out by a political clique; no time was given
for the opposition to organize itself, and the victors divided
among themselves a spoil of offices and titles. Hertford, who
became duke of Somerset, assumed the offices of treasurer and
marshal, vacant by the attainder of Norfolk; John Dudley,
Viscount Lisle, succeeded Somerset as great chamberlain, and
became earl of Warwick; the post of high admiral, which
Warwick vacated, went to Somerset's brother Lord Thomas,
who was created Lord Seymour of Sudeley. Most of the other
great offices remained in the hands of their former holders;
John, Lord Russell retained the privy seal, Sir William Paget
continued as chief secretary, Sir William Petre as secretary, and
Sir Anthony Browne as master of the horse; Sir Thomas Cheyney
and Sir John Gage remained as treasurer and comptroller of
the household. William Parr, earl of Essex, became marquis of
Northampton, Sir William Willoughby and Sir Edward
Sheffield were made barons. One prominent statesman gained
a title but lost an office; Thomas Wriothesley, who had been
led to take part in making Hertford protector, was created earl
of Southampton, but was deprived of the chancellorship and
removed from the council on 5 March on the ground that he
had of his own authority delegated his judicial functions to four
masters in chancery in order to leave himself more time to
attend the council. The seal was entrusted to William Paulet,
Lord St. John, who as 'great master of our house' (lord chamberlain of the household), seems to have acted *ex officio* as president
of the council; it was only in October that a new chancellor
appeared in the person of Sir Richard Rich—one of the 'assistants' of the will, who was given a barony. The protector
retained about him a great part of the administrative machinery
bequeathed to him by Henry VIII, and when on 20 February
the little king was duly crowned, it seemed that the government
of England might be carried on as before with hardly any sign
of a hiatus. In the words of King Edward's *Journal*: 'Also in this

time the late king was buried at Windsor with much solemnity, and the officers broke their staves hurling them into the grave; but they were restored to them again when they came to the Tower.'

In spite of all the efforts to produce an appearance of normality, and in spite of the actual continuity of government, the elevation of a protector evidently caused an uneasiness which the ruling clique endeavoured to dispel by the authority of parliament. Already there was in England a feeling that great matters of state should be established in parliament, and when parliament met on 4 November Somerset was at pains to regularize his position. As he had already exercised power for some months, especially as he had already summoned parliament, there could be no question of parliamentary appointment; but on the first day of the session, before business began, the clerk of the parliament formally read an instrument under the Great Seal, announcing the protector's commission, ostensibly in order to assign to him a special place and full prerogative in the sitting of parliament.[1] This procedure cannot have allayed all doubts, for on the last day of the session, 24 December, fresh letters-patent were issued[2] which, while recognizing the duke's position as protector and captain-general, and appointing him 'principal councillor and highest of our privy council', made his authority terminable at the king's pleasure. This instrument was signed by sixty-two persons including most of the peers present on the last day of the session, and the councillors; but there is no record of it in the *Lords' Journals*, and the one extant copy does not bear the Great Seal. The assertion that the protector's commission was ratified by parliament is therefore inexact; but the evidence seems to show that he either would not or could not exercise an authority over which parliament had no control.

Under the protector, so long as his power endured, and then directly under the king, the council was the mainspring of government. In theory it was still advisory rather than executive, but the process by which it was becoming a governmental

[1] *Lords' Journal*, i. 293; *Foedera*, xv. 164.
[2] There is a lacuna in the *Calendar of Patent Rolls* for this period. The document is known only 'in the copy in the possession of Mr. Staunton', published by J. G. Nichols in *Archaeologia*, xxx.

machine continued. During the reign of Henry it had grown steadily less aristocratic and more ministerial. On his death, of twenty-six councillors only one, Arundel, had a peerage of more than twelve years' standing; three were churchmen; and every one of the others held an office of state or household. It is worthy of note that in both the commissions to the protector the names of the councillors are severally mentioned, and that these names are identical with those of the sixteen executors and the twelve assistants of the will, save that Somerset lost his place by promotion, and Wriothesley his by degradation. Either because they had been named executors or because he preferred to have the law as an ally rather than as a critic, Somerset included the chief justices; otherwise his council was substantially that of Henry VIII, and in his day few changes were made. Very early Gardiner and Thirlby followed Wriothesley into retirement; in their places Thomas Seymour, Richard Southwell, and Edmund Peckham came in. In 1548 Wriothesley was readmitted and Thomas Smith, who succeeded Paget as secretary, along with the martial Shrewsbury, were admitted; but these three took the places of Seymour who was executed, and of Browne and Denny who had died.

While the size of the council was not increased, its organization was improved, particular days being allocated to particular business. It always sat in or near London, usually at Westminster, and from its central position it kept a vigilant eye upon the whole work of government. It determined matters of policy; it was there for example that the manner of proceeding against the lord admiral[1] was discussed. It issued proclamations which, though they lost the statutory authority conferred by the act of 1539, were still used to regulate all the life of the realm; they dealt, for example, with war, with religion, with inclosures, and with currency.[2] It exercised an overruling jurisdiction, usually, though not always, in the star chamber which in the short reign of Edward VI dealt with over 2,500 cases.

It is fair to say that in the day of Somerset it did not use its power tyrannically; the proceedings of the star chamber were, by Tudor standards, lenient, and the protector actually established a court of requests, to deal with poor men's causes in his own house. This fact is noteworthy; if the 'good duke' was

---

[1] *Supra* 489.   [2] Gladish, *The Tudor Privy Council* (1915), pp. 44 and 96.

not tyrannous he was certainly arbitrary;[1] he did all himself; when his overweeningness alienated some, and when his liberal measures alarmed others, he found himself without a party at all. The council which had been the vehicle of his power became the instrument of his destruction.

Its opposition, however, was factual rather than truly 'constitutional', and Somerset being fallen, it formed part of the apparatus with which his successor governed England in the name of the little king. Warwick at once set himself to obtain control, and by the end of 1551, he had added twelve new members all of his own faction. The little king, himself introduced into the council in August 1551, showed himself interested in its organization, and with precocious ability designed plans to improve its working. In 1552 he wrote with his own hand regulations[2] by which all the members, whose number he shows at forty, were assigned to five committees. As one of these was 'for the state', it has been thought that here was the origin of the cabinet. The opinion is not without some justification[3] though the essence of the cabinet is that it was (like the privy council in its relation to the council) an unofficial committee, and King Edward's plan takes its place in a whole series of expedients to reduce to workable size a council grown inconveniently large. Like some other projects of the same kind, it remained a project; the official records of the council give no sign of its use. This is significant; the council, even when the king ruled in name, moved at the bidding of the person or party who exercised real power. It was the instrument of authority, not a check upon it. On the other hand—as Northumberland found to his cost—it could not restore to a falling politician the 'auctoritye' which he had lost.[4]

Parliament, too, in one of its aspects, was a function of the royal power; but it was more than that; Henry, in making it a partner in his great enterprise, had given it an authority greater

[1] Paget in a letter of 8 May 1549 warned Somerset against 'those great cholerick fashions', which would seem unworthy in a king and were intolerable in a subject. Tytler, i. 174.

[2] *Literary Remains*, pp. 403, 489, 498.

[3] See on p. 531.

[4] Northumberland's power was plainly personal. Sir John Gates said to the duke on his way to the scaffold: '. . . you and your auctoritye was the onely originall cause of all together . . .' *Chronicle of Queen Jane and of Two Years of Queen Mary* (Camden Society, 1850), p. 21.

than he himself realized. Bishop Gardiner asserted that in parliament there was 'free speech without danger', and argued that it was 'prejudicial to the good order of the high court of Parliament' that a man 'being member thereof, might without cause be excluded, and so letted to parle his mind in public matters'.[1] Later, in Mary's reign, parliament in making its submission to Rome, and Gardiner in his sermon, referred to the three estates assembled in parliament as 'representing the whole bodye of the realme of Englande'.[2] It is to be remarked that the council, in the autumn of 1550, and again in November 1551, prevented the assembly of parliament lest it should support Somerset, and that on 31 March 1553 Northumberland dissolved a parliament which had sat only for a bare month, lest it should oppose his design for the succession. Plainly Sir Thomas Smith's conception that parliament was a grand jury of all England was already in English minds.

None the less the Crown had means of influencing both the composition and the conduct of parliament. The disappearance of the abbots had given the majority in the house of lords to the laity, many of whom were peers of new creation; the bishops had been made to take out new commissions equating them with other servants of the Crown, and among them were some who owed their position to the royal favour. Of the commons, partly as a result of the development of local government, very many held offices under the Crown. It has been shown that in the first parliament of Edward VI: 'Of the one hundred and eighty nine members, . . . whose names occur in the returns, at least a third either held some office about the court, or were closely related to ministers for the time being.'[3] The Crown had the power of creating new boroughs, and it had other means of packing parliament. Yet the packing of parliament must not be exaggerated; most of the seven new boroughs created in 1547 were thriving communities, and though the king might recommend his nominees, his recommendation might not always be accepted, while other magnates might nominate too. In 1547 the council's recommendation of Sir John Baker was rejected by the electors of Kent, and Gardiner in resenting his own exclusion from the lords referred also to the exclusion of

[1] Muller, *Stephen Gardiner and the Tudor Reaction*, pp. 163, 185.
[2] *Chronicle of Queen Jane and of Two Years of Queen Mary*, pp. 160, 163.
[3] A. F. Pollard, *England under Protector Somerset*, p. 74.

those whom he 'was used to name' from the nether house. When the second parliament was summoned in the autumn of 1553, Northumberland made the most strenuous efforts to pack it,[1] but these efforts were only an extension of the normal practice and they had no great result.

Once assembled, as appears from the *Lords' Journals*, and from the *Commons' Journals*, which, significantly, began in 1547, both houses debated freely, but many of the economic measures discussed by the commons did not get very far, and it can scarcely be doubted that most of the acts which emerged were sponsored by the government. Some of the acts of the first session, which lasted from 4 November to 24 December, restored to the houses part of their lost authority. The act[2] giving to proclamations the force of law was repealed, and so too, with certain limitations, was the act which permitted the king who was a minor to revoke all legislation passed before he completed his twenty-fourth year.[3] Much of the legislation, however, obviously reflected the mind of Somerset himself, who was inclined towards mildness in administration, and 'reform' in ecclesiastical affairs. The mildness must not be exaggerated. There were numerous exceptions to the act for a general pardon which was granted for all offences and forfeitures previous to 25 December 1547, and an act which was passed against vagabondage which limited the use of the gallows and the whip, provided for branding, forced labour, and absolute slavery, according to the degree of recalcitrance displayed by the able unemployed; 'infant beggars' might be forced to apprenticeship, girls till they reached the age of twenty and boys to the age of twenty-four; the only provision for the aged and impotent was provided by the charitable collections to be made on Sundays. A popular act[4] which swept away all treasons created by Henry VIII—except that it was still treason to alter the succession to the Crown as provided by statute and Henry VIII's will—abrogated, almost incidentally, the legislation against heresy, including the 'act of six articles', which was

---

[1] Dixon, iii. 508; Tytler, ii. 160.

[2] Incidentally included in the statute repealing the new treason laws. 1 Edward VI, c. 12.

[3] Ibid., c. 11.

[4] Ibid., c. 12. See comment in *Tudor Constitutional Documents*, ed. J. R. Tanner, pp. 378 ff. The act for vagabonds is 1 Edward VI, c. 3, and the act for a general pardon is ibid., c. 15.

specifically mentioned. The protector had no difficulty in having passed the various acts of his religious programme, and later the still more revolutionary programme of Northumberland was carried without demur. Generally speaking, the government could have its will in parliament if it redressed trade 'abuses' and made no great demands on the taxpayer.

With regard to finance the commons showed themselves less complacent. In the first session the subsidy of tunnage and poundage for the keeping of the seas was duly voted, but no other grant was made; in the second session, which lasted from 24 November 1548 to 14 March 1549, there was passed an act for a new kind of levy called a relief, which included taxes upon sheep and wool, in accordance with the protector's agrarian policy, along with a widely distributed personal tax. The imposition was payable in three years, and its yield was extremely small; the first payment was due only in May 1549, and produced something under £54,000; the second payment yielded about £47,500. By the time this payment was made Somerset had fallen, and his successor Warwick abandoned the tax on sheep and wool in return for a small subsidy. At the very end of the reign there was granted a subsidy with the accompanying two fifteenths and tenths, but none of this was paid until after the king's death, and the total direct taxation of Edward's day, including £120,000, granted to Henry VIII, brought in only about £300,000. The contribution of the clergy was likewise small.

This was very serious, for, throughout the whole reign, the government rested upon a very shaky finance. Henry VIII had bequeathed to his successor an almost empty coffer, a falling income, and discredited currency. The cost of the defences at Calais, at Boulogne, till it was lost, at Berwick, in Ireland,[1] and at sea made great demands; during the first five years of Edward his military expenditure amounted to nearly £1,400,000. As appears from the chronicles it was thought necessary to gratify the public by costly displays, and it was not only the public who must be gratified; Somerset helped himself largely from the royal lands, and the politicians who

---

[1] Dietz, p. 189, n. shows that though the net Irish revenue should have been about £4,700 a year in time of peace, England sent to Ireland annually sums which varied from £16,000 in 1547–8 to £42,000 in 1552–3.

acquiesced in his promotion were rewarded for their pains, the more easily since Paget produced a list of rewards which, he alleged, had been contemplated by Henry VIII before his death. Later in the reign Northumberland boldly said that there was no need to inform parliament of the king's generosity towards his faithful servants; but the commons must have known full well what was happening, and, armed with the knowledge, were less disposed to meet the royal necessities. That there was corruption in administration is not to be denied; Sir William Sharington, master of the Bristol mint, Lord Arundel, the lord chamberlain, Whalley, the receiver of revenues in York, Beaumont, the receiver-general of the court of wards and liveries, and even Paget himself as chancellor of the duchy of Lancaster were all guilty of embezzlement in one form or another.[1]

It is fair to state that there was no breakdown in machinery as a whole; that some of the malpractices were probably not without precedent, and that their exposure may well have been due to turns of the party wheel. Northumberland began to revive the financial methods of Empson and Dudley, but he certainly made other and more laudable endeavours to balance his accounts, perhaps under the guidance of William Cecil. He tried to clear off the foreign debt. In December 1552 there was a serious attempt at an audit; and in the parliament of 1553 were passed two acts which contemplated a complete reorganization of the financial machinery.[2] A patent of 1 January 1547 uniting the court of augmentations with that of general surveyors was declared to be valid,[3] and the king was empowered to dissolve or reorganize the other financial courts created by his father, and the duchy of Lancaster as well. Plainly the multiplicity of Henry's establishments had created some confusion, and in attempting a reform Warwick's government was more enlightened than that of Somerset; but its own administration had been faulty, and it had been in financial difficulty all

---

[1] Ibid., p. 181.

[2] Ibid., chap. xv. For the proposed audit see *Literary Remains*, ii. 468 and note.

[3] 7 Edward VI, c. 2. By a patent of 1 January 1547 (*Letters and Papers*, xxi. ii. 348, 408) Henry had dissolved the court of augmentations (created by statute 1536), and reconstructed it to include the court of surveyors general (established by statute 1542). There had been some question whether a patent availed against a statute. 7 Edward VI, c. 1, had regulated the procedure of receivers and treasurers and appointed penalties for failure to observe the rules.

along. Although the expenses of war ceased in 1550, the armed 'bands' had to be paid; there was little church wealth left to exploit;[1] the customs could not be raised, and parliament could not be trusted. In the summer of 1552 all the treasuries were empty, and at the following Michaelmas the *gens d'armes* had to be disbanded. Utterly unable to balance their accounts the governments of Edward VI followed the evil example of Henry VIII in debasing the currency. At first there was talk of reform, and in 1549 the amount of alloy was reduced; but there was a corresponding reduction in the weight of the individual coin, and between January 1547 and January 1551 the government made £537,000 out of the mint. In April 1551 there was a new debasement which brought in £114,500 before the council, alarmed by the unpopularity of its measures, and by their economic consequences, ordered a return to good currency in the following September.[2] The policy whereby an 'orgy of debasement' preceded the calling down of the vitiated coins to values more in keeping with their intrinsic worth appears to a modern financier as 'fatuous'. So in the long run it was, but the obvious design of the government was to issue the poor coins at the enhanced value in order to meet its own liabilities, and then to demand payment of money owed to itself on revenue accounts in the deflated currency.

It cannot be said that the Edwardian governments encountered opposition which was 'constitutional' in the ordinary sense, since they were able, by acts of power, to direct the fortunes of England into strange paths without formal protest from parliament or from any other body. Yet though the opposition was not formal, it was none the less real, and it involved, though indirectly, the principle that redress of grievance must precede supply. Because their arbitrary proceedings alienated not only the conservatives but also many of the virile elements most politically conscious, the government dared not ask the people of England for sacrifice, or parliament for money.

[1] In 1552 and in 1553 commissioners were sent round to collect church plate become 'superfluous', and something over £10,000 was got in. A beginning was made in the process of annexing the lands of the bishoprics; in 1550 Westminster was joined to London, which had to surrender some of its land. In 1551 Ponet, on translation to Winchester, gave up all his land and got a fixed stipend of 2,000 marks; in 1552 Gloucester was added to the see of Winchester, and its lands confiscated by the Crown. In 1550-1 the customs were less than £24,000, whereas in 1546-7 they were £40,000. Dietz, pp. 194, 200, 206 n.

[2] Ibid., p. 195.

Without money authority, whether wielded by idealist Somerset or unscrupulous Northumberland, was impotent, and that at a time when there was an obvious need to cope with two great questions which tormented England, as they tormented every country at the time—the economic problem and the problem of religion.

The economic problem was urgent. The brief reign of Edward VI was marked throughout by an economic stringency which appeared inexplicable to most observers, and, to the poorer members of society, unjust. The harvests, in the main, were good. How could there be dearth in the midst of plenty? A recent 'economic' explanation, based on the argument that the fields had been exhausted by continual cropping, is not supported by the evidence, and a contemporary theological explanation that want was due to the wrath of God was not entirely satisfying even to those who advanced it. John Hales might allege, in his famous *Defence*,[1] that 'scarcytie, famyn, siknes be plages of god' sent upon the people for their wickedness, but, as is plain from all he said and did, he understood that human error was partly responsible for the evils of the time, and most men agreed with him. It was generally felt that the whole economy of England was upset; the landlords were dissatisfied, raised the rents, and tried to inclose; local manufactures seemed to be declining; prices were high and wages were inadequate; in the country-side at least there was universal discontent. If these evils were due to human agency, they should be curable by human action, and various suggestions were made. The preachers, whose *naïveté* may have come nearest to the truth, held that the prime cause of misery lay in the covetousness of man, but the event was to show that some of the professed champions of the new faith were no better than the old abusers of God's truth. Some men thought that new laws should be made. Others, again, held that there were already laws enough, and that the fault lay in a lax administration.

As land was held to be, and indeed was, the principal source of English wealth, agrarian difficulties thrust themselves to the fore. There were sporadic disturbances in many parts of

---

[1] See *A Discourse of the Common Weal of this Realm of England*, ed. Lamond, p. lxvi. Hales defended himself from the charge that his plans for agrarian reform had produced disorder. For their occasion see p. 504 below.

England. Trouble spread like an infection from Somerset into Gloucestershire, Wiltshire, Hampshire, Sussex, Surrey, Worcestershire, Essex, Hertfordshire, and divers other places, and various petitions on the subject of inclosures were sent to the government.[1] The government, for its part, was not unresponsive. It would be rash to suppose that there already was a party known as the Commonwealth's Men,[2] for there is no evidence that the name was used before 1549, but there certainly existed an active group fired with an anxious resolution to improve the lot of the ordinary countryman. Its members were largely clerics, Latimer and Thomas Lever, of the vehement sermons, and Crowley of the fierce tracts; but there were laymen too, amongst whom John Hales was conspicuous. Hales, described as 'of Coventry' in a grant of 1545, was a substantial man who founded a free school at Coventry, and shared with Sir Ralph Sadler the office of clerk of the hanaper. He sat as a member for Preston in the first parliament of Edward VI, where he introduced three bills; one for the maintenance of tillage, one against regrating, and the other for compelling every man who kept sheep in inclosed pasture to keep also a proportionate number of cows for breeding as well as for milking. These bills reflect the views set forth in a paper which attributes the universal dearth of victuals to the lack of breeding of cattle and poultry, to regrating, and to the royal purveyance.[1] They do not, however, tally exactly with the opinions expressed in _A Discourse of the Common Weal of this Realm of England_, of date probably 1549, which has been generally attributed to Hales; broadly speaking, whereas Hales's paper attributed the dearth to inclosures, the _Discourse_ tends to attribute the inclosures as well as other evils to the debasement of the coinage. The famous _Dialogue_ has been credited to Sir Thomas Smith,[4] not without grounds, but the claims of Hales have not been finally discarded, and, at all events, he was extremely active in the campaign for agrarian reform. This campaign was not ineffective; Cranmer may have given his support, and Somerset,

[1] _A Discourse of the Common Weal of this Realm of England_, ed. Lamond, p. 148.

[2] _Calendar of State Papers (Domestic), 1547–1580_, p. 22.

[3] Lamond, p. xlii.

[4] Edward Hughes, 'The Authorship of the Discourse of the Commonweal', in the _Bulletin of the John Rylands Library_, xxi (1937), p. 167. The claims of Hales as the author, and Coventry as the scene, are maintained by Elizabeth Lamond in the edition published by the Cambridge University Press in 1893 (reprinted 1929).

whose intentions at least were good, was sympathetic. In the first session of parliament which met in November little advance was made. Bills to check inclosures, to confirm copyholders and leaseholders in their rights, and to bring up poor men's sons, all failed to pass. The act against vagabondage,[1] though it did distinguish between able-bodied and impotent poor, provided penalties of branding and slavery for the one and made no arrangements for the other, save the voluntary collections authorized by the act of 1546; vagabondage was still a crime, and the causes producing it were little regarded.

What could not be done in parliament, where landowners and justices abounded, was attempted by Somerset in an act of power. On 1 June 1548 he issued a proclamation condemning inclosures, and on the same day he issued commissions to ascertain how far the acts of Henry VII and Henry VIII against inclosures had been obeyed. It is not clear how many separate commissions were issued, but the only commission to operate actively was that of which Hales was a member; this dealt with the counties of Oxfordshire, Berkshire, Warwickshire, Leicestershire, Bedfordshire, Buckinghamshire, and Northamptonshire, and there is first-hand evidence as to the procedure adopted.[2] Sheriffs, justices of the peace, and others were ordered to command attendance of representatives of their counties who were confronted with a series of seventeen questions dealing with the making of inclosures, the decay of houses, and the multiplication of sheep. The answers to these questions might constitute a ground for the presentment of persons who had offended against the acts of Henry VII and Henry VIII, but the commissioners themselves were authorized to proceed only against those who hindered their inquiry.[3] Hales, in a letter of 24 July, represented that all was going well, and optimistically prepared bills for the forthcoming session of parliament; he hoped to gain his ends lawfully, and later he defended himself vehemently against the charge that he had stirred up trouble. It is plain, none the less, that men began to take things into their own hands, and the furrow drawn across Warwick's park may be a

[1] 1 Edward VI, c. 3.
[2] The relevant documents are conveniently printed in the *Ecclesiastical Memorials* of John Strype; the proclamation ii. 1. 145, the commission and Hales's charge under the commission ii. 2. 348 ff.
[3] Tytler, i. 114.

proof of local lawlessness rather than of the commissioners exceeding their powers. By the summer of 1549 men, even some of the watchmen themselves, were going about the country to organize agrarian unrest,[1] and by May the protector himself in repeating his proclamation against inclosures, warned men that he would punish outrages. Yet even when, on 8 July, he issued a fresh proclamation against disorders he was still urging the commissioners to proceed with their work, and the last instructions which he gave to them synchronized with the outbreak of Kett's rebellion.

The crushing of this revolt presaged a return to oppressive measures, and during the rest of the reign there was no further talk of agrarian reform;[2] indeed, it was the custom of the politicians to blame Somerset's mismanagement for all the evils of the time. A modern authority has suggested that the only real solution of the question was to confirm the agricultural tenants in rights which, if they had never been absolute under the feudal system, had yet been guaranteed by the economy of the times—the lord needed men to till his demesne. That the incursion of sheep into the arable land produced great hardship cannot be denied, but it may be questioned whether the fixation of peasant holdings would have produced a certain remedy. Not only would a peasant's holding have been too small for the breeding of sheep, but in the course of a few generations it must have become, in many cases, too small to support the expanding family. Soon the question would have been, whether, with the apparatus at its disposal, English agriculture could have brought more rough country under its sway, and whether the peasant, struggling to maintain himself, could have had the economic strength to undertake new ventures. That any sixteenth-century government would have supported him in these ventures is too much to expect. It cannot be disputed that the passing of the old economy brought misery to many; but its evils have been exaggerated; it brought opportunity to some, and prosperity to the nation as a whole. In any case there was no easy remedy in the hands of the government of the day.

In the realm of trade and manufacture, despite the dislocations resulting from the experiments with the currency, England did not fare too ill. The merchant and the capper in the

---

[1] Tytler, i. 187.

[2] The acts in restraint of inclosures were repealed by 3 & 4 Edward VI, c. 3.

Discourse set forth grievances which must have affected small towns and the country generally, but capitalists of enterprise were thriving, and London, despite a visit of the sweating sickness in 1551, was prosperous. The fall in the customs was due, not to lack of trade, but to the calculation of the duties according to an old 'Book of Rates' which gave the various commodities a low price quite obsolete. In the chronicles there is no hint of serious economic distress in the city. Townsmen were buying land and founding families. The great merchants of London gave the government the means to deal with its overseas borrowings, and when, in 1552, the government repaid its obligation by cancelling the privileges of the Steelyard, the enterprise of England was ready to exploit the opportunity.

The young king and his council were interested in geography, and it was with influential support that Sebastian Cabot, who had returned to England from Spain in 1547, promoted a 'mystery and company of Merchant Venturers' to seek for a North-East passage. In 1553 the London capitalists provided £6,000, and three ships were sent out under Sir Hugh Willoughby and Richard Chancellor 'for the discovery of the Northern part of the world'. Willoughby and two of his ships were lost near the North Cape, but Chancellor reached Archangel and went to Moscow where he laid the foundations of future commerce. Soon after his return in 1555 there was founded a company, later known as the Russia Company, with a regular constitution, a governor—Cabot was the first—four consuls, and twenty-four assistants. To this company was given a monopoly of Russian trade which was confirmed in the same year by the tsar. Next year Chancellor was drowned off Aberdeenshire on his way home from another adventure, but shortly afterwards Stephen Borough went to the Kara Sea, and in 1557 Anthony Jenkinson began the memorable journey which took him to Archangel, Moscow, Astrakhan, and even to Bokhara. When he returned to England in 1560 Elizabeth was on the throne, but it should be remembered that some of the splendour of her spacious days had its origin in a bold questing spirit which shone even in the difficult times of Mary and Edward VI.

Important as were the economic issues which presented themselves in the reign of Edward VI they have seemed, to most historians, less important than the religious issues, and to contemporary writers and thinkers religious questions—not always divorced from social and economic problems—were paramount. For centuries the presence of the universal church had loomed over all existence in western Europe; its unity had been broken by schism, its practice had been clogged by superstition, its ministers had not always been worthy, but it had always been there with the glory of its buildings and the regularity of its ritual to remind men that there was a life beyond the grave, and a heaven to which it held the only key. Now its authority was challenged; now men were denying that the pope held in his hands the power to bind or to loose eternally, and that his servants the priests could perform, by divine grace, the miracle of transubstantiation.

To a materialistic age the general preoccupation in the mystery of the Holy Communion may seem inexplicable, but to the sixteenth century the sacrament of the altar was the basis of priestly power, and as the character and conduct of priests came under criticism, certain vital questions presented themselves. Why should the priest communicate more fully than the layman, when Christ had died for all? Had his death given to the priesthood some peculiar privilege? Did that privilege involve the power of performing, through the grace of God, a miracle in the act of consecrating the Host? If Christ had died once for the sins of all men, why should the sacrifice be repeated? And how could the sacrifice be repeated, in any corporeal sense, when Christ's body was in Heaven? If a miracle there was, was the power to perform this miracle connected with the conferment of Holy Orders? To these questions many variant answers were given. Some held hard by the medieval doctrine of transubstantiation in all its simplicity; others denied that there was any miracle at all, asserting that the blessed elements underwent no change, and were, in fact, *nuda signa* to commemorate the great sacrifice.

Between these two extremes of belief various opinions were held, though Luther's doctrine of consubstantiation secured little support in England. Gardiner accepted transubstantiation, but he rejected the idea that the carnal and corporeal presence of his faith involved any 'grossness', in the sense that human teeth should tear the very flesh of the Redeemer. Cranmer accepted the Real Presence in some sense until he was converted by Ridley about the year 1546. It is sometimes said that he accepted the 'Suvermerianism' of Bucer, but 'Suvermerianism' is a latinized form of *Schwärmertum,* and as Bucer opposed the *Schwärmer* or Anabaptists he cannot be supposed to have held their doctrine. Cranmer's view was that there was Real Presence in the Eucharist,[1] according to the worthiness of the communicant, and though, at the very end of his life, he was converted, or nearly converted, to Zwinglianism by à Lasco, it is not clear that he ever accepted the elements as *nuda signa.* Even after his conversion he could say 'that Christ gave his body and shed his blood upon the cross for us, and that he doth so incorporate himself to us that he is our head, and we his members, and flesh of his flesh, and bone of his bones, having him dwelling in us, and we in him. And herein standeth the whole effect and strength of this sacrament.'[2] At first sight it might seem that between his views and those of his opponent, Gardiner, there was in essence no great difference. Yet there lay between the two men a gulf which may be described in Cranmer's own words:

'They teach, that Christ is in the bread and wine, but we say, according to the truth, that he is in them that worthily eat and drink the bread and wine.
'They say, that Christ is received in the mouth, and entereth in with the bread and wine. We say that he is received in the heart, and entereth in by faith.'[2]

Across this gulf no bridge could be found. The questions involved in the debate of transubstantiation affected the mind and the emotion of every man. Some were enraged at the criticism of long-recognized authority, and by the very suggestion that the saints of the past had lived and died in error; others were infuriated by the thought that men's souls had been enslaved, their minds deceived, and their possessions mulcted by a church which had unnecessarily interposed itself between man and his maker. As opinion hardened the quarrel became more definite, and soon western Europe was seething in open war. In Germany, Switzerland, Scandinavia, and France there had sprung up churches which had rejected the power of the pope and the old theology; some of these churches had wedded

---

[1] Constantin Hopf, *Martin Bucer.*
[2] *A Defence of the True and Catholic Doctrine of the Sacrament of the Body and Blood* (ed. of 1907), pp. 25 and vii.

themselves to existing political power, others were in rebellion against it. Everywhere there was conflict.

What was England to do? Under the direction of a powerful king she had repudiated the pope whilst she had retained the doctrine and, in large measure, the practice of the Roman church. Through the troubled waters Henry had steered the ecclesiastical ship with his own hands, setting the course by the star of the royal supremacy. None knew his plans; yet most men had been convinced that he must go either forward to continental protestantism or back to Rome. Now, when his strong fingers had left the wheel, the vessel plunged uncertainly, the officers quarrelled about the course, and the light of the supremacy was dimmed.

What was this supremacy? Was it inherent in the person of the king or in his office? No great acumen is required to guess that the party which controlled the prince would argue that the supremacy passed with the royal title, whereas an 'opposition' would assert that so transcendent a power could not be used either by an infant or a regent. The question was of immediate urgency, for two distinct parties disputed for control, and each claimed that it represented the inmost wish of the dead sovereign. Many symptoms seemed to declare that Henry was verging towards protestantism. He had given the education of his son to 'protestants', and to 'protestants' he had given virtual control of the council; the very survival of Catherine Parr, who favoured the men of the new way, seemed significant to contemporary observers; Cranmer asserted that in September 1546 Henry was going to change the Mass into communion, and it was Cranmer who held his hand when he died a few months later. On the other hand, King Henry's will conformed to catholic orthodoxy; the elements of the sacrament were referred to as the Body and the Blood, the help of the Blessed Virgin was invoked, and Masses were provided for the soul of the testator. It was Gardiner who sang the requiem at Westminster and celebrated the final Mass at Windsor, and it was he who acted as one of Cranmer's clergy in the Mass at the coronation of the new king.

It has been said, and truly, by Burnet, that on Henry's death Cranmer, 'being now delivered from that too awful subjection that he had been held under by King Henry, resolved to go on more vigorously in purging out abuses'. Certainly the protestants hoped that their hour had come and the archbishop felt free to express views upon the Mass, on which he had not hitherto been explicit. On the other hand, the protestants were not alone in feeling relief when the dread presence of Henry was removed. Gardiner, for his part, felt free to denounce Erasmus whom he had previously commended, and was ready to revise his views upon the royal supremacy (at least in so far as it affected his relations with Rome); he knew that, however vocal protestantism might be in the towns, there was a great weight of catholic sentiment in the country, and he may have hoped that with catholics like Wriothesley on the council conservative influences would prevail. He was aware that outbreaks of iconoclasm must be ungrateful to any government.[1] His 'doings and attempts' revealed in correspondence with Somerset, as recorded by Foxe, seem to indicate at least a faint hope that he might bring the new ruler of England to his own way of thinking; and it may be remarked that he accepted without audible comment the new patent which was enforced upon him and on all bishops in February 1547.[2] Somerset, however, was convinced of the need of reform,[3] and when he recognized this Gardiner stoutly maintained that the supremacy should not be exercised till the king came of age. He had, according to himself, a promise from the duke 'that he would suffer no innovations in religion, during the king's majesty's young age',[4] and he held that his 'late sovereign lord and master' was 'slandered' in the assertion that the changes proposed were in accordance with his intentions.

The issue between the two arguments was decided by force. The royal supremacy, bereft of Henry's conscience (which though peculiar was a very real thing), turned to sheer 'Erastianism'. Somerset knew that, to catholics, Mary, the issue of a catholic marriage, might be preferable to his own nephew. He had no mind, indeed he had no power, to restore the wealth of the church, rather did he hope to acquire more. He was aware of the strength of protestant opinion, with which he had a genuine sympathy; he was less advanced than Cranmer in his theological views, but in the main he followed where Cranmer led, and as the archbishop hoped to join hands with the continental reformers, England moved steadily towards protestantism.

[1] Powicke, *The Reformation in England*, p. 77. [2] *Acts of the Privy Council*, ii. 13.
[3] Powicke, p. 78. [4] Foxe, vi. 106.

When Warwick gained power the pace was accelerated. The need, and the desire, for spoliation of church wealth still continued, and the political necessity of extruding Mary was more urgent than ever; hence, although he afterwards professed that he was a catholic, he adopted the cause of reform with an enthusiasm which caused the injudicious Hooper to hail him as 'that most faithful and intrepid soldier of Christ', and hurried England along the path to Switzerland.

The great change was effected by divers means, by governmental pressure, by popular enthusiasm, by proclamation, by parliament; convocation played very little part. In the first months of Somerset's régime the council determined that bishops must take out new patents, and the decision was unopposed perhaps because each of the contending parties hoped that the supremacy would operate in its own favour—the order was issued on 6 February and it was only on 5 March that Wriothesley was ejected from the council. Thereafter for some months, whilst Somerset was engaged in Scotland, the reformation was advanced only by the vigorous sermons of Barlow and Ridley against images and ceremonies, by the failure of the government to deal severely against acts of iconoclasm in London and Portsmouth, and by the introduction of English into the offices of the church. Compline was sung in English in the royal chapel at Easter, and, soon after that, licence was granted to Richard Grafton and Edward Whitchurch to print books concerning the divine service 'in the English or Latin tongue'. Late in July, however, the government went definitely into action by issuing, under royal authority, Cranmer's book of *Homilies*, Nicholas Udall's version of Erasmus's *Paraphrases*, and a set of *Injunctions* which were to be enforced by a general visitation. Most of these were concerned with the conduct of the clergy and of the laity; services were to be conducted regularly, sermons were to be preached once a quarter; parishioners were to be instructed in the Paternoster, the Creed, and the Ten Commandments, and a register of weddings, christenings, and burials was to be kept. All this was regular enough; but the superstitious veneration of relics and images was condemned, and the English tongue was to be used, not only for the reading of the Epistles and the Gospels but for the singing of the Litany as well.

Some of the bishops resisted; Bonner surrendered after

brief visit to the Fleet, but Gardiner, who was more resolute, justified his opposition with great skill. He did not stand upon his argument that no innovations should be made whilst the sovereign was a minor; he did not, as has sometimes been stated, argue that the exercise of 'supremacy' had not the full effect of statute; what he did do was to plead that the supremacy, itself conveyed by statute, could not avail against an existing statute until parliament undid what parliament had done. The faith of the English church, he said, was set forth in the King's Book, and an act of 1543 condemned as heretics all who taught anything in contradiction to the King's Book. Cranmer's *Homilies*, teaching that faith alone justifies, were in manifest contradiction to the faith as by law established, and those who used it might find themselves liable to the penalties of Praemunire. The reference to Praemunire is of doubtful relevance, and it may be doubted whether Gardiner would have been so clear in his analysis had the supremacy been used to support his own aims. To these views, however, he strictly adhered; and he remained in prison until his defences were taken from him by the proceedings of the parliament which met on 4 November 1547.

The momentous acts of this parliament effected innovation almost incidentally. The first act of the session imposed fine and imprisonment upon all who spoke irreverently against the blessed sacrament, which it mentioned in terms of the greatest respect; but it provided that the justices of the peace should summon the bishops to be present at their quarter sessions, and, in a final clause, ordained that the laity should communicate in both kinds. The act which swept away most of the new treasons abrogated most of the legislation against heresy, and removed all restrictions upon printing, reading, and teaching the scriptures. On the ground that the *congé d'élire* involved elections 'which be in verie dede no elections', another statute provided that henceforth bishops should be appointed by letters patent. The act of 1545 confiscating the 'chantries', which had never been fully carried out, was renewed in a slightly different form, and, with various exemptions in favour of teaching colleges, certain dignified chapels, and all cathedral churches, pushed through a reluctant house. Both by what it did, and by what it failed to do, this measure was unhappy for the people of England. Many of the foundations, though secularized, had served useful ends, which now the deprived communities must provide from their own resources. In spite of promises very little indeed was done for education; the schools endowed from the sales of the chantries were mainly established by private donors, and Edward's fame as the founder of grammar schools rests mainly on the fact that his government failed to destroy some of those which already existed.

Whilst parliament was sitting convocation met, and the lower house which demanded not only communion in both kinds for the laity, but the right of marriage for the clergy, submitted to the bishops a serious claim that ecclesiastical business should be under ecclesiastical control. They suggested, amongst other things, that the clergy should sit in the house of commons. Their demands met with no response; the government kept all in its own hands. A proclamation against irreverent talk about the sacrament of 27 December, the release of Gardiner, under a general pardon, on 8 January, a proclamation enjoining the use of fasting for the benefit of the fishery, on 16 January, and a proclamation against unlicensed preachers and private innovations—all these things spoke of a conservative spirit. Yet, at the same time, the reformation went on apace. Latimer continued his vigorous preaching, and the council continued to issue proclamations in favour of the new ideas. On 18 January the Lord Protector and the council forbade the carrying of candles, the taking of ashes, the bearing of palms, the creeping to the cross, and other ceremonies; in February the removal of images was ordered; in March appeared a small pamphlet setting forth the order of communion for the laity, as enjoined by act of parliament; it was based on the *Simplex ac pia deliberatio* of Hermann von Wied, archbishop of Cologne, and was in English though the priest before using it still communicated himself in the Latin of the Sarum Use. In May Latin was excluded from St. Paul's and from the royal chapel. There is some evidence that the government was apprehensive lest innovation should go too far; a proclamation of April in denouncing talebearers, ordered unlicensed preachers to confine themselves to the *Homilies*, and in May even licensed preachers were subjected to the same restriction.

Yet the advance continued; at the end of June Gardiner was sent to the Tower because, in a public sermon preached upon St Peter's day, 29 June, he failed to give satisfaction to the council while Cranmer, fortified by the arrival in December 1547 c

two learned Italians 'Peter Martyr' Vermigli and Bernardino
Ochino, who had come to him via Strassburg, moved towards
continental ideas on the vexed question of the sacrament. He
moved slowly, however, seeking ever for a compromise which
would commend itself to all, and his various liturgical experiments were submitted to informal meetings of doctors and
bishops at Windsor and Chertsey. When parliament met in a
session which lasted from 24 November 1548 to 14 March 1549,
the right of marriage was conceded to the clergy, but the proposed new order of service was vigorously debated among the
peers, and considerably modified before it emerged as a schedule
to the Act of Uniformity. In the final voting it commanded the
support of the majority of the bishops, and of all the lay peers
but three. It could therefore be asserted that the new book,
though never submitted to convocation, carried the authority
of church as well as state, but it must be noted that the archbishop, in his endeavour at a true uniformity, had made concessions to the catholics which sorely disappointed the Zwinglians. His book, though couched in lovely English, was based on
the old Sarum Use and on the reformed Breviary of the Spanish
Cardinal Quignon, as well as upon the *Consultatio* of Hermann
von Wied. Gardiner, after study, declared that he could accept
it. His acceptance may have been easier because the uniformity
was to be enforced by sanctions which were, for the times, mild;
priests who used any other service were liable to a gradation of
penalties which might culminate in lifelong imprisonment, but
laymen might absent themselves from the new service without
any penalty, and even those who actively opposed its celebration escaped, for a first offence, with a fine of £10.

Compromise, especially compromise founded upon variant
interpretations, could not endure. The introduction of the new
liturgy produced the rising in the West Country of men to whom
the unfamiliar English, and the simplicity of the rite, made the
worship of God seem 'like a Christmas game', and among the
learned it soon became apparent that Cranmer could not
maintain his middle position. Towards a new advance he was
impelled by the return of exiles and by the advent of strangers.
In September 1548 there had arrived in England, from Friesland, John à Lasco or Laski, son of a Polish junker, whose views
of the sacrament were at least Zwinglian, and who opposed the
wearing of vestments; about the same time there came, on

Cranmer's summons, the noble Spaniard Francisco de Encin (Dryander) who had emerged from his wanderings throughou Germany as a somewhat radical Lutheran. Valerand Poullain, Calvin's successor at Strassburg, paid a fleeting visit; and although when he was disappointed of a post in Oxford he soor departed, the French tradition was maintained by Jan Utenhove of Ghent, who had been a pastor of the French church ir Strassburg. In the spring of 1549 fresh reinforcements appeared. Martin Bucer,[1] the Alsatian, who, rejecting the *Interim*, had tried to reconcile Lutherans and Zwinglians, arrived in Apri on Cranmer's invitation; along with him came the learned Hebraist Paul Fagius, reputedly even more eloquent thar Bucer himself, and about the same time Immanuel Tremellio, an Italian Jew, another scholar in Hebrew. There came toc John Hooper, once a Cistercian monk, whose resolute belief ir apostolic simplicity was quickened by contact with Laski, anc in August Poullain returned to become superintendent of a colony of Flemish weavers which Somerset had planted at Glastonbury.

The strangers were kindly received; 'Peter Martyr' wa: appointed professor of divinity at Oxford in March 1548, anc three years later startled opinion by taking his wife to live witl him in Christ Church. On 4 December 1549 Bucer was appointed regius professor of divinity at Cambridge, where first Fagius and after his early death Tremellio, held office as reader ir Hebrew. Some of the visitors were provided with livings, and for the others the Strangers' Church in London, of which Laski became superintendent in 1550, provided a natural centre With the opinion of protestant Europe all about him Cranme cautiously advanced towards the Zwinglian conception of the sacrament, but he did not accept it absolutely, and he remained entirely conscious that in the archbishop of Canterbury lay a rich repository of English tradition.

In the session of parliament which lasted from 4 November 1549 to 1 February 1550, catholic influences were still active but protestantism prevailed. A bill for the restitution of epis copal authority was rejected and, against the majority of the bishops, acts were passed for the removal of images and 'super-

---

[1] C. Hopf, *Martin Bucer and the English Reformation* (1946). For his arrival with Fagius see an article by Pierre Janelle in *Revue d'histcire et de philosophie religieuses*, i (1928), 163.

stitious' books, and for the reform of the ecclesiastical law. A third act, for the preparation of a new 'ordinal', which made no mention of minor orders, commanded the support of nine of the fourteen bishops present, but it must be observed that thirteen bishops were absent. Against the bill for the reform of the ecclesiastical law Cranmer himself voted, and this may seem surprising since the project was dear to his own heart. The terms of the act, however, explain his attitude. The king was to appoint a commission of thirty-two of whom only sixteen were to be clerics and of the clerics only four were to be bishops. The archbishop must have felt that the state was interfering in matters hitherto regarded as ecclesiastical, and that in so doing it was deliberately casting a slight on the episcopal bench. Moreover, in his desire for uniformity he may have welcomed an occasion on which he could side with the bishops of the old way. Whether his protest had any effect is uncertain, but when parliament had been dissolved it became plain that he was to have his will. No commission for the reform of the laws was issued until 6 October 1551; then the names of eight bishops occurred among the sixteen clergy appointed; and when, three days later, an effective committee was set up it included Cranmer himself and some of his stoutest supporters. This committee worked with diligence, but the code which they finally produced in a treatise called *Reformatio Legum Ecclesiasticarum* was never published in the days of Edward VI and was finally rejected in the reign of Elizabeth.[1]

Empiric England might leave the question of ecclesiastical jurisdictions to settle itself for the time being, but practical questions could not wait, and in their solution the ideas of the reformers prevailed. The new ordinal, which was included in the second act of uniformity, was distinctly 'protestant' in tone; no provision was made for the ordination of sub-deacons, acolytes, exorcists, lectors, and janitors; bishops, priests, and deacons alone remained. Throughout the years 1550 and 1551 the new ideas steadily made ground. Proceedings were instituted against well-known catholics, including Mary's chaplains, and even against Mary, who defied her would-be persecutors with Tudor arrogance and Tudor success. 'My father made the most part of you almost out of nothing', she said. For her Edward, born after her mother's death, was truly her father's son;

---

[1] *Literary Remains*, 397, n. 3.

whilst he was a minor he should not alter Henry's religious settlement, but she loyally acknowledged him as her king, and it was with his support that she continued her Mass un disturbed.

The catholic bishops were less fortunate. Bonner was deprived in October 1549, and his see, with which the new bishopric of Westminster was combined, was given, with its possessions much reduced, to Ridley, who was translated from Rochester Gardiner, since July 1548 a prisoner in the Tower, was deprived in February 1551, and his successor Ponet, who had succeeded Ridley at Rochester, was made to surrender the revenues of Winchester in return for a yearly salary of 2,000 marks. In August of the same year Voysey of Exeter was compelled to resign in favour of Coverdale; in the following October Day of Chichester gave place to Scory from Rochester; Hooper got the see of Worcester from Heath,[1] along with that of Gloucester, vacant through the death of Wakeman. Tunstall of Durham was imprisoned in his house in the spring of 1551 and, as the commons refused to attaint him, was deprived by a lay com- mission in October 1552; an act of parliament of 1553 which did not become operative divided the great diocese into two separate sees, Durham and Newcastle, both meanly endowed since the great part of the old revenues was to be seized by Northumberland. As Warwick, the duke had shared along with Bedford and Hoby in the spoils of Worcester, and the see of Exeter as given to Coverdale was worth only £500 instead of £1,566. 14s. 6½d.[2]

These changes, which cannot have been altogether grateful to Cranmer, were accompanied by circumstances which sorely tried him. Hooper, already involved in the vestiarian contro- versy, made trouble before he would accept installation, and in the winter of 1552 Northumberland offered the bishopric of Rochester to John Knox. His ostensible reason for doing so was to provide 'a whet-stone to quicken and sharp the Archbishop of Canterbury' and to strengthen the opposition to the ana baptists in Kent, but he may have wished to remove from Newcastle a masterful personality who caused unrest in the

[1] Day and Heath were deprived on 10 October 1551. See p. 502 n. *supra* for spoliation of lands and church plate.
[2] For acquisitions of ecclesiastical property made by Warwick and his friends see Dixon, iii. 252 and 536.

north, and might disturb his arrangements for the new see. Knox, however, who afterwards alleged that in a sermon preached before the king he had compared Northumberland and Winchester to Achitophel and Shebna,[1] proved 'neither grateful nor pleasable', and, doubtless to the relief of Cranmer who wished for no such whetstone, nothing came of the design of promoting him to the episcopal bench.

The attack on the conservative bishops was largely due to political and economic motives, but while it was going on truly religious men were urging further reform. There was a loud demand for the simplification of ceremonies and the revision of the Prayer Book, and shortly before he was made bishop of Gloucester Hooper pleaded for both these things in a sermon preached before the king. When Ridley took the see of London he at once began a campaign against the common manner of celebrating communion which made it appear 'a very Mass', forbade all rites not enjoined by the Book of Common Prayer, and urged churchwardens and curates to substitute tables for altars. In November 1550 the council ordered all altars to be removed throughout the land. The abolition of time-honoured ceremonies spread consternation among the country clergy; others beside the kindly Parkyn of Ardwick-le-Street, who loved the children's ceremonies,[2] grieved to see 'a little board set in the midst of the choir'; and their dismay was heightened when, following on the council's decision of 21 April 1552, the 'super-fluous' ornaments were removed from the churches. Yet, in the course of their attack, the protestants produced proof enough that some reformation was necessary. The evidence resulting from Hooper's visitation of 1551 is astounding. The questions put to the clergy were extremely simple.[3] They were asked to state the number of the commandments, to repeat the Creed and the Lord's Prayer, and to indicate the scriptural authority on which these essential things were founded. Of 311 clergymen confronted with these questions, only 50 answered them all, and 19 of these *mediocriter*; about 170 were unable to repeat the Ten Commandments, and 10 could not repeat the Lord's Prayer; some did not know who was the author of the Lord's Prayer, though good Mr. Dumbell of South Cerney, who could

---

[1] *Admonition to the Professors of God's Truth in England.*
[2] *English Historical Review,* lxii (1947), p. 72.
[3] Ibid. xix (1904), p. 98.

repeat it, knew that it was so called because 'it was given by our Lord the King', and was written in the royal Book of Common Prayer.

Fortified by such proofs that the old system was evil, the reformers urged the king and council to make further change, and by January 1551 Sir John Cheke was able to assure 'Peter Martyr' that if the bishops did not improve the Prayer Book Edward would do it himself and have his action ratified by parliament. Throughout 1551 Cranmer considered a revision. Though convocation met on 24 January 1552 there is no proof that it had any influence in the discussions which took place. Discussion was necessary, for all sorts of opinions were in the air; in May 1550 Joan Bocher was burnt for denying the humanity of Christ, and in April 1551 George van Parris, a Dutch physician, suffered the same penalty for denying his divinity. The moderate Bucer, who in January 1551 had delivered an elaborate *Censura* on the Prayer Book, died on 1 March following, and the archbishop was plied by the persuasions of enthusiastics like Laski. It may be that his cautious mind reacted less to the arguments of the foreigners than to those of Englishmen like Hooper and Coverdale, but, however persuaded, he moved further towards Zwingli, and the second Prayer Book which was authorized by parliament in the session of 23 January 1552 to April of the same year set forth a version of the communion which was essentially an act of remembrance, though the mention of the Body and the Blood as well as the bread and the wine may still have left a loophole for another interpretation. This may have been widened by the rubric bidding the communicants kneel, and Knox and his friends, who had probably introduced the sitting posture into their own services, were alarmed. In a sermon preached before the king in September 1552 Knox denounced the custom of kneeling and so alarmed the king that Cranmer, Ridley, and 'Peter Martyr' were ordered to reconsider the matter. The archbishop had his way; the custom of kneeling was retained, but the council added to the Prayer Book, as it passed through the press, the famous Black Rubric, which explained that the posture, though it expressed reverence and gratitude, implied no adoration. Not in the office of communion only did the second Prayer Book show the triumph of the new ideas; the word table was used throughout in place of altar, the General Confession supplanted the Auricular Confession which had been contemplated by the first book, vestments were simplified, the baptismal ceremony was simplified, and the ordinal of 1550 was, in substance, authorized. The service set forth was enforced under stringent penalties which were imposed not only upon the clergy who used any other form, but upon laymen who absented themselves from public worship or who attended services other than that now established. It has been argued that this coercion of the laity was an acceptance of the ideas of moral and social discipline imparted from the new churches of the Continent, but as against that, it must be remembered (*teste* Hunne), that laymen who had attended unauthorized services in earlier days had been apt to suffer for their irregularity.

The liturgy of the English church being thus established, it remained only to fix the dogma, and this too was done by lay authority. Previous confessions, the Ten Articles of 1536 and the Six of 1539, had been made by parliament, but parliament was not allowed any say in the statement published by Edward VI. In the last session of the reign Northumberland, who roughly rejected Cranmer's attempts to remedy abuses in the holding of church land, was at pains to put the bishops in their place, and there is no evidence that convocation, which sat at the same time as parliament, had any share in setting forth the articles of religion. They were in fact prepared by Cranmer himself and submitted to the royal chaplains, who were authorized to consider them by a letter of council of 21 October 1552. After further revision by the archbishop they were published along with a catechism by Bishop Ponet in June 1553, under a title which stated that they had been 'agreed upon by the bishops and other learned men in the synod'. When Cranmer protested against this assertion, he was told that they had been sitting when the publication was authorized; even this statement was false, although it was true, if irrelevant, that convocation was in session when the printing of the catechism was authorized.

Along with his statement of doctrine Cranmer is said to have prepared fifty-four articles for a uniformity of rites; but these were never issued, and his attempt at uniformity of doctrine had not the success for which he had hoped. Yet, despite the dishonesty which marked their publication, and despite their failure to attain the end desired, the Forty-Two Articles have great clarity, and in some ways a greater tolerance than the Thirty-Nine which took their place in Elizabeth's day. Their uncompromising protestantism was stated with the minimum of offence, and place was found for free will and for good works, as well as for justification by faith. Of the seven sacraments, however, only baptism and communion remained, and the doctrine of transubstantiation was specifically condemned.

Of the Forty-Two Articles not the least significant is that which defines the place of the civil magistrate. Article thirty-six, after explaining that the king of England is Supreme Head on earth next under Christ of the church of England and Ireland, and that the bishop of Rome has no jurisdiction in this realm of England, goes on thus: 'the Civil Magistrate is ordained and allowed of God: wherefore we must obey him not only for fear of punishment, but also for conscience sake'. Determined at once to keep the *Ecclesia Anglicana* in the old tradition, and at the same time to repudiate the pope, Gardiner had enunciated the same doctrine in *De Vera Obedientia*. As to the nature of the *Ecclesia Anglicana*, his views differed from those of Cranmer, but both men meant to keep the English church intact, and both found themselves in the same difficulty—they were compelled to exalt the existing power. In the year 1553 Gardiner had been long in prison; Cranmer's doom awaited him, but already the shadows fell around him. When he endeavoured in the parliament of 1553 to obtain his cherished *reformatio legum ecclesiasticarum*, Northumberland turned and rent him for traducing his betters: 'You bishops look to it at your peril that the like happen not again, or you and your preachers shall suffer for it together.'[1]

Northumberland had never been conspicuous for good manners, but his testiness at this juncture may readily be understood. He had done all in the name of the king, and the king's health was rapidly declining. With superb effrontery he endeavoured to maintain himself in power by playing upon the religious devotion of the dying boy. In June 1553 Edward was induced to 'devise' the crown to his first cousin once removed, the Lady Jane Grey, a girl of seventeen, and to her heirs male. Several acts of Henry VIII's reign had encouraged the idea that the king might fix the succession to the crown. Henry's

[1] R. W. Dixon, *History of the Church of England*, iii. 512.

arrangements, however, in the main followed the ordinary English rules of descent; his exclusion of his sister Margaret might be justified by her marriage to an alien, and his preference for the heirs of his nieces Frances and Eleanor over the ladies themselves, was acceptable as providing a possibility for heirs male. Northumberland, desperate and drunk with his own 'auctoritye', knew no restraints. His first thought was to marry his son Guildford to Margaret Clifford, the daughter of Eleanor, whose claim was the better since Eleanor was dead; on second thoughts he saw that he must pin his hopes on the senior line, that of Frances, and induced the king, in the second of two instruments, to leave the crown, not to 'the L' Janes heires masles' as he at first intended, but to 'the L' Jane and her heires masles'.[1]

This shocking arrangement was carried out so hurriedly that, while the Lady Jane was included in her own person, her sisters were included only in their heirs, and though this absurdity was rectified in the letters patent setting forth the king's 'devise', its essential incongruities remained. Its exclusion of Mary, Elizabeth, and Jane's mother, Frances duchess of Suffolk, could be justified only on the rejection of heirs female upon a sort of Salic law; yet, in the end, the crown was given to a female farther removed from Henry VIII than any of these three ladies.

It was a desperate throw made possible only because the ruling party in England, backed by French policy and protestant apprehension, was determined to exclude Mary at all costs. The council itself hesitated, Cecil stole away and was only recalled by the knowledge that Cheke would succeed him as secretary. When the lawyers were told to execute a will in the terms of the 'devise' they replied that even to draw up such a document would be treason. Browbeaten by Northumberland and comforted by the thought that obedience to a living king could never really be treason, they all surrendered except a brave justice of the common pleas, Sir James Hales. The document was completed, and on 21 June it was signed by over a hundred persons including councillors, peers, archbishops,

[1] Pollard emphasized the similarity between the change in this document and that in Edward's memorandum of 18 January 1552, wherein proceedings were to be instituted against 'the Duke of Somerset's confederates' and the text was altered to read 'the Duke of Somerset and his confederates'. *Political History of England*, vi. 64, 84. See *Literary Remains*, 561–76.

bishops, and sheriffs. Cecil signed, as he afterwards said, only as a witness; Cranmer was the last to sign, reluctantly perhaps but, upon his own assertion, unfeignedly. His plot prepared, Northumberland gathered his strength about him. Jane's sister Catherine was engaged to Pembroke's son, Northumberland's brother Andrew was engaged to Margaret Clifford, and his daughter was given to Lord Hastings, who, as a descendant of Edward III, had a claim upon the crown.

To the party about the court it may have seemed that the bold venture would succeed, but in fact its foundations were rotten. Henry II of France had no real desire to promote the ambitions of Northumberland; he merely wanted to extrude Mary of England in favour of his daughter-in-law the queen of Scots. The duke himself was disliked and dreaded even by his political allies, and, despite his loud-mouthed protestantism, was distrusted by Cranmer and the best of the churchmen. The government, compelled in 1552 by want of money to dissolve the mercenary bands created in 1551, had no striking force ready for the emergency. Mary, who had the Tudor courage, was not wanting to herself in the hour of crisis, and on Mary's side was ranged the opinion of England—if Henry's son Edward died without heirs everyone knew that the succession should pass to Henry's elder daughter.

These realities were unknown to the boy who was coughing away his life in the palace at Greenwich. He had no idea that he had been the tool of ambition. True Tudor, he took for granted that the royal conscience could not err, and that the royal will could not be disputed. For him religious truth was the protestantism he had learned from painstaking teachers, and he was not troubled with doubts. Long he was regarded by protestants as a young Josiah, but later ages have found him a bigot, and have seen in the clash of catholicism and protestantism only the struggle of rival intolerances. Foxe's story that the little king could hardly be induced to sign the death warrant of Joan Bocher has been discredited. Few people today would subscribe whole-heartedly to Knox's panegyric upon 'that most godly and most virtuous king that has ever been known to have reigned in England'. But the case against Edward and against protestantism has been overstated. There were few executions for religion in his day and England became a refuge for the persecuted of many lands; Gardiner was imprisoned; but although his circumstances were not easy he was able, in spite of a prohibition, to write six volumes of theological controversy, to collect Latin proverbs, and to write Latin verse.[1] In Mary's reign there were many executions for religion, and Cranmer was burnt.

[1] Muller, p. 204.

XV

THE REIGN OF MARY

EDWARD died on 6 July 1553 and Northumberland at once attempted the *coup d'état* which he had so carefully prepared. The king's death was kept secret for three days. The lord lieutenancies, which carried the control of the shire levies, had already been entrusted to the duke's partisans; now Windsor was fortified and the guns in the Tower made ready. About the duke were collected many nobles whose titles, if some of them were new, were high-sounding, and a close touch with the French ambassador was maintained. On 10 July the Lady Jane was brought by water from Isleworth to Westminster and on to the Tower where she was greeted with a tremendous roar of artillery. When, later in the same day, she was proclaimed queen throughout the city it seemed that the stroke had succeeded, and the imperial ambassadors wrote at once to their master that there was little chance of Mary's mounting the throne of England.

Mary herself had other views. Mistrustful of Northumberland's blandishments, and forewarned, probably, of his designs, she had declined an invitation to be present at Edward's death-bed, and when Lord Robert Dudley arrived to take her at her dwelling at Hunsdon he found her already gone. Ships had been sent to intercept her if she fled to the Netherlands, but flight was not in her mind. She meant to be queen of England, and when she left Hunsdon, perhaps as early as 4 July,[1] she had ridden hard for the Howard country. She encountered some perils in protestant East Anglia; the men of Cambridge assailed her company and even after she had gathered considerable strength her friends were refused admittance to Norwich; but she found refuge first at Kenninghall and later at Framlingham where her supporters rallied round her. Her party, as a Spanish observer noted, contained few persons of distinction, but its numbers increased surprisingly, and on the

---

[1] Antonio Guaras, *The Accession of Queen Mary*, ed. Richard Garnett, p. 89. Mary told the imperial ambassadors that she would declare herself queen on Edward's death; they had not approved, but on 7 July they knew that she had retired to Kenninghall on the pretext of illness among her servants; on 10 July they still thought her chances small. *Spanish Calendar*, xi. 73–9.

9th the resolute lady wrote to the council bidding them proclaim her queen. To this message which arrived on the 11th the
council replied in a letter composed by Sir John Cheke, and
Northumberland himself, writing in the name of Jane, drafted
a letter to the lords lieutenant bidding them resist the claim of
the Lady Mary 'basterd doughter to our said dearest cousen and
progenitor great unckle Henry the eight of famous memory'.[1]

Northumberland's own adherents distrusted him, fearing
perhaps with Arundel[2] that one who had already shown himself so imperious as a duke would be intolerable with the power
of the Crown at his disposal; his son's claim to the crown matrimonial bred fresh suspicions among the nobles, and among the
people at large he was, as Jane herself said, 'hated'. Ridley's
sermon asserting the illegitimacy of Mary and Elizabeth roused
deep anger, and the proclamation of Queen Jane evoked no
cheering. Few said 'God save her!' and a tapster named Gilbert
Potter[3] had the temerity to say that Mary should be queen; for
this he lost his ears next day, but public sympathy was with him
and the accidental drowning of his master, who had denounced
him, was hailed as a portent. When Mary's challenge appeared,
therefore, Northumberland found himself in a dilemma. A
good soldier and a good man-at-arms, he knew that his best
course was to strike boldly at his opponents, but he lacked the
strength at once to march out in force against Mary and to hold
London securely. Loath to quit the restless capital and the
doubting council he tried to entrust Suffolk with the expedition;
but Jane would not allow her father to go, and his advisers,
unfaithfully perhaps, urged him to take command himself,
reminding him that the shire of Norfolk still remembered the
weight of his arm. Persuaded against his will, he gathered a
force given as 3,000 horse and some foot, including the royal
guard, strengthened it with some thirty guns from the Tower,
and, bidding the lords reinforce him at all speed, set out to seek
his enemy.

The omens were bad. 'The people prece to se us', he said as

---

[1] *Chronicle of Queen Jane and Two Years of Queen Mary* (Camden Society, 1850),
p. 104.

[2] *Gentleman's Magazine* (1833), p. 12c.

[3] Potter afterwards received, from the grateful Mary, a handsome grant of lands
in Norfolk. The lands were, in part, derived from the spoils of two priories,
Blacborough and Wymondham. *Calendar of Patent Rolls*, Philip and Mary, i.
1553–4, p. 168.

he rode forth through Shoreditch, 'but not one sayeth God spede us';[1] and he met with no encouragement as he advanced. News came that Mary was proclaimed in Buckinghamshire, Berkshire, Middlesex, and Oxfordshire; that Sir Peter Carew had changed sides in the west; that Paget was mustering to march on Westminster and that the ships sent to intercept Mary had declared in her favour at Yarmouth. On the 16th the duke reached Cambridge where Dr. Sandys preached a sermon in his favour; but when, next day, he advanced to Bury St. Edmunds his dwindling and disheartened host found that it confronted a far superior array. The earl of Bath (John Bourchier) and other magnates had joined Mary; the earl of Sussex (Henry Radcliffe) was on his way; 'innumerable companies of the common folk' had appeared, and report gave her about 30,000 men. In the face of such power the duke recoiled at once to Cambridge and there on the 20th his sorry adventure ended.

Behind his back his party had collapsed; only his powerful will had held the waverers together, and he being gone each man began to shift for himself. At first they thought that they must act together at all costs, and the pliant marquess of Winchester, who had stolen away to his own house, was brought back by force; but even the semblance of unity could not be maintained. Secretary Cecil who, as he afterwards asserted, had never approved of the design, not only evaded his responsibilities but paralysed the action of the council, plotted desertion with some of its members, and practised for the seizure of Windsor castle. The reappearance of Paget, not unmindful of his grudge against Northumberland, so increased the forces of dissolution that weak Suffolk was powerless to resist. On the 19th there was a general *sauve qui peut*. Several of the councillors, on the pretext of seeing the French ambassador, left the Tower and met at Baynard's Castle where Pembroke was then living; Arundel, who had called down blessings on the head of the departing duke, now took the lead in denouncing him; Pembroke supported Arundel and the other councillors concurred; the lord mayor with his attendance was summoned and Mary

---

[1] *Chronicle of Queen Jane and Two Years of Queen Mary*, p. 8. This chronicle say that only 600 men went with the duke; these must have been his own retinue, bu the figure of 3,000 horse, which comes from Guaras, is possibly too high. The men of the guard were ill content and at first refused to march.

was proclaimed queen in the City. In the Tower Suffolk tore down the cloth of estate from above his daughter's head and, according to one account,[1] himself proclaimed Mary on Tower Hill; then he abandoned the Tower with his daughter in it, and went into the City. Next day, along with Cranmer, Lord Chancellor Goodrich, and the poor remains of the council, he dined with the lord mayor at the Guildhall, but about the diners London was mad for Mary in a tumult of bells and jubilation which had broken out as soon as she was proclaimed.

Apprised of these events Northumberland at Cambridge threw up his cap and proclaimed the new queen; but on the morrow, abject in defeat as he had been arrogant in triumph, he was arrested by Arundel, and a few days later was led to the Tower amid the jeers and stones of the beholders. With him went his brother Andrew and three of his sons, and Huntingdon with his son Lord Hastings. His luckless daughter-in-law was already in captivity, removed from the state apartments to the house of one Partridge within the precincts of the Tower. Within the next few days Lord Robert Dudley, the earl of Northampton, and the duke of Suffolk passed the grim portal, but of the other prisoners none was noble. Among the men of note were the two chief justices, Cholmley and Montague, Sir John Cheke the secretary, Bishop Ridley, and Dr. Sandys, along with Sir John Gates and Sir Thomas Palmer who were creatures of Northumberland and had been stout upholders of his 'auctoritye'. To such small compass was his party come; the dazzling train of magnates which had attended him had vanished. Arbitrary power had little solidity beneath its proud appearance, and in the face of stout opposition it withered away. History has witnessed the same spectacle on more than one occasion.

The paucity of the duke's following was emphasized when the victorious Mary entered London on 3 August. With her came Elizabeth who had ridden out to greet her and been well received; with her too came Anne of Cleves, the duchess of Norfolk, the marchioness of Exeter, and a splendid train of ladies and gentlemen whose names recalled the ancient glory of England. Coming in by Aldgate and receiving from the mayor, only to return it to him, the mace of the City, she was greeted with cheering, music, gunfire, and decorations in a welcome very different from that given to poor Jane. As she entered the

---

[1] Ibid., p. 12.

Tower four released prisoners sought her favour—old Norfolk who had been saved from execution by Henry VIII's death, young Edward Courtenay who had been imprisoned since 1538, the duchess of Somerset, Anne Stanhope, who had shared her husband's fall in 1551, and Bishop Gardiner who had been in durance since 1548. These illustrious captives had preferred to take their deliverance at Mary's own hand, and they had their pardon on the morrow. Next day Bonner was released from the Marshalsea, Tunstall from the King's Bench, and soon Heath and Day were freed from the Tower.

England had declared for a queen whose right seemed good according to the English way, and at first it seemed that England had been right in her choice. The new government established itself with little obvious change. Less blood was shed than in the disturbances which marked the fall of Somerset; the queen of her own accord showed a mercy deemed excessive even by the imperial ambassadors, themselves advocates of leniency. Only three men lost their lives. Northumberland died meekly, having vainly protested to Gardiner that he had been a catholic all along;[1] Gates resignedly, blaming the duke's 'auctoritye' as the cause of all; and Palmer stout-heartedly, repenting that his conduct had shown so bad an example of the reformed religion which he professed. Winchester and Pembroke were confined for a few days, and Pembroke had to surrender his gains from the diocese of Winchester; but both were free and sworn of the privy council by 13 August. Northampton was soon liberated though he was not restored to the council, and Suffolk himself was released after a short captivity. The Lady Jane, her husband, her brothers-in-law, and Cranmer were condemned in November and, though the sentences were not carried out, remained as prisoners in the Tower. The two chief justices, the chief baron, the master of the rolls, and the solicitor-general were deprived; and before long Hales, the only judge who had refused to sign Edward's 'devise', lost his office. Yet, the very fact that the lord mayor had entertained Cranmer and Suffolk after the proclamation of Mary argued that no radical change was expected in London; in fact Cranmer had been at liberty till 14 September, and no fewer than twelve of Edward's councillors found places at Mary's table, along with seven of

[1] Robert Parsons, *A Temperate Ward-word* (1599), pp. 43–4. Tytler, ii. 231.

Henry VIII's councillors whom she recalled, and twelve of her personal adherents.

The council so constituted was too big for the effective conduct of administration and it lacked unanimity. By this time, however, the practice of working in committees was well established; and even better established than the existence of committees was the existence of factions. There were rival cliques—it would hardly be right to call them parties—even in the king-ridden council of Henry, and in the day of Edward control had obviously passed from one aristocratic group to another. It seemed natural enough that with the advent of a new queen a fresh set of councillors should rise to eminence, and the changes which marked the beginning of the new reign were not unduly disturbing to the popular mind. The nobles of Northumberland's faction naturally lost their places and along with them went the two clerics, Cranmer and Chancellor Goodrich, the two chief justices, Mr. Secretary Cecil, and other skilled administrators including Sadler. Yet, in spite of the return of Norfolk, Winchester managed to keep his post as treasurer, and the household offices were given to Arundel, Cheyney, and Gage, who became master, treasurer, and chamberlain. Others of Edward's councillors who had gone over to Mary at the last moment still retained considerable authority, and the political experience lost by the expulsions was made good not by new men but by the tried servants of Henry VIII; Gardiner obtained the Great Seal, Bishops Tunstall and Thirlby reappeared, and along with them came the seasoned politicians Paget, Rich, and Southwell. Mary's personal followers were recommended by their fidelity rather than by their political experience and they did not play a conspicuous part in the conduct of affairs.

The quarrel between the various factions was very bitter; for long there was a struggle between Gardiner who, after opposing a foreign marriage decided to join the resolute catholics, and Paget who, having first supported the Spanish match, later joined hands with Pembroke and Arundel. At one time each of the rival parties plotted the imprisonment of the other or was said to do so,[1] and the imperial ambassadors alleged that the councillors forgot the service of their queen to think of their private quarrels. The queen herself alleged that she spent her days shouting at her council and all with no result. It seems,

[1] Tytler, ii. 392, 938.

however, that the imperial ambassadors drew their own profit from these unseemly quarrels.[1]

Yet behind the contendings may be detected a slow growth in constitutional machinery. In March 1554 the imperial ambassador, Renard, concerted with Paget and Petre a plan whereby affairs of state were to be conducted by an inner council composed of Gardiner, Arundel, Thirlby, Paget, Rochester, and Petre; next month he wrote that 'the last reduction and reform in the Council of State ought to be continued and become permanent'.[2] Soon after Philip's arrival in England he was advised by his father, who had been visited by Paget, to reduce the size of his council, retaining Gardiner, Paget, Thirlby, and Petre, whomever he discarded; and in fact when Philip went off to the Netherlands in September 1555 the government was entrusted to a group of seven councillors, the six already mentioned along with Pembroke. As a committee mainly of the same members afforced by a couple of lawyers was appointed in February 1554 'to consider what laws shall be established in this parliament', it is plain that one of the basic ideas of the cabinet system was already germinating. It might even be argued from the attempt made before the parliament of November 1554 to secure the return of members 'of the wise, grave and catholic sort' that there was some notion of a party system; but in fact no special efforts seem to have been made in 1553 and though for the later parliaments attempts to 'pack' were made there was really no great departure from customary methods.[3] The development was, like most constitutional development in England, the result of practical considerations rather than of any very exact theory. The government of Mary differed from that of her brother because the crown was now on the head of a resolute sovereign of mature age; but to outward seeming there was little change in its machinery throughout her reign, and at its outset none.

In ecclesiastical affairs, too, it seemed at first that few alterations would be made. The imperial ambassadors had not only urged the queen to moderation but let the council know that

---

[1] Muller, pp. 255, 262.   [2] Tytler, ii. 347, 372.

[3] Pollard, *Political History of England*, vi. 143, 147. The evidence is rather that an opposition was hardening; in 1555 Mary was trying to curb the independence of the commons by establishing a check upon attendance, and proposing to make the 'residence qualification' obligatory so as to ensure the election of humbler folk from the boroughs. The opposition replied with a 'place bill' (*infra*, p. 554).

they had done so, and for the moment she curbed her religious zeal. According to Foxe[1] she had gathered her first following in protestant East Anglia by promising that she would maintain the religious establishment of her brother; this story is supported by Knox, and it is true that Charles had advised his cousin to 'dissemble'. Certainly the new government gave the impression that nothing drastic would be done. Under imperial pressure Mary agreed that the obsequies of her brother should be celebrated in Westminster Abbey by Cranmer according to the protestant rite, though she herself, with many distinguished catholics, was present at a Mass sung by Gardiner in the Tower; and a proclamation of 18 August announced that the queen, though she adhered unflinchingly to the religion into which she was born, and hoped that her subjects would embrace it, would compel no man to do so until a new determination was made by parliament. This proclamation, which prohibited preaching throughout the realm, was to have grave consequences,[2] but for the moment it did not put a stop to the activities of Knox who was preaching in England, and who even prayed for 'our soveraigne ladye, quene Marie', though his prayer was a petition that the hearts of queen and council might be illuminated with the gifts of the Holy Ghost and inflamed with the true fear and love of God. There was perhaps no great probability that the queen would convince the preacher or that the preacher would convert the queen, but for the moment a truce was declared and the whole matter was remitted to the decision of parliament. Nothing could have been more correct. As regards finance, too, a sweet reasonableness prevailed. On 4 September one proclamation announced some reforms in the currency and another remitted the sums still due in respect of the subsidies voted in Edward's reign. The queen, it is true, demanded a loan from the City; but she was content with half the £20,000 she originally asked, and that she repaid within the month.

Despite the fair appearance, the hopes of a peaceful continuity in the affairs of England were fallacious. The mainspring of the sixteenth-century state was the prince, and there was that in Mary's character which made it impossible for her to compromise. She had the undaunted courage which had gained her the throne and a certain good-natured joviality which she could

---

[1] Cf. Knox: *Works*, iii. 295.
[2] See p. 542 *infra*.

use upon occasion. These things she had perhaps from her father, and from him too she had other and less attractive qualities of the Tudor. She had the clear conviction that the thing she wanted was right in itself and that because it was right she was entitled to seek it by all the means in her power without undue attention to scruple. From her mother she inherited a passionate devotion to the Roman church which remained utterly unshaken by suffering, by success, and by intellectual questionings. There is no suggestion that her good education under Udall and her reading of Erasmus's *Paraphrase* ever caused her to vary one hair's breadth from the path of truth as it had been shown her in her childhood. One other thing she took perhaps from her Spanish ancestry—her attitude to marriage. As a result perhaps of their long connexion with the Moors the Spaniards had acquired some oriental ideas about the place of woman. Was it not her grandmother the great Isabella herself who had said '*hombre d'armas en campo, obisbo puesto en pontifical, linda dama en la cama, y ladrón en la horca*'?[1] It was the duty of a woman to give herself to her husband and to bear him children; and from this duty she had been prevented by an unjust fate. From her infancy up there had been projects of marriage for her and she had been betrothed at one time to the dauphin, at another to her cousin the emperor; there had been talk of the duke of Milan, the duke of Orleans, of the old king of Hungary and the young king of Scots, of Don Luis of Portugal, of Philip of Bavaria, and of others besides. All these schemes had come to nothing and one great cause of their failure was the fact that Mary had been, since 1533, illegitimate.[2] To one so sure of her faith as was Mary the stigma itself can have meant little—she knew that she was the true daughter of Henry's true wife—but the cruel injustice must have rankled unceasingly. She had never been very beautiful though she was not ill-looking in youth, but now she was thirty-seven and her charms were already beginning to fade. In the first interview which she had with the imperial ambassadors after her accession, after modestly stating that she had never yet contemplated matrimony and as a private individual would not desire it, she let Charles understand quite plainly that she

[1] *Œuvres complètes de Brantôme* (Société de l'Histoire de France), ix. 297.
[2] She had even been coerced into making a written acknowledgement that her father's union with her mother had been incestuous.

expected him to find a husband for her, and that she regarded his alleged preference for a native husband as a mere device to drive time. She wanted a Spaniard. Charles was vain enough to believe at one time that he himself was the object of her hopes, but he had no mind to marry and wrote at once to put the possibility of an English match before his son Philip. The queen was set upon marriage and upon marriage with a husband to whom she could defer—in short upon marriage with a Spaniard. Yet the decisions of the prince must touch the fortunes of the whole state. Bound up with the royal marriage were the two urgent questions of the day, the questions of foreign policy and religion. These were so closely intertwined that they were in fact inseparable although for convenience they may be separately examined.

The death of Edward VI was of high consequence to the politicians of the Continent; Habsburg and Valois were still locked in their secular quarrel and to each the alliance of England was of great importance. In the year 1552 the fortunes of Charles, so flourishing in 1547, seemed to reach their nadir. His obvious determination to settle the political and religious affairs of western Europe in the interests of his own family had bred for him a host of enemies—the pope and the Italian princes, France and her ally the Turk, as well as the protestants of Germany. His introduction of 'Spanish' ideas had alienated some of the catholics of Germany, including his brother Ferdinand, and when his foes suddenly combined against him he had been unable to resist them; driven to ignominious retreat he had been compelled to sign, in August 1552, the treaty of Passau which, for the time being, recognized the *cuius regio eius religio* as far as Lutherans and catholics were concerned. Meanwhile France had overrun Lorraine with its three great bishoprics, Metz, Toul, and Verdun, and when the emperor had tried to recover Metz he had been foiled by the martial Guise in a siege which lasted from October till December. The year 1553 found him in the Netherlands defending his western borders and, though the death of Maurice of Saxony in July rid him of the ablest of his German enemies, he had been quite unable to regain what he had lost to the French. The war had degenerated into a position of stalemate. To Charles, therefore, the prospect that the old Anglo-Burgundian alliance might be

renewed was very agreeable, and he viewed with satisfaction
the accession of a princess who was of Spanish descent.

To France, on the other hand, the triumph of Mary was a
diplomatic reverse. Henry II had profited by the difficulties of
Charles not only in gaining territory in Lorraine but in establishing his influence in the British Isles. Scotland was already in his
grasp; the weak governments of Edward VI had recognized his
ascendancy and though, in order to oust Spanish Mary, he
supported the protestants, he had in Scottish Mary, a catholic
candidate for the English crown.

England therefore became the battleground of two skilful
ambassadors, Antoine de Noailles for France and Simon Renard
for the empire.[1] The failure of Northumberland's plot was an
imperial victory but, in spite of Mary's obvious preference for
Spain, Renard did not have things all his own way. Noailles
was quick to point out that Mary's marriage with a Spaniard
would put England under the heel of a foreigner. His professions
lacked sincerity since his own master hoped to establish an alien
on the English throne; but they gained for him a party, the
more easily since Henry II had manifestly aided the protestants
of Germany.

This was the diplomatic situation in which Mary was called
to find herself a husband and she did not hesitate. As early as
15 August Renard reported that the suggestion of Philip's
name would be welcome, and thereafter the task of the imperialists was not so much to persuade Mary to marry their prince
as to prevent her from alienating public opinion by rushing too
precipitately into a foreign match. Their skilful diplomacy was
not unopposed; many good catholics were nervous of Habsburg
ambition, and before August was ended there was some talk
that the queen might marry Cardinal Pole, who was only in
deacon's orders and was of the English royal house; Pole,
however, was 53 and had no mind to wed. Another possible
aspirant was Edward Courtenay whom Mary had restored in
blood and made earl of Devon; he too was of royal descent
and he enjoyed the support of Bishop Gardiner who had been
his fellow prisoner in the Tower. He was, however, some
ten years younger than Mary, he had spent half his life in
captivity, he lacked the stuff of which kings are made, and the
suspicion easily arose that 'Wily Winchester' was advancing

[1] See E. Harris Harbison, *Rival Ambassadors at the Court of Queen Mary* (1940).

him with the intention of gaining power for himself. Gardiner was supported by about one-third of the council but against him was the queen, and her will was decisive. She held secret interviews with Renard, and after she had been crowned amidst great enthusiasm on 1 October she no longer concealed her purpose. On the 29th she gave her solemn promise to the imperial ambassador that she would marry Philip; and when, at Gardiner's instigation, a delegation from both houses warned her against a foreign match she brushed ceremony aside and, disregarding the presence of the chancellor, bluntly asserted with her own royal lips that her marriage was her own affair. So unpopular was the match that, according to Noailles, rebellion was imminent and Plymouth was willing to put itself under French protection. No doubt the ambassador exaggerated, but England certainly looked askance at the prospect of a Spanish king; the Netherlandish alliance had generally been popular, but by this time, as London well knew, the Netherlands themselves were weary of paying for the ruinous wars of the Habsburgs.

On New Year's Day 1554 the street-boys pelted with snow-balls the forerunners of the imperial embassy, and when on the morrow the ambassadors themselves appeared in the capital they were met with silence and gloom. Gardiner, compelled to accept the alliance, had made the best bargain he could, and the terms concluded at Winchester House on 9 January safe-guarded the interests of England as much as possible. Philip was to have the title of king and to assist in government; but Mary alone was to have the conferring of all offices of church and state, which were to be held only by Englishmen. England was not to go to war against France. The laws and customs of England were to remain intact. Mary was to have a jointure of £60,000 a year and was not to be taken abroad without her consent. If she died childless Philip was to have no further interest in England; but if children came they were to inherit Burgundy and the Low Countries as well as England, and, in the event of Don Carlos's death without offspring, all the splen-did heritage of Spain. When the chancellor explained these terms on the 14th to an assembly of nobles and gentlemen at Westminster, and on the 15th to the lord mayor and the citizens, his eloquence roused no enthusiasm. England had no faith in the imperial promises; and some of the malcontents were in fact preparing for revolt.

The principal leader of the rising which occurred was Sir Thomas Wyatt; but Wyatt himself, son of the poet and the companion of the poetic earl of Surrey in the window-breaking escapade of 1543, was a hot-headed impecunious gentleman of the sort always to the fore when conspiracy is in the wind, and his ultimate intentions are not clear. Certainly he meant to prevent the Spanish match; probably he hoped to restore the religious settlement of Edward; possibly, as he dealt with Courtenay and made overtures to Elizabeth, he may have intended also to dethrone the queen, though of such a design there is no clear proof. He received help from the ambassadors of France and of Venice and prepared for simultaneous outbreaks to be led by the Carews in the west, by Sir James Crofts in the Welsh marches, by Suffolk in the midlands, and by himself in Kent. The insurrection was planned for 18 March but everything went wrong; Courtenay, when summoned by Gardiner on 21 January, betrayed his confederates; the sporadic risings came to nothing and Wyatt when he raised his standard at Maidstone on 25 January was disappointed of the help he expected from his own county. None the less he collected about 4,000 men; a force sent out from London under the octogenarian Norfolk went over to his side; on 3 February he arrived at Southwark. He was, however, already too late; London Bridge was held against him and London was in arms. Mary, once more equal to her hour, had appeared personally at the Guildhall two days previously and her appeal to the loyalty of the citizens had not remained unanswered. The threat of a bombardment from across the Thames forced Wyatt to leave Southwark on the 6th, but he crossed the river at Kingston by a bold manœuvre and came down to Knightsbridge before he halted. Next day he pushed on, beating off an attack by Pembroke's men near what is now 'Hyde Park Corner' and passed Winchester's men without fighting in Fleet Street, only to find in the end that Ludgate would not open to receive him. Retreat was impossible, for although some of his followers had caused great alarm in the palace of Westminster they had done nothing against Pembroke who had advanced as far as Temple Bar. Utterly discouraged, Wyatt surrendered without fighting, and by five o'clock he was on his way to the Tower along with a batch of his followers. There, on the 10th, he was joined by Suffolk and his brother who had been routed by the earl of Huntingdon

The government had had a fright; a conspiracy, forced into the open almost two months too soon, had failed by a very narrow margin; London itself had been in danger and the temper of its citizens uncertain. The queen's catholic friends had been ineffectual in the crisis and the battle had been won for her by men like Pembroke who had deserted Northumberland only at the last moment; of the very gentlemen-pensioners to whom her own person had been entrusted, some were protestant.[1]

Gardiner, who preached at court on 11 February, urged the over-gentle queen to show her mercy to the commonwealth by cutting off its 'hurtful members', and the measure of the council's alarm appears from the vengeance exacted. That the leaders should be executed was a matter of course; but forty-six of the common folk were hanged in London in a single day while others were sent to suffer in Kent *quoad terrorem populi*, and the government improved the opportunity by ridding itself of inconvenient persons whose complicity could not be proved. The Lady Jane Grey and her husband were beheaded on Tower Hill on 12 February, although the insurrection had not been made in their name and they had had no opportunity of taking part in it. As they had already been condemned the previous November no trial was necessary. Dudley died bravely enough, and the courage and dignity of his wife, a girl of sixteen years and five months, were conspicuous. In her death as in her life the Lady Jane illumines a sordid page of history; beautiful, gentle, and well-educated she was quietly firm in her religious faith. Perhaps, like many other girls, she was not entirely insensible to the glamour of a crown, but she fell a martyr to the dynastic ambitions of her kinsfolk and to the family discipline of a day in which a daughter did what a father bade her. Her father died on the 23rd, a victim rather of his own folly than of his crime; and before long her uncle Lord Thomas Grey mounted the scaffold. Wyatt was spared till 11 April in the hope that he would denounce his fellow-conspirators, and he may have told something, though possibly Courtenay had left him little fresh to tell. Courtenay was imprisoned in Fotheringhay for a while and later sent abroad; he might count himself fortunate. Gardiner had evidently suppressed some evidence

---

[1] e.g. Edward Underhill, *Narratives of the Days of the Reformation* (Camden Society, 1860), p. 161.

against him, but even if he had not been involved with Wyatt
at all he was still a danger to Mary's crown; Paget himself had
at one time suggested that he might be married to Elizabeth, and
Renard, thinking that he would be well out of the way, had
pressed for his execution. No less ardently did he urge that
Elizabeth too must be destroyed before Philip should appear
and at one time it seemed possible that his importunity would
succeed.

Elizabeth had managed to retain her sister's favour by attending Mass for the first time on the Nativity of the Blessed Virgin
(8 September); but Mary was not sure about her conversion,
and when, pleading indisposition, the princess disobeyed a
summons to come to Westminster at the time of Wyatt's
rising, suspicion was aroused. When she presented herself at
Westminster on 23 February she was refused audience; the
discovery of one of her letters to Mary in a French diplomatic
bag (obtained by highway robbery) suggested that she had
secret dealings with Henry II, and on 18 March she was sent
to the Tower. She entered tearful and reluctant, knowing that
great persons who passed through the Traitor's Gate seldom
returned alive, but in the end nothing could be proved against
her. Certainly Wyatt had written to her; but any answer she
had given to him had been verbal and on the scaffold he
exonerated her altogether. An ebullition of protestantism in the
City, the removal of Wyatt's head from the gibbet, and finally
the acquittal of Sir Nicholas Throckmorton by a jury (which
was very severely punished) convinced the government that it
must not go too far, and on 19 May Elizabeth was released
from the Tower and taken to Woodstock. There she was kept
under surveillance for almost a year, but her life was safe.

Mary may have been alarmed for a time by the threat to her
authority but she learned no permanent lesson; she was impressed less by the rebellion than by its speedy collapse, and the
only conclusion she drew—it would have been drawn by her
father too—was that the royal will was paramount. With un-
damped enthusiasm she went forward to a marriage which
was detested by many of her subjects. The parliament which
sat from 2 April to 5 May 1554 offered no opposition to the
treaty, though careful provision was made against Spanish
interference. It was declared by statute that the queen's regal
power was equal to that enjoyed by any of her progenitors, and

that Mary should not cease to rule when she took a husband; a bill for making offences against the queen's husband high treason was lost; Gardiner's proposals that Elizabeth should be excluded by name and Mary empowered to bequeath the crown by will met with no support. To the marriage itself, however, no objection was made. A papal dispensation, of no validity in English law, removed the canonical bar of consanguinity between the cousins. On 20 July the prince's 'navie of vii score saile' arrived at last in Southampton Water, and three days later Philip met Mary at Winchester.[1] On the 25th, graced by his father with the titles of king of Naples and king of Jerusalem, he was married to the queen of England in Winchester Cathedral by Bishop Gardiner.

The ceremonies were marked by the old English splendour; the queen's ring was 'a plain hoope of gold' because 'maydens were so married in olde tymes', and the 'faire ladyes and the moste beutifull nimphes of England' were in attendance. Philip and his retinue displayed the magnificence and dignity of Spain and the subsequent progress to Windsor was not lacking in the proper decorum. London, when the bridal pair made their formal entry on 18 August across London Bridge from Southwark, gave them a traditional welcome with pageants, orations, and artillery. It is significant that one of the pageants showed Henry VIII with a book marked *Verbum Dei* in his hand, and that Gardiner had the offending ornament replaced by a pair of gloves; significant too that twenty cartloads of Spanish gold were drawn with ceremony through the streets a few days after the festivities; more significant still that, on 20 November following, Cardinal Pole arrived in England.

Along with the preparations for her marriage the queen had hurried on the work of making England fit to receive the son of the catholic king. She did not feel herself bound by the promises made in her name by her friends at the time when her succession was in doubt;[2] from the very first she took it to be her

[1] See the letter of John Elder (a Scot who advocated union with England) printed as app. x to *Chronicle of Queen Jane and of Two Years of Queen Mary* Camden Society, 1850).

[2] It is doubtful if Mary made any definite promise to maintain the *status quo*, although Foxe's statement that she promised to make no 'innovation' when first she raised her standard is supported by Knox; the adherence of a somewhat 'protestant' country-side to a catholic queen might have been due to memories of Dussindale. Froude asserts that, from Kenninghall, she promised the council not to wed a

duty to restore England to the bosom of Rome. The proclamation[1] of 18 August in which the queen promised to exert no compulsion made her meaning plain enough. She held her hand for the time being because she knew that a reformation could best be accomplished with the aid of parliament, but she did not hesitate to promote the true cause by vigorous personal action before the estates assembled. Authority, she knew, would do much; she had, though she did not profess to use it, the prestige of the 'supremacy' attached to the Crown by her father; and she had in Gardiner a chancellor who, though he well recognized the authority of parliament, could not only distinguish between the executive acts done by Northumberland and the laws passed by the two houses, but could even shut his eyes whilst the laws themselves were ignored. With the approbation of the government catholicism at once began to gain ground.

On 13 August Dr. Gilbert Bourne, who preached at Paul's Cross by the queen's appointment, prayed for the dead and denounced the imprisonment of Bonner. A dagger was thrown at him and the bishop himself was in some danger from the mob. The queen turned the occasion to her advantage. For alleged complicity in the riot she arrested John Bradford (who had helped to save Bourne's life) and John Rogers, two prebendaries of St. Paul's, along with Thomas Becon, a preacher who had been made to recant in 1534; at the same time, by threatening to cancel the privileges of London, she made the mayor responsible for suppressing any further demonstrations in favour of protestantism. On St. Bartholomew's Day the Mass was reintroduced into half a dozen City churches 'not by commaundment but of the people's devotion', and on the following Sunday it reappeared in St. Paul's where the broken altar was being replaced by a structure of brick. Meanwhile the prohibition of preaching by the proclamation of 18 August had deprived the protestants of their principal weapon, and when on 29 August Gardiner was given power to license preachers throughout the country the initiative passed entirely into catholic hands. The

---

foreigner but he gives no proof; and the evidence seems to be that it was the imperial ambassador who made the promise and that Mary knew it was made only to meet the need of the hour. Her proclamation of 18 August condemns alike the opprobrious use of the epithets 'papist' and 'heretic' and it seems that the queen was at least prepared to allow her adversaries to deceive themselves as to her intention. Foxe, vi. 387; *Spanish Calendar*, xi. 104, 132.

[1] Gee and Hardy, *Documents Illustrative of English Church History*, p. 373.

attack was pushed with energy, but for the moment it was the
policy of the government, probably of Gardiner who was now
chancellor, to terrify rather than to destroy. There is no need
to accept the horrific picture of 'the doctor' as given by Ponet,[1]
but though he was affable with his friends he bullied his enemies
unmercifully and did not scruple to denounce as 'heresy' the
religion which was still established by law. The unwelcome
foreigners were allowed to depart; 'Peter Martyr', Ochino,
Poullain, Laski with his congregation, and the Glastonbury
weavers were all expelled or allowed to go their ways. Before
long they were followed by some English clerics of whom
Gardiner was glad to see the last and by John Knox. Bishops
Ponet and Scory left in 1554, Doctors Sandys and Cox in the
spring of that year. Laymen went too and the exodus was not
at first seriously hindered by the government.[2] This has been
described as a 'migration' rather than a 'flight' and it has been
suggested that it was part of a regular plan designed to ensure
the survival of English protestantism. The directors of this
enterprise are supposed to have been Suffolk's brother John
Grey, Francis Russell, later second earl of Bedford, Northampton, and William Cecil,[3] but though these men certainly
favoured 'reform' there is no need to postulate on their behalf
any such far-reaching scheme. The departures from England
are easily accounted for by a well-founded apprehension; what
was to come could easily have been foreseen. Gardiner, whose
bark was worse than his bite, did not go out of his way to make
martyrs unless he were constrained by higher authority; but
there was an obvious risk that Mary and Spain might persecute
and Gardiner himself was party to the arrest of some men whose
captivity ended at the stake. Ridley was already in the Tower
for his share in Northumberland's venture; Hooper was sent
to the Fleet on the ground that he owed money to the government; Barlow, who afterwards resigned his bishopric and was
released to go overseas, was imprisoned on the same accusation.
On 13 September Latimer, who had been warned to escape,

---

[1] Quoted by Froude in a note to chap. i of *Mary Tudor*. See 'The Troubles of
Thomas Mowntayne' in *Narratives of the Reformation* (Camden Society, 1860).
Mowntayne, who had gone with Northumberland to Cambridge, was browbeaten
and set in irons but in the end released.
[2] Only in the parliament of November 1555 was a bill introduced against such
as departed out of the realm without licence, and this bill failed.
[3] C. H. Garrett, *The Marian Exiles 1553–59* (1938), p. 16.

was ordered to the Tower for preaching in spite of prohibition. On the 14th Cranmer went to the same prison because his vigorous exposition of his views upon the sacrament (in which Ponet and Scory may have had a part) was found seditious; two months later he was condemned for his part in Northumberland's treason.

With the fall of the protestant bishops came the restoration of the catholics. By an exercise of authority which looked suspiciously like the supremacy, commissions were issued in August 1553 to deal with certain sees where irregularities had occurred. On various pretexts Durham was restored to Tunstall, Exeter to Voysey, Worcester to Heath, and London to Bonner; by some similar process Winchester was returned to Gardiner and Chichester to Day.

All these things were done while Edward VI's establishment still stood upon the statute book, but now all was ready for the meeting of parliament which assembled on 5 October 1553. The proceedings of its first session were marked by caution. The queen's legitimacy was established by law; various attainders were reversed; certain treason laws which had been renewed by Northumberland were repealed; the penalties for praemunire created since 1509 and for denial of the supremacy were abolished; an act which repealed nine statutes of Edward VI swept away the whole ecclesiastical settlement of the late reign and provided that, from the twentieth day of December following, public worship was to be conducted according to the practice of the last year of Henry VIII; another act punished irreverence against preachers and above all against the Mass. Beyond the establishment of Henry VIII, however, parliament would not go. Nothing was said of Rome; the queen was given to understand that she could not divest herself of her supremacy and that the church lands must remain with those who had them. As Renard later informed his master 'the Catholics hold more Church property than do the heretics'.[1] In the convocation which assembled at the same time the same spirit was obvious. Despite a stout opposition headed by Philpot, archdeacon of Winchester, the establishment of Edward VI was abolished, but nothing was said of the papacy and indeed the writs of summons had given to the queen the title of head of the church. Mary may have been disappointed by the attitude of parlia-

[1] 3 September; Muller, 261 (quoting Record Office Transcripts).

ment, but she went steadily upon her way. The restoration of
the former rites and ceremonies was hardly interrupted by
Wyatt's rebellion, and in the following March (1554) two
commissions were issued *tam auctoritate nostra ordinaria quam
absoluta* for the deprivation of seven bishops. Hooper, Taylor of
Lincoln, and Harley of Hereford were declared to have already
forfeited their sees because they had failed to show the 'good
behaviour' demanded by the letters-patent; and in the cases of
Hooper and Harley marriage was also condemned. Holgate of
York, Ferrar of St. David's (in prison owing to an accusation
under praemunire launched by his discontented chapter), Bush
of Bristol, and Bird of Chester, all professed religious, were condemned primarily on the ground that they had married. Barlow
had resigned Wells, Rochester had long been vacant, Goodrich
of Ely and Sampson of Coventry and Lichfield had died.
Canterbury, automatically made vacant when parliament had
attainted the already condemned Cranmer, and York were for
the moment left unfilled, as were two of the Welsh bishoprics;
but to most of the vacant sees catholic bishops were restored or
appointed, and soon work was found for them to do.

By injunctions issued on 4 March, just before the issue of the
commissions, the queen had laid upon the bishops the task of
carrying into effect the legislation just enacted though they were
not to use the style *regia auctoritate fulcitus* or to demand the oath
of supremacy. They were to suppress all naughty opinions, provide for Christian teaching and holy living, and above all to
deprive the married clergy. Professed religious were to be
divorced, deprived, and otherwise punished; seculars, on
abandoning their wives, might be granted a cure other than that
previously held. Not all prelates behaved with the enthusiasm
of Bonner,[1] who issued a commission of his own against irregularities about ten days before the royal injunctions appeared;
and some of the figures given for the number of deprivations
are preposterous. It seems probable, however, that throughout
the realm one-fifth of the clergy were removed. As the total
number of parishes was 'probably somewhere about 8,000', at
least 1,500 priests must, upon this computation, have been
liable to deprivation. Doubtless some of them found means
to avoid their fate—Chicken of St. Nicholas Cold Abbey

[1] London was the scene of some outrages by the gospellers and some pillorying of
outspoken critics of the government's proceedings.

forestalled his accusation by selling his wife to a butcher, and though he was carted about London by the indignant citizens he seems to have kept his living. None the less it is evident that many hapless families must have been cast into penury.[1]

This zeal for a return to the old way may have produced ill feeling among the poor, and among the rich it bred suspicion. In the second session of parliament, which was summoned to Oxford to avoid disturbance, though it actually met in London from 2 April to 5 May 1554, little was done to meet the queen's wishes in the matter of religion. As already stated, her marriage was approved but proposals to revive the heresy laws were defeated and even the restoration of the see of Durham was effected with great difficulty—Northumberland's spoliation had been excessive but the landed classes felt that a policy of restitution, once begun, might go very far.

Convocation, which met at the same time in answer to a summons wherein the queen omitted the title of supreme head, was more deferential to the royal opinion, but its main achievement was to appoint delegates to dispute with Cranmer, Ridley, and Latimer who had now been removed to Oxford. The debate which, if marred by some confusion of arrangement, displayed much acute reasoning, ended, of course, in the condemnation and excommunication of the three bishops who were found on the evidence of their answers to be heretical (20 April 1554). All were invited to recant; all refused, and Cranmer appealed from the sentence of the court to the just judgement of God Almighty. There being as yet no law under which this academic condemnation could be executed, the bishops were returned to separate prisons until some order could be taken. In May another group of protestants, Hooper, Ferrar, Rowland Taylor, Philpot, Bradford, Coverdale, Crome, Rogers, and others, issued a declaration which asserted not only steadfastness to their faith but loyalty to the queen. This profited them little; most of the signatories were already imprisoned and most were destined to the flames. The hour of their destruction was now at hand. Mary, as soon as she realized her first great objective by her marriage to Philip, hastened forward to the second and greater, the reconciliation of her realm with

---

[1] For a discussion of the figures see Dixon, *History of the Church of England*, iv. 144 ff.; also J. H. Pollen, *The English Catholics in the Reign of Queen Elizabeth*, pp. 39–46.

Rome. At once there became evident a new spirit of which the 'conforming' of Elizabeth may bear witness; persecution was in the air and the queen drew up with her own hand a direction to the council.[1] A legate was to come from Rome; Pole was to be the legate and along with the council was to choose delegates to conduct a visitation of churches and universities. Heretics were to be burnt with proper decorum—in London in the presence of some of the council—to the accompaniment of good sermons; their punishment was to serve for the edification of others, but that they should be destroyed Mary had no doubt. For her the executions were a necessary part of the process whereby thousands of her erring subjects might be brought back into the way of salvation, and it was with this great end in view that she impatiently awaited the arrival of the legate, whose aid she had already sought while the act of supremacy was still unrepealed.

Upon the news of Mary's accession the heart of Pole had burned within him and he had made it evident both to the pope and to the queen that he wished to go to England. His wish had been forestalled; Julius III had of his own accord appointed him legate, and indeed he was for every reason well fitted for the office. Of royal descent and favoured by Henry VIII, he had offered his prospects, his home, and his family upon the altar of his faith. He had been the associate of the great catholic reformers Contarini, Giberti, Caraffa (who became pope as Paul IV), and the others. He had played a great part in the first session of the council of Trent. He had come near to being made pope on the death of Paul III in 1549— indeed had he been a competent self-seeker he would probably have worn the tiara. His rejection he had accepted as a proof that God had other plans for him, and now, he thought, his opportunity was coming.

Before he could execute his task, however, he had to overcome the suspicions of England and the suspicions of the emperor. Since Wolsey's day Englishmen had disliked the title of cardinal and they had some reason to be doubtful of Cardinal Pole. Reginald Pole had owed much to Henry VIII; he had gone to Paris to collect evidence for the divorce and as late as October 1535 had assured Cromwell of his readiness

[1] Dixon, iv. 236.

to do the king service at all times; until 1537 he had enjoyed a good income from English benefices in which he had never resided, and he had remained in safety abroad to launch the fierce diatribe which brought his brother, and later his mother, to the scaffold.[1] To the English protestants he was their 'crafty' enemy, and even to many English catholics a dangerous friend. Crafty he was not. Efforts have been made to show that he was politically minded, and truly when he insisted that the church must be reformed from within he was wiser than the clever politicians; but though his ideals were noble his methods were ineffective and his judgement was bad. It is significant that Mary, despite her desire for his presence, did not press him to come before all was ready;[2] his idea was that she should not marry at all, and she meant to marry Philip. The emperor, no less keen than Mary for the marriage, was wrong in his suspicion that Pole would marry Mary himself, but he was quite right in his belief that the single-minded cardinal might easily plunge the queen into ill-considered action if he attempted to direct her policy before she was firmly established upon her throne. Charles therefore contrived by various devices to postpone the cardinal's journey to England until Mary was safely wed, and it was only in the autumn of 1554 that he set foot at length upon his native soil.

By this time all was prepared. Parliament had been summoned for 12 November 1554. The catholic bishops had been restored; doubtless the glamour of the Spanish gold had not been without influence on the lords, and care had been taken to secure the return of reliable commons. First Renard and then Paget had been sent to Brussels to explain that the holders of church lands must be confirmed in their rights; and though Pole was shocked, rightly feeling that penitents should not bargain, and that the Church should not seem to sell her mercy, he was convinced that there was no remedy. It was with full knowledge of the situation that he passed from Calais to Dover on 20 November. A bill reversing his attainder was hurried through on the 21st; three days later he was conducted to Whitehall with great ceremony to be greeted by the queen who, in her ecstasy, believed

---

[1] W. Schenk, *Reginald Pole Cardinal of England* (1950), pp. 64, 71, 158 *et passim*. The Treatise *Pro Ecclesiasticae Unitatis Defensione*, generally known as *De Unitate*, long and vituperative, was not published till 1555, but the author sent a copy to Henry VIII in 1536.

[2] Letters of 28 October and 14 November 1553. *Spanish Calendar*, xi. 323-4, 357

that she had felt her babe quicken, and ordered *Te Deums* to be sung. On the 28th lords and commons, sitting together at White- hall, heard an address by the legate; next day a petition to be received once more into the church was passed with two dissent- ients in the commons and none in the lords, and on St. Andrew's Day at Whitehall, the king and queen and both houses being on their knees, the realm was absolved from the sin of schism and received back into mother church. The legislation which fol- lowed an outbreak of pious rapture was noteworthy. The treason laws were extended to include a mere assertion that any person had a better title than Mary and to cover Philip's person as well as that of his wife. To Philip was given the regency in the event of Mary's death. The heresy laws of Richard II, Henry IV, and Henry V were revived and all the statutes passed against papal authority since 1528 were repealed. In order to facilitate pious generosity the acts of mortmain were put into abeyance for twenty years; but the abbey lands and the impropriated livings and tithes were placed under the protec- tion of the English law, with a threat of praemunire against those who sought any clerical adjudication in regard to their status. Even the act of absolution was given parliamentary auth- ority. Pole was disgusted; but his advisers, if not he himself, realized that while the majority of Englishmen would follow the direction of the queen in matters of religion, they were not so devout that they would sacrifice their lands upon the altar of faith.

Now the queen had in her hand the means of dealing a direct attack upon heresy. Heresy had been an offence at common law all along—witness the execution of Joan Bocher and George van Parris in the days of Edward VI[1]—but the Crown had never, even in the days of Henry VIII, been persistent in persecution. Now the church courts had regained their right to condemn. Yet even so they could not, of their own authority, burn those whom they convicted; burning was a matter for the secular arm and could be carried out only after the issue of a royal writ. It was therefore still within Mary's power to use sparingly the means at her disposal; but she did not spare. Pole was at her side to issue commissions for the trial of heresy; ecclesiastics, notably Bonner, were ready to try and to condemn; she herself would see that the condemnation was carried into effect. The

---

[1] *Supra*, p. 520.

Spaniards, whose record in Spain and in the Netherlands shows that they had no regard for toleration, counselled caution as a matter of policy; Gardiner, though he joined in the first attack, probably in the hope that a few examples would suffice, was half-hearted and in fact himself condemned no single heretic in the whole of his great diocese. Yet the queen did not falter. She had the salvation of her people at heart and she knew her duty; perhaps her religious zeal was tinged with human animosity especially against the clergy who had married. She herself was the prime cause of the persecution which sullied her reign.

The signatories of the manifesto of May 1554 were the first to suffer; Pole granted a commission for their trial about a week after the dissolution of parliament, and on 28 January 1555 the accused were brought before Gardiner and his colleagues. Coverdale was in hiding or abroad; Crome recanted; the other five who were brought to trial were at once condemned. They were degraded by Bonner on 4 February and executed without delay. It was John Rogers, the editor of Tyndale's translation, who 'broke the ice'—the phrase is Bradford's; he was burnt at Smithfield in the afternoon of the day on which he was degraded. Hooper was burnt at Gloucester on 9 February. Both men were offered life in return for recantation and both refused. Taylor, Saunders, and Ferrar died each at the scene of his labours; Ferrar on 30 March after he had appealed in vain to Pole. All died valiantly and the effect of the first executions was not that which the government had expected; so great was the popular reaction that Renard dreaded a rising. The moment of hesitation passed, however, and the church courts resumed their offensive by condemning five laymen of Bonner's diocese. This was in March, just before Ferrar's execution; a priest was burnt at Colchester in the same month and thereafter the persecution continued in a grim progression interrupted occasionally for political reasons, as when parliament was sitting in the autumn of 1555. Just before the meeting of this parliament, on 30 September and 1 October Ridley and Latimer were tried at Oxford and condemned, though they might have had pardon on submission since neither had been a monk and neither was married. They were burnt in the dry ditch outside the walls of Oxford on 16 October and Latimer's words to Ridley have rung like a trumpet down the ages: 'We shall this day light such a candle, by God's grace, in England as shall never be put out.' Ridley was long in dying; for Cranmer a still longer torture was reserved. As an archbishop he was denounced at Rome by Philip and Mary; Paul IV referred the case to the inquisition, founded in 1542 by his predecessor, which in turn delegated the investigation to three English clerics, the bishop of Gloucester, the dean of St. Paul's, and the archdeacon of Canterbury. Disregarding Cranmer's previous condemnation by the doctors of the university which was purely academic the delegates began their inquiry *ut de novo*, advancing charges of perjury and adultery[1] as well as of heresy. Cranmer repudiated the jurisdiction of a papal sub-delegate and made a bold appeal to Mary, disputing the papal authority in England but professing that he was willing to answer a summons to Rome. To Rome the results of the inquiry had already been forwarded and there on 25 November Cranmer, pronounced contumacious for not appearing, was excommunicated; on 4 December he was deprived of his archbishopric; he was burnt in effigy and a commission was issued for his degradation and delivery to the secular arm in England. Thither the news came swiftly, and the sentence was soon carried out. Disregarding Cranmer's appeal to a general council the papal commissioners Bonner and Thirlby degraded him on 14 February 1556—Bonner with a coarse relish—and ten days later the queen signed the warrant for his execution. Then, for almost a month, he was kept in suspense; there was no intention of sparing him, but his enemies knew that they might succeed in dragging from him a recantation which would be of great value to their cause.

Their hopes were not idle, for Cranmer was no bigot but one who understood the views of others; he was, moreover, uneasy in his own mind on the matter of the royal supremacy. That supremacy he had accepted in the day of Henry; he had exalted it in the day of Edward; could he deny it now when Mary was queen? Had not Elisha condoned the bowing down in the house of Rimmon? Upon the doubts of a naturally gentle man the accusers played, giving him hope of mercy by removing him from the prison of Bocardo to the comfort and the half-liberty of the deanery of Christ Church. During this period he made

[1] Cranmer was married twice. Little is known of his first wife. Sixteen years after her death he married Margaret the niece of Osiander (the Lutheran divine of Nuremberg). This lady he kept in seclusion (though not in the ventilated box of Harpsfield's story) after the passing of the statute of six articles, and in Mary's reign he sent her abroad.

four 'submissions', and finally two abject recantations in which
he abandoned his views upon the supremacy and upon the
sacrament, and recognized in his own iniquity the prime cause
of the woes which had afflicted England.[1] This confession secured,
his enemies were ready to make an end, and he was ordered
to prepare for death on 21 March. On the night before his
execution he prepared a seventh recantation; but the crowd
which assembled in St. Mary's church for the final ceremony
heard not an ignominious surrender but a triumphant challenge.

Cranmer had now taken a firm stand upon the verities
approved by his conscience; he repudiated entirely all that
he had written 'contrary to truth' and 'for fear of death' and
promised that the hand which had penned the recantations
should be the first to feel the flames. They hurried him to the
stake forthwith, but victory came riding to him on the wings of
death. Not only did he vindicate his own cause but he struck a
shrewd blow at the cause of his enemies; all men could judge
the value of recantations like that of Northumberland and
that soon to be made by Sir John Cheke. Certainly Cranmer
must bear some responsibility for the deaths of Joan Bocher
and van Parris; it should be remembered that he had saved
Joan Bocher in Henry's day, and, though the fact may have no
particular significance, he was not among the councillors who
signed the warrant for her burning. Almost alone he had inter-
ceded with Henry VIII for Fisher, for More, for the monks of
Sion, for the Princess Mary and for Bishop Tunstall, for
Anne Boleyn, and for Cromwell. It seems probable that he
concealed from Henry his real views upon the sacrament and
upon the marriage of priests; on the other hand, he resisted the
statute of six articles openly and he dared to tell the king that
he offended God. He had not the blind courage that knows no
fear nor the uninformed conviction that is impervious to doubt;
in the hour of trial he wavered, and those who will may cast
stones. Today his life is remembered by his death and his
memory survives in the lovely cadences of his liturgy.

His death did not end the burnings and before the reign was
done nearly 300 victims perished in the flames,[2] mainly in

[1] A. F. Pollard, *Cranmer*, chaps. xii and xiii.
[2] For the essential accuracy of Strype's list see Pollard, *Political History of England*
vi. 154. Burghley, by adding the number of those who perished by torment and
famine, raised the total to 'near 400'.

London and the south-eastern midlands. War was waged not only on the living but on the dead: at Cambridge the bodies of Bucer and Fagius were exhumed and burnt at the stake; the body of 'Peter Martyr's' wife, Catherine Cathie, was cast upon the dunghill of the dean of Christ Church; the body of John Tooley, who had been executed for robbery, was dug up and burnt because the malefactor had denounced the bishop of Rome as he stood by the gallows. It was Pole, made archbishop of Canterbury on the day after Cranmer's death who most felt the necessity for ensuring that the bodies of those who had died in heresy should be burnt after death.

Some justification for the severities may be found in the fierce diatribes published by the refugees; Knox's *Faithful Admonition to the Professors of God's Truth in England*, published in July 1554, was an all but open incitement to tyrannicide, and Knox was not the only pamphleteer. It may be admitted that some of those accused of heresy recanted and that a few—a very few— of Foxe's heroes were undesirables. The essential fact remains that the great majority of the victims were plain men and women—there were sixty women—who died for their religion. They did not die in vain. It was a hard age and English crowds witnessed spectacles of cruelty with callous indifference; but Englishmen had not the wanton joy in sheer cruelty that disgraced some other peoples of the period; they knew courage when they saw it and they could not but regard with respect a faith which endowed its believers with unconquerable fortitude.

Mary avenged her mother; with the support of Pole she did what she believed was her duty to her infected realm. So doing she cast away the goodwill of her people by which she had been raised to the throne in the face of great political odds.

While the fires of persecution burnt all went ill with the unhappy queen; indeed her persistence in burning may represent some blind attempt to placate the fates that were so hard to her. The hope of offspring, loudly proclaimed to the world in 1555, was found to be a delusion; her symptoms were not those of pregnancy but of the disease which was destroying her. On the summons of his father her unloving husband left her in September; and from the Netherlands, where he seasoned his politics with gaiety, pestered her with demands for the crown matrimonial; when she explained that she could not grant this by her sole authority, he attributed her failure to fulfil his desire to lack of effort and made it an excuse for prolonging his absence. In October the regency of the Netherlands was given to him and in January 1556 he became, by his father's abdication, king of Spain; thereafter he had no further interest in England save as a source of aid towards the realization of his continental designs.[1]

To his abandoned wife the parliament which met on 21 October 1555 brought little comfort. Care had been taken over the elections; the dying Gardiner was at his best in the opening oration; the pope's Bull confirming the abbey lands was read aloud, and the government could claim that it had made a real effort to restore the finances; but still the commons were recalcitrant. They voted only one subsidy instead of the two fifteenths and the subsidy, and they would do little towards restoring to the church its lost revenues. At length it was conceded that the clergy should be free from the burden of first-fruits and should pay their tenths to the legate instead of to the Crown; but the money was to be used to relieve the Crown of its liability for monastic pensions and to the Crown the laity were still to pay their tenths. The queen was authorized to surrender her revenues from first-fruits and tenths if she wished to do so; but lay impropriators were left to their own consciences which, in the event, proved to be insensitive.

Even this very limited concession to the church was not gained without difficulty; the third reading of the bill was forced through by locking the door of the commons' house upon the members as if they were a jury which had failed to agree. When the commons returned the compliment by locking themselves in in order to reject a bill against refugees overseas, the government promptly sent to the Tower both Sir Anthony Kingston who had held the door and the serjeant-at-arms who had given him the key; and in the committal of another member on the same day star chamber showed little regard for the privileges of the house. The Crown endeavoured to secure for itself the right to compel the attendance of members, and failing in this attempted to enforce the old qualification of residence in order to confine representation of boroughs to innocuous townsmen; the commons replied by adding to the bill a clause

---

[1] From an early date a note of all matters of state which passed from the council was to be made in Latin or Spanish. Pollard, *Political History of England,* vi. 158

excluding from the house any person in the pay or in the employment of the Crown, and in the end it was thrown out. The good understanding between Crown and commons, which had been so sedulously cultivated by Henry VIII, was destroyed by a high-handed action which may have been due to foreign promptings.

For while parliament was still sitting, on 12 November 1555, Gardiner died. Pole expressed the hope that his successor would be a chancellor 'less harsh and stern', but despite his bluster 'wily Winchester' was far less severe to heretics than the narrow-minded legate. If Gardiner was not so cruel as the old writers alleged, he was perhaps not quite so much a pure 'English' patriot as some of his modern admirers have asserted. He opposed the Spanish match, it is true; but from the Spanish state papers it is evident that when he clearly understood the queen's resolution he made himself more acceptable even than Paget to the imperialists. He may not have approved of the wholesale burnings, but he made no great effort to stop them. Power, it seems, was his darling and his own life was perhaps something worldly.[1] Yet, endowed with learning, intellect, courage, common sense, and humour he was a stout-hearted servant of England and his death robbed the queen of her most reliable counsellor. Heath, who received the Great Seal on 1 January 1556, was not of his calibre and the direction of affairs passed more and more into the hands of the cardinal. His hands were not equal to the task.

The zeal and sincerity of Pole are not to be doubted. The scheme of reform which he laid before a legatine synod which met on 2 December 1555 was bold and far-reaching; it was, in fact, a restatement of the *Consilium de Emendanda Ecclesia* which he had helped the catholic reformers to produce at Rome in 1536. In a frank allocution he told his hearers that the spiritual ill health of the church was largely due to the abuses, and especially the covetousness, of the priests, and reminded them of their responsibilities. Under his guidance the synod was led to produce a series of salutary decrees aimed at curing the radical evils. In the preceding June an English Prayer Book had already been

---

[1] Muller, *Stephen Gardiner*, p. 354. There is no need to accept the story that he travelled about with 'two lewd women in men's clothing'. He was given to hectoring his opponents and accused them of 'heresy' while the legislation of Edward VI was still unrepealed.

published and now preparations were made for a new book of homilies, a catechism, and an English translation of the New Testament. These projects were not carried into effect. The synod adjourned in February 1556 and it never met again, for the cardinal received a stab in the back from a former ally.

Long ago Cardinal Caraffa had feared that the *spirituali* who had gathered round Pole at Viterbo came near to the Lutheran view of justification, and his fears had been proved well founded, as he thought, when, in August 1542, 'Martyr' and Ochino, 'religious' both,[1] quitted the catholic fold. Possibly he may have been jealous of the Englishman; certainly he held him in suspicion; and when, in May 1555, he mounted the papal throne at the age of seventy-nine, political considerations heightened his ill will. He was a Neapolitan and he desired to expel the Spaniards from southern Italy. Accordingly, in December 1555, he made a secret alliance with France and though his manœuvre was halted for a time by the truce of Vaucelles between Henry and Philip in February 1556, he soon found occasion to renew it. In July 1556 he concluded a formal treaty with France whereby Naples was to be given to a son of Henry II; that France was an ally of the Turk did not disconcert him. Philip in reply bargained with the protestants of Germany and in September 1556 launched Alba against the papal states. Although the general behaved with moderation, and refrained from occupying Rome, the furious pope did not hesitate to excommunicate the catholic king, describing him as 'the son of iniquity' who endeavoured to 'surpass even his father Charles in infamy'. In April 1557 the pope withdrew all his representatives from Philip's dominions and deprived Pole of his authority *a latere*, though for the moment he left him *legatus natus*; in June he went farther, appointed the senile William Peto as legate and summoned Pole to Rome on a suspicion of heresy.

While the *Reformatio Angliae* designed by the cardinal came thus to an abrupt end the plight of his protectress, who would not desert him in his evil day, was pitiable. She was compelled to choose between the deference she owed to her husband and the obedience due to the Holy Father. Her own inclination,

---

[1] Peter Martyr Vermigli had been head of the Augustinian house at Spoleto, and visitor-general of his order; Bernardino Ochino had been vicar-general of the Capuchins.

stimulated perhaps by Paul's treatment of Pole, placed and kept her on the side of Philip.

Philip returned to England on 20 March 1557, and despite the fond hopes of his wife, he came solely to impel England into war with France. So doing, he sought only his own advantage, but in fact France had already supplied England with more than one *casus belli*. Henry had done his best to induce Scotland to attack England. He had given to the English exiles in his land a support which he hardly troubled to conceal; they haunted his court; they behaved as pirates in the Channel; they threatened the Isle of Wight and maintained relations with the governor of Yarmouth castle; they fomented several conspiracies aimed at setting Courtenay or Elizabeth, or both together, on the English throne. All these ventures failed; but though their failure proved a dislike of France it certainly indicated no liking for Spain. England was alienated by the personal arrogance of the Spaniards, by their failure to give any privileges to English traders, and by their refusal to open up to them the treasures of the golden west. America was not unknown to English venturers, and in 1556 Field and Tomson penetrated to Mexico (where, incidentally, they found a Scotsman already settled), but, although subjects of Philip, they met with an ill reception from the Spaniards who were there. Field died and Tomson was sent in fetters to Seville.[1]

In these circumstances England was little inclined to offer military help to Spain and, in truth, she had little help to offer. Mary had tried to maintain the number of her ships but she had difficulty in equipping them for sea; she held two musters of her gentlemen pensioners, gallant in their green and white, but although every pensioner had three well-mounted attendants with him the total force was little over 200 men and except for the Yeomen of the Guard, increased during the reign to 440 and splendidly attired, she had no other regular troops. Philip, it is true, had 100 Englishmen in his guard which contained a like number of Spaniards, Germans, and Swiss, but to what extent this corps moved with the king is not clear.[2] No

---

[1] *Voyages and Travels* (from Hakluyt's *Principal Navigations*; Arber's *English Garner*, i. 7.)

[2] Samuel Pegge, *Curialia*, pp. ii, iii, 27, 47. The navy was less inactive than has been supposed; in 1557–8 particularly there was considerable activity in the dock-yards and considerable expenditure (M. Oppenheim, *A History of the Administration of the Royal Navy and of Merchant Shipping in Relation to the Navy*, i. 109–14).

doubt the queen retained a number of gunners to serve her artillery, but she did not command the enthusiastic support of the shire forces, now organized under the lords lieutenant, from which the main strength of an English army was normally drawn. There was, moreover, little money available for war, since the queen, though she had improved her revenues from her lands and her customs, had been able to obtain only a small parliamentary grant, and that in 1555.

Yet Philip's importunity and the insolence of France in supporting Stafford's[1] attempt on Scarborough in April 1557 availed to bring England into war, which was formally declared in the following June. In July a force of 7,000 men under Pembroke crossed to the aid of the Spanish king who besieged St. Quentin and on 10 August utterly routed a relieving army under Montmorency. Paris was at the mercy of the conqueror but *Philippus cunctator* hesitated and before long the tables were turned. Guise came back from Italy; strong forces were concentrated at Abbeville and the famous gunner Piero Strozzi reconnoitred the walls of Calais. The English commanders at Calais were by no means unaware of their danger though they did not realize its imminence; their defences were in need of repair; they were short of men; among their soldiers was discontent due partly to lack of pay, and there may have been treachery among the civil population. Yet all their pleas for assistance were in vain. The council, confident that there could be no serious campaign in the winter, had dismissed what troops it had; and there was no fleet at sea; England could not send help and what help Philip gave was too little and too late. When, on 1 January, the storm broke upon the doomed fortress, something of the old English valour was seen, but Lord Wentworth had to yield Calais on the 7th and on the 20th Lord Grey of Wilton, after beating off eight assaults, surrendered Guisnes upon honourable terms. So England lost the bridgehead which she had held for 220 years.

It was in the logic of history that the expanding power of France should recover a French stronghold, and the French gain was real. England's loss from a material point of view was less serious than was believed at the time; as the case of the cloth staple showed, English trade did not depend upon the

---

[1] Stafford was a grandson of the Buckingham executed in 1521 and a descendant of Thomas of Woodstock, the youngest son of Edward III.

possession of territory overseas and Calais was very expensive to maintain. Yet, if the queen did not actually say that after she was dead the name of Calais would be found graven upon her heart, the story was true in essence for she was stricken by the blow. The damage to English prestige was very great, and it became greater still when it was realized throughout Europe that she could not even attempt to regain her lost domain.

Mary could do nothing and Philip would do nothing. The parliament which was summoned for 20 January 1558, though the elections had been carefully handled, voted only one subsidy and one fifteenth; having shown itself in other ways unsympathetic to the royal desires it was prorogued on 7 March. A forced loan was levied from shire and town; the export duty on beer and the import duty on wine were raised, and on 28 May was published a new book of rates designed to bring the value set upon amounts of goods into true relation with the inflated prices. The harvest of this reform, however, could not be reaped at once, and it was Elizabeth who benefited. Yet somehow or other money was raised to put some ships into commission and to send an army to the Low Countries. It was partly to English aid that Philip owed his victory at Gravelines in July 1558, but though the concentration of English ships off Flanders enabled the French to seize the island of Alderney, England had neither reward for her pains nor compensation for her loss. There had been some suspicion that Philip had deliberately let Calais fall because he thought that he could easily retake the town for his own use; but Philip had no such idea in his mind. He meant to acquiesce in the loss of Calais in the hope that France would acquiesce in the gains that his subjects had made elsewhere— England should pay for the Spanish conquests in Italy. For the protracted quarrel between Habsburg and Valois was now drawing to an end. The need for a joint attack upon heresy furnished a motive for peace; Guise and his brother met Granvelle[1] at Marcoing, and in October *pour-parlers* were opened at Cercamp. There a truce was signed on the 17th, though the English ambassadors did not arrive until the 21st. The interests of England were not considered at all, and when the peace-negotiations reached their conclusion in the treaty of

[1] Antoine de Granvelle, bishop of Arras, trusted minister of Philip II, son of Perrenot, the chancellor of Charles V, later cardinal.

Cateau-Cambrésis in April 1559, Philip and Granvelle made no serious effort to regain Calais. Calais was lost.

At home, meanwhile, Mary was hastening unhappily to her end. Still she hoped for an heir; still she burnt heretics; still she stripped the Crown of the clerical wealth it had appropriated; still she strove to put the offices of church and state into the hands of good catholic men; but her doom was upon her. In August she came from Richmond to Westminster where she 'took her chamber and never came abroad again'. As she lay there dying, comforted by good dreams 'seeing many little children, like angels, play before her', all she had striven for was fading away. Her cousin the emperor died in his retreat at Yuste on 21 September. Her husband did not come near her; he was turning his eyes elsewhere and in November the Spanish ambassador Feria paid a secret visit to Elizabeth. Englishmen began to court the rising sun, and Mary expressed herself content that her sister should succeed. Parliament had been recalled on 5 November; on the 14th the commons were summoned to confer with the lords for weighty affairs; early on the morning of 17 November Mary died and twelve hours later Cardinal Pole was also dead.

Behind her Mary left an empty treasury and a considerable debt abroad, a country depressed by the loss of Calais, and a people disgusted with the faith which had kindled the fires of Smithfield. It is by these fires that she is chiefly remembered. Yet, her religious bigotry apart, she was not a cruel woman. She was brave herself and like her father she 'loved a man'; she it was who increased the soldier's pay from 6d. to 8d. a day, and in her will she made provision for a house in London for the relief and help of poor, impotent, and aged soldiers. Though she burnt heretics without pity she showed towards traitors a lenity which was sometimes dangerous. To her sister, an obviously dangerous rival, she was not ungenerous. She was loyal to her friends; for the sake of Pole she defied the pope she forbade Paul's nominee Peto to accept the honours given him and she did not hesitate to stop a papal messenger.

What she thought right she did; what she saw she saw clearly though passion sometimes distorted her vision; but she did not see very far. For all her zeal she cannot be reckoned one of the makers of the counter-reformation; secure in the faith of her childhood she was utterly unconscious of the intellectual doubts and spiritual dissatisfactions which vexed more sensitive spirits. For her, heresy was a sin, to be combated not by reason but by faith and by a relentless application of the medieval discipline. In her deference to Spain and to the pope she flouted the spirit of English nationality which was one of the dominant forces of the age. Striving against the spirit of the time she failed in what she attempted, and the note of her reign, it has been said, is sterility.

Yet all was not failure. It was something that England, at a time when the old order was changing throughout Europe, kept, despite the changes, her old tradition in church and state, her old discipline, and her old virility. The essential institutions of her governmental system steadily developed, and, in Mary's day, wasteful administration was checked and an attempt was made to balance revenue and expenditure.[1] England saw neither civil nor religious war, and beneath the appearance of stagnation was the stirring of a new life. Under the ashes of the martyrs an unquenched protestantism was glowing, behind the failure at Calais the courage of a strong race stood unimpaired. The economies of town and country were adapting themselves to new conditions, and in spite of their grumbles the merchants were flourishing. The Russia company was founded in 1555; next year an ambassador from Ivan the Terrible came to Mary's court and a treaty of commerce was made. The explorations of Chancellor, Stephen Borough, and Anthony Jenkinson penetrated to Nova Zembla and to Bokhara. In spite of the Portuguese, captains like John Locke and William Towerson made voyages to Guinea. Field and Tomson went to Mexico. The spirit of adventure, far from being dead, was seeking new outlets. Although to the politicians of western Europe the English state presented a picture of impotence, the English people were active and strong. Below the arid soil the roots of nationality were quick, and ready to shoot forth the stems which were to bear a great harvest.

[1] Dietz, chap. xvi. It is fair to add that under Northumberland Cecil had already made plans for reforms in the financial administration. Ibid., p. 197.

XVI

THE ACHIEVEMENT OF THE AGE

Not the least achievement of the age was the creation of a nation-state. This state was, after the manner of the times, a monarchy. The manifestations of the royal power were various, but behind the steady purpose of Henry VII, the massive strength of Henry VIII, the prim piety of Edward VI, and the bigoted devotion of Mary may be discerned the essential truth that the prince had authority simply because he was the prince. To the world of today the reverence given to the monarch is astonishing, but in the sixteenth century the royal office was invested with the sacrosanctity of the middle ages and at the same time accorded the deference due, in a practical world, to the actual head of affairs. The divinity which hedged the king did not come from the unction alone; the kings were kings and the queen was a queen before the ceremony of coronation. It might be said to come by descent were it not that the rules of descent were uncertain, and subject, with ill-defined limitations, to the adjustments made by parliament. It could not be pretended that the kings owed their authority to parliament; it was the king who summoned parliament and he summoned it, or omitted to summon it, according to his will. The first of the Tudors gained the crown on the field of Bosworth and he kept it because he could; his son succeeded because he was his father's son; and the children of Henry VIII held the crown in succession because they were the king's children, the son first and then the two daughters in order of seniority. On a last analysis it would seem that the essentials of monarchy in England were simply some title by descent, a power to wear the crown and—this is vital—an ability to gain recognition from the people of England. It is not to be disputed that the strong personalities of the Tudors developed the royal authority in England, but the crown picked up on Bosworth field had a magic of its own.

Writ all over the period is the compelling force of the royal sovereignty. Of an act passed in the first parliament of the newly enthroned Henry VII a contemporary wrote 'there were many gentlemen against it but it would not be for it was the king's pleasure', and in each successive reign the king's pleasure, expressing itself in various ways, commanded obedience. There was sometimes resistance, which occasionally took the form of rebellion, but, in the main the king's will prevailed to an extent which is surprising when one considers the actual force by which it was maintained. The very rebels hesitated to push their attacks home against the person of the prince; the very victims of royal injustice, standing on the scaffold, expressed their submission to the master who destroyed them; Cranmer no less than Gardiner acknowledged the overriding majesty of the king.

This reverence for the Crown was not peculiar to England and it was not peculiar to England that the royal power created a machinery to make its power effective throughout the land. The machinery developed in England, however, was of peculiar competence. The council, as in other lands, dealt with every branch of government, but in England more quickly than in most other lands high birth retreated before ability; great nobles continued to come, the archbishop kept his position, but more and more influence passed to the 'heads of departments', and more and more seats were given to men who had done their apprenticeship in the service of the state. As elsewhere the council was divided into committees; what distinguished the English council was that dependent organizations, whether charged with the administration of particular areas or of particular departments, were manned partly by members of the council as well as by hard-working officials. So was the central power kept in close touch with subordinate institutions. In their treatment of the justices of the peace those competent men-of-all-work whose utility had been evident long before the sixteenth century, the Tudors adopted the same policy. On the one hand, from the day of Henry VII onward, the justices were made directly responsible to the council; and on the other, by the inclusion of the names of councillors in every commission for the peace, the central government armed the local officialdom with some of its own authority. With the rise of the justice of the peace the sheriff lost much of his old prestige, though he was still 'pricked' after a meeting of councillors wherein the treasurer took precedence, and in the reign of Edward VI he lost his control of the forces of the shire to the lord-lieutenant, an officer directly appointed by the king.

What is true of administration is true of finance. Some of the new courts were directly connected with the royal revenue and on them too members of the council served. Moreover, by the development of the chamber at the expense of the formal exchequer, the whole financial system was brought more directly beneath the supervision of the Crown. In one way and another the whole machinery of the state was kept under the constant and precise surveillance of the king and his chosen advisers.

In an age which did not distinguish clearly between good law and good government, justice no less than the administration was under the authority of the Crown. The law was the king's law. The council had its judicial functions which it exercised both as a court of first instance and as a court of appeal, and the jurisdictions used by subordinate bodies, which were in most cases administrative as well as judicial, came directly from the royal power. Star chamber, chancery, council of the north, court of requests, stannary courts, all alike drew their authority from the prerogative justice of the king. The law of the church which no less than civil law might be learnt in the universities, until Henry VIII abolished its study in 1535, was in the hands of trained ecclesiastics, but the church courts, being limited to their own field, did not greatly conflict with the courts of the Crown; indeed they contributed to the lay courts, and especially to the court of chancery, some of the principles of Roman law upon which canon law was founded. When Henry VIII took to himself the jurisdiction hitherto enjoyed by the pope in England the Crown gained a new access of power which it kept until 1555; canon law was made subservient to the 'prerogative royal' and appeals from church courts were taken to chancery.[1]

There was, however, another extremely important form of justice which was not entirely *ad arbitrium regis*. The common law was the king's law, its judges were the king's judges; but in fact the law of England and its officers were in some sense outside the royal control. What was the origin of this law of England was not very clear. On the lips of formal orators it was, of course, an example of that municipal law which, like the law of nature and the law of nations, was a formalization of the divine law. But where had it its origin? Fortescue had based it upon

[1] Act for submission of the clergy, 25 Henry VIII, c. 19.

the customs of England established by 'Brute and his fellow-ship', and most men seemed to believe that it represented the way in which things had always been done in England. Yet, however old in its origin it was a living thing; it could be declared and formulated in parliament, and, although this was not clearly admitted, it could be augmented too. Parliament might pretend that it was only explaining or emphasizing existing right, but it was constantly applying old principles to entirely new situations. It was in fact making new law. So doing it acted by the authority of the realm and not in the name of the Crown—the fact that a monarch was found to be a usurper did not invalidate the legislation of his reign.

This living law, expressed in English and French as well as in Latin, and recorded much in precedents, was not to be studied in the universities. It must be studied in practice. As a member of one of the ten inns of chancery the novice learned the rudiments of his business; then he passed to one of the four inns of court as a member of which, besides taking part in mock trials and hearing discourses, he attended the great courts in Westminster Hall and saw how things were done. Graduation there was none; but in due course the student would become an 'utter barrister' and after sixteen years from his entrance to his inn he might find himself one of a batch of barristers whom the chancellor selected at intervals to become serjeants-at-law. His promotion, which involved the providing of a feast and the giving of gold rings, was an expensive affair, but he gained the right of pleading in common pleas and the chance of eventually becoming one of the half-dozen justices who presided in each of the two great courts[1] and might sit upon assizes throughout the land. The men who rose to the bench knew their law and, as they had behind them the weight of a strong legal corporation, their word was not lightly regarded. Members of the bench were not as a rule members of council, but they were much consulted; along with other lawyers they regularly discussed the legislative proposals to be put before parliament and, by 'writs of assistance', they were summoned to sit in the upper house. Strong in its ancient descent and administered by men of long experience, the common law was in some measure independent of the Crown; yet, by the methods recounted, it, no less than

[1] The chief baron, who presided over the court of exchequer, was not necessarily a serjeant till 1579, but he was, in fact, usually a man of law.

the prerogative law, was integrated with the royal administration.[1] It was part of the Tudors' wisdom to use and not to defy the things which stood for the English tradition.

Nowhere is their policy more obvious than their handling of parliament. They were not averse from using the strong hand occasionally, but, in the main, they took parliament into partnership and made it an ally of the royal power, first against the over-mighty subject and later against the pope. It was in the day of Henry VIII that the alliance between Crown and commons became most manifest; but this alliance had begun in the day of Henry VII and it continued almost until the end of Mary's reign. More than one speaker was elevated to the royal council; and there is a whole 'line of Speakers who, as under-treasurers or chancellors of the exchequer, received a useful training before having to deal with financial grants by the commons . . .'.[2] In Mary's day a committee of council considered the legislative business of an impending session;[3] care was taken to see that councillors found seats in the commons, and in the reign of Edward VI the 'King's Privy Council in the Nether House' appears as a recognized element in parliamentary business. In March 1549 it was concerned with the conduct of a member, and in November of the same year it joined with twelve other members of the house to inquire of the council the king's pleasure concerning a tax. In Mary's reign the 'Queen's Privy Council in this House' appears in the *Commons' Journals*.[4] The liaison between executive and parliament was well established and the result was noteworthy. For long the executive would have thought scorn to suppose that it was in any way responsible to parliament, but parliament became used to discussions of policy and began insensibly, by use of the financial weapon, to exercise control. It would not, for example, pay for

---

[1] Edmund Dudley in the *Tree of Commonwealth* (written 1509, published 1859) exposed some of the Crown's methods of interfering in legal processes.

[2] A. F. Pollard, *Parliament in the Wars of the Roses*, p. 21. The speaker was paid by the Crown. Apparently a fee of £100 and £100 in respect of expenses (Gladish, 115). The two chief justices, the king's attorney, the king's solicitor, and the clerk of the parliament, with their clerks, all received payment for attendance at parliament (*Acts of the Privy Council*, ii. 37).

[3] Ibid. iv. 398. Ibid. vii. 28 for the same procedure under Elizabeth. For councillors in parliament see Wallace Notestein, *The Winning of the Initiative by the House of Commons* (Proceedings of the British Academy, 1924).

[4] *Journals of the House of Commons*, i. 11, 12, 30. The commons, however, retained control over their own membership, cf. ibid. i. 17, 25, for their scrutiny of returns from Sandwich and Maidstone.

the foreign policy of Mary. Parliament did not cease entirely to be an instrument of opposition; and it did not become solely an instrument of the royal power; in alliance with the Crown it served an apprenticeship towards the day in which it could itself direct the affairs of state. Whereas in other lands diets or states-general were meetings called occasionally when the royal power was in difficulties and were more apt to criticize than to assist, the English parliament was tending to become a permanent and essential part of the machinery of government. Under the Tudors it was much controlled by the Crown who took the initiative hitherto sometimes used by discontented barons; but, though attempts were made to pack it, it was never subservient and the day was not far distant when it would claim the initiative for itself. The great achievement of this period was the development of an institution which was later to be copied throughout the world. This development was due not to profound theory but to practical common sense.

Common sense was the note of the Tudor accomplishment. Doubtless the monarchs gave the lead and set the pace, but they moved in general accordance with the English habit of thought and the greatness of the change which they effected may not have been obvious to the majority of their subjects. To the men of 1558 it may well have seemed that the state as they knew it did not differ essentially from the state of 1485 which had been known to their forebears. Perhaps after the manner of most generations they felt that things were not quite so good as they used to be; but England was England still.

The achievement of the Tudors in their dealings with the church was parallel to their achievement in the realm of the state and was in some ways even more remarkable. In the early days of Henry VIII the church in England had its being as part of a world-church whose authority, though far from universal, was recognized at least throughout western Europe; when he died the *Ecclesia Anglicana* was part of the realm of England of which the king was Supreme Head, and though under Mary it returned for a brief space to the old way, it reappeared under Elizabeth, though England was now a protestant country. The transference of the 'supremacy' to the Crown was accompanied by great alterations; the abbeys lost their lands, the bishops became, though not entirely, the king's servants; a succession of new and different *formulae* was enunciated in the Bishop's Book, the King's Book, and the Books of Common Prayer. The treasures of the past were taken from the churches and although ecclesiastical ornament was not always of a high art, much of beauty must have vanished; the mural paintings,[1] which in the day of Edward VI gave place to plain walls and scripture texts, were at least lively and colourful, and some, if we may judge from recent discoveries, artistic too. Yet the great changes were accepted, if without enthusiasm, at least without very violent protest; there were martyrs—who did not die in vain—and there were ebullitions of rebellion, but there were no wholesale massacres and there was no civil war.

The complacence of the subjects has been attributed to religious indifference in an age of common-sense materialism, but this explanation, though not without force, will explain neither the deaths of the martyrs nor the reaction which followed the martyrdoms; it has been attributed, and again with justification, to the compelling power of the Crown; but it was due very largely to the instinctive wisdom of the Crown which contrived to invest the revolution which it made with a cloak of English precedent and to persuade the people of England that the action it took was not outside the English tradition and not irrevocable. None the less it may be doubted if the Crown could have succeeded in its venture had its policy not awakened a responsive chord in the heart of the English people; when Henry denounced dogmas and practices which seemed to rest upon man-made tradition and mistranslations of the Greek, he enlisted the support of a common-sense people already affected by the realism of the day.

It is true enough that the royal action was not without precedent. As has been pointed out[2] the whole mechanism of the church in the middle ages depended upon a working compromise with the state, and in England that compromise had been carried out with an illogical efficiency. The church contributed to the necessities of the Crown; the bishops, though they had their Bulls from Rome, were often royal ministers; the Crown shut its eyes to the systematic disregard of the statute of praemunire; the career of Wolsey has shown in outstanding

---

[1] For example, the mural paintings in the Church of St. Peter and St. Paul, Pickering (*The Times*, 16 August 1950).

[2] Sir Maurice Powicke, *The Reformation in England.*

fashion how the authority of the pope's special legate might reinforce the power of a monarch. It would be idle to pretend that the magnitude of Henry's venture escaped the notice of his contemporaries; but it is easy to see that many Englishmen supposed that the king was asserting the rights of England against the foreigner, that the changes he made were not so great after all and that England would in the end return to something very like the old way. The 'foreign' extravagances of Edward VI's reign did not commend themselves; the 'foreign' executions of Mary's reign excited disgust; but the main fabric of Henry VIII's establishment was generally, if not enthusiastically, accepted, and when, in the day of Queen Elizabeth, the hoardings of demolition and the scaffoldings of reconstruction were removed the *Ecclesia Anglicana* which emerged bore, at least outwardly, a marked resemblance to the church which it had supplanted. To the church of Elizabeth about nine-tenths of the English clergy adhered.[1]

In spite of all the changes of the official religion the ordinary rules of conduct remained unaltered. The old disciplines still held, and the restraint of the various bodies of insurgents is remarkable. There was no general slackening of morality. The chronicles take it for granted, not only that murder and theft should be severely punished, but that bawds should be carted through the city, apprentices whipped for introducing wenches into their masters' houses, and dishonest traders exposed to public infamy with their rotten goods strung round their necks.

The practical age of the early sixteenth century was not a great age of literature; its learning and its letters were largely concerned with the controversies of the day and save for *Utopia* and the English Bible there were few books of outstanding merit. The infiltration of renaissance ideas was spasmodic; the first flourish of criticism and speculation was nipped by the frost of theological suspicion, and the contribution of Italy to English letters afterwards appeared in a gift of added grace and poetic form rather than of soul-stirring experience.

The universities suffered from the difficulties of the times.[2] The keener spirits in the church had already realized that

---

[1] J. H. Pollen, *The English Catholics in the Reign of Queen Elizabeth*, p. 40; A. O. Meyer, *England and the Catholic Church under Queen Elizabeth*, p. 29.
[2] See C. E. Mallet, *History of the University of Oxford*, and J. B. Mullinger, *The University of Cambridge*.

'heresy' must be repelled by learning, and the colleges founded under the early Tudors owed much to the bishops. At Oxford, Brasenose College was founded by Smyth of Lincoln (1509); Corpus Christi College was founded by Fox (1517); and Cardinal College, the forerunner of Christ Church, by Wolsey (1525). At Cambridge, Jesus College was established by John Alcock, bishop of Ely (1496); Christ's College (1505) and St. John's College (1511) owed more to Fisher than even to the Lady Margaret; Magdalene College founded by Lord Audley (1542), and Trinity College, reorganized by Henry VIII (1546), were lay foundations. Usually the erection of these new colleges represented an attempt to put to good use foundations which had fallen into decay; Henry, Margaret, and to some extent Wolsey, admired good learning; Fox and Fisher were real scholars; but the enthusiasm of the church for reform was checked by the dread of heresy, and as the mendicants, who were particularly strong at Oxford, had a real hatred for Greek the advance of the new learning in that university encountered a stiff opposition. There the conservatives, taking the title of 'Trojans', did their best to terrorize the followers of Fox, who had established a lecturership in Greek. They were defeated because King Henry, stimulated no doubt by More and Pace, intervened effectively, and, although the forces of reaction remained strong, some reforms were made. Oxford produced good Grecians like John Stokesley, later bishop of London, William Tyndale, and Thomas Starkey, the pamphleteer, all of Magdalen Hall, and Richard Morison the propagandist, but her classical tradition owes not a little to Wolsey who, despite his many interests, still found time to patronize learning. For his school at Ipswich he designed a liberal scheme of instruction (founded perhaps on his experience in Magdalen College School, where he had been 'informator' for six months); it is noteworthy that members of his household like John Clement and Richard Pace, who never were undergraduates, as well as Richard Morison, were all scholars of note.

In 1520 the cardinal, into whose hands Oxford had placed its statutes in 1518, founded a readership in Greek, and when, in 1527, he revised the constitution of the new Cardinal College he provided for six professors who, probably by arrangement with Fox, gave public lectures in the new humanist fashion. One of the teachers, Nicholas Kratzer, turned his attention to

astronomy; another, Vives, to education; others, like Edward
Wotton and John Clement, gave themselves to medicine; but,
partly because the cardinal reinforced his college with men
like John Frith and Richard Taverner from Cambridge, the
classical tradition did not die, and produced its fruits in the
persons of Richard Reynolds, the Carthusian martyr, Nicholas
Harpsfield, George Etherege, Alexander Nowell, and William
Whittingham.

At Cambridge, thanks to the memory of Erasmus, and to the
practical aid of Fisher, the classical tradition was stronger.
Thomas Lupset, the most brilliant of the pupils of Erasmus, left
England for Paris in 1514, and though he returned to succeed
Clement at Oxford in 1520, he went abroad again in 1523. He
being gone, the banner of Greek at Cambridge was borne by
Richard Croke of Eton and King's College, who, in 1518, was
formally appointed reader, and in his inaugural lecture next
year exhorted his hearers to emulate Oxford in the matter of
Greek. His admonition was probably unnecessary; Cambridge
produced, in the course of the next few decades, a brilliant
galaxy of scholars.[1] From St. John's College, in 1524 remodelled
by Fisher according to the system of Fox, there came forth
George Day, later bishop of Chichester, Richard Croke, John
Cheke, 'who taught Cambridge and King Edward Greek',
John Redman, later master of Trinity, Roger Ascham, Thomas
Lever, and William Cecil to mention no others; from Queens'
came John Aylmer, bishop of London, Thomas Smith, and
John Ponet, later bishop of Rochester and Winchester; from
Pembroke Hall Nicholas Ridley, later bishop of London, John
Bradford the martyr, and Nicholas Heath, later archbishop of
York; from Jesus, Cranmer; from King's College Richard Cox,
bishop of Ely, and the martyr John Frith; from Corpus Christi
Richard Taverner, who went to Cardinal College; from Trinity
Hall came Stephen Gardiner, while from the new foundation of
Trinity College, graced by Redman and William Bill from St.
John's, came John Christopherson, later bishop of Chichester,
and John Dee, the astronomer suspected of being a magician.

As the names of these scholars suggest, the progress of learning was intimately connected with the progress of the reformation. The study of Greek was now encouraged, now suspected,

[1] See A. Tilley, 'Greek Studies in England in the Early Sixteenth Century', in
English Historical Review, vol. liii (1938).

according to the royal attitude in ecclesiastical affairs. In the early twenties a group of young enthusiasts at Cambridge who met at the White Horse Inn,[1] where 'little Bilney' of Trinity Hall was prominent, became imbued with Lutheranism, and it was for this reason, perhaps, that Cambridge, following the example of Oxford, placed her statutes in Wolsey's hands in 1524. The cardinal, alarmed by the appearance of the Cambridge infection in his own new college, had the leaders arrested, and while he had power kept a close eye upon both the universities.

On his fall learning was not released from the grip of politics; on the contrary it was made to serve the ends of whatever interest dominated the state. How close was the connexion between the universities and the great world of affairs appears from the very names of the academic officers. Oxford had among its chancellors Warham, Longland, Cox, and Pole, whilst Cambridge could boast Fisher, Cromwell, Gardiner, Somerset, and Pole; Sir Thomas More was high steward of both universities. As long as Henry VIII lived his overriding personality swayed the fate of scholarship and scholars; the decision to obtain the opinion of the learned on the 'divorce' at once robbed Oxford and Cambridge of their independence and directed their efforts to an unprofitable end; men like Croke and Pole were sent abroad to collect evidence in favour of the king. Later each change in the royal policy brought a reaction in the academic field. Occasionally the universities benefited. Cromwell's commissions of 1535 were not inspired by academic enthusiasm, but none the less they did something to reform learning, though less than was envisaged in the injunctions issued in the same year. In his foundation of Corpus Christi College Fox had attempted the establishment of 'university' as opposed to 'college' teaching and the Cromwellian commissioners evidently felt that his idea was right. At Oxford they founded new public lectureships in Greek, Latin, and civil law—the study of canon law was virtually abandoned—and it was ordained that student of 'physick' were to be examined by the 'physick professor' before they were allowed to practise. Cambridge was ordered to establish public lectures in Greek or Hebrew as well as college lectures in Latin and Greek. The king who, in 1536, relieved the universities from the payment of firstfruits and tenths

---

[1] See pp. 257 and 343 *supra*.   [2] 27 Henry VIII, c. 42.

considered that they would be able to maintain one 'discrete and larned personnage to reade one opyn and public lectour ... which lecture shalbe called perpetually Kyng Henry the Eight his lecture'. Experience showed, however, that the colleges either would not, or could not, subscribe enough to maintain university teachers, and in 1540 the king founded five regius chairs at Cambridge in divinity, civil law, physic, Hebrew, and Greek, each endowed with the handsome salary of £40 per annum. It is said that he intended to do the same thing at Oxford at the same time; but it was only in 1546 that regius professorships in theology, medicine, civil law, Hebrew, and Greek came into being, and for the maintenance of three of the chairs—Hebrew, Greek and theology—the newly reconstituted Christ Church College was made responsible. Behind the appearance of progress were signs of a decline though the decline has perhaps been exaggerated by historians.[1] Much of the attention of scholars was given to theological polemics and parties rose and fell with the turn of the political wheel. The theology of the Bishop's Book of 1537 was developed at Cambridge; during the reaction presaged by the 'six articles', Gardiner prevailed against good scholars like Thomas Smith and John Cheke and ousted innovation by having regents expelled and undergraduates birched; towards the end of the reign the political pendulum swung once more in favour of reform.

Again, the destruction of the old religious foundations caused general uneasiness and broke valuable relationships between schools and churches on the one hand and colleges on the other. Some of the colleges like Christ Church benefited by the suppression of old foundations, but the monastic colleges—Gloucester, Durham, Canterbury, St. Bernard's, and St. Mary's at Oxford—disappeared as did the houses of the mendicants. There were fewer of the poor 'unattached' students struggling on towards a career in the church, and their places were taken by gay young men of fashion.[2] Few students proceeded even to the degree of bachelor; very few indeed went on to a doctor's degree; and the newly established public lectures were not well attended. The attack on the chantries caused great alarm, and the commission sent to Cambridge in January 1546 found that very few colleges

---

[1] The account given by Lever in his sermon of 1550 (*Lever's Sermons*, ed. Arber, 1870, p. 120) may be too pessimistic.

[2] See the animadversions by Ascham and Latimer cited in Mullinger, ii. 88–99.

were living within their incomes. The happy result of this inquiry was the foundation, in December 1546, of a great 'modern' college on the ruins of some medieval foundations, but, though the university was thus enriched in the gain of Trinity College, its anxieties were by no means at an end. In an age of confiscation the position of the colleges was far from secure and in 1549 it was proposed to erect a new college of civil law by the amalgamation of Clare Hall and Trinity Hall. This proposal was part of a scheme of reform which was inaugurated at the start of a new reign, and for a while it seemed that the progressive party must have things its own way, since Somerset, in pursuance of his theological programme, interested himself in the universities. It was not, however, till the summer of 1549 that the commissioners sent to the universities issued their injunctions; these were far-reaching and obviously aimed at recasting altogether the old *trivium* and *quadrivium*. New curricula, new text-books, new time-tables were introduced and, though Cranmer obviously tried to avoid wholesale ejections, new teachers were provided too. In 1548 'Peter Martyr' became professor of divinity at Oxford, and in 1549 Bucer was given the corresponding chair at Cambridge; at Cambridge, too, first Fagius, and on his death Tremellio, were made professors of Hebrew.

The path of the reformers, however, was not easy; as their proposals with regard to Clare Hall showed they had perhaps too little regard for existing institutions and existing scholarship. 'Peter Martyr' found himself for ever contending with antagonists and those of a most pertinacious kind, while the gentler Bucer soon found things to criticize and men to criticize him; he complained that poor scholars were excluded by fellows who grew old in indolence, and that secret adherents of catholicism publicly disparaged him in their lectures. The atmosphere of controversy was not congenial to good letters and with the fall of Somerset any chance of great improvement disappeared. Northumberland was at some pains to cultivate the good opinion of Cambridge, and Edward himself was said to have cherished great designs for that university; probably, however, 'Alcibiades', as Ponet called him, had no real interest in academic reform and certainly he had no money to spare. Nothing was done to promote good learning and the proposal to make admission to the degree of M.A. conditional upon sub-

scription to the forty-two articles speaks rather of the difficulties of the time than of academic liberty. In the reign of
Mary a resolute attempt was made to return to the old way.

Gardiner had no opinion of the recent changes; he said to
Thomas Mowntayne 'wheras yow have set upe one begarlye
howse, yow have pulde downe an C. prynsly howsys for yt;
puttyng owte godly, lernyd, and devoyte men that sarvyd God
daye and nyghte, and thurte [thrust] yn ther plase a sorte of
scurvye and lowsye boyes'.[1] He soon made his authority felt.
At Cambridge, where he recovered his position of chancellor,
new heads were given to all the colleges save Gonville, Jesus,
and Magdalene. At Oxford, where, as bishop of Winchester,
he at once instituted a visitation of New College, Corpus, and
Magdalen, the effect of his influence was speedily seen; the
unfortunate Edward Anne, an undergraduate who had written
a lampoon upon the Mass, was publicly flogged in the hall
of Corpus, receiving one stroke for every line of his poem. On
Gardiner's death Pole became chancellor of Cambridge; and
soon he was chancellor of Oxford, where a Spanish theologian
was introduced into the chair of divinity. Under the cardinal's
supervision the universities became rigidly orthodox and it has
been stated that, after the reconciliation of England with Rome,
no college, either at Oxford or Cambridge, bought any save
service-books.[2]

It is a curious commentary upon university conservatism
that Cranmer, even after the prolocutor had pronounced
against him in his academic trial, took part in a ceremony
wherein John Harpsfield (brother of Nicholas) disputed for the
degree of D.D., and was thanked for his conduct by the president
while 'all the doctors gently put off their caps', although on the
morrow he was condemned as a heretic.[3]

As with the universities, so with the schools.[4] Boys received
their early education at home, or in the household of some
gentleman to whom they were sent. As the story of Lady Jane

---

[1] *Narratives of the Reformation*, p. 183, *Autobiography of Thomas Mowntayne.*
[2] Pollard, *Political History of England*, vi. 173, n. 3.
[3] Pollard, *Cranmer*, p. 345.
[4] See A. F. Leach, *English Schools at the Reformation*; Foster Watson, *The English Grammar Schools to 1660*; Beatrice White, Introduction to *The Vulgaria of John Stanbridge and The Vulgaria of Robert Whittinton* (Early English Text Society, Original Series, 187, 1932). Horman's *Vulgaria* (Roxburghe Club, 1926).

Grey shows, girls might be educated as well as boys, and, as is clear from her own case and from that of Ascham who wrote of her, among the tutors of the young there were good scholars with a genuine love of teaching. In the *Boke named the Governour* by Sir Thomas Elyot (1531), in Vives's[1] *De Ratione Studii Puerilis* (1523), and in the works of Ascham, sound and liberal ideas were set forth, and, as appears from Colet's foundation, some of the good theories were carried into effect. But the atmosphere of the early sixteenth century was not favourable to the endowment of learning. In London, St. Antony's School preserved its reputation for some time though it was 'sore decayed', and its scholars fought a losing battle in the streets with the boys of St. Paul's—St. Antony's 'pigs' versus St. Paul's 'pigeons'. The more official debates between the scholars of 'divers grammar schools' which were wont to take place in the churchyard of St. Bartholomew disappeared; in 1553, however, Christ's Hospital made a modest beginning when 'a great number of poore children' were taken in at the citizens' charges to the 'late dissolved house of the Gray Friers'.[2] Westminster was sorely impoverished by the dilapidation of the see; the revenues, even of Eton and Winchester, were at one time in danger, and though Shrewsbury was founded in 1552, the country grammar schools benefited far less from the munificence of 'King Edward' than has often been supposed.

The *Commons' Journal* shows that there were bills to found schools, but although foundations were made or improved at Berkhampstead, St. Albans, Stamford, and Pocklington, no general programme of endowment was ever carried out.[3] In the eyes of parliament education was perhaps less important than 'the true stuffynge of Fetherbeddes, Bolsters, Mattresses and Cuysshions'.

None the less, the teaching of Latin was improved. The old interest in disputations may have declined, but it was still important that boys, whether they meant to be 'clarkes' or not, should know Latin—above all, that they should be able to

---

[1] Joannes Ludovicus Vives was a Spaniard brought by Wolsey to Oxford where he lectured on rhetoric; he became tutor to Princess Mary. Besides the work mentioned he wrote *De Disciplinis*, full of sound wisdom, and some of his *Colloquia* were included along with those of Erasmus in books for the use of young students. See *Tudor School-boy Life*, Foster Watson, 1908.

[2] For the London schools see Stow's *Survey of London*, i. 71, ed. C. L. Kingsford.

[3] *Journals of the House of Commons*, i, 4, 5, 7, 9, 22. *Statutes of the Realm*, iv. 1; ix. 156.

converse in Latin. Hence came the use of *Vulgaria*, collections of sentences in Latin and English dealing with the events of everyday life. The first *Vulgaria* to be used in England (1519) were those collected by John Stanbridge who, after education at Winchester and New College, taught in Magdalen College School under Anwykyll, and from 1488 to 1494 himself held the the office of 'informator'. Stanbridge's work, which was largely based upon Terence, was printed only in 1519; it formed the basis of a much more elaborate collection published in 1520 by Robert Whittinton, a good though vainglorious grammarian, who had been a pupil at Magdalen College School and who, under the pseudonym of Bossus, engaged in a literary controversy with Lilly of St. Paul's and William Horman of Eton. Horman, however, had already produced an excellent set of *Vulgaria* for use at Eton which was printed in 1519; by that time Stanbridge's method was already in use at St. Paul's, and it is reasonable to suppose that he had made use of his collection in teaching at Winchester, where he was headmaster from 1494 to 1503; certainly Stanbridge's method had been introduced to St. Paul's by 1519.

It was not for his method but for his selection of authorities that Whittinton was criticized, and the *Vulgaria* remained in general use for many years even though Roger Ascham in the *Scholemaster* (published only in 1570 and then unfinished) advocated an abandonment of the endless rules, and an approach to the Latin language through the medium of classical texts properly taught by a system of double translation.[1]

Meanwhile Greek had taken its place in the curriculum of some English schools. St. Paul's under William Lilly, John Rightwise, and their successors maintained the tradition of Colet. Greek was taught at Eton, if not by William Horman when he was headmaster, at least whilst the *Vulgaria* was being written,—perhaps before 1497, certainly before 1519. It is possible that Winchester introduced some teaching of Greek whilst Horman was master there, and certain that when Alexander Nowell became headmaster of Westminster in 1543 he read parts of the New Testament with the elder boys. The statutes of East Retford (1552), which ordered that Greek should be taught to the highest forms, are evidence of a good intention though there is no proof that the intention there, or in other grammar

---

[1] He had been anticipated in his ideas by Vives.

schools, was carried into effect. For most boys, it may be surmised, education meant a training in Latin grammar and in the
stoical endurance of pain.[1] Yet, though Holofernes still held
sway, the tradition of study was not lost; the grammar was
sound; the discipline was real; here and there were sown the
seeds of a better approach to learning, and the education of a
Tudor schoolboy might easily become the foundation of a true
scholarship.

A sign of the intellectual awakening of England appears in
the increasing demand for books. For a time it seemed as if a
profitable English market would be exploited by foreigners,
and an act of 1484 had facilitated the importation of books
from abroad. Caxton had learned his trade overseas; in 1487 he
commissioned a foreigner to print an edition of the Sarum
Missal for him, and amongst his apprentices and successors
were many aliens by birth. Latin was still used for works which
were meant to have universal appeal and Latin books, even
after England could produce them, were often printed overseas.
Erasmus's books were produced abroad, as were More's *Utopia*
and Vergil's *Anglica Historia*; indeed, except for some Cicero
and Terence for the use of schools, no 'classic' was produced in
England before 1535.

None the less, the trades of printing, book-binding, and book-
selling advanced steadily in England, partly because they
secured royal patronage. In December 1485 a Savoyard named
Peter Actors was made stationer to the king; his successor,
after 1501, was the Norman William Faques who, since he was
also a printer, took the style of 'printer to the king'; in 1506
Faques was succeeded by Richard Pynson, also a Norman, who
introduced the use of roman type; on Pynson's death in 1530
the office went to Thomas Berthelet who was deprived of his
position on the accession of Edward VI. His successor was
Richard Grafton who, in 1543, not only reprinted Hardyng'
Chronicle but added a continuation of his own. Grafton was
deprived when Mary ascended the throne and was succeeded
by John Cawood. In 1547 Reyner Wolfe was appointed king'

---

[1] In *Educational Charters and Documents, 1589–1909*, ed. A. F. Leach, may b
studied the curricula and time-tables of various schools. Of the severities o
school life some examples are given in *Social Life in Britain from the Conquest to th
Reformation*, G. G. Coulton. The *Vulgaria* of Whittinton abound in references to th
rod and its terrors, some meant to be jocular.

printer in Latin, Greek, and Hebrew; but in a steady development much work was done by men who did not enjoy the royal
patronage at least officially,—among them, for example, Caxton,
who was far more than a printer, and his apprentice Wynkyn
de Worde. The contract for the printing of Horman's *Vulgaria*
by Pynson in 1519, which is still extant, shows that the business
of publishing was very well understood, and an act of 1534
compelling foreign printers to sell their editions entire in sheets
to English stationers shows that the English book-trade was
alive to its own interest. There have been traced no fewer
than ninety-one signs of stationers in London of date prior
to 1558.[1] Many of the shops were stationed about St. Paul's
churchyard; but there were others elsewhere in the City and
there were a few presses outside London, at Oxford, for example,
at St. Albans, and later at Cambridge. Early in the sixteenth
century, the stationers' company, which claimed some existence
dating from 1404, gained importance as the defender of the
native traders, and in 1557 it received a formal charter from the
king and queen.

There was loss as well as gain. It is a sad commentary on the
times that the library in the Guildhall founded by 'Dick
Whittington', and much used by students of divinity, disappeared in the year 1549—Somerset took the books and manuscripts away and they were never returned.

A study of the list of books published is informative. It
reveals both an increasing use of English and a slow development in English taste. Caxton stuck in the main to the old
favourites, translations from Virgil and Cicero, stories from
the Greek mythology, works of edification, and, most important
of all, English classics from Chaucer and Gower, and from the
Arthurian cycle. The publication of statutes and of law books
was dictated by utilitarian considerations, but there was a
genuine interest in history for its own sake. More's *Historie of
Kyng Rycharde the Thirde*, though it was not attributed to its
author till 1557, was written about the year 1514, and throughout the reigns of Henry VIII and Edward VI there were issued
fugitive pieces dealing with the great events of the day. Some
of these were used to enrich the chronicles which rose to the
status of literature in the works of Thomas Fabyan, Richard

---

[1] See E. Gordon Duff, *A Century of the English Book Trade*, 1457–1557 (1905), and
the *Catalogue of the Caxton Celebration* (1877).

Grafton, and Edward Hall. Fabyan, an alderman and sheriff of London, wrote a chronicle which ended with the year 1485 and which was mainly a repetition of *The Great Chronicle*, though it included borrowings from Gaguin's *Compendium super Francorum Gestis* of 1497; he himself called it a *Concordance of Chronicles*, but after his death it was printed by Pynson in 1516 with the more imposing title of *The New Chronicles of England and of France*. When, in 1533,[1] it was reprinted by Rastell, it included some continuations to 1509; these were not by Fabyan whose name has gained an unwarranted importance because Stow attributed to him the whole of *The Great Chronicle*, of which he himself made great use.

None the less, Fabyan has the distinction of having made a real endeavour to produce history instead of the common-place books and chronicles hitherto compiled by citizens. Grafton's work is not of outstanding merit; it is founded on a reprint of Hardyng, but it also uses the work of More and contains a continuation provided by the printer himself. Hall's *Chronicle* entitled *The Union of the two Noble and Illustre Famelies of Lancastre and Yorke . . . beginnyng at the Tyme of King Henry the Fowerth . . . and so successivly proceadyng to the Reign of the High and Prudent Prince Kynge Henry the Eight*, is at the outset only a dull repetition of earlier narratives; for the reign of Richard III, however, it uses the work of More, and for the reign of Henry VII the work of Polydore, while for the reign of Henry VIII it embodies a number of fugitive pieces already mentioned, as well as the author's own knowledge and his own lusty spirit. Its English is clear and wholesome, and it expresses the confident pride of the Englishmen of the day. Other contemporary chronicles such as those of Wriothesley, Machyn, and the anonymous chronicles[2] dealing with the history of London, one from 1523 to 1555 and the other from 1547 to 1564, were not published until the nineteenth century. As records of fact they are useful to the historian of today, but they reflect the caution necessary at a time of religious and political change and they have no pretensions to style. The *Journal* of Edward VI, though it is remarkable as the product of a boy's mind, and valuable

---

[1] In 1529 John Rastell had printed *Pastyme of the People* containing the histories of diverse realms, including England.

[2] 'Two London Chronicles, from the Collections of John Stow,' ed. C. L. Kingsford (*Camden Miscellany*, 3rd series, vol. xii (1910)).

for the information it contains, is devoid of literary grace.[1] That there was an interest in the history of other countries and other ages may be guessed from the publication in 1493-4 of the story of the siege of Rhodes, and of Lord Berners's beautiful translation of Froissart (1523-5) which has become an English classic in its own right.

Histories were no doubt considered valuable as teaching the art of life and government; these topics were considered complementary one to the other by the men who believed that 'mastery' in a man was a good thing and that no man could rule others unless he could rule himself. There is a series of books dealing with the whole upbringing of youth and the direction of the mature mind into the art of government. The *Bokys of Hawkyng and Huntyng and also of Coatarmuris,* translated for the French by Juliana Barnes or Berners, prioress of Sopwell in Hertfordshire (1486), the *Boke of Courtesye* of 1478 and 1491, and Barclay's *Myrrour of Goode Maners* (1523), though their themes are medieval, may be taken as evidence that a full life was valued for its own sake. The *Tree of Commonwealth,* written by the captive Dudley in the vain hope of moving the pity of the arrogant young king, is in the main medieval too. It is rather a dreary allegory showing that the principal root of the state is the love of God, and that the ground in which the root is fixed is the prince who, even more than the clergy, must set an example to all his subjects. Subsidiary roots are justice, truth, concord, and peace, each of which yields an appropriate fruit, but the greatest fruit of all is the honour of God; this gives its own virtue to the virtues which might in themselves be culled from pagan trees, and can even turn to good the 'perilous cores' which are in themselves dangerous: vain glory may become perfect glory and lewd enterprise may become perfect enterprise. Behind the imagery of the middle ages, however, and the familiar parable of Menenius Agrippa, the renaissance spirit shows itself in frank realism. To modern readers the great value of the book is the implicit condemnation of administrative despotism. The author is plainly denouncing abuses which he well knows—interference with justice by letters from the privy council sent ostensibly as replies to special petitions, and sometimes as mere directives; interference by the king in his own

---

[1] Published in the *Literary Remains of Edward VI* (Roxburghe Club) and also in Burnet's *History of the Reformation of the Church of England.*

suits, promotion of churchmen for political reasons and their use as temporal officers, non-residence of clergy; neglect of study whereby the universities decline and the divines 'become fast', —these are among the evils whose operation Dudley must have witnessed when he was in the service of Henry VII.

A more gracious side of the renaissance influence appears in the *Boke named the Governour* of Sir Thomas Elyot (1531) which will bear comparison with *Il Cortigiano* as well as with the old English books of manners. After explaining that in a *respublica* there must be a single sovereignty, and that under the authority of the prince there must be inferior governors or magistrates, the book sets out a scheme of education for the child of a gentleman who is to have authority in the public weal. The wealth of citation from the classics and from the fathers is surprising, but it is evident that the author is writing of England and for Englishmen. Due attention is devoted to ordinary learning in the nursery and under a tutor. Some instruction in music, in painting, and in carving is well enough, provided that the pupil has aptitude. Dancing is a legitimate exercise and there is no impropriety in men and women dancing together, indeed the practice may inculcate circumspection and other branches of prudence and so help to develop morality. Great attention is paid to the development of the body; some forms of sport are condemned as giving too little exercise, others as giving too much. Of sedentary games chess is the best, card-playing has some value in training the wits, dicing is utterly bad. Some forms of bowls are too gentle to benefit the body, but football is a 'beastly fury' which should be avoided. Wrestling, riding, hunting, hawking, and swimming are all good; best of all, however, is archery which can be practised by a single sportsman and will not only make a man able but fit him to serve the commonwealth in arms.

Subsequent chapters explain the qualities which should be developed in a governor and distinguish between real virtues and those which are merely apparent; there is some comprehension of distributive justice but the author shows that the qualities which make a good governor are those which make a good man. It was not only in one book that Elyot showed himself as a child of the renaissance. He translated the *Rules of a Christian Lyfe* by Pico and made translations both from the Greek and from the Latin. His humane view of the meaning of education was adopted by the gentle and practical Roger Ascham who, characteristically, wrote a whole book—*Toxophilus* (1545)—upon the use of the bow, and, no less characteristically, used the commendation of archery both to condemn less worthy things and to inculcate moral principles. His book on the *Scholemaster* was written in the reign of Queen Elizabeth but it dealt with an experience acquired by him in earlier days and makes famous reference to the education of the Lady Jane who found her tutor 'Mr. Elmer' so helpful and her parents so hard.

Long before Ascham wrote, however, the interest of England had turned to the controversies of the day, and the polemical writings of the time gave little heed to beauties of form. The use of dialogue[1] exemplified in the *Toxophilus* and in the *Discourse of the Common Weal of this Realm of England*, conduced to clarity rather than to grace, and although the invective of Brynklow, Latimer, and their fellows sometimes produced a lively picture or a telling phrase, the writers did not concern themselves particularly with the literary art. From the flames of the controversies, truly, there emerged golden treasure in the English Bible and the English liturgy which have done so much to shape the prose of England, and have trained the ears of countless generations to power of expression and beauty of sound; but the beauty which was here wedded to holiness seems to have been the result rather of innate genius than of conscious effort. Conscious effort, however, there must have been elsewhere; the increasing recourse to the classics was not without its effect and in the *Arte of Rhetorique* (1553) Thomas Wilson roundly condemned the 'counterfeiting of the Kinge's Englische' by the use of French and Italian phrases. The works of Sir Thomas More, however, published by his nephew William Rastell in 1557, though they may be reckoned as controversial pieces, must be counted also as a great monument of good English.[2]

It was in poetry rather than in prose that the use of continental models was eventually carried to excess, but, in fact, the gifts of the renaissance came only tardily to English verse.

English poetry was at a low ebb; the successors of Chaucer had been ponderous and dull, and during the fifteenth century inspiration did not come from without. Among the classical

---

[1] Stimulated by the study of German examples; C. H. Herford, *Studies in the Literary Relations of England and Germany in the sixteenth Century* (1886).

[2] R. W. Chambers, *Thomas More*, p. 20.

books acquired by Englishmen, the poets are almost unrepresented; Theocritus was unknown and as a writer of eclogues 'the Mantuan' who was reverenced was not Virgil, but Baptista Spagnuoli Mantuanus, whose poems were read in grammar schools even in Shakespeare's day. The works of the two great poets of the early sixteenth century, Alexander Barclay and John Skelton, are not conspicuous either for good form or for happy thought; in invective alone do they excel, and they may claim a connexion with the renaissance only by the vigour of their criticism. It is significant that both authors were Latin scholars of repute and that both were churchmen. Barclay translated Sallust and some eclogues of the Mantuan, and it was probably from the Latin version of Locher that he took the *Ship of Fools* on which his reputation mainly rests. Of its content something has already been said;[1] its form is commendable in that the author experimented in the seven-lined stanza and had the courage to add to the original 'envoys' of his own. He had the sense, too, to have his work enriched with lively woodcuts, and his personality appears in attacks upon Skelton made both in the *Ship of Fools* and in the *Eclogues*. His reputation stood high in his day; he was employed to 'devise histoires and convenient raisons', to ornament the buildings prepared for the 'field of cloth of gold' and as *Preignaunt Barclay* he was complimented along with Chaucer, Lydgate, and Skelton in a work of 1521. Yet neither in spirit nor in form did he achieve real greatness. Despite his condemnations, he had no clear vision of reform; often his thought was flat and his verse halting. He contemplated the impending 'ruyne of the holy fayth' in the impressive lines:

> My dolefull teres may I nat well defarre
> My stomake strykynge with handes lamentable.[2]

and proposed for remedy the old specific of the grand crusade in which England and Scotland should sink their differences:

> If the Englysshe Lyon his wysdome and ryches
> Conioyne with true loue, peas and fydelyte
> With the Scottis vnycornes myght and hardynes
> Than is no dout but all hole christente
> Shall lyue in peas welth and tranquylyte

[1] p. 253 *supra*.
[2] *The Ship of Fools*, translated Alexander Barclay (Edinburgh, 1874), ii. 193.

And the holy londe come into christen hondes
And many a regyon out of the fendes bondes.[1]

Barclay, as might be guessed, was a Scotsman by birth. His contemporary Skelton was all English, and directed some of his most trenchant verses against the Scots, whose disasters under James IV and Albany he recorded:

But for the special consolacion
Of al our royall Englysh nacion.[2]

He had a good classical education at Cambridge and probably at Oxford. He was employed as a tutor to the young prince who later became Henry VIII, and was patronized not only by the royal family, but by the countess of Surrey. He was versatile and had a rude strength which commended itself to a critic like Southey, but his execution was far inferior to that of the Scot, William Dunbar. His pose was to criticize with a bluff directness the men and manners of his time. The *Bowge of Courte* was a satire upon court life, and in *Colyn Cloute*, written about 1519, he made his rustic hero attribute all the woes of England to the misdemeanours of the clergy. The powerful Wolsey did not escape unscathed, and in *Speke, Parrot* and *Why come ye nat to Courte* (1522-3) he followed up his attack upon the great cardinal with a fierce if unfair satire. Not surprisingly, he was compelled to take sanctuary at Westminster with Abbot Islip, though he seems to have emerged before his death in 1529. He had a good control of assonance, and his 'breathless rime' is sometimes effective, but though he had some humour he had little music and little nobility of thought. The only dramatic work of his which has survived, the interlude *Magnyfycence*, is intolerably dull; there is almost no plot and the characters who speak their parts are personified qualities devoid of true humanity.

Drama was little affected by the new thought. The English had always been fond of spectacles and the pageants continued to be a colourful feature of town life; but their gallant themes, while they provided opportunities for fine clothes and feats of chivalry, cannot be regarded as 'plots'. It was noise and splendour which attracted the multitude. In *A Discourse of the Common Weal of this Realm of England*, it is alleged that many popular festivities as 'stage playes, enterludes, Maye games', and so on,

[1] Barclay, ii. 209.
[2] John Skelton, *Poetical Works* (London, 1843), i. 184.

which were occasions of 'muche expenses', have been abandoned; but it is plain from the chronicles, and from the king's 'Journal', that Londoners were not stinted in the matter of public spectacles. The accounts of the master of the revels, show that the visit of the French ambassador, the Marshal de Saint André, who brought the order of St. Michael to Edward in June–July 1551, was the occasion of an astonishing display. Two 'bankett houses' were erected, one in Hyde Park and another in Marylebone Park. They were decked with flowers, roses, 'lavender spyke, honysokilles, marygoldes and gillofloures' on two separate occasions, and the total cost was upwards of £440.[1] Roses were bought by the bushel, from Hackney, Hoxton, Islington, and Shoreditch, for the first festival; for the second, presumably because the roses were passed, 'gillofloures' had pride of place. Probably the prevalence of sweating sickness in London may have suggested recourse to the open fields, and it seems that in the end even they had to be abandoned in favour of Hampton Court. On this occasion the king must have wasted outright £475 of the £500 he spent, and generally it seems the case that on occasions of public display expense was little regarded.

For ordinary celebrations there was less pageantry and more drama. Miracle plays were common—as late as 1517, a new processional play in honour of St. Anne was inaugurated at Lincoln where the scholarly Atwater was bishop—and they went on in the old way until the development of the reformation rendered improper plays dealing with saints; but in them, too, there was little art.

The moralities, which were to outlive the reformation, continued to deal with the familiar topic of man subjected to the warring influences of good and evil. Besides *Magnyfycence*, already noted, may be remembered the interlude of the *Nature of the Four Elements* by John Rastell, brother-in-law of Sir Thomas More who, like Barclay, was employed on the decorations for the field of cloth of gold. Interludes were popular and, as appears from Roper's *Life*, More[2], when he was a member of Cardinal Morton's household, sometimes acted parts which he himself improvised. It is possible that, with the growth of

[1] See *Loseley Manuscripts*, ed. John Kempe (1836). I owe this information to Professor Arnold Edinborough of Queen's University, Ontario, who is making fresh study of these accounts. Roses usually cost 1s. 4d. a bushel.

[2] R. W. Chambers, *Thomas More*, p. 59.

criticism, elements of fun and realism presented themselves more strongly, and the devil or the vice was made a subject for merriment; but the idea of a comic devil was quite familiar to the medieval stage. In the drama of Henry VIII's day there was nothing that was new.

During the reign of Henry VIII England drew from France the idea that the making of poetry was part of knightly accomplishment; the courtiers began to write verses, which, if they commended themselves, were copied out and circulated, often with accompanying music, though it was not until 1557 that a considerable body of 'songes and sonettes' was printed in the volume generally known as *Tottel's Miscellany*. Two poets stood out conspicuously, Sir Thomas Wyatt (1503?-1542) and Henry Howard, earl of Surrey (1517?-1547). In both the pulse of the renaissance beat strongly; each took a full share in the active life of his day, experiencing to the full its vicissitudes of triumph and defeat; each saw the inside of a prison and Surrey at last paid for his rashness with his head. It cannot be claimed that these men imbued their verse with very deep feelings drawn from their experience of life; rather they were artists who added to their other accomplishments the joy of poetic creation. Yet they did introduce into English verse the personal note, and together they restored a sense of music and of form which had been absent from it since Chaucer died. Wyatt it was who first rendered into English the Italian sonnet, copying Petrarch both in spirit and in form, yet daring to depart from his master by finishing with a rhyming couplet; and besides his sonnets he wrote lyrics, epigrams, and satires, many of them based upon Italian models. In the hands of Surrey the cadence of the sonnet was immensely improved, though often he contented himself with three quatrains and a couplet, and with his translations from the *Aeneid* England first heard the majestic blank verse which was to be carried to supreme triumph by Shakespeare and Milton. It was only, however, in the second half of Henry's reign that Surrey wrote; in its first burst of splendour English poetry made no advance. Men's eyes were set upon the old and the far; things near and present seemed of little value in the world of letters; Erasmus never learned English.

The music of the early Tudor period was not distinguished by great inspiration, but none the less it reflected in its advance and in its competence the movement of the age; like all other arts it had been much under the patronage of the church in the middle ages, and its development during the renaissance period exhibited the changes which were proper to the time. On the one hand, ecclesiastical music became more scholarly and more artistic; on the other hand, there was a great advance in secular music, for dances, for songs, and, finally, with the beginning of chamber music, for its own sake. Yet, as in literature, architecture, 'constitution', and all else, an active influence from without did not destroy the accepted English form.

The England of the early Tudors was a music-loving land. In church music, particularly, there had arisen, under royal patronage, a very definite school. The chapel royal, established early in the fifteenth century, had become a nursery where English composers developed their technical skill and their creative power. John Dunstable, who died in 1453, spent much of his time abroad, but early in the sixteenth century the tradition which he had left behind him in the Netherlands came back to inform the art which had been developing in England, and William Cornyshe (1465?–1523) was no mean expositor of the English form. Even under Henry VII the technique of church music had begun to change; the Eton College manuscript, of date between 1490 and 1504, shows an attempt at regular counterpoint and smoothness unknown in the Old Hall manuscript which represents the period from 1430 to 1480.[1] Robert Fairfax is distinguished for his pure and definite counterpoint and when, in the reign of Henry VIII, the English school was enriched by Flemish influences, marked progress was made. A manuscript of about 1516[2] contains an anonymous piece of great beauty, *O My Deare Sonne*, as well as the famous *Quam Pulchra Es* of Richard Sampson (1470–1554). Richard Sampson and his contemporary John Taverner, best remembered by his *O Splendor Gloriae*, may not have excelled Fairfax in grandeur but they certainly outstripped him in artistry and in the fitting of the music to the words. This fitting of the music to the words became of supreme importance with the triumph of protestantism. The masses and motets disappeared with the images and the ornaments, and to the chapel royal with its organ and choir

---

[1] This contains 138 compositions representative of the period, and is at Old Hall near Ware.
[2] Included in British Museum. Royal MSS., no. 11 e 11.

was left the task of accommodating the great English tradition to the simpler forms demanded by the new religion. In this heavy task the composers brilliantly succeeded, and with their triumph are associated three famous names. Christopher Tye (1497–1572/3), who was first a chorister at King's College and later master of the choir-boys at Ely, published in 1553 the *Actes of the Apostles*, a collection of part-songs some of which, for example *Winchester*, are still used as hymns. Another Cambridge man, Robert Whyte, Tye's son-in-law and pupil, succeeded him at Ely, and later became organist and master of choristers at Westminster; he wrote, for example, *Peccatum Peccavit* and *O Praise God in His Holiness*, both fine works; but as he published nothing himself, it was only after his death that his reputation rose to its full height. Thomas Tallis, who was organist at Waltham Abbey, continued the tradition, and made for himself a great name familiar to all lovers of music; his achievement, however, belongs mainly to the great days of Elizabeth.

It was not only by the maintenance of the chapel royal, and not in church music alone that the Tudor princes showed their interest. All were patrons of music. The privy purse expenses of Henry VII show payments for flutes, for a pair of clavichords, for lutes, for making a case for a harp; and rewards are given to child singers, choristers, musicians and waits in various towns, bagpipers, the king's piper, harpers, singers, and others. It seems clear that William of Newark (1450–1509) and William Cornyshe, to whom sums were paid, were commissioned to write songs and carols.

Henry VIII had a keen interest in music, and he maintained a considerable establishment. Besides the masters of his chapel royal there were luters, minstrels, virginal-players, fifers, and viols; he had an organ-maker and a keeper of the king's instruments whose inventory shows the strength of the royal equipment.[1] He himself was a competent performer; a foreign observer remarked that he 'plays well on the lute and the harpsichord and sings from the book at sight'.[2] He was a composer too. It may be doubted whether Hall is quite correct in asserting that Henry composed 'two goodly masses, each in five parts',

---

[1] *Songs, Ballads and Instrumental Pieces composed by King Henry the Eighth.* Reproduced from British Museum MS. 31922 in Oxford for the Roxburghe Club (1912).
[2] *Letters and Papers*, ii. 117.

and an unfriendly critic has stigmatized him as 'an eclectic of the feeblest kind';[1] but it seems probable that he wrote the songs attributed to him. These are in three or four parts, and if not of superlative merit they are happy and pleasant, expressing the royal opinion that as

> Idillnes is cheff mastres of vices all
> Then who can say
> But myrth and play
> Is best of all.

Hall reports that during one of his progresses Henry exercised himself daily in sports and also in '... singing, dancing ... playing at the recorders, flute, virginals and in setting of songs and in making of ballads'.[2] He also saw to it that his children were educated in music; Edward himself recounts how the French ambassador 'heard me play on the lute',[3] while Mary, it is said, played both the virginals and the lute.[4]

Like the kings, the great nobles encouraged musicians. The evidence of the *Northumberland Household Book* shows how much was done by a great family, and men like Buckingham and Wolsey obviously thought that music was part of their high estate; plainly an education in music was, by this time, part of the necessary instruction of a gentleman. From the accounts of the court ceremonies preserved in the Herald's manuscripts[5] it is clear that part-singing was a recognized accomplishment; the famous story that Thomas Cromwell[6] advanced to greet Pope Julius II (to whom he presented jellies of his own making), singing, with two supporters, a three-part song, may not be true, but it attests the English reputation for music to which Erasmus himself paid tribute.[7] Wynkyn de Worde's *Song Book*[8] which appeared in 1530—the first book of music printed in England—contains many secular pieces, some of which, though they were ornamented by counterpoint, and written in various parts, were probably based on folk-songs. The rising middle class, it seems, cultivated music, and even the ordinary people, though

---

[1] Ernest Walker in *History of Music in England*, 3rd ed. (1932).
[2] Hall, i. 19.
[3] Edward VI's *Journal*, ed. J. G. Nichols (Roxburghe Club), pp. 332–3.
[4] Letter of Queen Catherine to Mary, *Letters and Papers*, vi. 472.
[5] Leland's *Collectanea*, vol. iv. 235 et seq.
[6] Foxe, Bk. viii.   [7] Erasmus, *Encomium Moriae*.
[8] Only the bass part of this book survives, now in British Museum.

their tastes may have been simple, were evidently appreciative of tuneful sound; the public ceremonies of state, which were designed largely to captivate popular goodwill, were always graced by music, often in the form of lusty trumpet peals. Trumpet playing, it may be remarked, is in itself a difficult art.

Good architecture is usually the sign of a prosperous age, and in the reigns of Henry VII and Henry VIII there was much good building in England. Relatively little was owed to renaissance influences, for the native style—a developed perpendicular—was virile and more than competent. England produced both fine conception and good workmanship; her craftsmen knew what they wanted to do and were able to do it. As they experimented, the gothic scheme of pointed arcade, buttress, and stone vault was gradually discarded and the arch became little more than a frame to fill in designs which were fundamentally four-square. In some buildings, like that of the church of Thaxted in Essex, the use of uprights and horizontals, long used for mullions, was so much developed that the artist seemed almost to be awaiting for the advent of the steel girder which would permit him to apply his system to the main fabric itself. By skilful technique the vault was rendered extremely flat. The fan-vaulting at Sherborne abbey and Bath abbey, it is true, follows the tradition of the Gloucester cloister, but that in Henry VII's chapel and in the cathedral at Oxford, though it gives an appearance of immense depth, is really produced by an extremely ingenious arrangement of slightly wedged stones supported on largely concealed transverse arches from some of whose voussoirs the fans spring.

During the fifteenth century good craftsmen had been working at Oxford and Windsor,[1] and Henry VII had inherited artists and artificers of great skill. Most of these were English. His three master-masons, in the year 1509, were Robert Vertue, Robert Jenins, and John Lebons; his chief carpenters, his painters, and his brassfounders all had English names, so too had most of the 'kervers' and sculptors. Famous among the image-makers were Lawrence Imber and Thomas Drawswerd, the latter one of the family in York which had a great reputation. The great glazier of the day, Bernard Flower, was

[1] W. R. Lethaby, *Westminster Abbey and the King's Craftsmen*, see Bibliography, v. Architecture.

apparently English, and when he died the contractors who undertook to finish his work were described as 'glaziers of London'. One of them, however, Gaylon Hoon, has a foreign-sounding name; some of the carvers, as Derek van Grove and Giles van Castel, must have come from abroad, and certainly the stalls in Henry VII's chapel are German in feeling.

For the refinements of ornament Italians were employed. Henry VII's tomb with its recumbent statues was the work of a hot-headed Florentine, Pietro Torregiano, who had once broken the nose of Michelangelo. With him a contract was signed in 1512, and the work was completed in 1518. Meanwhile, he had not only constructed the tomb of Henry's mother, the Lady Margaret, in the south aisle, but had signed, in 1517, the indenture for the making of the altar to the west of the royal sepulchre which, in fact, was finished only in 1526 by Benedetto di Rovezzano. Wolsey employed Italians, notably Giovanni di Majano, for some of the ornaments—the terra-cotta busts of the emperors, the medallions, and the plaque of his arms—at Hampton Court, and later contracted with Antonio Cavallari to construct a tomb for him; but the surveyor, 'Mr. Williams, a priest', who was possibly the architect, James Betts, 'master of the workes', and Nicholas Towneley, clerk-controller, obviously were Englishmen. Holbein, who may be reckoned Italian in his architecture, designed two gatehouses at White-hall, and a few traces of Italian work appear in churches and in private houses; but in the main English architecture was in the hands of Englishmen.

The English tradition expressed itself both in churches and in domestic architecture. The idea that under the impending reform pious effort entirely ceased is false; besides the chantry chapels, which were a mark of the early Tudor period, there was some ambitious church-building on a grand scale. It was a great period for towers. The central or 'Bell Harry' tower at Canterbury was built between 1495 and 1503, the tower of Fountains abbey about the same time; the bell-tower at Evesham was built in 1533, and the west tower of Bolton priory was still unfinished when the house was dissolved. Magdalen tower wrongly associated with the name of Wolsey, will serve as a reminder that some of the ecclesiastical artistry of the day was given to the universities; the stalls in King's College chapel which have classical detail throughout, were erected between

1531 and 1535, and a chapel was a necessary feature of each of the colleges erected at the time. The abbey church of St. Peter and St. Paul at Bath, erected during the reign of Henry VII on the site of an earlier Norman church, is a superb example of late perpendicular with its great west window, its fan-vaulting, and the carved angels of Jacob's ladder; but the glories of this 'lantern of the west' pale before those of King Henry VII's chapel at Westminster wherein English architecture reaches a supreme height. The king's design was to erect at the east end of the abbey a chapel to express his 'synguler and specyall devocyon' to the glorious Virgin, and to provide a splendid resting-place for himself and his wife. It was his intention to place in the central radiating chapel the body of his saintly predecessor Henry VI, and to commemorate in a magnificent chantry the piety of the house of Lancaster. For its valour the chantry of Henry V, just opposite to the entry of the new chapel, already provided a memorial. His plans were not completely carried out—Henry VI still lies at Windsor; but the building which arose on the site of the old lady chapel is, as Bacon said, 'one of the statelyest and daintiest monuments of Europe both for the Chappel and for the Sepulchre'. The king had at his disposal the craftsmen who had worked, and some of whom continued to work, on St. George's chapel at Windsor; and as, unlike some other kings, he had money at his disposal, his enterprise succeeded well. In 1503 the foundation-stone was laid in the presence of Abbot Islip, and the shell of the building was complete by 1509, though it was many years before the decoration was finished. In the end, however, there arose a work which was a tribute not only to his own piety and wisdom, but also to the skill of English craftsmen and the possibilities of English architecture. It is worthy of note that much of the credit for the architecture of Henry's chapel is given, by common report, to Sir Reginald Bray,[1] one of the king's men-of-all-work and essentially a 'new man'. Here and there, as in Gardiner's chantry at Winchester, classical features are combined, rather oddly, with the perpendicular style; but the essential feature of the ecclesiastical building of the period is the persistence of the English tradition.

---

[1] It is of relevance that Bray commissioned, probably in the nineties, a magnificently illuminated antiphoner for use, it may be presumed, in a chantry of his foundation. (Now in the possession of J. C. Thomson, Esq., late of Charterhouse School.)

Of domestic architecture the same thing may be said.[1] The English tradition proved itself capable of meeting the requirements of an age in which the Englishman's house, except in the figurative sense, ceased to be his castle—it is characteristic of the English way that the battlements survived as an ornament. The English dwelling-house of any consequence had always had as its essentials a hall set opposite to the kitchen, buttery, and pantry, from which it had come to be divided by screens; from the farther end, where a dais soon made its appearance, the hall gave to the 'chamber', which represented the family residence. As the years passed the chamber developed into a complex of rooms, usually on more than one story, but the hall itself retained its commanding position, and, both in the abbot's house and in the gentleman's manor, was frequently graced by the addition of a great bay-window thrown out from the side of the dais. The insistence upon a lofty hall meant that the extensions were made in latitude rather than in elevation; and partly for convenience, partly because big dwelling-houses were usually associated either with monasteries or with castles, these extensions tended to group themselves in a quadrangular form which commended itself also to the later Renaissance architects.

This form had applied itself naturally to the building of colleges, and in the earlier Tudor period both Oxford and Cambridge were enriched with noble examples of the English builder's art. Conspicuous among these is St. John's College at Cambridge with its fine gate-tower bearing the arms of the Lady Margaret, her badges—the Tudor rose and the Beaufort portcullis—and the daisy which represents her name; but the greatest academic monument of the age is at Oxford, where Christ Church bespeaks the glory and the ambition of the proud cardinal.

In an age when the royal power was paramount, the skill of the architect was naturally put at the service of the Crown, and during the early sixteenth century the increase of the kingly power is expressed in the development of the royal palaces. Henry VII rebuilt Baynard's castle in the City, created from burnt-out Sheen the palace of Richmond, made extensive alterations at Greenwich, and added new chambers to the palace

[1] T. Garner and D. Stratton, *The Domestic Architecture of England during the Tudor Period* (see also Bibliography, s.v. Architecture).

of Westminster. Henry VIII built new houses for himself at Bridewell and St. James's, and added to Windsor, Richmond, and Greenwich. He also acquired from others and turned to his own use, New Hall or Beaulieu near Chelmsford, Hunsdon House in Hertfordshire, Nonsuch in Surrey (never completed), and, on the cardinal's fall, York House and Hampton Court.

The residence of the archbishops of York at Westminster, glorified by additions which included the famous gates by Holbein, became, as Whitehall, one of the favourite palaces of the English kings; Hampton Court, though now graced by the addition of Wren's splendid wing, stands as a living monument of Wolsey's vaulting ambition and the competent greed of his master.[1] On a site which he had leased from the Hospitallers in 1514, choosing it for its gravel soil and healthy air, the cardinal began to erect a country retreat which he fitted with all the amenities of the age. It is characteristic of the man that, spurning the doubtful waters of the Thames, which others drank, he had his domestic supply brought in leaden ducts from the salubrious Coombe Springs more than three miles away, and on the other side of the river. By 1520 the main outline of his work was complete, and although its exact extent is obscured by later additions, it was considerable. It seems that he built the west-front, the base-court, and the clock-court, bounded on the north by the magnificent great hall, in which he kept open table, and on the south by the panelled suite reserved for his own use. So magnificent a house excited the cupidity of the king. It is said that the astute cardinal baulked the royal jealousy by formally surrendering the lease of the manor in 1525, though for some years afterwards he wrote 'from my manor of Hampton Court'; at all events, when the cardinal fell Henry entered into possession without delay, erected tablets bearing his coat of arms in many places, and made vast additions of his own. Wolsey had employed some Italian workmen; there are traces of renaissance ornament in Henry VIII's buildings; and it was a sign of the changing age that the vast palace was made of brick. Yet, for all that, the great building with its fine hall, its courts, its high-pitched gables, its bay-windows, and its gracious chimneys is essentially a development of the English style.

What the king did was done by his nobles and gentlemen.

[1] E. Law, *Short History of Hampton Court*.

Sometimes ecclesiastical buildings were turned into dwelling-houses by their new owners. From the church and cloisters of Netley abbey Sir William Paulet made a great quadrangular house; at Titchfield Sir Thomas Wriothesley built a great gate-house tower in the nave of the abbey church, turned the refectory into a hall and housed his family and servants in the rest of the building; at Lacock in Wiltshire Sir Edward Sharington made a fine new house out of a nunnery. Some of the gentlemen's houses built at this time were executed in brick, among them Layer Marney and Sutton Place, which exhibits signs of Italian influence; but the generality adhered to the English manner, though by no means all achieved the complete quadrangle. It is a curious fact that the builders often avoided a southern aspect from a belief, maintained by the redoubtable Andrew Boorde (an ex-Carthusian protégé of Cromwell's), that a south wind was very unhealthy. In building as in other things, the protector had a royal ambition, and to him was due Somerset House, described as 'the first building of Italian architecture erected in England', and attributed to John of Padua, 'deviser of his majesty's buildings'. Somerset erected his magnificent palace on the ruins of the town houses of the bishops of Worcester and Lichfield and of the church of St. Mary at Strand; to obtain material for his building he blew up the priory of the Hospitallers at Clerkenwell, and even despoiled the north aisle of St. Paul's ornamented with Holbein's famous *Dance of Death*. His work, however, was not completed at the time of his fall, and by a kind of Nemesis it vanished beneath the hands of the eighteenth-century improvers.

In the towns there was little building of great distinction, perhaps because the wealthy merchants were in process of becoming country gentlemen, and because the fortunes of the city companies were so uncertain. The Guildhall at Cirencester, however, is a beautiful example of 'late perpendicular', and some of the town houses in Bristol, visible till recently, were well made, with windows panelled in stone. The same embellishments appear in good houses in Glastonbury, Salisbury, Norwich, and Sherborne, for example, and in other towns, especially those of the midlands, fine half-timbered houses may still be seen. Even in counties where stone was abundant half-timbered houses were popular. Where stone was used for street houses the building was carried straight up, but where timber was employed an upper story was sometimes projected upon corbels.

The furniture for the houses, even of the palaces, was something simple, though Wolsey seems to have introduced into Hampton Court a magnificence unknown before his day. In the homes of the rich, arras or tapestry was freely used, some of the chambers were panelled, some of the floors were tiled and occasionally there was a carpet; but even in great dwellings the floor of the hall was still strewn with rushes which collected dirt as the weeks passed. For long, beds, even for the courtiers in the palace, were mere trundle-beds introduced at night into the chambers of presence, while lesser folks fared as best they might in the hall and elsewhere. The beds which appear in medieval wills were probably articles of furniture which served for couches by day, but by the middle of the sixteenth century the bed which was bequeathed in a testament was sometimes a handsome piece heavily carved in oak, and indeed the progress of the bed is the sign of an improving civilization. The box-chair of the period, sometimes heavily carved, must have lost in comfort what it gained in dignity. In the accounts of feasts there is frequent reference to 'cupboards' on which the plate not in use was displayed, and it was a mark of the royal dignity that the king's 'buffetiers' or 'beefeaters', who guarded his plate, should find employment even when a feast was in progress. If, however, the cupboard displayed in the Victoria and Albert Museum[1] really was made for Arthur prince of Wales, the craftsmanship of the period, though not unpleasing in design, was rather rough in execution; a cupboard from the West Country, of date about 1520, is, however, much more finished. Teak was used for the best furniture, tables and forms were carved with scalloping on the lower edges, and by 1550 a good draw-table had been produced. From 1540 on some fine carving in Renaissance style appeared on oaken panels, but the carving of the early years of the century still retained a medieval stiffness. Altar-pieces of English alabaster from Chellaston, Tutbury, and Nottingham which, in all probability, portrayed scenes familiar to the public from their occurrence in miracle plays[2] were very popular. They

---

[1] See O. Brackett, revised H. Clifford Smith, *English Furniture Illustrated*, for photographs and descriptions.

[2] Dr. W. L. Hildburgh in *Archaeologia*, cxiii (1949).

had a very wide market which extended from Brittany to the Baltic and penetrated inland as far as Burgundy. They portrayed, almost invariably, scenes from the life and death of Christ—though occasionally the centre-piece displayed the Trinity—and in their presentation they adhered to a strict canon; the risen Jesus, for example, is always shown as stepping upon the body of a sleeping soldier. The figures are not ill grouped, but none of them displays the life which is shown by some of the continental woodwork of the same date. England continued to use conventional types, and the demand for altarpieces waned before Renaissance influence was felt.

Pictorial art for its own sake may not have made a strong appeal to a practical age which liked clear representation. The illustrations on manuscripts, for the most part, continue the tradition of the fifteenth century, and the illustrations of the early books are usually woodcuts of a very rude kind. The 'devices' which were used for ornament on special occasions seem to have been more conspicuous for colour and clarity than for interpretation; but it is significant of the period that the human figure, suitably clad, was very often portrayed. It is in portraiture that the drawing and painting of the period excels, and the artists were, for the most part, foreigners. The representation of Henry VII which appears in the manuscripts of the *Book of Hunting* is a conventional picture of no particular merit, but the portrait of Margaret Beaufort in the National Portrait Gallery, though conventional in its pose, shows signs of the new art. The artist, however, is unknown. Perhaps he was a foreigner, for at Arras there are some fine line-drawings of English subjects, notably Henry VII, Perkin Warbeck, and Margaret Tudor, presumably from the pencil of some Netherlander; and it now seems certain that the admirable portrait of Henry VII, of date 1505, was done by Master Michiel (Miguel Zittoz?[1]), the court painter to Isabella of Spain, who after that queen's death returned to his native Flanders to work for Margaret of Austria. This Master Michiel was a good artist; he had painted the Catherine of Aragon as a girl now to be seen in the Kunsthistorisches Museum in Vienna, and it has been suggested that Catherine called attention to his skill when her father-in-law was seeking a portrait of a lady who might become his second wife.

[1] See *Burlington Magazine*, lxiii (1933), 104.

Henry VIII was interested in art but there was not in England a native school strong enough to benefit from the breath of the Renaissance. The king's serjeant-painters, John Browne and his successor after 1532, Andrew Wright, were mainly occupied in supplying decorations for his majesty's ships, or for the temporary structures required for fêtes; it is significant that the gild of the English painters was called the 'painter-stainers' company. The English monarch was unable to secure the services of great Italians like those who graced the court of his rival Francis, and the foreigners who painted the scenes like the Battle of Spurs and the Field of Cloth of Gold, now to be seen at Hampton Court, were at best mediocre. More distinguished were the members of the Ghent family whose name was anglicized as Hornebolt in a patent of denization of 1534; Luke, Gerard, and Susanna were all active in England and probably produced some of the many portraits and miniatures of the king. Another Netherlander was Johannes Corvus (Jan Raf) from Bruges, at one time master of the painters' gild there, who produced the portrait of Fox, now in Corpus Christi College, Oxford, and that of Mary Tudor in 1532 when she was duchess of Suffolk. To him have been attributed portraits of the Princess Mary who was later queen, and of Henry Grey, duke of Suffolk; certainly if these works were not actually by him, they came from a brush trained in the same school.

To an unknown artist, perhaps German rather than Netherlandish, is assigned the remarkable portrait of Henry VIII recently acquired by the National Portrait Gallery, which differs from the Holbein portraits though it cannot be very much earlier in date. It shows the king with chestnut hair, blue eyes, and somewhat aquiline nose; the face with its faintly knitted brows seems to indicate a quick temper, and the small eyes are calculating; but imperious dignity is not lacking. Like many portraits of the age, it was attributed to Holbein, to whom were ascribed at one time pictures done by his imitators and apprentices; but it was not by Holbein. Holbein needs no borrowed plumes; to him was due one of the greatest contributions ever made to the art of portraiture in England.

Born in 1497 at Augsburg,[1] the gateway through which the

---

[1] N. Wornum, *Life and Works of Holbein*, and A. B. Chamberlain, *Hans Holbein the Younger*.

Italian Renaissance entered Germany, he began in 1515 to practise his art in Basel; there he met Erasmus of whom he painted three portraits, and it was with an introduction from Erasmus to Sir Thomas More that he came to England in the winter of 1526–7. During a stay which lasted till 1528, he made drawings and paintings of the More family, and of other persons of distinction with whom More was associated; of Warham and Fisher, for example, of Sir Henry Guildford and his lady, of Sir Thomas Godsalve, of Sir Henry Wyatt, and of Sir Thomas Elyot. In 1532 he returned to England to find that Warham was dead, More in disgrace, and the old friendships gone. He found appreciative subjects, however, in the merchants of the Steelyard, some of whom he painted, while he supplemented his earnings by embellishing the walls of the Guildhall, and designing the title-page of Coverdale's Bible. It was not long before he found influential patronage; he painted Thomas Cromwell, and from 1536 until his death in 1543 he was the paid servant of the king. To this period belongs the great series of portraits of the king, queen, and courtiers, which have made the age real to us, and established for ever the artist's name. He must be reckoned as one of the great portrait painters; in Lambeth Palace there are portraits of all the archbishops of Canterbury since Matthew Parker, each done by the leading artist of the day, but none surpasses the brilliant William Warham which represents Holbein, and even that is inferior in some ways to the lovely drawing, now at Windsor, on which the picture was presumably based.

There was nothing 'impressionistic' in Holbein; every line meant something, and every line was firmly drawn—though a magnifying-glass shows that the seemingly firm line was really a series of very small lines. Even in his roughest sketches he contrived to give individuality to his faces and figures. It has been suggested that he brought with him from Germany a simple apparatus whereby the artist, looking through a peep-hole, was able to outline the main features of his sitter on a vertical sheet of glass set upright a couple of feet away, and some have said that because he deprived himself of the stereoscopic effect of two-eyed vision, his work is rather flat.[1] Be that as it may, it is not to be denied that he gave to his portraits a

[1] K. T. Parker, *The Drawings of Hans Holbein in the Collection of His Majesty the King at Windsor.*

life and a character to which few artists have attained. His women, it must be confessed, are not always attractive; most of them are mean-mouthed, some of them as Margaret Wyatt, Lady Lee, with her long nose and sidelong eyes, look sly as well as rather prim; but the drawings are sometimes kinder than the pictures, and the members of More's family particularly are most sympathetically done. The costume of the women lends itself to good portraiture; the gable-shaped head-dress sets off the face well, and the round French cap, which succeeded it in fashion, is not unduly assertive. The dresses and the jewellery, even of the queens, are rich rather than gaudy; Anne of Cleves is perhaps the most ornate, though the painting in the Louvre is less kind than that of the unknown artist in St. John's College, Oxford.[1]

As for the men, they show the characteristics which brought the sitters to eminence in a realistic age. Almost every face bears the stamp of competence and command. Henry VIII, with his small eyes set far apart and his rather thin beard, may not be a figure of manly beauty, but he has majesty; and the somewhat porcine features of Thomas Cromwell show both ability and power. As for the king's servants who helped to make their master great, every one of them shows strength, sometimes tinged with craftiness. Sir Thomas More's father, however, appears a kindly old man, and Sir Thomas Elyot has the face of a dreamer.

Holbein's fellow workers and his rivals were mainly foreigners like himself. In 1543 Andrew Wright the serjeant-painter was succeeded by Antonio Toto, who had come to England with Torregiano, and Toto held the office throughout the reign of Edward VI. Edward had been painted by Holbein in youth, and when, in 1553, he was painted again as presenting the palace of Bridewell to the citizens of London, the work was done by Guillim Stretes, a Dutchman. Mary was painted by Lucas Heere and by Anthonis Mor, both Netherlanders; Mor had been sent from Madrid to paint a portrait of Mary for Philip, and he it was who later produced a picture of Philip and Mary together. The famous picture of Cardinal Pole was done by Titian. But though foreign artists were apparently preferred, there is evidence of an increasing interest in art. Henry VIII is said to have had a gallery of which he kept the only key, and

[1] It is very doubtful if this is really a portrait of Anne of Cleves.

that he and so many of his servants sat to good artists is proof that there was appreciation, at least, of portrait-painting. Moreover, though Holbein founded no school—his assistants were mainly foreigners to whom he gave only temporary employment—a few Englishmen did rise to eminence as painters of miniatures, notably John Shute and John Betts who seems to provide a link between Holbein and Nicholas Hilliard. So, even in the field of art, the early sixteenth century presaged dimly the great day of Elizabeth.

What of the people of England during this pregnant age? They too were preparing for the great destiny which was to come upon them. Behind the outstanding figures of princes, prelates, and ministers, each playing on the great stage the part proper to his rank and status, may be discerned a lusty folk. The English were a healthy stock, but there seems to have been but little advance in medicine; Andrew Boorde published his treatises, and the unification of the company of barbers and the guild of surgeons by act of parliament in 1540 bespeaks at least some interest in the fight against disease. Yet Holbein, who painted the king in the act of granting a charter to the barber-surgeons, himself died of the plague in 1543; throughout the whole period there are references to epidemics generally described as 'the sweating sickness';[1] and against this illness there was no specific save seclusion or flight. During the epidemic of the summer of 1528 the king shut himself up quite alone; and the cardinal, who was at Hampton Court with a very small company, received from his still affectionate master the caution to 'keep out of the air, to have only a small and clean company about him, not to eat too much supper, or to drink too much wine', and to take some pills which the royal apothecary had made up for him. In later years Hampton Court became a convenient refuge for the king and his entourage when the dread visitant came to London, as for example in the summer of 1551. For king and commoner alike the established way of combating epidemics was to avoid them, but there is some evidence of a growing realization that prevention was better than cure. The various regulations for the keeping clean of London, the exact precautions taken to preserve the health of the infant Edward VI, and the cardinal's interest in his

---

[1] Apparently a kind of influenza.

water-supply and drains, all attest a consciousness that, with the growth of the towns and the loss of the old simplicities of life, there was lost some of the old security of the country-side. In spite of ignorance, carelessness, and the crudity of sanitation, the health of England was not ill preserved. Foreign observers not infrequently admired the physique of the islanders; Fortescue had contrasted the strength of the well-fed English commons with the feebleness of the half-starved French peasants, and one of the Spaniards who came to England with Philip remarked that though the English lived in hovels, they ate like lords. In the English chronicles the people appear as a sturdy folk, cold-blooded, perhaps, in their attendance at hangings and floggings, but boldly indifferent to danger and death, rejoicing in the noise and the colour of public spectacles, keen in business, competent in agriculture, adventurous at sea, and, when occasion called, hardy in arms.

Sound in her stock as competent in her institutions, instinct with life and energy, England awaited the arrival of Elizabeth.

APPENDIX

TUDOR COINAGE FROM HENRY VII
TO ELIZABETH

DURING the Tudor period English coinage developed from a medieval into a modern currency. For the first stage of the development the practical Henry VII was responsible. The early issues of his reign were distinctly medieval, no farther advanced in art or technique than the coins of the preceding century and a half. His later coins, on the other hand, were of true Renaissance style, showing signs of classical influence in the fine, dignified portrait and in the decorative detail.

A comparison of Henry VII's first issues of coins with his last issues shows that during the course of his reign almost every denomination—from the gold sovereign (20s.), ryal (10s.), angel (6s. 8d.), and half-angel (3s. 4d.) to the silver testoon (1s.), groat (4d.), half-groat (2d.), penny, halfpenny, and farthing—completely altered its appearance. The only exceptions were the angel and half-angel, and even these showed modifications of detail which gave greater elegance to the design.

Two of the denominations, the gold sovereign and the silver testoon, were new; now, for the first time, England had coins which corresponded to the monetary values long used in computations. The gold sovereign, with a weight of 240 gr., and a fineness of 23 carats, 3½ gr., made its first appearance in 1489. Its currency value was 20s. Its large flan, measuring 1.6 inches across, invited magnificence and spaciousness of design, and the early issues of the sovereign do in fact rank among the finest products ever to come out of the English mint. The testoon or shilling, issued in the last years of the reign, was no less beautiful in its own way. It bore, along with contemporary groats and half-groats, for the first time in the history of the English coinage, a portrait of the king which was a true likeness.

Henry VII used the mint of London only for the issue of his gold coinage, and of his testoons and groats. The smaller denominations in silver were issued from time to time at the royal mints at Canterbury and York, and at the ecclesiastical mints of the archbishops of Canterbury and York and of the bishop of Durham. Towards the close of his reign, however, he imposed certain restraints on these ecclesiastical mints.

In the last fifteen years of his reign Henry VII struck into coin at the London mint 31,000 lb. of gold and 164,000 lb. of silver, and on his death he left to his son a coinage which, for fineness and artistic merit, was unsurpassed by any contemporary European currency. Seventeen years later Henry VIII had largely dissipated his father's wealth, and had started on a deliberate policy of debasing the currency. Once begun, the easy descent could not be stayed; the impetus increased, and by the end of the reign his coinage had deteriorated into one of the most shameful and shabby currencies ever seen in England. Both his gold and his silver coins were heavily alloyed, the silver coins so much so that they often showed discolorations where the base metal had worked out to the surface.

The process of debasement began in 1526, under the direction of Cardinal Wolsey, with the revaluation of gold coins at 22s., and later 22s. 6d., to the sovereign, with the introduction of 'crown' gold of only 22 carats fineness for two new gold coins, the crown and half-crown, and with the reduction in weight of the silver coins from 12 gr. to 10⅔ gr. to the penny. The 'Wolsey' coinage of 1526–44 consisted of the following gold coins: the sovereign (22s. 6d.), the angel (7s. 6d.), the half-angel (3s. 9d.), the George noble (6s. 8d.), and half-George noble (3s. 4d.), so called because they bore on their reverse side the figures of St. George and the Dragon, all these struck of 'standard' gold of 23 carats, 3½ gr. fineness, and the crown and half-crown of 22 carats fineness. The contemporary silver coins were the groat, the half-groat, penny, halfpenny and farthing, all of 11 oz., 2 dwt. fineness.

In the coinage of the last years of Henry VIII's reign, 1544–7, the gold coins—sovereign, half-sovereign, angel (now valued at 8s.), half-angel and quarter-angel, crown and half-crown—had their weights severely cut down, and their fineness reduced to 23, 22, and finally 20 carats. The silver coins—testoon, groat, half-groat, penny, and halfpenny—were reduced in fineness from 11 oz., 2 dwt. to 9 oz., then to 6 oz., and finally to 4 oz., with an accompanying loss in weight.

The extent of the debasement is even more clearly seen when its effect on four of Henry VIII's denominations, the gold sovereign and angel, and the silver groat and penny, is presented in tabular form, as follows:

SOVEREIGN

	1509–26	*1526–44*	*1544–7*	
Weight . .	240 gr.	240 gr.	200 gr.	192 gr.
Currency value	20s.	22s. 6d.	20s.	20s.
Fineness . .	23 ct., 3½ gr.	23 ct., 3½ gr.	23 ct.	22–20 ct.

ANGEL

	1509–26	*1526–44*	*1544–7*
Weight . .	80 gr.	80 gr.	80 gr.
Currency value	6s. 8d.	7s. 6d.	8s.
Fineness . .	23 ct., 3½ gr.	23 ct., 3½ gr.	23 ct.

GROAT (currency value 4d. throughout)

	1509–26	*1526–44*	*1544–7*
Weight . .	48 gr.	42⅔ gr.	40 gr.
Fineness . .	11 oz., 2 dwt.	11 oz., 2 dwt.	6–4 oz.

PENNY

	1509–26	*1526–44*	*1544–7*
Weight . .	12 gr.	10⅔ gr.	10 gr.
Fineness . .	11 oz., 2 dwt.	11 oz., 2 dwt.	6–4 oz.

One result of Henry VIII's ruthless debasement of the silver coinage was that his bad silver money drove out of circulation good silver coins of earlier date, as well as contemporary gold coins. Gold coins in particular were exported or hoarded, and the baseness of the coins left in circulation contributed to a sharp rise in prices.

Towards the close of his reign Henry VIII used, besides the Tower mint at London, other royal mints at Southwark, Canterbury, York, and Durham. The closing of the mints of the archbishops of Canterbury and York, and of the bishop of Durham, marked the end of the ecclesiastical issues. These issues had been confined to small silver coins from the half-groat downwards. Wolsey, characteristically, struck coins of a higher denomination—the York groats—in his palace mint. This usurpation of the royal prerogative is recorded in the articles of his impeachment in 1529.

In the first years of Edward VI's reign there was no improvement in the coinage. His councillors continued to strike base money from Henry VIII's coin dies, using his dread name to avert criticism. By 1549, however, Edward VI had begun to issue coins in his own name, and had taken the first step on the long and difficult road towards a restored currency. In 1549 and 1550 his coinage consisted of gold sovereigns and half-sovereigns, crowns and half-crowns of an improved fineness of 22 carats, and silver testoons or shillings of 6–8 oz. fineness, with smaller denominations in silver remaining at the old base level of 4 oz. fineness. By 1551 standard gold of 23 carats, 3½ gr. had been reintroduced for a new sovereign valued at 30s., and for an angel, now valued at 10s., while crown gold of 22 carats fineness was used for a sovereign of 20s., for a half-sovereign, for a crown and half-crown. At the same time silver of 11 oz., 1 dwt. fineness was used for the silver crown, half-crown, shilling, sixpence, threepence, and penny. It will be noted that the silver crown, half-crown, sixpence, and threepence have now made their first appearance. After the issue of a very base shilling in 1551, the use of base silver seems to have been limited to the penny, the halfpenny, and the farthing, with the rose design.

The principal source for the supply of base silver was found in the base testoons of Henry VIII. These were demonetized by Edward VI, and re-coined into base silver coinage in order to provide revenue to cover the cost of the better silver coinage. The mints of Canterbury and York were chiefly occupied in the conversion of testoons into new currency. Edward VI's other mints were the Tower mint, the mints at Southwark and at Durham House in the Strand, and the mint of Bristol.

Although Edward VI aimed seriously at restoring the coinage, his successive attempts at raising its standard resulted in the issue of coins of the same denomination, but of varying quality and weight, within a very short period of time. The currency must have been most confusing to the public. Moreover, the proportion of improved coins in circulation was very small compared with the mass of debased coins still current, and, as before, bad money drove out good. Only the complete demonetization of the base issues could effect a true rehabilitation of the coinage, and this was not finally achieved until the reign of Elizabeth.

Meanwhile, Mary Tudor maintained her own issues and those of her husband Philip at a high level. All her gold coins—the sovereign of 30s., the ryal of 15s., the angel and half-angel—were of standard gold of 23 carats, 3½ gr. Her silver coins—shilling, sixpence, groat, half-groat, and penny—were of 11 oz. fineness, and a base penny of 3 oz. which circulated along with them may have served as a token currency. All the coins of Mary, and of Philip and Mary, were minted at London.

At the beginning of Elizabeth's reign much of the debased money of Edward VI, and even of Henry VIII, was still in circulation. It was allowed to remain current at reduced values for a short time— for example, testoons of the two worst classes were countermarked at values of 4½d. and 2¼d.—but was gradually gathered in to the mint and there melted down. In 1561 the base money was finally demonetized.

The coinage of Elizabeth included gold coins of either standard fineness, 23 carats, 3½ gr.—the sovereign of 30s., the ryal of 15s., the

angel, half-angel, and quarter-angel—or of crown gold, 22 carats—
the pound, the half-pound, the crown, and half-crown. The silver
coins were all of 11 oz. fineness until 1560, and after that were of
11 oz., 2 dwt. fineness. The denominations in silver were the crown,
half-crown, shilling, sixpence, threepence, half-groat, three-half-
pence, penny, three-farthings, and halfpenny. The only mint in
operation was the Tower mint in London.

Among the great achievements of Elizabeth's reign was the restoration of a currency which was both honest and distinguished in
design and style. She even made an attempt to introduce machinery
for its production. A Frenchman, Eloye Mestrell, who had formerly
worked at the Paris mint, was employed for a time at the Tower mint
on the manufacture of silver coins struck with a screw press. These
'mill' coins were far superior to hammered coins, but the opposition
of the other mint employees to the new machinery was too strong,
and the experiment had to be abandoned. It was not renewed until
the reign of Charles I.

It is to the credit of Elizabeth that she recognized so early the
need for machinery in the issue of a perfect coinage. It is even more
to her credit that, rejecting low devices, she restored the English
coinage to the high standards of her grandfather, Henry VII, and
gave it the characteristics of a good modern currency, artistic merit
and integrity.

BIBLIOGRAPHY

THE bibliography here presented does not profess to be complete, but it mentions most of the works which have supplied the information used in the text.

The material has been arranged under heads which, it is hoped, will be convenient, but the categories used are not quite mutually exclusive, and occasionally there is reference from one section of the bibliography to another.

GENERAL

BIBLIOGRAPHIES AND WORKS OF REFERENCE

The Bibliography of British History, Tudor Period, 1485–1603, by Conyers Read (1933), is the best guide to the authorities for the period. Small but useful pamphlets and leaflets published by the Historical Association deal with particular subjects, e.g. *A Short Bibliography of English Constitutional History* by Helen M. Cam and A. S. Turberville (Historical Association Leaflet no. 75, 1929); for recent work the *Annual Bulletin of Historical Literature* is very useful.

Of very great convenience to students are the *Handbook of British Chronology* (ed. F. M. Powicke with assistance from Charles Johnson and W. J. Harte, 1939), and the *Handbook of Dates for Students of English History* (ed. C. R. Cheney, 1945); the former supplies lists of kings, of officers of state, of bishops of England, Scotland, and Ireland, of sessions of parliament (to 1547), and of English church councils (to 1536), as well as notes on calendars and lists of the regnal and exchequer years. The latter is particularly useful as containing tables arranged to show at a glance the date of Easter and of the great church festivals for any given year; it also supplies a convenient list of saints' days and festivals. A list of the sessions of parliament is also to be found in the index volume (vol. xiii) of the *Cambridge Modern History*; this contains, too, genealogical tables for all the ruling houses of Europe which are of great importance for an age when the prince was identified with the state.

M. S. Giuseppi, *Guide to the Manuscripts preserved in the Public Record Office* (2 vols., 1923–4), does far more than show where

3720.7   R r

the archives are. *A Repertory of British Archives*, part 1, by Hubert Hall (for the Royal Historical Society, 1920), provides a conspectus of the most important documents both in London and the provinces, and though the writer was mainly interested in the earlier period, he gives, for the Tudor period, useful references to documents not easily found otherwise. The *List of Record Publications* (List Q, 1938) contains, besides lists of Calendars and Registers, and the *Lists and Indexes* issued by the Public Record Office, a catalogue of the volumes in the Rolls Series and references to other publications undertaken by the Government.

PERIODICALS. Among the periodicals those of most value are the *English Historical Review, History,* the *Bulletin of the Institute of Historical Research,* and the *Economic History Review* founded 1927. Along with them may be reckoned the *Transactions of the Royal Historical Society;* this society, in 1897, was amalgamated with the Camden Society, which had been publishing original documents and writings since 1838, and in 1925 Hubert Hall edited a *List and Index of the Publications of the Royal Historical Society 1871–1924 and of the Camden Society 1840–1897,* which is extremely helpful.

A Bibliography of Early English Law Books, by J. H. Beale (1926), contains, amongst other things, a notice of most publications prior to 1800, and an accurate list of the *Year Books.*

Amongst works of bibliography should be mentioned the works of C. L. Kingsford, *The Chronicles of London* (1905), and *English Historical Literature in the Fifteenth Century* (1913), which trace the relationship of the various continuations whereby sixteenth-century writers endeavoured to bring the *Brut* down to their own day.

The *Dictionary of National Biography* is a mine of information; corrections appear from time to time in the *Bulletin of the Institute of Historical Research,* but most of the Lives are reliable and founded upon good authorities.

The *Victoria History of the Counties of England* (1900– still in progress; so far 101 volumes have appeared, besides indexes) provides information on local history of all sorts.

POLITICAL HISTORY (DOMESTIC)

(i) COLLECTIONS OF DOCUMENTS

The documentary materials for the reign of Henry VII are contained in *Memorials of King Henry VII* (1 vol., ed. James Gairdner, Rolls Series, 1858); in *Letters and Papers Illustrative of the Reigns of Richard III and Henry VII* (2 vols., ed. James Gairdner, 1861, 1863; Rolls Series, 1861); and in *Materials for a History of the Reign of Henry VII* (2 vols., ed. William Campbell, 1873, 1877). The *Memorials of King Henry VII* contains the *Vita Henrici VII* and the *Annales Henrici VII* by Bernard André. André, who was a native of Toulouse, was an Augustinian friar. He probably came to England with Henry VII, who made him his poet laureate and later, 1496, appointed him tutor to Prince Arthur. He enjoyed the patronage of Fox and was well treated at court. He is described as being blind, but it is impossible to believe that he was born blind. His Life of Henry VII, begun in 1500, carries the story down to the capture of Perkin Warbeck; the *Annales*, which may have been partly a continuation, are fragments dealing with the years 1505 and 1508. Gairdner's volume also includes a poem, 'Les Douze Triomphes de Henri VII', probably written in 1497, and very possibly the work of André. In addition it includes Journals and Reports of Ambassadors dealing with Brittany, Spain, Portugal, and Aragon. In the *Letters and Papers* Gairdner drew on the resources of the Record Office, as well as the Harleian and the Cottonian collections, the Vatican transcripts, the Egerton MSS., the Royal MSS., and other collections in the British Museum. There are also some pieces from the Archives of Lille and Ghent and from the Advocates' (now in the National) Library in Edinburgh. There are a few manuscripts from the College of Heralds and from Lambeth. In *Materials for a History of the Reign of Henry VII* William Campbell used Polydore Vergil and the London Chronicles, but his great service was that he presented many useful excerpts from patent rolls, close rolls, French rolls, privy seal writs, and various financial records, including those of the wardrobe and the duchy of Lancaster. The patent rolls for this reign appear in two volumes of the *Calendar of Patent Rolls*. A useful selection of documents and literary extracts is provided in *The Reign of Henry VII from Contemporary Sources* (3 vols., ed.

A. F. Pollard, 1913–14); the texts supplied should be compared with those in the authorities which are exactly cited.

For the reign of Henry VIII almost all the English documentary evidence is summed up in the monumental *Letters and Papers of Henry VIII*, which covers not only the State Papers from the Record Office but many manuscripts from the British Museum, and presents also relevant pieces in foreign repositories as well as digests of the Patent Rolls.

For the reigns of Edward VI and Mary the printed evidence is scantier. Vol. i of the *State Papers (Domestic Series)* provides only a brief catalogue of documents, which is supplemented by some pieces in vol. xii (Addenda). The volumes of the *State Papers (Foreign)* for Edward and Mary are more adequate, though they confine themselves to state papers proper in the Record Office. For both reigns the *Calendar of Patent Rolls* is available. In *The Reigns of Edward VI and Mary*, P. F. Tytler published, in 1839, a collection of documents which is still of use, though many of the pieces are now to be found in the printed calendars, notably the *Calendar of State Papers (Spanish)*, which, in seventeen volumes, covers the 'letters, despatches and state papers' dealing with Anglo-Spanish affairs preserved in Vienna, Brussels, and Simancas for the whole period 1485–1558. Scarcely less valuable is the *Calendar of State Papers (Venetian)*, of which seven volumes deal with the period 1509–58; and there are a few pieces in the *Calendar of State Papers (Milan)* (vol. i), which covers the period 1485–1618.

Foedera, the great collection of treaties and political documents, was made between 1704 and 1713 by Thomas Rymer and continued after his death, the entire twenty volumes having appeared by 1735. This first edition covered the period 1101–1654, but in the Record Office edition (ed. Adam Clarke, 1818–67) the period covered was 1066–1645. Though the *Foedera* has lost some of its importance because many of the important pieces are now to be found in calendars, it is still of great value. Thomas Duffus Hardy's *Syllabus of Documents in Rymer's Foedera* (1869–85) provides an admirable register of its contents and also acts as an index to the various editions. The *Corps universel diplomatique* (8 vols., ed. J. Dumont, 1725) contains texts of many treaties; most of those which concern Britain are to be found also in the calendars and in *Foedera*.

The publications of the Historical Manuscripts Commission

are reports on many collections invaluable for national and local history, among them the *Calendar of the Manuscripts at Hatfield House* (sometimes known as the Cecil MSS.), vol. i, which presents, in summary form, papers some of which were published *in extenso* by Samuel Haynes in 1740. The numerous *Lists and Indices* issued by the Public Record Office direct attention to various manuscript sources, and are extremely useful where printed calendars are not available.

For other collections of documents see the sections on Constitutional History, Ecclesiastical History, and Economic History.

(ii) CONTEMPORARY CHRONICLES AND NARRATIVES

The best contemporary history is that of Polydore Vergil, *Anglica Historia*, of which the first draft was ready in 1513 though the first edition did not appear until 1534. This took the history of England down to 1509, and it was not until the third edition of 1555 that an additional book (Bk. xxvii) took the story down to 1537. The portions dealing with the Tudor period have not yet been all translated into English, but a new edition with a translation of the books which cover the period 1485–1537 has been made for the Camden Series of the Royal Historical Society by Denys Hay (1950), who has written a valuable and analytical biography, *Polydore Vergil, Renaissance Historian and Man of Letters*, (1952). Vergil's work, however, is the basis not only of Bacon's book (see below, p. 617) but of the corresponding portions of Edward Hall's *Chronicle*. This history, correctly styled *The Union of the Two Noble and Illustre Famelies of Lancastre and Yorke*, &c., was the work of a serjeant-at-law who died in 1547. The author wrote as a profound admirer of Henry VIII, and his account of that monarch's 'triumphant reign' has been published separately in two handsome volumes under the editorship of Charles Whibley (1904).

The Life of Cardinal Wolsey, by George Cavendish (2 vols., ed. S. W. Singer, 1815), was first printed in 1641 as *The Negotiations of Thomas Woolsey the Great Cardinall of England*, but it was evidently known in manuscript in Shakespeare's day.

The chroniclers of the period, except for Hall, for the most part prudently confined themselves to fact, and their works were not printed till the nineteenth century, when they were given to the public by the Camden Society. During the period of John Gough Nichols particularly, this society was much interested in the Tudor period, and amongst its publications were the *Chronicle of Calais in the Reigns of Henry VII and Henry VIII to the Year 1540* (ed. J. G. Nichols, 1846), which includes a useful plan of the Marches of Calais; the *Diary of Henry Machyn, Citizen and Merchant-Taylor of London, from 1550–1563* (ed. J. G. Nichols, 1848), very full; *Chronicle of Queen Jane and of Two Years of Queen Mary and especially of the Rebellion of Sir Thomas Wyat* (ed. J. G. Nichols, 1850), written by some person of consequence who was actually in the Tower and dined with the unfortunate Lady Jane; *Chronicle of the Grey Friars of London* (1189–1556) (ed. J. G. Nichols, 1852), useful for London, and *A Chronicle of England during the Reigns of the Tudors from 1485 to 1559, by Charles Wriothesley, Windsor Herald* (ed. William Douglas Hamilton, vol. i, 1875, vol. ii, 1877), which is well informed about public affairs. In the various Miscellany volumes there are smaller narratives of great importance. Along with these Camden publications, which are well annotated, may be mentioned a Spanish account of the *Accession of Queen Mary* by Antonio de Guaras (ed. with an English translation by Richard Garnett, 1892). The *Narratio Historica Vicissitudinis Rerum*, &c., describing the fall of Northumberland, by Petrus Vincentius, a Lübeck schoolmaster who was in England, shows the conflict between political and religious issues in the mind of a German (1553, only three copies known, reprinted, London, 1865) (see Denys Hay in *English Historical Review*, vol. lxiii, 1948). The precocious *Journal* of Edward VI may be found in Burnet's *History of the Reformation* and in the *Literary Remains of Edward VI* (Roxburghe Club, ed. J. G. Nichols, 2 vols., 1857). Many narratives were collected by Arber in his 'English Garner' (8 vols., 1877–90); these were re-edited in a new edition (1903–4). The volume entitled *Tudor Tracts* (ed. A. F. Pollard, 1903) contains, among other things, *The Expedition into Scotland*, &c., by William Patten, which gives the best account of the battle of Pinkie, illustrated with contemporary maps; two volumes of *Voyages and Travels* (ed. Beazley, 1903), in the same series, contain excerpts from Hakluyt's *Principall Navigations, Voiages and Discoveries of the English Nation* (1598–1600).

(iii) GENERAL HISTORIES

Sub-contemporary writers did not distinguish between political and ecclesiastical history, and to them the period of the early Tudors was essentially the period of the reformation. For several generations of Englishmen the *Actes and Monuments* of John Foxe, often called the *Book of Martyrs*, though it dealt with many countries and many centuries, was regarded as the standard history of a vital epoch. Its tone is fiercely protestant, but though its narratives have been subjected to much criticism, Foxe, who used original authorities when he could, has on the whole come out well. The best modern edition is that issued over the name of George Townsend (1843-9). A defence of Henry VIII's achievement was made by Lord Herbert of Cherbury in the *Life and Reign of Henry VIII* (1649), which made use of original documents; but meanwhile a violent attack on the monarch had been made in *De Origine ac Progressu Schismatis Anglicani* (1585), begun by Nicholas Sander and completed by Edward Rishton. There is an English translation with notes by David Lewis (1877). The book was the less 'dangerous' in England because it remained so long untranslated, but in 1676 a translation into French was made by Maucroix, and its appearance during the doubtful days of Charles II stimulated Gilbert Burnet to produce, between 1679 and 1714, his *History of the Reformation of the Church of England*. Moderate in expression and admirably documented, though faultily arranged, Burnet's book held the field in the day of the whigs, and it is still valuable today. The best modern edition is that of Nicholas Pocock (7 vols., 1865). But in the reaction which followed the French Revolution the Roman catholic case was presented by John Lingard in his *History of England* (1819-30), and by M. A. Tierney, whose edition of Dodd's *Church History of England* (1839-43) was well supported by original documents, some of them not easy of access to protestants. In reply to these books and to the Oxford Movement with which they were connected, James Anthony Froude published the book which must be regarded as the classic for the period— *The History of England from the Fall of Wolsey to the Defeat of the Spanish Armada* (12 vols., 1856-70). A follower of Carlyle in his denunciation of 'shams', Froude was a strong protestant; but he was also a staunch Englishman and his work is that of a convinced nationalist. He has been represented as an historical romantic, but though his bias is revealed in his argument, his facts, in spite of some errors, are mainly correct. He was of great industry and used not only the contemporary literary authorities but the English and Spanish State Papers, as well as material from the Bibliothèque Nationale. The excellence of his style commended him to a wide public, and he did much to fix the opinion of England. He deserves his reputation in spite of his critics, amongst whom may be numbered James Gairdner who, in *Lollardy and the Reformation in England* (4 vols., 1908–13), wrote the history of England from 1485 to 1558 from what may be called an Anglo-catholic point of view. Conspicuous among recent writers are H. A. L. Fisher, whose volume in *The Political History of England*, vol. v, 1485–1547 (1906), is marked by a liberal humanity; and A. F. Pollard, who covered the whole period in several works of deep insight and sound scholarship marked by special attention to constitutional history, namely *The Reign of Henry VII from Contemporary Sources* (3 vols., 1913–14), *Henry VIII* (1920), *England under Protector Somerset* (1900), vol. vi of *The Political History of England* (1910), and *Factors in Modern History* (1907). Shorter works worthy of attention are *The Making of the Tudor Despotism*, by Professor C. H. Williams (1928), and the lively volume on *Tudor England* by Mr. S. T. Bindoff (1950). *England under the Tudors*, by Mr. G. R. Elton (1955), pays particular attention to the development of the machinery of government and emphasizes the importance of Thomas Cromwell. The chapters upon England in the *Cambridge Modern History* (vol. i, chap, xiv; vol. ii, chaps. xiii, xiv, and xv) are valuable and equipped with good bibliographies. Worthy of special attention is *England under the Tudors*, vol. i, *Henry VII*, by W. Busch (translated by Alice M. Todd and A. H. Johnson with an introduction by James Gairdner, 1895); though only part of a larger work which was projected, the book examines all the sources available to the writer with Germanic thoroughness, and uses them with admirable judgement.

The Pilgrimage of Grace, by M. H. and R. Dodds (2 vols., 1915), is a very well documented study written with moderation. *The Western Rebellion of 1549* by F. Rose-Troup (1913) makes use of much original material. *Rival Ambassadors at the Court of Queen Mary*, by E. Harris Harbison (1940), makes excellent use of unpublished manuscripts belonging to the family of Noailles.

Tudor Studies (ed. R. W. Seton-Watson, 1924) contains several articles important for constitutional and ecclesiastical history.

BIOGRAPHIES

In an age when personality meant so much, the biographies of princes, ministers, and prelates are hardly distinguishable from the histories of the times.

The Life of Henry VII (1622), by Francis Bacon, though written in great haste, largely from Polydore Vergil (and marred by the famous misreading of *latenter* instead of *laetanter* in the description of Henry's entry into London), none the less is a great biography marked by the true insight of a man of the age. James Gairdner contributed a volume on *Henry VII* to the English Statesmen Series, 1889. *Henry VII*, by Gladys Temperley (1918), is a good study which contains a useful itinerary of the king.

J. S. Brewer, who edited the *Letters and Papers*, produced a solid history of Henry VIII, *The Reign of Henry VIII from his Accession to the Death of Wolsey* (ed. James Gairdner, 2 vols., 1884), in which he took the view that the glory of the reign departed with Wolsey. But the standard modern life is *Henry VIII*, by A. F. Pollard, in the sumptuous Goupil Series (1902), based largely on the *State Papers*. *Henry VIII*, by Francis Hackett (1929), resting much on originals, though without notes, is of value as a psychological study; and in 1936 Conrado Fatta published *Il Regno di Enrico VIII d'Inghilterra*, meant to direct the attention of Italy from *l'aspetto scandalistico del regno* to its real importance; his work, which is well annotated, is particularly strong on the diplomatic side.

The fullest account of Edward VI appears in the *Literary Remains of Edward VI* (ed. J. G. Nichols, Roxburghe Club, 1857).

Queen Mary has found several biographers. *Mary Tudor*, by Beatrice White (1935), is well balanced, and *Spanish Tudor*, by H. F. M. Prescott (1940), is a sympathetic study. *Queen Elizabeth*, by J. E. Neale (1934), is a scholarly work admirably written. The solid biography, *Anne Boleyn*, by P. Friedmann (1884), which insists that the mainspring of Henry's difficulty was his connexion with Mary Boleyn, abounds in information.

The best modern biography of Wolsey is that of A. F. Pollard (1929). This emphasizes the cardinal's domestic work rather than the foreign policy which had attracted the attention of Mandell Creighton in *Cardinal Wolsey* (1898). *The Life and*

Letters of Thomas Cromwell, by R. B. Merriman (2 vols., 1902), is still of great value. Sir Thomas More has many biographers, including William Roper, his son-in-law, and his great-grand-son Cresacre More. Notices of these and other lives appear in *Thomas More*, by R. W. Chambers (1935), which, admirably documented and written with profound sympathy, is now the standard biography.

As long ago as 1694 John Strype published his *Memorials of the Most Reverend Father in God, Thomas Cranmer* (republished in 2 vols., 1812 and 1853), but the best modern biography is in *Thomas Cranmer and the English Reformation*, by A. F. Pollard, 1904. *Stephen Gardiner and the Tudor Reaction*, by James Arthur Muller (1926), is extremely valuable; the author is inclined to exagger-ate the consistency of his hero as an 'English' statesman and to minimize his ambition, but an admirable documentation presents possible critics with the means of forming their own opinions. *Reginald Pole Cardinal of England*, by W. Schenk (1950), is valuable as emphasizing the connexion between Pole and the fathers of the counter-reformation. A sober, documented bio-graphy is that of *William Tyndale*, by J. F. Mozley (1937). *The Life of Dean Colet*, by J. H. Lupton (1887), himself sur-master in St. Paul's, is a recognized standard work. *The Life of Jane Dormer, Duchess of Feria*, by H. Clifford (ed. J. Stevenson, 1887), is the biography, by one of the servants, of the English-woman who married the Spanish ambassador at the court of Queen Mary.

POLITICAL HISTORY (FOREIGN)

(i) COLLECTIONS OF DOCUMENTS AND SOURCES

Much of the evidence from foreign repositories has been made available in the calendars published by the British Govern-ment, but some foreign collections merit attention. *The Calen-dar of State Papers, Spanish*, may be checked in the *Colección de Documentos inéditos*, and the Venetian Papers in the various calendars from *Relazioni degli Ambasciatori Veneti al Senato durante il Secolo Decimo Sesto* (ed. E. Albèri, 15 vols., 1839–63): the first six volumes are concerned largely with English affairs. Among the 'Collection de documents inédits sur l'histoire de France' may be mentioned the *Négociations diplomatiques entre la France et l'Autriche (1491–1530)* (2 vols., ed. A. J. G. Le Glay, 1845–7;

no. 44), and the *Papiers d'état du Cardinal Granvelle* (9 vols., ed. C. Weiss, 1841–52), which, since they contain the papers of Granvelle's father Chancellor Perrenot, cover the period 1500–65. Among the publications of the Société de l'Histoire de la France is the *Correspondance de l'Empereur Maximilien et de Marguerite d'Autriche (1507–19)* (2 vols., ed. A. J. G. Le Glay, 1839), and the *Mémoires de Martin et Guillaume du Bellay* (4 vols., V. L. Bourrilly and F. Vindry, 1908–19). In the Archives de l'Histoire Religieuse de la France may be found the *Ambassades en Angleterre de Jean du Bellay (1527–29)* (ed. V. L. Bourrilly and P. de Vassière, 1905); and in the collection called the 'Inventaire analytique des Archives des Affaires Étrangères' are the *Correspondance politique de MM. de Castillon et de Marillac, ambassadeurs de France en Angleterre (1537–42)*, ed. J. Kaulek (1885) and the *Correspondance politique de Odet de Selve, Ambassadeur de France en Angleterre (1546–9)*, ed. Lefèvre-Pontalis, 1888. The important *Chroniques de Jean Molinet* have been re-edited by Georges Doutrepont and Omer Jodogne for the Académie Royale de Belgique, 3 vols. (1935–7).

Works of an earlier date still of value are the *Lettres du Roy Louis XII et du Cardinal G. d'Amboise* (4 vols., ed. Godefroy, 1712), the *Ambassades de Messieurs de Noailles en Angleterre* (5 vols., ed. R. A. de Vertot, 1763). A very full bibliography of the materials for French history is supplied in *Les Sources de l'histoire de France* (3 vols. in 18, ed. Molinier, Hauser, Bourgeois, and André, 1901–35); the first and second volumes of the second part cover the period 1494–1559.

The Correspondence of the Emperor Charles V (ed. William Bradford, 1850), though a small empiric collection, is still very useful. Documents relative to England's dealings with the papacy are found in *Vetera Monumenta Hibernorum et Scotorum Historiam Illustrantia* (ed. A. Theiner, 1864) and in *Römische Dokumente zur Geschichte der Ehescheidung Heinrich's VIII von England* (ed. S. Ehses, 1893). For Mary's negotiations with Rome see *Hardwick State Papers* (1778), i. 62–102.

(ii) Later Works

The chapters in the *Cambridge Modern History* (vols. i and ii, 1904) are of great value and are equipped with very full biographies. The best modern history of France is that edited by E. Lavisse, *Histoire de France* (18 vols., 1900–11), of which volume v, parts i and ii, by H. Lemonnier, cover the period 1492–1559. The best English book is *A History of France*, by J. S. C. Bridge (5 vols., 1921–36), which regrettably ends with the death of Louis XII.

The standard German life of Maximilian is that by H. Ulmann (1891). In English, *Ferdinand and Isabella, History of the Reign of Charles the Fifth* (reproducing and continuing the work of William Robertson), and *History of the Reign of Philip the Second*, by W. H. Prescott, were standards, but they have been supplanted by *The Emperor Charles V* (2 vols., 1902–13), by Edward Armstrong, and to some extent by *Philip II of Spain*, by Martin A. S. Hume (Foreign Statesmen Series, 1897), though the *Spanish Calendar* is perhaps the better guide. *The Rise of the Dutch Republic* (3 vols., 1855, and many later editions), by J. L. Motley, is still useful though only a part deals with the period prior to 1559, and throughout its general tendency is to over-emphasize the importance of the religious issue. A brief but well-balanced presentation of Netherlandish history is *A History of the People of the Netherlands* (5 vols., 1898–1912), which is a translation of P. J. Blok's *Geschiedenis van het Nederlandsche Volk* by O. A. Bierstadt and R. Putnam; only vol. i deals with our period.

The best English history of Venice is that of Horatio F. Brown (1895), *Venice: An Historical Sketch of the Republic*. The standard work in Italian is *Storia Documentata di Venezia*, S. Romanin (10 vols., 2nd ed. 1913). *La Storia di Venezia nella Vita Privata* (3 vols., 1927–9), by Pompeo Molmenti, admirably illustrated, is valuable for general life, architecture, and art. For the renaissance in Italy the standard work is J. A. Symonds's *Renaissance in Italy* (7 vols., 1875–86).

SCOTLAND

(i) RECORDS, COLLECTIONS OF DOCUMENTS, AND CONTEMPORARY HISTORIES

The official records in H.M. Register House, Edinburgh, are described in *A Guide to the Public Records of Scotland* (M. Livingstone, 1905), in the *Public Records of Scotland*, by J. Maitland Thomson, 1922—a humane and informative book—and in *The Scottish Records, their History and Value*, by Henry M. Paton—an admirable pamphlet (no. 7) in the series of The Historical Association of Scotland, 1933; *The Acts of the Parliaments of Scotland*, *The Exchequer Rolls of Scotland*, *The Register of the Great Seal*, though all the series are not yet complete, have been published in full for the period 1485–1559; *The Register of the Privy Seal* is printed to 1548. All these have been published by H.M. Register House and Stationery Office.

The judicial records are less straightforward. *The Acts of the Lords of the Council in Civil Causes, 1478–1495* (1839) are the record of a supreme court which had ceased to be parliamentary and become conciliar, though the transition must have seemed unimportant when parliament and council were so closely allied. Only in 1918 did a second volume (for 1496–1501) appear from Register House, furnished with a valuable introduction; and in 1943 the publication was continued for 1501–3 in the *Acta Dominorum Concilii* (published by the Stair Society). For the period 1513–45 the political business of the council was included in the record of its judicial business, and from this record have been extracted and digested the portions which deal with political affairs; these are published in a volume under the title *Acts of the Lords of Council in Public Affairs* (ed. R. K. Hannay, 1932). With the year 1545 begins the record proper of the privy council which is being published by the Register House in successive series, and is now printed to the year 1689. *The Hamilton Papers* (of the Council of the North), now in the British Museum, also printed in two volumes by H.M. Stationery Office, cover the period 1532–90.

Much material for Scottish history is found in the publications of the Historical Manuscripts Commission. This may be approached through *An Index to the Papers relating to Scotland*, by C. S. Terry (1908), which most regrettably has not been continued beyond 1908. *The Scottish Historical Review* (25 vols., 1904–28), with the New Series begun in 1947, is a mine of information of all sorts. *The Sources and Literature of Scottish Church History*, by M. B. MacGregor (1934), is a slight but convenient catalogue.

For the political history of Scotland the English State Papers are essential. *The Calendar of State Papers relating to Scotland* (2 vols., ed. M. J. Thorpe, 1858) is a brief catalogue which deals only with the contents of the State Paper Office for the years 1509–1603. The Scottish material, however, not only from the Record Office but from the British Museum and other sources, is included in the *Letters and Papers of Henry VIII*, and for the period after 1547 in the *State Papers relating to Scotland and*

Mary Queen of Scots, which have been published by the Stationery Office in a series which now comes down to the year 1595.

The purely judicial records of Scotland have not been used in the making of this book, but the judicial work of the central court may be studied with the aid of the *Indexes* published by the Register House, and a valuable summary of the available evidence is provided in *The Sources and Literature of Scots Law* (Stair Society, 1936).

Much contemporary evidence, both documentary and literary, has been published by the great Scottish Historical Clubs; to their publications an excellent guide has been furnished in *A Catalogue of the Publications of the Scottish Historical and Kindred Clubs and Societies, 1780–1908,* by C. Sanford Terry; this has been continued to 1927 by Cyril Matheson. Among the works valuable for the period 1485–1559 are *A Diurnal of Remarkable Occurrents that have Passed within the country of Scotland, since the death of James the Fourth till the year MDLXXV* (published by the Bannatyne Club and Maitland Club, 1833); Father James Dalrymple's translation of *The Historie of Scotland,* by John Leslie, bishop of Ross (vol. ii, ed. E. G. Cody, Scottish Text Society, 1895), *The Historie and Chronicles of Scotland,* by Robert Lindesay of Pitscottie (3 vols., ed. A. J. G. Mackay, Scottish Text Society, 1899–1911); published by the Scottish History Society are the *Balcarres Papers, Foreign Correspondence of Marie de Lorraine, Queen of Scotland 1537–1557* (2 vols., ed. Marguerite Wood, 1923–5), and *The Scottish Correspondence of Mary of Lorraine* (ed. Annie I. Cameron, 1927). For the works of John Knox see p. 623.

A selection from manuscripts in the Bibliothèque Nationale is provided in *Papiers d'état, pièces et documents inédits ou peu connus relatifs à l'histoire de l'Écosse au XVIe siècle* (ed. A. Teulet, 3 vols., for the Bannatyne Club, 1852–60). A later and somewhat different selection appears in the *Relations politiques de la France et de l'Espagne avec l'Écosse au XVIe siècle* (5 vols., 1862).

The only sub-contemporary histories of importance not published by the clubs are George Buchanan's *Rerum Scoticarum Historia* (1582), translated and continued by James Aikman (6 vols., 1827–9), and *The History of Scotland, 1423–1542,* by William Drummond of Hawthornden (1655), published in several eighteenth-century editions, sometimes as *A History of the Five Jameses.*

(ii) Secondary Authorities

Most of the recent research upon Scottish history has appeared in introductions to club and government publications. Among general histories that of P. Hume Brown (3 vols., 2nd ed., 1911) is still the best, though the *History of Scotland*, by Andrew Lang (4 vols., 1900–7), is a necessary though often exaggerated criticism of the standard presentation. *A History of Scotland*, by C. S. Terry (1920), is a brief but clear narrative enriched by genealogical tables of the great Scottish families (for the Scottish aristocracy see *The Scots Peerage* (9 vols., 1904–14)).

For constitutional history *The Parliaments of Scotland*, by R. S. Rait (1924), and *The College of Justice*, by R. K. Hannay (1933), are extremely important. *The Archbishops of St. Andrew* (5 vols., J. Herkless and R. K. Hannay, 1907–15) is based very largely upon original materials, and supplements the restrained account given in *A History of the Church in Scotland*, by A. R. MacEwen (2 vols., 1913–18). *The Reformation in Scotland*, by D. Hay Fleming (1910), is the work of a convinced protestant based almost entirely upon original material, and the same author's *Mary Queen of Scots* (1897) contains in its voluminous notes references which cover the history of Scotland from 1542 to 1568. *The Scottish Presbyterian Polity in the Sixteenth Century*, by J. G. Mac-Gregor (1926), is a useful reminder that the Scottish kirk did not draw its institutions entirely from Geneva. *John Knox: a biography*, by P. Hume Brown (2 vols., 1895), is still a standard work, but *John Knox*, by Lord Eustace Percy (1937), is a generous appraisal which makes good use of Knox's minor writings. *The Works of John Knox* (6 vols., ed. David Laing, Bannatyne Club, 1846–64) are very important. The first two volumes, comprising *The History of the Reformation in Scotland*, were also issued by the Wodrow Society (2 vols., 1848). The text of the History, re-spelled on the ground that the author used several amanuenses, has been edited by Professor Croft Dickinson (2 vols., 1949), with a most valuable apparatus of introduction, notes, and index.

IRELAND

(i) Original Sources

The most useful general bibliography of Irish history is Constantia Maxwell's *Short Bibliography of Irish History* (Historical Association Leaflet no. 23, 1921). The best guides to the public records of Ireland previous to 1919 are H. Wood's *Guide to the Records Deposited in the Public Record Office of Ireland* (1919) and R. H. Murray's *A Short Guide to the Principal Classes of Documents Preserved in the Public Record Office, Dublin* (1919); unfortunately the value of these guides is impaired by the fact that some of the records described were destroyed by fire in 1922, and they cannot be used without reference to H. Wood's 'The Public Records of Ireland before and after 1922' (*Transactions of the Royal Historical Society*, 4th series, xiii, 1930). The Statutes at Large passed in parliaments held in Ireland from the 3rd year of Edward II, 1310, to 40th year of George III, 1800 (20 vols., 1785–1801), provided only an incomplete record. Of the forty-nine acts of the Drogheda parliament only twenty-three were printed.

The volumes of *Irish Historical Studies*, begun in 1936, contain, besides much useful bibliographical material, a number of well-informed articles dealing with the period 1485–1558; among them are 'The Early Interpretation of Poyning's Law 1494–1534' (David B. Quinn, ii, 241); 'The History of Poyning's Law' (R. Dudley Edwards and T. W. Moody, ii, 415); Edward Walshe's 'Conjectures Concerning the state of Ireland, 1552' (David B. Quinn, v, 303), and 'The Vindication of the Earl of Kildare from Treason 1496' (by G. O. Sayles, viii, 39).

Much material for the history of Ireland is found in the *Calendar of State Papers, Ireland* (24 vols., covering the period 1509–1670, 1860–1910); only vol. i, 1509–73, edited by H. C. Hamilton, however, is concerned with the period of the early Tudors, and the content of that volume is a jejune catalogue. A more fruitful source is the *Calendar of State Papers, Carew* (6 vols., ed. J. S. Brewer and W. Bullen, 1867–73); the volume which covers the period 1515–74 contains a miscellaneous collection of documents and letters and provides much valuable information as to governmental methods and affairs of state. Even more valuable is vol. vi (1871), which contains miscellaneous extracts and the Book of Howth now in the Lambeth Library. This, though written in several hands and in the rudest of English, has great merits; it is contemporary, and it is free from obscurity of expression; it is full of life and provides excellent anecdotes and descriptions of the personalities of the time.

Two contemporary chronicles much used by later writers

are the *Annals of Loch Cé, 1014–1590* (ed. and trans. by W. M. Hennessy, 2 vols., 1871), of which part of vol. ii deals with the years 1485–1558, and the *Annals of Ulster* (otherwise Annals of Senat), *431–1541* (4 vols., ed. with translation and notes by W. M. Hennessy and B. MacCarthy, 1887–1901), of which only vol. iii (1379–1541) and vol. iv, which contains introduction and index, are useful for the times of the early Tudors.

A sub-contemporary authority much used by all later writers is James Ware's *Rerum Hibernicarum Annales Regnantibus Henrico VII–Elizabeta* (1664).

(ii) SECONDARY AUTHORITIES

The standard history for the subject is R. Bagwell's *Ireland under the Tudors* (3 vols., 1885–90). The best short account is R. Dunlop's *Ireland from the Earliest Time to the Present Day* (1922). *A History of Medieval Ireland*, by E. Curtis (2nd ed., 1938), deals mainly with the earlier period, but is particularly interesting for the 8th earl of Kildare. *The Making of Ireland and its Undoing, 1200–1600*, by Alice Stopford Green (1919), is useful but enthusiastic. Of particular value is *Henry VII's Relations with Scotland and Ireland*, by A. Conway (1932), a scholarly piece of work based on original sources and enriched by an article on Poyning's Parliament by E. Curtis.

WALES

The standard history of Wales, *A History of Wales*, is that of Sir John Lloyd (1911). An account of the position of the Welsh chiefs and the Marcher lords is provided in *Wales and the Wars of the Roses*, by H. T. Evans (1915). 'Welsh Nationalism and Henry Tudor', by W. Garmon Jones (*Transactions of the Cymmrodorion Society*, 1917–18), is extremely useful, though critical of some popular opinions. Lord Herbert of Cherbury in his *Life of Henry VIII* justifies that monarch's policy towards Wales. The best-known modern study is the brief *Tudor Policy in Wales*, by Sir J. F. Rees (Historical Association Pamphlet no. 101, 1935). Professor William Rees has contributed a paper on 'The Union of Wales with England', illustrated with an excellent map, to the *Transactions of the Cymmrodorion Society* (1937); these *Transactions*, which begin in the year 1877, contain much material for Welsh history.

CONSTITUTIONAL

(i) Printed Records and Collections of Documents

PARLIAMENT AND PRIVY COUNCIL. The legislation of the period is to be found in vols. ii, iii, and iv, pt. i, of the *Statutes of the Realm* (11 vols., 1810–28). Up to the year 1489 it was the practice to engross the output of each parliament on the statute roll in a single statute written in French, whose content did not always agree with the version given in the rolls of parliament. From 1491–2 on the acts were enrolled under separate titles on the parliament roll itself and the statute roll terminated. For the parliaments 1485–9 the *Statutes of the Realm* provide both versions of each statute.

The rolls of parliament published as the *Rotuli Parliamentorum* (6 vols., 1771–83: index 1832) are interesting as showing the whole business of parliament (the House of Lords). The series ends with the parliament of 1503–4, but the rolls of parliament for the years 1513–36 are printed in vol. i of the *Journals of the House of Lords* (n. d., ? 1742). The *Journals* proper begin in 1509; they are short and jejune but contain the orations with which parliament was opened, references to the presentation of the speaker, and useful lists of names.

The *Journals of the House of Commons* begin with the year 1547; they too are brief, but have interest in showing the legislation which was proposed but not carried.

The official records of the privy council are not extant for the period between 1435 and 1540, though odd fragments survive. On 10 August 1540 it was decided that an official record should be kept by the clerk, who was also to be the king's secretary (then becoming an officer of state). The register thus established was printed for the period August 1540 to April 1542 in vol. vii of the *Proceedings and Ordinances of the Privy Council of England* (ed. Harris Nicholas, 1837). The succeeding years up to 1558 are contained in the first six volumes of the *Acts of the Privy Council of England* (ed. John Roche Dasent, 1890–3).

The *List of the Proceedings of the Court of Star Chamber* (vol. i, 1901, *Lists and Indexes* no. xiii) covers the period 1485–1558. For a list, with brief digests, of proclamations see R. R. Steele, *Tudor and Stuart Proclamations, 1485–1714* (2 vols., 1910). See E. R. Adair in *English Historical Review*, vol. xxxii, 1917, for an interpretation of the statute of proclamations.

Year Books and *Reports*: the *Year Books*, unofficial and anonymous reports of actual cases written in French, which survive in an almost unbroken series from Edward I to Richard III, become intermittent for the reigns of Henry VII and Henry VIII; the last surviving *Year Book* is for 1536. Richard Tottel and various others began to print abridgements in the sixteenth century. The best collection is found in *Les Reports des cases en les ans des Roys Edward V, Richard III, Henri VII et Henri VIII touts qui par cydevant ont este publies*, by John Maynard (1678–80). The *Reports* which supplant the *Year Books* were written in English by known reporters, of whom the most important for the early Tudor period are Sir James Dyer for 1513–82, and Edmund Plowden for 1549–80, in *The Commentaries or Reports* (first published 1578; London, 2 vols., 1816). For the list of *Year Books* see Holdsworth, *A History of English Law* (vol. v. 358), and the Bibliography by J. N. Beale (p. 610 *supra*).

(ii) Secondary Sources

The Constitutional History of England, by Henry Hallam (1827), the first history defined by the adjective 'constitutional', though the phrase had been used in the dedication to C. J. Fox which precedes *A Historical View of the English Government*, by John Millar (1787), represented that England had a constitution and the Tudors broke it. The simple Whig hypothesis has been much modified since Hallam wrote, but the book still has value. An admirable survey is contained in *The Constitutional History of England*, by F. W. Maitland (posthumous; edited by H. A. L. Fisher, 1908, 10th impression 1946). The best modern study of the constitutional history of the Tudor period in its relation to British constitutional history in general is to be found in Sir David Lindsay Keir's *Constitutional History of Modern Britain, 1485–1937* (3rd ed., 1947). Of much value is *Tudor Constitutional Documents, 1485–1603*, by J. R. Tanner (1922), with admirable commentaries. There are useful documents in *England under the Early Tudors, 1485–1529*, C. H. Williams (1925).

For legal history the standard work is that of W. S. Holdsworth, *A History of English Law* (12 vols., 1922–38; index to first 9). See also *The Influence of the Legal Profession on the Growth of the English Constitution*, by W. S. Holdsworth (Creighton Lecture, 1924).

For the constitutional history of the period 1485–1547 the

standard work is K. Pickthorn's *Early Tudor Government* (2 vols., 1934); the first volume is arranged strictly according to subject, the second is more in narrative form.

For Parliament see *The Evolution of Parliament*, by A. F. Pollard (2nd ed., 1926), the same writer's *Parliament in the Wars of the Roses* (1936), and the numerous contributions made by him to the *English Historical Review, History*, and the *Bulletin of the Institute of Historical Research*. In these periodicals there are many articles of great importance, some of them most illuminating as to the Tudor method of legislation. Special mention may be made of the following articles in the *Transactions of the Royal Historical Society*: Isobel D. Thornley, 'The Treason Legislation of Henry VIII' (vol. xi, 3rd series), 'Some Proposed Legislation of Henry VIII', by T. F. T. Plucknett (vol. xix, 4th series). G. R. Elton in *English Historical Review*, vol. lxiv, 1949, and vol. lxvi, 1951, makes a valuable contribution.

The multifarious activities of the Tudor council may be studied in the various periodicals and transactions already named. A. F. Pollard on 'Council, Star Chamber and Privy Council under the Tudors' (*English Historical Review*, vol. xxxvii, 1922) and also 'The Privy Council under the Tudors' (ibid., vol. xxxviii, 1923) demand special attention.

A useful statistical account of the activities of the council (marred by errata in printing) is found in *The Tudor Privy Council*, by D. M. Gladish (1915).

Many of the council's special activities have been made the subjects of special monographs. See *The Privy Council under the Tudors*, by Lord Eustace Percy (1907), *A Treatise of the Court of Star Chamber*, by William Hudson (*fl.* 1630), which has been the basis of much subsequent writing, *Select Cases before the King's Council in the Star Chamber Commonly called the Court of Star Chamber, 1477–1509* (2 vols., 1903, 1911), and *Select Cases in the Court of Requests* (1898), both ed. by I. S. Leadam, *The King's Council in the North*, by Rachel R. Reid (1921), *The Council in the Marches of Wales*, by Caroline A. J. Skeel (1904), 'The Council of the West', by the same author (*Transactions of the Royal Historical Society*, 4th series, vol. iv, 1921). *The Channel Islands under Tudor Government, 1485–1642*, by A. J. Eagleston (1949), is a study based on French as well as English sources.

For the administration of the Borders see *The Lord Wardens of the Marches of England and Scotland*, by Howard Pease (1913), and *The Last Years of a Frontier*, by D. L. W. Tough (1928). *The Border History of England and Scotland*, by George Ridpath (1776), is still useful. *Tudor Cornwall: Portrait of a Society*, by Mr. A. L. Rowse (1941), gives a reminder that 'society' was not uniform throughout England.

For officers of state see *Handbook of British Chronology* (*supra*, p. 609). For the office of secretary, which developed about this time, see *The Principal Secretary of State*, by Florence M. G. Higham (Manchester Univ. Pub., Historical Series, xliii, 1923).

The machinery of local government may be studied in Gladys Scott Thomson, *Lords Lieutenant in the Sixteenth Century* (1923), in C. A. Beard, *The Office of Justice of the Peace in England in its Origin and Development* (1904), in Bertha H. Putnam, *Early Treatises on the Practice of the Justices of the Peace in the Fifteenth and Sixteenth Centuries* (vol. vii of Oxford Studies in Social and Legal History, ed. P. Vinogradoff, 1924), which has a bibliography of contemporary writings on the justices of the peace.

The organization of the boroughs is described in Alice Stopford Green, *Town Life in the Fifteenth Century* (1894), to which there is no sixteenth-century equivalent; the introductions to various municipal records are also useful.

For the royal household *Curialia*, by Samuel Pegge (1791), is still valuable.

FINANCE

The standard work on taxation is *A History of Taxation in England from the Earliest Times to the Present Day*, vol. i, by S. Dowell (1884). A detailed account of English finance with useful statistics is found in F. C. Dietz's *English Government Finance, 1485–1558* (University of Illinois Studies in the Social Sciences, ix, 1920); the figures supplied in this pioneer work may here and there be subject to revision. Evidence of the supplanting of the king's chamber by the exchequer is to be found in an article by A. P. Newton, 'The King's Chamber under the Early Tudors' (*English Historical Review*, vol. xxxii, 1917), and also in *Tudor Studies* (*supra*, p. 616). *The Early English Customs System*, by N. S. B. Gras (Harvard Economic Studies, 1918), deals mainly with an earlier period, but is valuable because it contains the 'Book of Rates'.

Information about the coinage can be found in *English Coins*, by George C. Brooke (3rd ed., 1950), and in Sir Charles Oman, *The Coinage of England* (1931).

THE NAVY

Three volumes published by the Navy Records Society contain material important for the history of the early sixteenth century: *Naval Accounts and Inventories of the Reign of Henry VII, 1485–88 and 1495–97* (ed. M. Oppenheim, vol. viii, 1896), *Letters and Papers relating to the War with France, 1512–13* (ed. A. Spont, vol. x, 1897), and *Fighting Instructions, 1530–1816* (ed. J. S. Corbett, vol. xxix, 1905). There is much evidence as to the musters of fleets in the *Letters and Papers of Henry VIII*. In *A Miscellany* (presented to Professor J. M. Mackay of Liverpool, 1914) is printed a contemporary account, with good drawings, of Henry VIII's navy. In the *English Historical Review*, vol. xxxiii (1918), 'The Navy under Henry VII', by C. S. Goldingham, is an important article on Henry VII's naval construction. The standard modern work is *A History of the Administration of the Royal Navy, 1509–1660*, by M. Oppenheim (1896). Useful information may be obtained from the introductory chapters of *Drake and the Tudor Navy*, by J. S. Corbett (2 vols., 2nd ed., 1899). In the *Transactions of the Royal Historical Society* for 1907 (3rd series, vol. i) was published an article and a contemporary drawing of the burning of Brighton by the French in the time of Henry VIII.

For English maritime expansion *The Voyages of the Cabots and the English Discovery of North America under Henry VII and Henry VIII*, by J. A. Williamson (1929), is important; (see *The Voyages of John and Sebastian Cabot*, by the same writer, in Historical Association Pamphlet no. 106, 1937, and 'An Early Grant to Sebastian Cabot', by A. P. Newton, in the *English Historical Review*, vol. xxxvii, 1922). For the effect of the development of ocean navigation upon English history see *The Ocean in English History*, by J. A. Williamson (1941), and *Tudor Geography*, by E. G. R. Taylor (1930), which is extremely useful.

THE ARMY

There is abundant material for the history of the British army in the *Letters and Papers of Henry VIII*, in *Foedera*, and in contemporary correspondence, but the history has remained unwritten.

A History of the British Army, by J. W. Fortescue (vol. i, 1910), though very useful, passes lightly over the Tudor period, and the military history included in *The Sixteenth Century*, by Charles

Oman (1936), is mainly concerned with continental armies. In the *Miscellany* of the Scottish History Society (vol. viii, 1950) are published the accounts of the campaign of Flodden with an introduction examining the English method of organizing a field-army.

ECCLESIASTICAL HISTORY

(i) RECORDS AND DOCUMENTS

Amongst ecclesiastical records pride of place goes to the *Valor Ecclesiasticus* (Record Commissioners, 6 vols., index in vol. vi, 1810–34), which is 'no less than an entire Survey and Estimate of the whole ecclesiastical property of England and Wales' on the eve of the reformation. (For a satisfying analysis see *English Monasteries on the Eve of the Dissolution*, by A. Savine, in Oxford Studies in Social and Legal History, vol. i, 1919). *Concilia Magnae Britanniae et Hiberniae* (ed. David Wilkins, 4 vols., 1737) contains, besides notices of convocations—often extremely brief—a great collection of documents including papal bulls, royal injunctions and articles, and episcopal letters. The works of John Strype (1643–1737) (reprinted in 19 vols., 1812–24, with an index of 2 vols., 1828) rest so much upon manuscripts, some of which the author seems to have acquired rather oddly, that they may well be reckoned as documentary sources. Burnet's *History of the Reformation* (*vide supra*, p. 615) is extremely well documented.

Valuable documentary evidence is to be found in the publications of the Lincoln Record Society (well edited), though relatively few of these touch the period 1485–1558. Three volumes of *Lincoln Wills*, which cover the period from 1485 to 1532, are mainly important for social history, but the two volumes of *Chapter Acts of the Cathedral Church of St Mary of Lincoln (1520–1547)* serve to show the continuity of the administration, and the three volumes of *Visitations of the Diocese of Lincoln, 1517–31* (ed. Hamilton Thompson, 3 vols. 1940–7) contain valuable information as to the state of the religious houses. *The Visitations of the Diocese of Norwich, 1492–1532* (ed. A. Jessop, Camden Society, 1888) and the *Collectanea Anglo-Premonstratensia, 1291–1505* (3 vols., Camden Society, 1904–6) provide useful evidence. *The Register of Richard Fox while Bishop of Bath and Wells* (1492–4), ed. by E. C. Batten (1889), *The Register of Richard Fox, Lord Bishop of Durham* (ed. M. P. Howden, Surtees Society, 1932), and the *Letters of Richard Fox* (ed. P. S. and H. M. Allen, 1929) are all valuable. *Visitations and Articles and Injunctions of the period of the Reformation* (3 vols., Alcuin Club, 1910) supplies information for the later visitations. Vol. i, edited by W. H. Frere, provides a good introduction, and vol. ii, edited by W. H. Frere and W. M. Kennedy, supplies information about the visitations of 1536 and 1558. See also the *English Historical Review*, vols. xxxix–xli (1924–6) for additional information in articles by W. P. M. Kennedy and J. M. Wilson. 'The Deprived Married Clergy in Essex, 1553–61', by H. E. P. Grieve (*Transactions of the Royal Historical Society*, 4th series, vol. xxii, 1940) seems to show that the dislocation of clerical life was less than might have been supposed.

(ii) SECONDARY AUTHORITIES

The general works in which ecclesiastical and political history are closely intermingled have already been noted.

In many ways the most useful history of the Anglican church is R. W. Dixon, *History of the Church of England, from the Abolition of the Roman Jurisdiction to 1570* (6 vols., iii and iv being useful for this period, 3rd ed., 1895–1902), which is definitely Anglican in outlook and very well documented. *The English Church in the Sixteenth Century*, by J. Gairdner (1902), is a work of learning enriched by good notes, good bibliographies, and a valuable map of the religious houses in England at the time of their dissolution; it is critical of the protestant movement which it seems to regard as not particularly spiritual in its origin, and it presaged what may be called an Anglo-Catholic tendency obvious in excellent biographies such as R. W. Chambers's *Thomas More* (1935) and James Muller's *Stephen Gardiner and the Tudor Reaction* (1926). In *Pre-Reformation England* (1938) Canon Maynard Smith concluded that a reformation was necessary; in his second volume, *Henry VIII and the Reformation* (1948), he commented upon the haphazard way in which the reformation was accomplished. *The Reformation in England* (vol. i, *The King's Proceedings, 1517–35*), by P. Hughes (1950), is a critical study of the royal conduct. The Abbé Constant in *La Réforme en Angleterre* (2 vols., 1930–9, English translations by R. E. Scantlebury, 1934, and E. I. Watkin, 1941) offers a well-documented account from the standpoint of a foreign catholic whose views, though moderate, are quite definite; and a contribution to the discussion of the religious issue is found in C. H. Smyth's *Cranmer and the Reformation under Edward VI* (1926). A definite expression of the protestant view is found in numerous works of G. G. Coulton, notably *Ten Medieval Studies* (3rd ed., 1930) and *Five Centuries of Religion* (4 vols., 1923–50). Coulton has been denounced as 'a man with a muck-rake', seeing only the evils of the old church. He was, however, conscious of Froude's 'exaggerations', and his work was necessary because writers who glossed over the defects of the medieval church were able to represent the reformation as an act of power on the part of the state. A valuable study of the reformation considered as an act of state by an author who was well aware that it has other aspects, is found in Sir F. Maurice Powicke's *The Reformation in England* (1941). This traces the development, out of the old world-order, of an ecclesiastical system which was accepted as an integral part of a national church, and emphasizing the long tradition of cooperation between Crown and church in England, shows that the resemblance between the new church and the old was not unnatural. *Studies in the Making of the English Protestant Tradition*, by E. G. Rupp (1947), a student of Lutheranism, is a protest against the tendency whereby 'the Reformers are in a fair way to becoming peripheral annotations to the story of the English Reformation'. *The Marian Reaction in its Relation to the English Clergy*, by W. H. Frere (1896), is a useful statistical summary of the fortunes of the English clergy, with a handy chart of the changes which occurred in the sees. *English Monks and the Suppression of the Monasteries*, by Geoffery Baskerville (1937), is important, especially for the fate of the ejected religious. *Martin Bucer and the English Reformation*, by Constantin Hopf (1946), is useful for the development of the Edwardian Prayer Books. A. W. Pollard's *Records of the English Bible* (1911) is still a standard, and *A Brief Sketch of the History of the Transmission of the Bible*, by H. Guppy (1926), is enriched by some excellent facsimiles.

ECONOMIC AND SOCIAL

RECORDS AND DOCUMENTS

England is particularly rich in local records, those of towns, gilds, manors, and parishes. These appear in all sorts of forms—

many are found among the publications of the Historical Manuscripts Commission, many are published locally by societies or individuals. Conspicuously well produced are the records of Northamptonshire with which the name of Miss Joan Wake is associated; of the fourteen volumes already published, three, ix, xii, and xiii, dealing with Peterborough, its monastery and its cathedral, fall within the early Tudor period. Of *The Records of the Borough of Leicester* (4 vols., 1899–1923), vols. ii and iii, admirably edited by Mary Bateson, deal with the period 1485–1558.

Hall's *Repertory* (p. 610 *supra*) gives a general description of records of this class; see also *The Parish Registers* (1910) and *Churchwardens' Accounts* (1913), both by J. C. Cox. A standard work is *A Bibliography of British Municipal History*, by Charles Gross, Harvard Historical Studies, 5 (1897).

A useful selection, from many sources, is provided in *Tudor Economic Documents* (ed. R. H. Tawney and Eileen Power, 3 vols., 1924). A collection of documents dealing with one aspect of economic history is found in the *Domesday of Inclosures* (ed. I. S. Leadam, 2 vols., 1897, for the Royal Historical Society): see article by E. F. Gay and I. S. Leadam in *Transactions of the Royal Historical Society* (1900).

For the social life of the country there is much evidence. The domestic economy of royalty may be studied in the *Privy Purse Expenses*: Henry VII's in Bentley's *Excerpta Historica* (1833); his wife Elizabeth's (ed. Harris Nicholas, 1830), Henry VIII's (ed. Harris Nicholas, 1827), the Princess Mary's (ed. Frederic Madden, 1831), and Princess Elizabeth's in the *Camden Miscellany* (vol. ii, 1853); inventories of the wardrobes of the duke of Richmond and of Princess Catherine as dowager are found in the *Camden Miscellany* (vol. iii, 1855). Detailed accounts of court festivities appear in John Leland's *De Rebus Britannicis Collectanea* (ed. T. Hearne, 6 vols., 1715).

The 'Ordinances of Eltham', setting forth the discipline of the royal court in 1526, are found in the *Proceedings of the Privy Council*, vii (see *English Historical Review*, 1922). Some accounts of the master of the revels are to be found in the *Loseley Manuscripts* (ed. John Kempe, 1836. A new edition is promised by Professor Arnold Edinborough, of Queen's University, Ontario).

The economy of a great noble is set forth in *The Regulations and Establishment of the Household of Henry Algernon Percy, the fifth*

Earl of Northumberland, at his castles of Wressle and Leckonfield (ed.
Bishop Percy, 1770; new ed. 1905), and in *The Household Books
of John, Duke of Norfolk and Thomas, Earl of Surrey, 1481–1490*
(ed. J. Payne Collier, Roxburghe Club, 1844). *The Rutland
Papers* (Camden Society, 1842) contain materials for the reigns
of Henry VII and Henry VIII, and the *Plumpton Correspondence*
(Camden Society, 1839) reveals the domestic history of a
Yorkshire family which depended to some extent upon the
earl of Northumberland. *The Paston Letters* (ed. James Gairdner,
6 vols., 1904), which cover the period 1422–1509, cannot be
neglected. *The Cely Papers* (Camden Society, 1900), which tell
of the fortunes of a family of merchants of the staple, largely
bound up with the Cotswold wool-trade, go down only to 1488.

For life in the towns see the reference to local records. For
life in the country Manorial Records, some quoted by Tawney
and Power, are illuminating.

CONTEMPORARY LITERATURE

ECONOMIC AND SOCIAL

The economic problems of England attracted the interest of
the times. Two accounts of England as seen by foreigners are to
be found among the Camden volumes; one is *A Relation, or
rather a True Account, of the Island of England* (ed. with a translation by C. A. Sneyd, 1847), which probably contains the
material on which an official *Relazione* was prepared by Andrea
Trevisano, the first resident Venetian ambassador in England.
In the translation of the first eight books of Polydore Vergil's
Anglica Historia (ed. Henry Ellis, 1846) there is a brief description of the whole island of Britain, which although it is introduced as a preface to very ancient history, is evidently the fruit
of the writer's own observations. The *Itinerary* of William of
Worcester (William le Botoner), who died in 1482(?), describes
the England of the fifteenth century, but is available only in a
rare and ill-edited volume. A very full contemporary description
of England is contained in the *Itinerary* of John Leland (1535–43)
(the best edition is that edited by Lucy Toulmin Smith, 5 vols.,
1906–10). *The Historical Geography of England before 1800*, ed. H. C.
Darby (1936), contains a chapter, ix, on Leland's England by
E. G. R. Taylor. *A Survey of London*, by John Stow, though not
published till 1598 and concerned with Elizabethan London, is

extremely useful (best edition that of C. L. Kingsford, 2 vols., 1908).

The religious and economic changes of the age produced much literature, some of it very bitter. Amongst the publications of the Early English Text Society may be noted *A Supplicacyon for the Beggers*, by Simon Fish, and *The Decaye of England by the Great multitude of Shepe* (*Four Supplications*, 1871); the *Complaynt of Roderyck Mors*, by Henry Brinklow, and the same author's *Lamentacyon of a Christen agaynst the Cytye of London* (1874). *A Caveat or Warening for Commen Cursetors, vulgarley called Vagabones* (1869), deals with the growing question of vagabondage. The famous *Discourse of the Common Weal of this Realm of England* (ed. Elizabeth Lamond, 1893: reprint 1929) is significant as showing a comprehension of the results of inflation.

AGRICULTURE

The Boke of Husbandry, by Master Fitzherbert (1534) (reprinted by the English Dialect Society and ed. W. W. Skeat, 1882), is a very practical book on farming written in all probability by one who was a justice of the common pleas. His *Boke of Surveyinge and Improvements* (1525), mainly concerned with legal issues, was reprinted in 1767 in a *Collection of Certain Ancient Tracts concerning the Management of Landed Property*. Equally practical, though the work of an unsuccessful farmer, are the *Hundreth Good Pointes of Husbandrie* and the *Hundreth Poyntes of Good Husserie* (both 1557), by Thomas Tusser; in 1573 the two together were enlarged into *Five Hundreth Pointes*.

ECONOMIC

The best general history is *Economic History of England*, by E. Lipson (3 vols., 9th ed. 1946), which presents clearly the evidence for the development of English agriculture, industry, and trade. *A History of Agriculture and Prices in England*, by J. E. Thorold Rogers (vols. iii–vii, 1882–1902), though its figures have sometimes been questioned, is still useful. It is supplemented to some extent by *Prices and Wages in England*, by Sir William Beveridge and others, the first volume of which appeared in 1939 (still in progress). Evidence as to the extent of the inclosures is provided by E. F. Gay in 'Inclosures in England in the Sixteenth Century' (*Quarterly Journal of Economics*, vol. xvii, 1903), in 'The Inquisition of Depopulation', by E. F. Gay and I. S. Leadam

(*Trans. R.H.S.*, N.S., vol. xiv, 1900), and in 'The Inclosure of Common Fields in the Seventeenth Century', by E. M. Leonard (*Trans. R.H.S.*, N.S., vol. xiv, 1905). *The Agrarian Problem in the Sixteenth Century*, by R. H. Tawney (1912), is a full and exact study of the position of the English peasantry in the face of economic changes. Equally important is Dr. Tawney's *Religion and the Rise of Capitalism* (originally the Holland Memorial Lectures of 1922; published 1926; reprinted with new preface 1937), which applies to English society the theories developed by Max Weber in *Die protestantische Ethik und der Geist des Kapitalismus* (1904–5): translated by Talcott Parsons, 1930, and by E. Troeltsch in *The Social Teaching of the Christian Churches* (2 vols., trans. Olive Wyon, 1931). For a critique of Dr. Tawney see 'Capitalism and the Reformation', by P. C. Gordon Walker (*Econ. Hist. Rev.*, vol. viii, 1937). See also H. M. Robertson's *Aspects of the Rise of Economic Individualism* (1st ed. 1933, reprinted 1935), which asserts that the rise of capitalism in England and Holland in the sixteenth century was due not to religious but to geographical and economic factors. To part of Robertson's argument J. Brodrick, S.J., replied in *The Economic Morals of the Jesuits* (1934). *The Early History of English Poor Relief*, by E. M. Leonard (1900), is still very useful. For the revenues of the different strata of society see 'Income from Land in England in 1436', by H. L. Gray (*E.H.R.*, vol. xlix, 1934). 'The Village Population in the Tudor Lay Subsidy Rolls', by S. A. Peyton (*E.H.R.*, vol. xxx, 1915), though mostly concerned with the period after 1558, seems to show that the rural population was less stable than might have been expected.

For English foreign trade see G. Schanz, *Englische Handelspolitik gegen Ende des Mittelalters* (2 vols., 1881), which contains valuable tables of the customs returns. *Studies in English Trade in the Fifteenth Century* (ed. E. Power and M. M. Postan, 1933), 'Commercial Trends and Policy in Sixteenth Century England', by F. J. Fisher (*Econ. Hist. Rev.*, vol. x, 1940), and 'The Internal Organization of the Merchant Adventurers of England', by W. E. Linglebach (*Trans. R.H.S.*, N.S., vol. xvi, 1902), are all useful. *Bronnen tot Geschiedenis van den Handel met Engeland, Schotland en Irland, 1485–1585*, by H. J. Smit (2 vols., 1942, 1950), available only in Dutch, contains many documents of importance for Dutch trade, many of them in Latin or French; it is largely concerned, however, with Scottish merchants. *The Evolution of the English*

Corn Market, by N. S. B. Gras (Harvard Economic Studies, 1915), is supported by much documentary evidence, and *The Early English Customs System* (Harvard Economic Studies, 1918), by the same author, though largely concerned with an earlier period, contains much material very valuable for the Tudor period, especially the Book of Rates for 1507. The so-called *Chronicle* of Richard Arnold (1502), ed. J. Douce (1811) as *The Customs of London*, is really a commonplace-book but contains a useful list of prices, and a table of the prices of various loaves according to the varying price of wheat. *The Sports and Pastimes of the People of England*, by Joseph Strutt (new and improved edition with index by W. Hone, 1838), and his *A Complete View of the Dress and Habits of the People of England* (new and improved edition by J. R. Planché, 2 vols., 1842), are still of use, and *English Social History*, by G. M. Trevelyan (1944), covering the period from Chaucer to Queen Victoria, provides a good general review of the subject.

POLITICAL THOUGHT
CONTEMPORARY WRITINGS

The works of Sir John Fortescue, though written before 1485, did much to formulate English thought and had a lasting effect. The most modern edition of *The Governance of England* is Plummer's edition of 1885, which is still good; of the *Commendation of the Laws of England—De Laudibus Legum Anglie*, the best version is that edited and translated by S. B. Chrimes (1942). *The Tree of Commonwealth*, by Edmund Dudley (1509), is a medieval parable whose importance lies in its revelation of the abuses which flourished under the mantle of the law. More's *Utopia*, published in Latin at Louvain in 1516, presented platonic ideas in a manner which commended them to a wide public, although the first English translation was made by Ralph Robinson of Corpus Christi College, Oxford, as late as 1551. Gilbert Burnet published in 1684 a translation which is somewhat closer to the text. *The Boke Named the Governour*, devised by Sir Thomas Elyot, Knight (1531), was edited by H. H. S. Croft in 1883, and is interesting as showing the realization of the Greek idea that education was vitally concerned with the production of good government. During the stress of Henry VIII's quarrel with the papacy, there was much debate upon the nature of sovereign power; William Tyndale's

Obedience of a Christian Man (1528) (ed. H. Walter, Parker
Society, 1848) laid stress upon the powers of a Christian prince,
'he that judgeth the king, judgeth God', and Stephen Gardiner
in his *De Vera Obedientia Oratio* (1535) (best ed. by P. Janelle
and published with two other tracts by Gardiner in *Obedience in
Church and State*, 1930) defended the royal supremacy. Sir John
Cheke in 1549 produced a treatise on *The Hurt of Sedicion*,
maintaining the thesis that 'a rebel is worse than the worst
prince' (inserted by Holinshed in his *Chronicles*, reprinted 1641).
The Pilgrim, by William Thomas (1546, edited J. A. Froude,
1861), is a vehement defence of Henry by one who knew Italy
well, was clerk of the council, and was eventually executed in
1554 for participation in Wyatt's rebellion. The catholic point
of view appears well in the *Dialogue between Cardinal Pole and
Thomas Lupset*, by Thomas Starkey (in *England in Henry VIII's
Time*, E.E.T.S., 1871–8). With the advent of the Swiss theology
came the doctrine of the duty of resistance which became
definite when Mary ascended the throne, and in the hands of
Knox went outside the ideas of Calvin, and even of Bullinger
whom Knox consulted. In the pamphlets written during his exile
(*Works*, ed. D. Laing, 6 vols., 1846–64) Knox made the duty
of resisting idolatry into an almost open justification of tyrannicide; his ally Christopher Goodman published in 1558, at
Geneva, a tract named *How Superior Powers Oght to be Obeyed*,
which justified resistance to ungodly princes. Meanwhile, in
1556, another refugee, John Ponet, in his *Shorte Treatise of
Politike Power*, defended resistance on grounds 'more far-reaching than those of Knox and Goodman'. In other protestant
divines like Cranmer and Hooper non-resistance is still commended.

LATER WORKS

Amongst modern books the standard work is J. W. Allen's
A History of Political Thought in the Sixteenth Century (1928), which
is important as setting English ideas into the background of
continental thought. *English Constitutional Ideas in the Fifteenth
Century*, by S. B. Chrimes (1936), is a well-documented book
which makes good use of the cases from the *Year Books*. *The
Early Tudor Theory of Kingship*, by Franklin le van Baumer (1940),
and *Foundations of Tudor Policy*, by W. G. Zeeveld (1948), are
both useful; the latter lays stress on the work of Sir Richard
Morrison as a propagandist.

LEARNING AND EDUCATION

Useful studies on humanism in England can be found in chapter xxv ('The Renaissance in Europe', by A. A. Tilley) of the last volume of the *Cambridge Medieval History* (1936), and in the chapters on the Classical Renaissance by Sir Richard Jebb, and on the Christian Renaissance by M. R. James in vol. i of the *Cambridge Modern History* (1907). *Humanism in England during the Fifteenth Century*, by R. Weiss (1941), though short is comprehensive. *The Oxford Reformers: John Colet, Erasmus, and Thomas More*, by F. Seebohm (1867, revised 1869), and the *Life of John Colet, D.D., Dean of St. Paul's*, by J. H. Lupton (1887), though valuable, no longer retain the position which they once occupied. Not only does modern criticism tend to discount the nascent protestantism of the humanists, but much of their narrative must be corrected from a modern work of supreme importance. This is the edition of the *Opus Epistolarum Desiderii Erasmi* (11 vols., ed. P. S. and H. M. Allen and H. W. Garrod, 1906–47); the letters up to 1518 are translated in three volumes with good notes by F. M. Nichols, as *The Epistles of Erasmus* (1901–18). *The Age of Erasmus* (1914) and *Erasmus: Lectures and Wayfaring Sketches* (1934), both by P. S. Allen, are important parerga, and the *Life of Erasmus* by J. A. Froude (1894), though subjected to modern criticism, is valuable for its liveliness and spirit. The *Moriae Encomium* (ed. by Mrs. P. S. Allen, and trans. by J. Wilson, 1913) is well done. The *Colloquia* of Erasmus were translated somewhat freely by M. Bailey in 1725; the best edition is that provided by E. Johnson (1878) as *The Colloquies of Erasmus*; the *Colloquy on Rash Vows and Pilgrimages for Religions Sake* was translated with an excellent apparatus by J. G. Nichols as *Pilgrimages to St. Mary of Walsingham and St. Thomas of Canterbury* (1849). In 'Pietro Griffo, an Italian in England (1506–12)', attention is called to a little-known humanist who wrote on English History—*De Officio Collectoris in Regno Angliae* (Denys Hay in *Italian Studies*, 1939).

The *Vulgaria* (1519) of William Horman (Roxburghe Club, 1926) set forth a practical way of teaching Latin according to the humanistic ideas; *The Vulgaria of John Stanbridge and the Vulgaria of Robert Whittinton* (ed. Beatrice White, Early English Text Society, 1932) shows the development of the 'direct' method of teaching Latin, and shows also the mind of Orbilius. *Vives: On*

Education (1913) is a translation by Foster Watson of the *De Trandendis Disciplinis* (1531) of the Spaniard Juan Luis Vives who came to England in 1522 and acted as a tutor to the Princess Mary. His *Linguae Latinae Excercitatio* (1538), a series of school dialogues, used by Eton in 1561, are also translated by Foster Watson as *Tudor School-boy Life* (1908).

The *Utopia* and the *Boke Named the Governour* also set forth schemes of education. *Toxophilus*, by Roger Ascham (1545), which is more than an essay upon archery, and *The Scholemaster* (1570), which contains the famous passage upon the education of Lady Jane Grey, are both to be found in the series of *English Reprints* (ed. E. Arber, 1869–71).

Foundation statutes and curricula of various schools are to be found in *Educational Charters and Documents, 598–1909*, by A. F. Leach (1911), whose book on *English Schools at the Reformation (1546–48)* (1896) shows Edward VI as a despoiler rather than a founder. *The English Grammar Schools to 1660*, by Foster Watson (1908), gives a valuable survey of school education during the early Tudor period.

For university education recognized standards are *A History of the University of Oxford*, by C. E. Mallet (vol. ii, 1924), and *The University of Cambridge*, by J. B. Mullinger (vols. i and ii, 1873–1911). For the lives of Oxford men see *Athenae Oxonienses, 1500–1697*, by A. Wood (1721), (ed. Philip Bliss, 4 vols. (1813–20)). The series of 'College Histories' published between 1898 and 1906 is useful for both universities, and much detailed information is to be found in the numerous volumes of the Oxford Historical Society begun in 1884.

LITERATURE

Much of the literature of the period, being directed to practical ends, has already been noted in other sections.

The *Shyp of Folys, by Alexander Barclay* (1509), edited by T. H. Jamieson (2 vols., 1874), enriched with contemporary woodcuts, contains the life of Barclay and a bibliography of his works. *Certayne Egloges* and the *Mirrour of Good Maners* have been reprinted by the Spenser Society (1855) and *The Eclogues of Alexander Barclay* have been edited by Beatrice White for the Early English Text Society (1928). *The Poetical Works of John Skelton* were edited with useful notes by Alexander Dyce

(2 vols., 1843); a more modern edition is *The Complete Poems of John Skelton*, edited by P. Henderson (2nd ed., 1949). A life of the poet is provided in *Skelton*, by H. L. R. Edwards (1949). *The Poems of Sir Thomas Wiat* have been edited by A. K. Foxwell (2 vols., 1913), and E. M. W. Tillyard published a selection in *The Poetry of Sir Thomas Wyatt* (1929). *The Poems of Henry Howard, Earl of Surrey* are found in the Aldine edition of the British poets (n.d.). The works of both writers and of some other contemporaries appear in Tottel's *Miscellany* (2 eds., 1557), which was included by Edward Arber among his English Reprints (1870). In *Deux gentilshommes poètes de la cour de Henri VIII* (1891), which signalizes the borrowings from France, Edmond Bapst endeavoured to show George Boleyn, Lord Rochford along with Surrey as a harbinger of the renaissance.

Songs and Ballads chiefly of the Reign of Philip and Mary (ed. T. Wright, Roxburghe Club, 1850) contains a selection of poems mostly by little-known authors, including the 'Ballad of Chevy Chase'. An article on 'The Ballad History of the Reigns of Henry VII and Henry VIII', by C. H. Firth (*Transactions of the Royal Historical Society*, 3rd series, vol. ii, 1908), seems to show that these works are interesting for their political allusions rather than for their artistic merit.

Early Tudor Drama, by A. W. Reed (1926), contains much information about John Rastell, brother-in-law of Sir Thomas More, and his various publications. A standard work is *The Medieval Stage*, by E. K. Chambers (2 vols., 1903), which includes material up to the reign of Henry VIII. *Studies in the Literary Relations of England and Germany in the Sixteenth Century*, by C. H. Herford (1886), shows that the debt of English letters to Germany was not confined to the field of theology.

Full accounts of the literature of the period are given in the *Cambridge Modern History of English Literature* (14 vols., 1907–16, Index, 1927, of which vols. iii and v deal with the period 1485–1558), and in the *Oxford History of English Literature*, especially the volume *English Literature at the Close of the Middle Ages*, by E. K. Chambers (reprinted 1947).

For the printing of books in England see *A Century of the English Book Trade, 1457–1557*, by E. Gordon Duff (1905); the *Catalogue of the Caxton Celebration, 1877* (ed. George Bullen) is very useful.

MUSIC

Songs, Ballads and Instrumental Pieces of Henry VIII (Roxburghe Club, 1912) not only produces the original scores but provides information about Tudor music and musical instruments. *Dances of England and France from 1450 to 1600*, by Mabel Dolmetsch (1949), explains how the dances were done and prints the actual music used. Discussions of Tudor music appear in *A History of Music in England*, by E. Walker (3rd ed., 1952), and a *History of English Music*, by H. Davey (2nd ed., 1921), in *The Oxford Companion to Music*, by P. A. Scholes (1938), and Grove's *Dictionary of Music and Musicians*, 4th ed. (6 vols., ed. H. C. Colles, 1940). *The Oxford History of English Music* (vol. ii, by H. E. Wooldridge, 1905) is a recognized standard. Particularly important for the period 1485–1558 are *Early English Composers*, by A. H. G. Flood (1925), and the monumental *Tudor Church Music*, by P. C. Buck and others (10 vols., 1925–30). *English Chamber Music*, by E. Meyer (1946), is informative, but rests on the doubtful theory that the rise of chamber music was due to the rise of the middle class. *Popular Music of the Olden Time*, by W. Chappell (2 vols., 1855–9) and a collection of *National English Airs* (2 vols., by the same author, 1838–40, re-edited by H. E. Wooldridge as *English Popular Music*, 1893) are useful reminders that the English had an international reputation for music, and that all their music was not 'high-brow'.

ARCHITECTURE

A sub-contemporary work much used by later writers is *The First and Chief Groundes of Architecture used in all the Auncient and Famous Monyments*, by John Shute (1563). The standard modern work is Sir Reginald Blomfield's *History of Renaissance Architecture in England* (2 vols., 1897), which sets forth the work done by Italians and Germans in the sixteenth century. *Early Renaissance Architecture in England*, by J. A. Gotch (2nd ed., 1914), has useful illustrations and a good bibliography. For the ordinary reader the most useful books are *The Domestic Architecture of England during the Tudor Period*, by T. Garner and A. Stratton (2 vols., 2nd ed., revised and enlarged 1929), which is well illustrated; *English Homes*, by H. A. Tipping (in a 2nd ed., 1936, a single volume covers the period 1066–1558), which is also well illustrated; and a brief but good brochure on *The English House* (Historical Association Pamphlet 105, 1936), by A. Hamilton

Thompson. Special studies are *Westminster Abbey and the King's Craftsmen* (1906) and *Westminster Abbey Re-examined* (1925), both by W. R. Lethaby, and *A Short History of Hampton Court*, by E. Law (1924).

A useful bibliography of English architecture is to be found in *A Short Bibliography of Architecture*, by A. Stratton (Historical Association Leaflet 54, 1923).

The simplicity of English furniture is made evident by the few exhibits in the Victoria and Albert Museum. Excellent photographs and descriptions of some of these and of other pieces are found in *English Furniture Illustrated*, by O. Brackett and H. Clifford-Smith (1950).

ART

Reproductions of the portraits in which the art of the period was successful may be found in the National Portrait Gallery (illustrated catalogue and many reproductions obtainable), in *Historical Portraits*, by C. R. L. Fletcher and E. Walker (4 vols., 1909–19), which displays in vol. i (covering the period 1400–1600) many uncoloured reproductions. *Hans Holbein the Younger*, by A. B. Chamberlain (2 vols., 1913), contains valuable information about the careers of Holbein and his less-known contemporaries in England and good coloured reproductions. *The Drawings of Hans Holbein in the Collection of His Majesty the King at Windsor* (ed. K. T. Parker, 1945), splendidly illustrated, is useful for Holbein's technique. *English Miniatures*, by J. de Bourgoing (ed. G. C. Williamson, 1928), and *British Miniaturists*, by B. S. Long (1929), both amply illustrated, make it clear that there was no miniaturist of very great merit in England except Holbein before the day of Elizabeth. The standard biography of Holbein is still *Some Account of the Life and Works of Hans Holbein, Painter of Augsburg*, by N. Wornum (1867), which is much used by later writers. 'The Henry VII in the National Portrait Gallery' by Gustav Glück in the *Burlington Magazine* (vol. lxiii, 1933) calls attention to the work done in England for English subjects by Miguel Zittoz, court painter to Isabella of Spain and Margaret of Austria.

Michael Bryan, *Dictionary of Painters and Engravers* (5 vols., ed. G. C. Williamson, 1926–34), is a useful illustrated reference-book arranged alphabetically. Of special importance is the astonishing *Allgemeines Lexikon der bildenden Künstler*, by Ulrich Thieme and Felix Becker, with collaborators (37 vols., 1907–50).

LIST OF HOLDERS OF OFFICES

There are excellent lists of English officers of state in the *Handbook of British Chronology* issued by the Royal Historical Society in 1939. Other lists, which include the offices of Marshal and Admiral, are to be found in the *Complete Peerage*, by G. E. C. (vol. ii, Appendix D., ed. Vicary Gibbs, 1912). A list of the king's secretaries and one of the secretaries of state are to be found in *The Principal Secretary of State*, by Mrs. Higham (1923). Lists of officers for the end of the period are furnished by J. G. Nichols in preface xv of the *Diary of Henry Machyn* (*Camden Society*, 1848). Useful lists of all officers are included in Hadyn's *Book of Dignities* (1851), though these must be treated with some caution; the chancery lists, however, are founded on the excellent catalogue of Thomas Duffus Hardy (1843).

The lists here presented have been checked against original authorities whenever possible. Where they are founded upon the *Handbook of British Chronology* and upon the *Complete Peerage*, they may be regarded as authoritative; for some of the less important officials, however, it has been impossible to establish exact dates of appointment, and even to be sure, for some dates, whether the office was occupied at all. It may be remarked that in the case of Charles Somerset, later earl of Worcester, J. E. Doyle in his *Official Baronetage of England* (3 vols., 1886) cites an entry in the *Patent Roll* which has not been identified in *Letters and Papers*, but which seems likely to be correct. No doubt further research will close existing gaps.

The presentation of the titles along with the names of the officers of state will make identification simpler. Incidentally, it will serve to show how many *ministeriales* were 'new men'; no doubt they came from substantial families, but most of them had not the honour of simple knighthood at the start of their careers.

The outlines of a *cursus honoris* appear; it may be remarked that the early masters of the rolls, who were clerks, resigned on obtaining a see, and that when a commoner became lord keeper or chancellor he was given a title.

The brief biographies supplied with the list of speakers will serve to show that the speaker was usually a man experienced in the service of the Crown.

CHANCELLORS

Henry VII

1485, 18 Sept.	Thomas Rotherham, *alias* Scott, bishop of Rochester 1468–72, bishop of Lincoln 1472–80, archbishop of York 1480–1500.
1485, 7 Oct.	John Alcock, bishop of Rochester 1472–6, bishop of Worcester 1476–86, bishop of Ely 1486–1500.
1487, 6 Mar.	John Morton, bishop of Ely 1479–86, archbishop of Canterbury 1486–1500; cardinal 1493.
1504, 21 Jan.	William Warham, bishop of London 1502–3, archbishop of Canterbury 1503–32. He had been keeper of the

seal since 11 Aug. 1502, succeeding Henry Deane, archbishop of Canterbury, who had been keeper from 13 Oct. 1500 to 27 July 1502.

Henry VIII

1509.	William Warham.
1515, 24 Dec.	Thomas Wolsey, bishop of Lincoln 1514, bishop of Bath and Wells 1518–24, bishop of Durham 1524–9, bishop of Winchester 1529–30, archbishop of York 1514–30; cardinal 1515; legate *a latere* 1518.
1529, 26 Oct.	Sir Thomas More, kt. 1521.
1533, 26 Jan.	Sir Thomas Audley, kt. 1532, cr. Baron Audley 1538. He had been keeper of the seal from 20 May 1532.
1544, 3 May.	Thomas, Lord Wriothesley. Wriothesley had acted as keeper from 22 Apr. 1544, kt. 1540, baron 1544.

Edward VI

1547.	Thomas, Lord Wriothesley, cr. earl of Southampton 1547. He was deprived 7 Mar. 1547, when William Paulet, Lord St. John, afterwards marquis of Winchester, was made keeper.
1548, 23 Oct.	Richard Rich, kt. 1533, cr. Baron Rich Feb. 1547.
1552, 19 Jan.	Thomas Goodrich, bishop of Ely 1534–54. Goodrich had been keeper since 22 Dec. 1551.

Mary

1553, 23 Aug.	Stephen Gardiner, bishop of Winchester 1531–51, 1553–5.
1555, 14 Nov.	Sir Nicholas Hare and others held the seal in commission pending the choice of a successor to Gardiner.
1556, 1 Jan.	Nicholas Heath, bishop of Rochester 1540–3, bishop of Worcester 1543–51, 1553–5, archbishop of York 1555–9.

TREASURERS

Henry VII

1486, 14 July.	Sir John Dynham or Dinham, Lord Dynham 1466, (Lieutenant of Calais under Richard III).
1501, 16 June.	Thomas Howard, earl of Surrey 1483, duke of Norfolk 1514.

Henry VIII

1509.	Earl of Surrey remained in office.
1522, 4 Dec.	Thomas Howard, earl of Surrey 1514, and duke of Norfolk 1524, cr. earl marshal 1533, son of the former treasurer.

Edward VI

1547, 10 Feb.	Edward Seymour, kt. 1523, earl of Hertford 1537, 1st duke of Somerset, 16 Feb. 1547, Protector Jan. 1547–Jan. 1550.
1550, 3 Feb.	William Paulet, kt. before 1525, Lord St. John 1539, earl of Wiltshire 1550, first marquis of Winchester 1551.

Mary

1553.	Marquis of Winchester remained in office.

KEEPERS OF THE PRIVY SEAL

Henry VII

1485, 8 Sept.	Peter Courtenay, bishop of Exeter 1478–87, bishop of Winchester 1487–92.
1487, 24 Feb.	Richard Fox, bishop of Exeter 1487–92, bishop of Bath and Wells 1492–4, bishop of Durham 1494–1501, bishop of Winchester 1501–28.

Henry VIII

1516, 18 May.[1]	Thomas Ruthall, bishop of Durham 1509–23.
1523, 14 Feb.	Sir Henry Marny, 1st Lord Marny Apr. 1523, K.B. 1494, K.G. 1510, chancellor of the Duchy of Lancaster, vice-chamberlain of the household and captain of the Yeomen of the Guard 1509–34.
1523, 25 May.	Cuthbert Tunstall, bishop of London 1522–30, bishop of Durham 1530–52, 1553–9.
1530, 24 Jan.	Thomas Boleyn, cr. earl of Wiltshire and Ormond 1529, K.B. 1509, Viscount Rochford 1525.
1536, 29 June.	Thomas Cromwell, 1st Lord Cromwell, cr. earl of Essex 1540.
1540, 14 June.	William Fitzwilliam, kt. 1513, 1st earl of Southampton 1537.
1542, 3 Dec.	John Russell, kt. 1513, cr. Lord Russell 1539, cr. earl of Bedford 1550.

Edward VI

1547, 21 Aug.	Lord Russell reappointed.

Mary

1553, 3 Nov.	Earl of Bedford reappointed, d. 14 March 1555.
1555.	Sir Robert Rochester, controller of the household (temporary; no patent), 'servant' of the Lady Mary, kt. in Feb. 1554, K.G. in July 1557, chancellor of the Duchy of Lancaster by May 1554.
1555, 31 Dec.	William Paget, 1st Lord Paget 1549, kt. by 1544, K.G. 1547.

[1] Date when his salary began.

MASTERS OF THE ROLLS

1485, 22 Aug.	Robert Morton, bishop of Worcester 1487.
	Robert Morton and William Eliot, jointly.
1487, 26 Feb. (? 26 Nov.) David William.	
1492, 5 May.	John Blythe (later bishop of Salisbury 1494).
1494, 13 Feb.	William Warham (later bishop of London, 1502, archbishop of Canterbury 1503, and Lord chancellor 1504.
1502, 1 Feb.	William Barnes or Barons (later bishop of London 1504).
1504, 13 Nov.	Christopher Bainbridge (later archbishop of York 1508).
1508, 22 Jan.	John Yonge (later dean of York 1514).
1516, 12 May.	Cuthbert Tunstall (later bishop first of London 1522, then of Durham 1530).
1522, 20 Oct.	John Clarke, archdeacon of Colchester.
1523, 9 Oct.	Thomas Hannibal.
1527, 26 June.	John Taylor, prebendary of Westminster 1518.
1534, 8 Oct.	Thomas Cromwell, executed 1540.
1536, 10 July.	Christopher Hales, solicitor-general 1525, attorney-general 1529.
1541, 1 July.	Sir Robert Southwell, kt. 1537.
1550, 13 Dec.	John Beaumont (deprived for abusing his office).
1552, 18 June.	Sir Robert Bowes, warden of the east and middle marches 1550.
1553, 18 Sept.	Sir Nicholas Hare, master of requests 1537, and again 1552, speaker of the House of Commons 1539, kt. 1537.
1557, 5 Nov.	Sir William Cordell, solicitor-general 1553, speaker of the House of Commons 1558, kt. by 1557.

KING'S SECRETARIES
(*before* 1540)

1485.	Dr. Richard Fox, bishop of Exeter 1487.
1487.	Dr. Oliver King (acting 1489 and 1492), bishop of Exeter 1493.
1500.	Dr. Thomas Ruthall (till May 1516).
1516.	Dr. Richard Pace.
1526.	Dr. William Knight.
1528.	Dr. Stephen Gardiner.
1533.	Thomas Cromwell.

PRINCIPAL SECRETARIES OF STATE
(*from* 1540)

1540.	Sir Thomas Wriothesley, cr. earl of Southampton 1547.	Sir Ralph Sadler.
1543, 23 Apr.		Sir William Paget.
1544, Mar.	Sir William Petre.	
1548, 17 Apr.		Sir Thomas Smith.

1549, 15 Oct.	Dr. Nicholas Wotton.
1550, 5 Sept.	William Cecil, kt. 1551, cr. 1st Lord Burghley 1571.
1553, 2 June.	Sir John Cheke (additional).
1553, Aug.	Sir John Bourne (to Mar. 1558).
1557, 30 Mar.	John Boxall
1558, 20 Nov.	William Cecil (only one secretary).

GREAT MASTERS OR LORDS STEWARDS OF THE HOUSEHOLD

(1531, *designation of the office changed from Lord Steward to Great Master*; 1553, *returned to the title of Lord Steward*)

1485.	John Radcliffe, 1st Baron Fitzwalter, ex. 1496.
1488.	Robert Willoughby, Lord Broke.
1490.	Sir William Stanley.
Before 20 July 1506.	George Talbot, 4th earl of Shrewsbury, d. 1538.
1540, Oct.	Charles Brandon, duke of Suffolk.
1545.	William Paulet, Lord St. John of Basing (later earl of Wiltshire and marquis of Winchester).
1550.	John Dudley, earl of Northumberland.
1553.	Henry Fitzalan, 12th earl of Arundel.

LORDS CHAMBERLAINS OF THE HOUSEHOLD

1485.	Sir William Stanley (beheaded 1495).
1508 or 1509.	Sir Charles Somerset (later Lord Herbert and earl of Worcester). Appointed 3 May 1508, cr. chamberlain 22 April 1509.
1526.	William, Lord Sandys of 'the Vyne', held post till his death, 1540.
1543, May.	William Paulet, Lord St. John of Basing (later earl of Wiltshire and marquis of Winchester).
1546, July.	Henry FitzAlan, 12th earl of Arundel.
1550, 2 Feb.	Thomas, Lord Wentworth, d. Mar. 1551.
1551.	Thomas, Lord D'Arcy of Chiche.
1553.	Sir John Gage, d. Apr. 1556.
1556.	Sir Edward Hastings (later Lord Hastings of Loughborough).
1558.	William, Lord Howard of Effingham.

TREASURERS OF THE HOUSEHOLD

1485, Oct.	Sir Richard Croft.
1502.	Sir Thomas Lovell.
1522.	Sir Thomas Boleyn (later Viscount Rochford and earl of Wiltshire).

By 1537.   Sir William Fitzwilliam (earl of Southampton 1537).
1537, Oct.–Mar. 1539. Sir William Paulet (later Lord St. John of Basing, earl of Wiltshire and marquis of Winchester).
1541.   Sir Thomas Cheyney, d. 1558.

COMPTROLLERS OF THE HOUSEHOLD

1485.   Sir Richard Edgecumbe.
In 1492.   Sir Roger Tocotes.
In 1498.   Sir Richard Guildford.
1509.   Sir Edward Poynings (probably appointed before death of Henry VII).
1520?   Sir Henry Guildford.
1526   William Fitzwilliam (cr. earl of Southampton 1537).
1532, May.   Sir William Paulet (1st Baron St. John of Basing, 1st earl of Wiltshire, and 1st marquis of Winchester).
1537, 18 Oct.   Sir John Russell (cr. 1st earl of Bedford 1550).
1540.   Sir John Gage.
1547.   Sir William Paget (resigned on being summoned to parliament as Lord Paget of Beaudesert, Dec. 1549).
1550.   Sir Anthony Wingfield (d. Aug. 1552).
1552, 27 Aug.   Sir Richard Cotton.
1553, Aug.   Sir Robert Rochester (appointed by Mary at her accession).
1557, 25 Dec.   Sir Thomas Cornwallis.
1558.   Sir Thomas Parry.

LORDS ADMIRAL

1485, 21 Sept.   John de Vere, earl of Oxford, till his death 10 Mar. 1513.
1513, 4 May.   Sir Thomas Howard (later duke of Norfolk).
1525, 16 July.   Henry Fitzroy, duke of Richmond, till his death 22 July 1536.
1536, 16 Aug.   William Fitzwilliam, earl of Southampton.
1540, 18 July.   John, Lord Russell (later earl of Bedford).
1542, 28 Dec.   Edward Seymour, earl of Hertford (later duke of Somerset).
1542–3, 26 Jan.   John Dudley, Viscount Lisle (later earl of Warwick and duke of Northumberland).
1547, 17 Feb.   Thomas, Lord Seymour.
1549, 28 Oct.   John Dudley, Viscount Lisle.
1550, 4 May.   Edward, Lord Clinton (later earl of Lincoln).
1553, 26 Oct.   William, Lord Howard of Effingham.
1558, Feb.   Edward, Lord Clinton (till his death in Jan. 1585).

EARLS MARSHAL

1483, 28 June.   John, Lord Howard (cr. duke of Norfolk and earl Marshal), d. 22 Aug. 1485.
1486, 19 Feb.   William Berkeley, earl of Nottingham (later marquis of Berkeley), d. 14 Feb. 1492.

1494, before 31 Oct.	Henry Tudor, afterwards Henry VIII.
1510, 10 July.	Thomas Howard, earl of Surrey (later duke of Norfolk).
1524, 21 May.	Charles Brandon, duke of Suffolk.
1533, 28 May.	Thomas Howard, duke of Norfolk, in tail male.
1547, 17 Feb.	Edward Seymour, duke of Somerset.
1551, 20 Apr.	John Dudley, earl of Warwick.
1553, 3 Aug.	Thomas Howard, duke of Norfolk, restored, d. 25 Aug. 1554.
1554, 25 Aug.	Thomas Howard, duke of Norfolk, grandson of the above, executed 2 June 1572.

LORDS LIEUTENANT AND LORDS DEPUTY OF IRELAND

Date	Lords lieutenant	Lords deputy
1486.	Jasper Tudor, duke of Bedford.	Gerald Fitzgerald, 8th earl of Kildare.
1492.		Walter FitzSimons, archbishop of Dublin, and Sir Thomas Ormond (governor).
1493.		Robert Preston, Lord Gormanston. William Preston, his son.
1494.	Prince Henry, duke of York.	Sir Edward Poynings.
1496.		Henry Deane, bishop of Bangor (appointed 1 Jan.). Gerald, 8th earl of Kildare (appointed 6 Aug. and in office till 1513).
1513.		Gerald, 9th earl of Kildare. [Sir Piers Butler.] [Sir Maurice Fitzgerald.]
1520.	Thomas Howard, earl of Surrey.	
1522.		Piers Butler, pretended earl of Ormond.
1524.		Gerald, 9th earl of Kildare.
1527.		Richard Nugent, Lord Delvin.
1529.	Henry Fitzroy, duke of Richmond and Somerset (d. 1536).	
1530.		Sir William Skeffington.
1532.		Gerald, 9th earl of Kildare.
1534.		Sir William Skeffington.
1536.		Leonard Lord Grey.
1540–7.		Sir Anthony St. Leger.
1548.		Sir Edward Bellingham.
1550.		Sir Anthony St. Leger.

Date	Lords lieutenant	Lords deputy
1551.		Sir James Croft.
1553–6.		Sir Anthony St. Leger.
1556–60.		Thomas Radcliffe, Lord Fitzwalter, earl of Sussex.

SPEAKERS

The lists of speakers has been obtained, for the most part, from the *Rolls of Parliament*, the *Lords' Journals*, and the *Commons' Journals*; in a few cases the date of appointment is not given by any of these authorities. For the earlier period it was the practice to obtain leave to elect a speaker on the second day of parliament and to present him on the third or fourth day. In the last parliament of Edward VI, however, the speaker was appointed on the second day of parliament, and during Mary's reign upon the first day of parliament. It seems legitimate to conclude that the speaker (who was paid by the Crown), was a government nominee, and this conclusion is supported by the facts that both Dyer and Pollard were 'nominated' by the treasurer of the household and that Higham was 'brought to the chair by Mr. Treasurer and Mr. Comptroller'. When, therefore, official evidence fails it may safely be surmised that the speaker was appointed within a few days of the formal opening of parliament which, it may be remarked, was not infrequently upon a Monday.

Date of office	Name	Other positions, &c.
1485, 9 Nov.	Thomas Lovell, Northamptonshire	Fought at the battle of Bosworth for Henry Tudor. Chancellor of exchequer for life. Knighted after Stoke. Treasurer of the Household 1502. Constable of the Tower 1509. Surveyor of the court of Wards and Liveries. Rise of Wolsey affected his position, retired from public life.
1487, 12 Nov.	John Mordaunt, Bedfordshire	Commander at Stoke. Serjeant-at-law. King's Serjeant. Chief justice of Chester. Knighted Feb. 1503. Chancellor of Duchy of Lancaster. Member of P.C.
1489, 15 Jan.	Sir Thomas Fitz-william, Yorkshire.	Apparently conspicuous only for marrying a wealthy woman, niece of the 'Kingmaker'.
1491, 19 Oct.	Richard Empson, Northamptonshire.	Knighted 1504. High steward of Cambridge. Chancellor of Duchy of Lancaster. Conspicuous for tax exactions and fines. Executed on charge of conspiracy 1510.

Date of office	*Name*	*Other positions, &c.*
1495, 17 Oct.	Robert Drury, Sussex.	Work in connexion with Scotland and Border policy. Present at marriage of Princess Mary, and Field of Cloth of Gold. Collector of Subsidies. P.C.
1497, 19 Jan.	Thomas Englefield, Berkshire.	Knighted on marriage of Prince Arthur to Catherine of Aragon.
1504, 29 Jan.	Edmund Dudley, Staffordshire.	Member of P.C. 1485. Under-sheriff of London 1497. With Empson exacter of fines and taxes. Suspected of corruption. Wrote *Tree of Commonwealth*, advocating absolute monarchy. Executed for constructive treason 1510.
1510, 23 Jan.	Sir Thomas Englefield.	
1512, 6 Feb.	Robert Sheffield.	Fought at Stoke, knighted after battle. Recorder of London.
1515, 8 Feb.	Thomas Neville, Kent.	Member of Henry VIII's household and of P.C. On commission for inclosures 1517. Member of star chamber 1519. K.G.
1523, 18 Apr.	Thomas More, Middlesex.	M.P. 1504. Under-sheriff of London 1510. Envoy to Flanders 1515. *Utopia*, published 1516. Master of requests. Member of P.C. 1518. *Persona grata* at court. At Field of Cloth of Gold. Knighted 1521. As Speaker opposed Wolsey. High steward of Cambridge. Chancellor of Duchy of Lancaster, Lord chancellor 1529, resigned 1532. Executed, 1535.
1529, 6 Nov.	Thomas Audley, Essex.	Succeeded More as chancellor. 'Understood *business* well and *men* better.' Kept in favour with Henry VIII throughout all changes of policy.
1533, 9 Feb.	Humphrey Wingfield, Great Yarmouth.	Commissioner for Wolsey 1529. Member of P.C. Knighted when made Speaker. First burgess to be Speaker.

Date of office	Name	Other positions, &c.
1536, 12 June.	Richard Rich, Essex.	Gave evidence against More. Supporter of Cromwell but deserted him and testified against him. Baron 1547–8. Chancellor Oct. 1548.
1539, late Apr. or early May.	Nicholas Hare, Norfolk.	Master of requests, 1537, and of Rolls 1553. Justice of Chester. Member of P.C. Keeper of Great Seal. Knighted 1537. In prison for part of his term as Speaker.
1542, 18 Jan.	Thomas Moyle, Kent.	Chancellor of court of Augmentations. Knighted Oct. 1537.
1545, 23 Nov.	Thomas Moyle may have been Speaker in this last parliament of Henry VIII, but no proof in *Lords' Journals*.	
1547, 4 Nov.	John Baker, Huntingdonshire.	Joint ambassador to Denmark 1526. Attorney-general. Member of P.C. Chancellor of exchequer 1545–58.
1553, 2 Mar.	James Dyer, Cambridgeshire.	Serjeant-at-law, knighted 1552. Judge of the queen's bench, later president of court of common pleas.
1553, 5 Oct.	John Pollard, Oxfordshire.	Serjeant-at-law. Vice-president of council of Welsh Marches 1550. Knighted 1553.
1554, 2 Apr.	Robert Brooke, London City.	Serjeant-at-law. Recorder of London 1545. Chief justice of common pleas 1554. Knighted by Philip Jan. 1555.
1554, 12 Nov.	Clement Higham, West Looe, Cornwall.	Joined Mary at Kenninghall. Made member of P.C. Chief baron of exchequer. Knighted by Philip Jan. 1555.
1555, 21 Oct.	John Pollard (again).	
1558, 20 Jan.	William Cordell, Suffolk.	Solicitor-general. Master of Rolls 1557. Knighted by 1557. Member of P.C.

KEY TO GENEALOGICAL TABLE I

m. — murdered.
ex. — executed.
k. — killed.
d. — died.
A — Anne Mortimer: if female descent is recognized York would take precedence over Lancaster.
B — Margaret Beaufort: if female descent is not recognized the Lancaster claim is defeated on the death of Henry VI, 1471, his son Edward having predeceased him (K. Tewkesbury) 1471.

Descendants of Edward IV

1. Lady Jane Grey (1537–54).
Married to Guildford Dudley as part of a plot for altering the succession. Proclaimed queen, 1553. Executed by Mary Tudor after Wyatt's Rebellion, 1554.

2. Henry Courtenay, marquis of Exeter (1496?–1538).
Through his mother Catherine, heir to the English Crown should Henry VIII die without lawful issue. Powerful in Devon and Cornwall. Well known to stand for the old religion. Had intrigued with Chapuys. Discovered to have had communication with Montague and to have spoken against the king and his advisers to Sir Geoffrey Pole. Executed, as an aspirant to the Crown, 9 December 1538.

Descendants of George, duke of Clarence

3. Edward, earl of Warwick (1475–99).
Imprisoned in Tower for fourteen years. Finally executed, 28 November 1499, for alleged complicity with Perkin Warbeck.

4. Margaret, Lady Salisbury (1473–1541).
Compromised by corresponding with her son Reginald Pole. Henry VIII determined to destroy her whole family, therefore she was attainted 1539, imprisoned for two years, and executed 1541.

5. Henry Pole, Lord Montague (1492?–1538).
Took great care not to offend the government but was convicted of treason on the evidence of certain fragments of conversation in which he was said to have anticipated the king's death, to have regretted the abolition of the monasteries and the slavishness of parliament; his correspondence with Reginald Pole also told against him. Executed 9 December 1538.

Descendants of Elizabeth and John de la Pole, duke of Suffolk

6. John de la Pole, earl of Lincoln (1464?–87).
Killed at Stoke, fighting for Lambert Simnel, 1487.

7. Edmund de la Pole, earl of Suffolk (1472?–1513).
Fled to Flanders, because he was told that the Emperor Maximilian

would gladly help one of Edward IV's blood to gain the throne of England. Outlawed 1501. Seized by duke of Guelders on way to Friesland, 1504, and delivered to Henry VII by Philip of Castile, 1506. Put in Tower, exempt from general pardon at accession of Henry VIII, executed 1513.

8. Richard de la Pole (d. 1525).

Killed at battle of Pavia, 1525, fighting in the army of Francis I.

Descendants of Thomas of Woodstock, duke of Gloucester and earl of Buckingham

9. Edward Stafford, 3rd duke of Buckingham (1478–1521).

Owing to his position of High Constable, his wealth, his descent from Thomas of Woodstock and other connexions, one of the most powerful men in England. As early as 1503, when Henry VII was ill, was talked of as a possible heir to the throne. Charges of disloyalty, such as wishing king's death and having designs upon the throne, trumped up against him on evidence of some discontented servants. Executed on these charges, 1521.

10. Thomas Stafford (1531?–1557).

Opposed Spanish marriage, and claiming royal descent on both sides, sailed from Dieppe by connivance of Henry II of France, and with two ships made a landing at Scarborough, 1557, taken and hanged.

TABLE II

THE ANGLO-HABSBURG CONNEXION

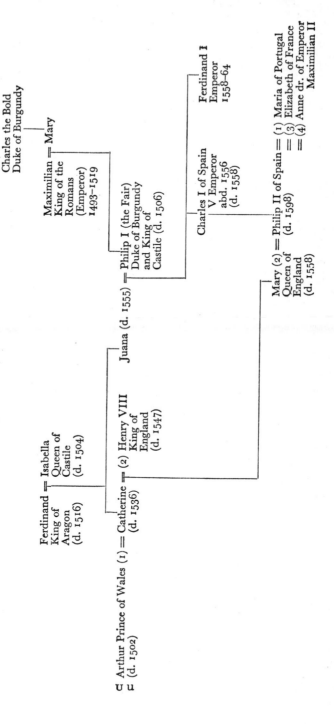

INDEX

PRINTED IN
GREAT BRITAIN
AT THE
UNIVERSITY PRESS
OXFORD
BY
CHARLES BATEY
PRINTER
TO THE
UNIVERSITY